PROPAGATION CHANNEL CHARACTERIZATION, PARAMETER ESTIMATION AND MODELLING FOR WIRELESS COMMUNICATIONS

PROPAGATION CHANNEL CHARACTERIZATION, PARAMETER ESTIMATION AND MODELLING FOR WIRELESS COMMUNICATIONS

Xuefeng Yin

Tongji University, Shanghai, China

Xiang Cheng

Peking University, Beijing, China

WILEY

IEEE PRESS

This edition first published 2016

Registered office
John Wiley & Sons Singapore Pte Ltd, 1 Fusionopolis Walk, #07-01 Solaris South Tower, Singapore 138628

For details of our global editorial offices, for customer services and for information about how to apply for permission to reuse the copyright material in this book please see our website at www.wiley.com.

Library of Congress Cataloging-in-Publication data applied for

ISBN: 9781118188231

A catalogue record for this book is available from the British Library.

Set in 10/12pt, TimesLTStd by SPi Global, Chennai, India.
Printed and bound in Singapore by Markono Print Media Pte Ltd

1 2016

Contents

Preface

The investigation of the propagation channel is becoming more and more important in modern wireless communication. The demand for spectral efficiency motivates exploitation of all channels that can possibly be used for communications. Nowadays, a common trend for designing physical layer algorithms is to adapt the transceiving strategy, either by maximizing the diversity gains or by utilizing the coherence of the channels to improve the signal-to-noise power ratio.

Dr. Xiang Cheng and I have been working on topics relevant to channel characterization for years. My major research has been focused on measurement-based stochastic channel modeling using high-resolution estimates of the channel parameter from real measurement data. Xiang's work concentrates more on accurate yet easy-to-use channel modeling and simulation based on geometry-based stochastic channel modeling approach. This book is intended to cover both theoretical and experimental studies of channels by merging Xiang's and my own study results, obtained in the last decade. Most of the content has been published in journals and conference proceedings. New results that are still under review for publication are also addressed in order to give a complete presentation of specific topics. In general, the book can be viewed as a collection of the latest results in the field of theoretical and experimental channel characterization. The contributions of Dr. Xiang Cheng and myself to this book are equivalent.

There are already several books dedicated to channel investigations (Durgin 2003, Koivunen 2007, Parsons 2000, Saunders 1999, Pätzold 2002). Our book includes more of an introduction to the methods used for the steps of channel characterization than these earlier studies, instead of presenting only the final results. From this point of view, our book tells more complete stories about channels, linking the methods applied in the different stages of channel analysis. Furthermore, combining the description of theoretical and the empirical methods in one book helps the reader conceive more clearly the merits of these methods. Another feature of the book is that we also cover the methods used for extracting the parameters of generic models. Normally, books address channels in one or two chapters and do not describe and comment on the methods used for characterizing them. An example is the book by (Correia 2001). However, since readers usually want to know how realistic the models are, it is important to give a clear view of the underlying methods. Although some of the methods described in this book have also been described in the literature for general cases (McLachlan and Krishnan 1997, Stoica and Moses 1997), we focus on the adaptation of methods applied for analyzing the propagation channels.

This book can be used as a textbook for the courses dedicated for propagation channel characterization, or for parts of courses that focus on wireless communication systems and networks. We organize the book in such a way that the chapters are self-contained and can be selected individually for specific topics. We start in Chapter 2 by introducing the phenomena of propagation in wireless communication channels and the terminologies, and also the parameters used to characterize their properties. Then, in Chapter 3, the generic parametric models applied for representing multiple components in channel impulse responses are introduced. For stochastic behaviors of channels represented by these model parameters, statistical models are needed. We, therefore, review the approaches adopted in channel characterization and modeling; from their theoretical aspects in Chapter 4 and by using measurements in Chapter 5. The impacts of measurement equipment on the observations and model accuracy are also discussed in Chapter 5. Chapters 6 and 7 introduce the high-resolution channel-parameter estimation methods for extracting the parameters of the generic channel models from measurement data, based on deterministic specular-path models and statistical models, respectively. Chapter 8 elaborates the modeling procedure and key techniques for constructing stochastic models based on parameter estimates. At the end of the book, Chapter 9 illustrates specific channel models for different communication systems as examples of the methods and techniques introduced in this book.

Acknowledgements

This book is based on the publications written by the authors and their colleagues in recent decades. We are indebted to colleagues and students who have made valuable suggestions and comments leading to many important changes. In this regard, we are particularly grateful to the co-authors of the publications relevant to this book: Prof. Bernard Fleury, Aalborg University; Prof. Troels Pedersen, Aalborg University; Attaphongse Taparugssanagorn, Asian Institute of Technology; Dr. Li Tian, ZTE Technology; Stan X. Lu, Huawei Technology Company; Dr. Zhimeng Zhong, Huawei Technology Company; and Nicolai Czink. We also wish to acknowledge the valuable comments of the manuscript reviewers: Prof. Bo Ai, Beijing Jiaotong University and Prof. Jianhua Zhang, Beijing University of Post & Telecommunication.

References

Durgin GD 2003 *Space-Time Wireless Channels*. Pearson Education.

(ed. Correia L) 2001 *Wireless Flexible Personalised Communication (COST 259: European Co-Operation in Mobile Radio Research)*. John Wiley & Sons.

(ed. Pätzold M) 2002 *Mobile Fading Channels*. Wiley.

Koivunen J 2007 Characterization of MIMO propagation channel in multilink scenarios PhD thesis Helsinki University of Technology.

McLachlan GJ and Krishnan T 1997 The EM Algorithm and Extensions vol. 1. Wiley Series in Probability and Statistics.

Parsons JD 2000 *The Mobile Radio Propagation Channel* 2nd edn. John Wiley and Sons.

Saunders SR 1999 *Antennas and Propagation for Wireless Communication Systems*. John Wiley & Sons.

Stoica P and Moses RL 1997 *Introduction to Spectral Analysis*. Prentice Hall.

List of Acronyms and Symbols

List of Acronyms

3GPP	Third Generation Partnership standards bodies
5G	Fifth generation wireless communications
AEE	average estimation error
AFD	average fading duration
AGV	autonomous guided vehicles
AoA	azimuth of arrival
AoD	azimuth of departure
ARA	acceptance rejection algorithm
ARIMA	autoregressive integrated moving average
AS	azimuth spread
ASA	array size adaptation
CBM	correlation-based model
CCDF	complementary cumulative distribution functions
CLEAN	an iterative beam removing technique
COMET-EXIP	covariance matching estimator-extended envariance principle
CoMP	cooperative multipoint
COST	Commission of Science and Technology
CRLB	Cramér–Rao lower bound
DECT	Digital European Cordless Telephone
DER	direction estimation range
DFER	Doppler frequency estimation range
DI	diffuse component
DML	deterministic maximum likelihood
DoA	direction of arrival
DoD	direction of departure
DRAG	dynamic range of array gain
ECM	environment characterization metric
EM	expectation-maximization
Emp.	empirical
EoA	elevation of arrival
EoD	elevation of departure
ESPRIT	estimation of signal parameters via rotational invariance techniques

Est.	estimated
E-step	expectation step
F2M	fixed to mobile
FB	Fisher–Bingham
FFHMA	fast frequency hopping multiple access
FHMA	frequency hopping multiple access
GAM	generalized array manifold
GBDM	geometry-based deterministic model
GBSM	geometry-based stochastic modeling
GR	Gerschgorin radii
HFB	higher frequency band
HRPE	high-resolution parameter estimation
IMT	international mobile telecommunications
IS-GBSM	irregular-shaped GBSM
ISI	improved initialization and search
ISM	industrial, scientific and medical bands
JADE	joint angle and delay estimation
LCR	level cross rate
LoS	line-of-sight
MD	mobile scatterer
MEA	method of equal area
MEDS	method of exact Doppler spread
METIS	Mobile and Wireless Communications Enablers for the Twenty-Twenty Information Society
MIMO	multiple-input, multiple-output
ML	maximum likelihood
MMEA	modified method of equal area
MODE	method of direction estimation
M-step	maximization step
MUSIC	multiple signal classification
NC-ML	non-coherent-maximum-likelihood
NFD	Newton forward difference
NGSM	non-geometric stochastic models
NLoS	non-line-of-sight
NSL	normalized side-lobe level
OLOS	obstructed line-of-sight
OMUSIC	orthonormal-basis MUSIC
OSM	orthogonal stochastic measure
PDF	probability density function
PDP	power delay profile
PE	pseudo-envelope
PMM	propagation-motivated model
PN	pseudo-noise
PSD	power spectral density
PSM	parametric stochastic models
RF	radio frequency

RIMAX	Richter's maximum likelihood estimation
RMSEE	root mean square estimation error
RS-GBSM	regular-shaped GBSM
Rx	receiver
SAGE	space-alternating generalized expectation-maximization
SCM	spatial channel model
SCME	spatial channel model enhanced
SD	static scatterer
SIOD	space-invariance of determinant
SISO	single-input, single-output
SML	stochastic maximum likelihood
SNR	signal-to-noise ratio
SoS	sum of sinusoids
SS	specular-scatterer
ST	space time
SVD	singular-value decomposition
SW	switch
TDL	tap-delay line
TDM	time-division-multiplexing
TEM	transverse electric and magnetic wave
Tx	transmitter
ULA	uniform linear array
V2V	vehicle to vehicle
Vec-MUSIC	vector-MUSIC
vMF	von-Mises–Fisher
VTD	vehicular traffic density
WINNER	wireless world initiative new radio
WSS	wide-sense stationary
WSSUS	wide-sense stationary uncorrelated-scattering
XPD	cross-polarization discrimination

List of Symbols

$\mathbb{R}$	real line
$\mathbb{C}$	complex plane
$\mathbb{S}^p$	p-dimensional sphere
$\mathbb{D}$	domains
$\otimes$	Kronecker product
$\odot$	Hadamard – that is, element-wise – product
$\Re\{\cdot\}$	real part of the complex number given as an argument
$\mathcal{I}\{\cdot\}$	imaginary part of the complex number
$\|\cdot\|_{\mathrm{F}}$	Frobenius form of the vector or matrix given as an argument
$\|\cdot\|$	absolute value of the given argument
$\det(\cdot)$	determinant of the matrix given as an argument
$\mathrm{tr}(\cdot)$	trace of the matrix given as an argument

$(\cdot)^{\mathrm{H}}$	Hermitian of the vector or matrix given as an argument
$(\cdot)^{\mathrm{T}}$	transpose of the vector or matrix given an argument
$(\cdot)^*$	complex conjugate of the scalar given as an argument
$(\cdot)^\dagger$	pseudo-inverse of the matrix given as an argument
$(\ \cdot\ )$	scalar product of the given arguments
$\delta_{(\cdot)}$	the Kronecker delta
$\delta(\cdot)$	Dirac delta function
$\boldsymbol{I}_{(\cdot)}$	an identity matrix of dimension given as an index
$\mathrm{diag}(\cdot)$	diagonal matrix with diagonal elements listed as argument
$\lambda_d(\cdot)$	the dth eigenvalue of the matrix given as an argument
$\Pi_{(\cdot)}$	the projection operator onto the column space of the matrix given as an argument
$\boldsymbol{c}(\cdot)$	array response
$\boldsymbol{c}'(\cdot)$	the first derivative of the array response
$\Gamma(\cdot)$	the gamma function
$I_n(\cdot)$	the modified Bessel function of the first kind and order n
$\sigma_{(\cdot)}$	standard deviation of random variable given as an argument
$\boldsymbol{\theta}$	parameter vector
$\boldsymbol{\theta}_{\mathrm{S}}$	parameters with indices specified in a set S in a subset of $\{1, \ldots, p\}$
$\boldsymbol{\theta}_{\tilde{\mathrm{S}}}$	parameters with indices listed in the complement of S intersected with $\{1, \ldots, p\}$
X^{S}	hidden-data space for $\boldsymbol{\theta}_{\mathrm{S}}$
S^i	index set in the ith iteration of the SAGE algorithm
$\boldsymbol{Y}(t)$	output signal from an antenna array
$\boldsymbol{W}(t)$	noise vector
D	total number of path components
$\mathcal{R}_1$	the region where the Tx array is confined
$\mathcal{R}_2$	the region where the Rx array is confined
$\boldsymbol{\theta}_\ell$	parameter vector associated with the ℓth path component
$\boldsymbol{A}_\ell$	the polarization matrix of the ℓth propagation path
$f_{k,m,p}(\boldsymbol{\Omega})$	the field pattern of the mth element of array k for polarization p
$\boldsymbol{u}(t)$	the input signal vector
$\boldsymbol{\Omega}$	a unit direction vector
T_t	a sounding period
T_s	a sensing period of a Rx antenna
T_r	a period separating two consecutive sensing intervals
T_{cy}	separation between the beginnings of two consecutive measurement cycles
T_g	the guard interval
σ_w^2	noise variance
λ	wavelength
M_1	the number of antennas in the transmitter site
M_2	the number of antennas in the receiver site
$\boldsymbol{r}_{k,m}$	the location of the mth element of array k

α_{ℓ,p_2,p_1}	weight of the polarization component with Tx polarization p_1 and Rx polarization p_2
$\Lambda(\cdot)$	log-likelihood function of the parameter(s) given as an argument
ν	Doppler frequency
$\breve{\phi}$	estimation error of nominal azimuth of arrival
$\sigma_{\breve{\phi}}$	azimuth spread
$h(t;\boldsymbol{\theta})$	the spread function of the radio channel
$P_d(\boldsymbol{\theta})$	power spectrum of the dth path component with respect to $\boldsymbol{\theta}$
$P(\boldsymbol{\theta})$	power spectrum of the correlator output in the Rx with respect to $\boldsymbol{\theta}$
$\boldsymbol{H}(t)$	channel transfer matrix
$\tilde{\boldsymbol{H}}(t)$	channel matrix distorted by noise
$R_{(\cdot)}(\tau)$	autocorrelation function of the process given as the argument in the subscript
$R_{(\cdot)(\cdot)}(\tau)$	cross-correlation function of the two processes given as arguments
κ	concentration parameter
$\bar{\phi}$	nominal azimuth
σ_Ω	direction spread
q	differential order in the Newton forward-difference formula
h	step size in the Newton forward-difference formula
Δ	a forward shift operator in the Newton forward-difference formula
β	ovalness parameter in the Fisher-Bingham 5 distribution

1

Introduction

1.1 Book Objective

The characteristics of the propagation channel are of great importance for designing wireless communication systems, analyzing communication qualities, and simulating the performance of networks. However, in most books on wireless communications, propagation channels are usually presented in only one or two chapters, which describe the fundamental characteristics of channels – for example path loss, shadowing, and multipath fading – and present some standard models. Since the procedures for measuring the wireless channels, the methodologies adopted for parameter estimation, and the modeling approaches implemented are neglected in these books, it is impossible for readers to understand how the models are established for specific scenarios. This also results in suspicions about the applicability of models, and questions also arise about the appropriateness for implementation in channel simulations.

Furthermore, fast-growing wireless communication networks and services bring greater demands for high spectral efficiency. Numerous techniques have been used, all essentially exploiting the resources from propagation channels. For example, parallel spatial channels are resolved and utilized by multiple-input, multiple-output (MIMO) techniques for diversity or multiplexing. Similar MIMO techniques in other domains, such as in polarizations and in wavefronts, have been developed. It is of no doubt that future wireless system design will be more and more adaptive to the environments in which they are used. Network architecture design is also becoming increasingly complicated in order to make the most use of specific channels. For example, the techniques of distributed antennas, massive MIMO, relay, cooperative transmission, and joint processing all require detailed knowledge of channels in both a stochastic sense and in site-specific scenarios. Therefore, channel characterizations based both on theoretical approaches and real measurements are going to become critical in the future.

Considering the multiple aspects of a channel, it is actually a "mission impossible" to write a book that is sufficiently comprehensive that every topic of channel studies is included. This book is written with the aim of covering only some aspects of the propagation channel:

- the high-resolution approach of analyzing channels based on measurement data
- stochastic channel modeling either using empirical parameters or based on simulation of scattering.

Propagation Channel Characterization, Parameter Estimation and Modelling for Wireless Communications, First Edition.
Xuefeng Yin and Xiang Cheng.

The objectives of this book are threefold. First, the book provides the fundamentals of both empirical measurement-based and theoretical-scattering-based channel modeling. The topics covered are widely spread, touching on the fields of wideband channel measurements, model parameter extraction, stochastic model generation, and theoretical channel modeling. Second, the book provides some updated channel models, which can be used for practical simulations. Engineers in the wireless communication industry can therefore use them to evaluate their system performance. Thirdly, this book highlights ongoing trends, revealing some fresh research results that might be interesting for researchers when designing new systems.

1.2 The Historical Context

1.2.1 Importance of Channel Characterization

The statistical characteristics of channels can significantly influence the design of wireless communication systems. For example, the path-loss model, based on the measurements in specific regions, can be used to determine the appropriate value of the separation between cells, in order to keep the interference below a certain threshold. Shadowing models can be used to determine the maximum and the minimum transmission power in order to avoid blindspots in the coverage. Multipath fading models, which include the fading rate and fading-duration characteristics, can be used to determine the packet length and the transmission rate. Delay spread models can be used to evaluate the frequency selectivity of the environment, so as to determine the coherence frequency bandwidth or the separation of the orthogonal channels in the frequency domain. Doppler frequency spread models can be used to calculate the coherence time of the channel, and therefore determine the cycle duration to renew the estimate of channel coefficients. The models in the spatial domains, for example the cluster-based bidirectional models, can be applied to determine the antenna beamwidth in beamforming applications, or to calculate the degrees of freedom for channels with MIMO configurations. Stochastic models themselves are based on extensive measurements in many environments categorized into specific types, such as outdoor, indoor, urban/suburban, and so on; they are therefore valid in similar environments.

The model parameters can be used to determine the many thresholds used in communication systems. For example, for frequency hopping multiple access systems, the frequency offsets due to the Doppler effect of the channel, and the timing problems due to the multipath arrivals at different time instants, can cause a certain portion of the desired signal's energy to appear in spurious adjacent frequency bins; consequently the detection of the desired signal becomes difficult [Joo et al. 2003], and the detection matrix may have erroneous entries [Yegani and McGillem 1993]. With the knowledge of the delay-Doppler frequency dispersion behavior of channels in certain environments and scenarios, the threshold level of envelope detectors can be appropriately selected. Furthermore, if the instantaneous knowledge of the channel dispersion characteristics is available, the channel can be equalized accordingly.

1.2.2 Single-input, Single-output Channel Models

Channel investigation started at the end of the 1960s [Okumura et al. 1968]. At that time, wireless systems were built for voice communications using frequency division multiple access technique. The channel characteristics of interest when the single-input, single-output (SISO) system was considered was therefore the fading distributions at particular frequencies.

For outdoor scenarios, it has been found that the fading distribution is Rayleigh in a local geographical area with diameter of less than a few hundred wavelengths, and lognormal over large geographical areas [Lee and Yeh 1972, Okumura et al. 1968, Schmid 1970, Turin et al. 1972]. Suzuki [1977] considered more distributions, including the Nakagami and lognormal distributions, to fit the empirical data. It was found that the Rayleigh distribution is not always a good fit for most data, and that the lognormal distribution is often better. A possible reason for this observation is that the distribution – actually a mixture of Rayleigh distributions with a lognormal mixing distribution – is an intermediate distribution between the Rayleigh and the lognormal distributions [Suzuki 1977].

For indoor propagation environments, the SISO channel models have been established for the line-of-sight (LoS) and obstructed LoS (OBS) scenarios, as in the factory and open-plan office cases [Kozlowski et al. 2008, Rappaport and Seidel 1989, Rappaport et al. 1991, Saleh and Valenzuela 1987, Seidel et al. 1989, Yegani and McGillem, 1989a,b, 1991]. The motivation for investigating indoor SISO channels is to provide models for indoor deployment of radio systems that accommodate data rates up to 1 Mb/s. Such systems include the Digital European Cordless Telephone (DECT) 802.41 and WLAN (IEEE 802.11) standards, as well as the communication systems for autonomous guided vehicles (AGVs) [Rappaport et al. 1991]. The interesting characteristics of the channels in indoor environments include the path loss and delay spread. It has been found that the delay spread can be several times greater in unpartitioned factory buildings than in partitioned office buildings [Hawbaker and Rappaport 1990b]. Besides the large-scale parameters, the detailed wideband characteristics of the channel, for example the dispersion of the channel in the delay domain, has been examined. For example, Hawbaker and Rappaport [1990a] found the so-called "pulse overlapping" phenomenon, which revealed that even in the LoS scenario the OBS path components can be added to the LoS path components within the resolution of transmitted pulse, resulting in so-called multipath fading.

Furthermore, resolvable rays in the time domain have been applied to modeling channels. This kind of model was called a discrete model. For outdoor environments, discrete channel models consist of discrete rays or discrete peaks of the power-delay profiles [Cox and Leck 1975, Turin et al. 1972]. The magnitude of each ray can be set to follow the lognormal distribution [Suzuki 1977]. The correlation bandwidth is also applied as a model parameter for channels [Cox and Leck 1975], and this is large when the channel-delay profile exhibits several dominating discrete peaks, but small when multipath is severe [Cox 1972]. Since the channel-impulse response in the delay domain is available, the distribution of the number of paths, and the mean and standard deviations of logarithmic path strength are considered for channel characterization [Cox 1972]. Furthermore, by using multiple observations of the channel, the Doppler frequency spectrum has also been computed and used for modeling the channel [Cox 1973]. In addition, the trend of describing the channel properties in two dimensions has appeared in the literature [Cox 1973]. The Doppler spectra versus delay and the distribution of path strength versus delay have been studied for outdoor channels in urban environments [Cox 1973]. The small-scale characteristics of the channel – the channel property at specific delays – have become important for modeling.

Some important observations have been obtained through measurement. For outdoor urban environments, the excess delay of a channel at 900 MHz can be up to 9–10 μs [Cox 1973];

the delay spread, defined as the square root of the second central moment of the power-delay profile, is 2–2.5 μs. The path with 0.1 μs resolution exhibits a Rayleigh distribution, inferring that the fading coefficients for the first arrival path can be modeled as a Gaussian random process. For paths with different delays, uncorrelated scattering is confirmed by the observation that their Doppler frequency power spectra are quite different. The conclusion that paths with different delays are uncorrelated seems more useful for urban environments. Some authors have proposed to use correlated paths to construct discrete models, but this contradicts the observations of Cox [1973].

For indoor manufacturing environments, Yegani and McGillem [1991] provided the statistics for channels in different sites in a factory under four scenarios with different settings of LoS/OBS, and sparsely or densely distributed scatterers. It was found that the interarrival times of the paths were well modeled by the Weibull distribution, the number of paths by the modified beta distribution, and the path-gain coefficients by the Rayleigh, Rician, and lognormal distributions. The values of the parameters of these distributions were reported by the authors. It is interesting to observe that the average number of paths for different sites at a fixed threshold of signal strength is about the same, an indication that the statistics of the number of paths arriving at the receiver is not very sensitive to the topography of the factory site. Furthermore, the geometry of the factory and the layout of the working area have a strong influence on the distribution of the path-gain coefficients. There are also some new findings, for example when the dynamic range is not selected the path-gain coefficients follow the lognormal distribution regardless of the LoS, OBS, or how densely distributed the scatterers are. As for the threshold, when this is greater than −10 dB the path-gain distribution follows the Rician PDF, but when lower than −10 dB, the Rayleigh distribution provides a better fit. Thus, the estimated PDF for gain coefficients depends on the level of the dynamic range set at the receiver.

The research into channels for SISO has evolved into multiple areas. For example, polarization characteristic have been investigated since the 1970s, when polarization diversity was used to combat multipath fading. Employing orthogonally polarized channels over the same microwave link for satellite communications can result in twice the system capacity as when using single-polarized antennas [Lee and Yeh 1972]. In 2001, Andrews et al. [2001] pointed out that six channels without any correlation can immensely improve the transmission rate and system capacity of a wireless communication system, by polarization in a scattering-rich environment. Channel models have been proposed that can be used to generate the channel responses with an arbitrary pair of vertical and horizontal polarizations at both transmitter and receiver sites [3GPP 2007, Jeon et al. 2012]. Besides the cross-polarization discriminations (XPDs) of individual propagation paths, these models also involve the responses of antennas in different polarizations.

Jiang et al. [2007] studied the correlation coefficients for both copolarized and crosspolarized channels. They found that:

- polarization decorrelation outperforms spatial decorrelation in the strong LOS scenario
- horizontally polarized channels are more correlated than vertically polarized channels
- the correlation of copolarized channels increases as the Rician K factor increases
- channels have much higher correlation in the elevation domain.

A strong conclusion was that the crosscorrelation of crosspolarized channels is not affected by the environment, while the performance of copolarized channels is scenario dependent.

1.2.3 Spatial Channel Models (SCMs)

Estimating the direction or bearings of incoming signals has been a research topic for years. The original objective was for signal detection and estimation, including radar target tracking and component separation. The methods used for estimating direction of arrival are similar to those used in time-series spectral analysis and they are applied specifically with the samples obtained from spatially distributed arrays of sensors, including antennas for receiving electromagnetic waves and microphones for acoustic signals.

The study of the arrival angles of signals as part of the design of communication systems can be traced back to the 1970s. For example, Lee and Brandt [1973] found from field measurements of mobile radio signals that the signal arrival is concentrated at elevation angles lower than 16°. Based on this finding, an omnidirectional antenna with vertical directivity is usually selected to increase the average received signal strength.

There are also many practical concerns that require a knowledge of the spatial characteristics of a channel. For example, when MIMO techniques are used in communication systems, the spatial diversity and/or multiplexing gains need to be evaluated based on realistic modeling of the covariance of the spatial channels. Furthermore, in the case where the beamforming technique is used in a base station, it is necessary to know the distribution of the energy in the direction of arrival; in other words how the energy is concentrated and what its spread is in the dominant path. Additionally, with directional parameters, the propagation of the waves can be easily visualized when the actual constellation of the scatterers is presented for specific environments. Geometry-based channel modeling (GBSM) has flourished in the last decade. One major reason is that channel dispersion in the directional domains can be obtained by spectral analysis of the measurement data.

Spatial–spectral analysis methods can be categorized into two classes: spectral-based methods and parametric-model-based methods. Theoretically, conventional methods such as the periodogram [Schuster 1898] and the correlogram [Chatfield 1989], which belong to the category of spectral-based methods, are not applicable in many cases due to the limited spatial aperture of the sensor array and the responses of the sensors. Eigenstructure-based methods have therefore been widely adopted. These include the MUltiple SIgnal Characterization (MUSIC) algorithm and its many variants [Asztély and Ottersten 1998, De Jong and Herben 1999, Friedlander 1990, Jäntti 1992, Kaveh and Barabell 1986, Krim and Proakis 1994, Krim et al. 1992, Rao and Hari 1993, Rao 1990, Salameh and Tayem 2006, Stoica and Nehorai 1989, Wang et al. 2001] and other subspace-based methods, such as the propagator method [Marcos et al. 1994; 1995, Tayem and Kwon 2005], and Estimation of Signal Parameters by Rotational Invariance Techniques (ESPRIT) (which does not result in a spectrum, but provides analytically the solutions for parameter estimates) [Asztély and Ottersten 1998, Jäntti 1992, Paulraj 1986].

In the 1990s, algorithms based on parametric models of channels were used to extract channel-model parameters from measurement data. The maximum-likelihood estimator and the approximation of it with an iterative estimate updating procedure can be used to estimate both the deterministic parameters and the statistical parameters of channels depending on the generic model applied. These algorithms are also called super-resolution methods, as they may achieve higher resolution than conventional spectral-based methods. Typical examples of these algorithms are the expectation-maximization (EM) algorithm [Frenkel and Feder 1999, Moon 1997, Nielsen 2000, Zhang et al. 2001], the space-alternating generalized expectation-maximization (SAGE) algorithm [Fessler and Hero 1994, Fleury et al. 1999,

Taparugssanagorn et al. 2007, Yin et al. 2007; 2006a], Richter's Maximum likelihood (RiMAX) algorithm [Richter, 2004; 2005, Richter et al. 2000; 2003], and the variants of the SAGE algorithm produced by adopting models different from the widely used resolvable specular path model [Bengtsson and Ottersten 2000, Yin et al. 2006b]. The papers cited here analyse and compare these algorithms, and also cover aspects such as the impact of the antenna arrays used for data collection and the influence of the model mismatch between the usually applied resolvable specular-path model and the true scattering effect.

These algorithms are applied to extract multidimensional parameters of channels from measurement data. The parameters include the direction of arrival, direction of departure, delay, Doppler frequency and polarization matrices of individual propagation paths. The estimates are used to construct the stochastic geometry-based or ray-based channel models, such as:

- the well known 3GPP TR 25.996 models [3GPP 2007]
- the Wireless World Initiative New Radio (WINNER) II spatial channel model-enhanced [IST 2007]
- the IMT-advanced channel models [ITU 2008].

In the spatial channel model, clustering of multiple paths is a necessary step for generating the small-scale parameters of the channel, such as the cluster delay spread, cluster angular spread, and the time-variant behavior of the clusters. How to appropriately cluster the multiple propagation paths has been discussed in literature [Czink et al. 2005a,b]. At first, visual-inspection-based clustering methods were proposed [Czink 2007a], but this is impractical for a large amount of measurement data. Moreover, the clustering results may not be unique when users have different opinions about the clusters. Automatic clustering methods, requiring minimum interactions of users, were designed as an alternative. These methods make use of the so-called multipath component distance measure (or environment characterization metric) to group the paths into a cluster [Czink et al. 2005b; 2006]. Readers are referred to Czink 2007b for the detailed description of various clustering methods and their performance.

The multipath clustering concept has also been extended to the modeling of time-variant channels [Czink et al. 2007a,b, Xiao and Burr 2008, Xiao et al. 2007]. The objective of introducing time-variant clusters is to reduce the computational complexity when generating spatial-correlated time-variant channel realizations or channel matrices. The parameters of the clusters, especially the centroid of clusters, are tracked through consecutive channel snapshots [Czink and Galdo 2005, Czink et al. 2007b]. It was found by Czink et al. [2007b] that for both outdoor and indoor scenarios clusters can be easily tracked. The histogram of the logarithmic cluster lifetime follows an exponential distribution of the cluster lifetime for outdoor scenarios.

1.2.4 Channel Models for 5G

Recently, the fifth generation (5G) of wireless communication has attracted tremendous research attention. The increasing demand for high-data-rate communications for 5G has motivated use of signals transmission at higher frequency bands (HFB) beyond 6 GHz [Andrews et al. 2014].

The European Seventh Framework project "Mobile and wireless communications enablers for the twenty-twenty information society" (METIS) proposed for 5G communications a frequency band ranging from 450 MHz to 85 GHz [Jämsä et al. 2014]. Channel characterization for HFBs has been focused on 60 GHz millimeter (mm-) wave propagation [Collonge et al. 2004, Correia and Frances 1994, Daniels and Heath 2007, Moraitis and Panagopoulos 2015, Piersanti et al. 2012, Smulders 2009, Weiler et al. 2015, Yang et al. 2006]. The large-scale characteristics, such as path loss, shadow fading, frequency selectivity, and the influence of human bodies and different materials on channels, have been investigated for the 60 GHz frequency [Collonge et al. 2004, Correia and Frances 1994, Piersanti et al. 2012, Yang et al. 2006]. High-resolution parameter estimation (HRPE) algorithms, such as space-alternating generalized expectation-maximization (SAGE) [Fleury et al. 1999] and Richter's maximum likelihood estimation (RiMAX) [Richter and Thomä 2005], have been used to extract multipath components (MPCs) from the outputs of virtual linear or planar arrays [Gustafson et al. 2014b; 2011, Martinez-Ingles et al. 2013]. Multipath clusters were identified and their statistics have been reported as stochastic channel models for various propagation scenarios [Gustafson et al. 2014a,b]. Besides the mm-wave frequency bands, more channel measurement studies for other HFB frequencies have been carried out recently, such as

- 10–11 GHz [Belbase et al. 2015, Kim et al. 2015, Weiler et al. 2015]
- 28–38 GHz [Azar et al. 2013, Rappaport et al. 2013, Wu et al. 2015]
- 70–73 GHz [Karttunen et al. 2015, Nie et al. 2013a, Semkin et al. 2015, Zhang et al. 2014]
- 81–86 GHz [Semkin et al. 2015].

A common setup in these studies is that antennas with narrow half-power beamwidth (HPBW), such as pyramidal horn antennas, are used [MacCartney et al. 2013, Samimi et al. 2013, Zhang et al. 2014]. One motivation for using narrow-HPBW antennas in such measurements is that the large antenna gain resulting can counteract the significant path-loss in HFB propagation. Furthermore, if the antenna's HPBW is narrow enough, a direction-scan-sounding (DSS) method can be used by rotating the antenna's axis towards different directions: the channel is "scanned" in multiple directions. Based on these measurement studies, power delay profiles (PDPs) and path-loss models for omnidirectional channels have been synthesized from directional observations, and channel models for HFB wave propagation have been proposed for outdoor cellular, backhaul, and indoor propagation scenarios [Hur et al. 2014, MacCartney and Rappaport 2014, Nie et al. 2013b].

1.2.5 Other Kinds of Channel Model

Future generations of wireless communication systems will employ new techniques that rely on channel modeling for more complex network constellations. For example, design of the distributed antenna, cooperative relay and joint processing systems, or algorithms require models of co-existing multilink channels. Some preliminary works have been done on multilink correlation channel models [Yin et al. 2011; 2012a,b,c,d].

Other channel models exist for the non-stationary scenarios and distributed scenarios. Some models focus on the specific behavior of the channels, such as their reciprocity behavior with respect to time and frequency, their polarization characteristics, interference, and the LoS and NLoS probabilities and so on.

1.3 Book Outline

This book contains ten chapters. It starts by introducing the phenomena of propagation considered in wireless communication channels and defines the terminologies and parameters used to characterize their properties. Then generic parametric models for channel multiple components are introduced. The approaches adopted in channel characterization and modeling from theoretical and experimental aspects are elaborated respectively. A focus of the book is on high-resolution channel parameter estimation methods for extracting deterministic specular-path components and statistical dispersive components from measurement data. Next, the general procedures and key techniques adopted for constructing stochastic models based on parameter estimates are described. Finally, some specific channel models are presented as examples of implementing the methods and techniques introduced in this book. Below is a detailed description of the content in individual chapters.

Chapter 2 “Characterization of Propagation Channels” begins by introducing three phenomena of fading in wireless channels: path loss, shadowing, and multipath fading. Then the stochastic characterizations of these phenomena are described. Following that, we emphasize the duality relationship between the selectivity and dispersion of multipath fading, and also explain the definition of the wide-sense stationary uncorrelated-scattering (WSSUS) assumption and its applicability in practice. In this chapter, a review of propagation channel modeling is provided, describing the different approaches that aim to accurately and/or efficiently generate channel impulse responses with desired channel characteristics. These channel models are categorized into two main classes: MIMO channel models and vehicle-to-vehicle (V2V) channel models.

Chapter 3 “Generic Channel Models” introduces the basic mechanisms of radio propagation, the representation of channels in terms of multidimensional spread functions, and the generic models usually applied for channel parameter estimation. These generic models include the specular-path model, dispersive-path model, time-evolution model for the path parameters, and power spectral density models for individual components. Furthermore, the influence of system configurations – for example the amplify-and-forward relay systems – on the format of the generic models is also described. Finally, the model of the received signal in, for example, the channel sounding context is given.

Chapter 4 “Geometry-based Stochastic Channel Modeling” describes the difference between the geometry-based deterministic modeling approach and the geometry-based stochastic modeling approach, the details of the latter for the regular-shaped and irregular-shaped scenarios. Furthermore, the simulation methods of the theoretical/mathematical reference model in reality are introduced.

Chapter 5 “Channel Measurements” introduces the methodologies, equipment, and procedures of measuring the impulse responses of propagation channels. Besides the general description of channel measurements, we go one step further to discuss the influence of imperfections that occur during the calibration of the equipment on the measurement results. For example, we analyze the impact of the existence of time-variant phase noise and the inconsistency between the measured and real radiation-pattern measurements. We also, for the first time, provide experimental analysis on how the directionality of the radiation of antennas influences parameter estimation results. All of these studies are based on experimental data. We focus on introducing the phenomena, and only briefly describe several solutions available at present. The readers are encouraged to discover more solutions for these important issues.

Chapter 6 "Deterministic Channel Parameter Estimation" focuses on the high-resolution parameter estimation algorithms based on the generic specular-path model. These are used for extracting the parameters of individual path components from measurement data. Besides the traditional SAGE and RiMAX algorithms, which have been used extensively for parameter estimation in MIMO measurement-based channel modeling, we also introduce some newly developed estimation methods, which are expected to be used in the very near future for accurate channel modeling. For time-varying scenarios, where the path parameters evolve with respect to time, a tracking scheme based on the particle filter concept is elaborated. For each method introduced, we include the results of performance evaluations that were carried out by processing real measurement data.

Chapter 7 "Statistical Channel Parameter Estimation" describes another category of parameter estimation methods, which use generic models for the power spectral density of the channel to estimate the statistical parameters, such as the second moments of the channel/channel components. Two methods are introduced: the generalized array manifold (GAM) model-based approach and the power spectral density-based approach. These methods, although not been widely adopted, can result in more accurate estimates of channel statistics. We also describe the practical limitation of the methods when used in practice.

Chapter 8 "Measurement-based Statistical Channel Modeling" systematically describes modeling procedures that are based on measurements in detail. Both the common issues in the modeling – clustering algorithms and data segmentation – are discussed, along with specific approaches and recent new topics in channel modeling, such as non-stationarity modeling, relay, and cooperative multipoint (CoMP) channel modeling.

Chapter 9 "In Practice: Channel Modeling for Modern Communication Systems", as the last chapter of the book, provides the examples of models established using the methods introduced in earlier chapters. These examples cover many common scenarios that have been popular for channel modeling in recent years. The readers can use this chapter as a collection of models recently developed for the MIMO, vehicular, relay, CoMP, and multilink channels. Students can undertake trials for other scenarios based on the procedures presented in these examples.

Bibliography

3GPP 2007 TR25.996 v7.0.0: Spatial channel model for multiple input multiple output (MIMO) simulations (release 7). Technical report, 3GPP.

Andrews J, Buzzi S, Choi W, Hanly S, Lozano A, Soong A and Zhang J 2014 What will 5G be?. *IEEE Journal on Selected Areas in Communications* **32**, 1065–1082.

Andrews MR, Mitra PP and deCarvalho R 2001 Tripling the capacity of wireless communications using electromagnetic polarization. *Nature* **409**, 316–318.

Asztély D and Ottersten B 1998 The effects of local scattering on direction of arrival estimation with MUSIC and ESPRIT *Proceedings of IEEE International Conference on Acoustics, Speech, Signal Processing (ICASSP)*, vol. 6, pp. 3333–3336.

Azar Y, Wong G, Wang K, Mayzus R, Schulz J, Zhao H, Gutierrez F, Hwang D and Rappaport T 2013 28 GHz propagation measurements for outdoor cellular communications using steerable beam antennas in New York City *Proceeding of IEEE International Conference on Communications (ICC)*, pp. 5143–5147, Budapest.

Belbase K, Kim M and Takada J 2015 Study of propagation mechanisms and identification of scattering objects in indoor multipath channels at 11 GHz *Proceedings of European Conference on Antenna and Propagation (EuCAP)*, pp. 1–5, Lisbon.

Bengtsson M and Ottersten B 2000 Low-complexity estimators for distributed sources. *IEEE Transactions on Signal Processing* **48**(8), 2185–2194.

Collonge S, Zaharia G and Zein G 2004 Influence of the human activity on wide-band characteristics of the 60 GHz indoor radio channel. *IEEE Transactions on Wireless Communications* **3**(6), 2396–2406.

Correia L and Frances P 1994 Estimation of materials characteristics from power measurements at 60 GHz *Proceedings of 5th IEEE International Symposium on Personal, Indoor and Mobile Radio Communications*, vol. 2, pp. 510–513.

Cox D 1973 910 MHz urban mobile radio propagation: Multipath characteristics in New York City. *IEEE Transactions on Communications* **21**(11), 1188–1194.

Cox D and Leck R 1975 Correlation bandwidth and delay spread multipath propagation statistics for 910-MHz urban mobile radio channels. *IEEE Transactions on Communications* **23**(11), 1271–1280.

Cox DC 1972 Delay doppler characteristics of multipath propagation at 910 MHz in a suburban mobile radio environment. *IEEE Transactions on Antennas and Propagation* **AP-20**(5), 625–635.

Czink N 2007a *The random-cluster model – a stochastic MIMO channel model for broadband wireless communication systems of the 3rd generation and beyond* PhD thesis Technische Universitat Wien, Vienna, Austria, FTW Dissertation Series.

Czink N 2007b *The random-cluster model – a stochastic MIMO channel model for broadband wireless communication systems of the 3rd Generation and beyond* PhD thesis Technology University of Vienna, Department of Electronics and Information Technologies.

Czink N and Galdo GD 2005 Validating a novel automatic cluster tracking algorithm on synthetic IlmProp time-variant MIMO channels. Technical Report TD-05-105, COST273.

Czink N, Bonek E, Yin X and Fleury BH 2005a Cluster angular spreads in a MIMO indoor propagation environment *Proceedings of the 16th IEEE International Symposium on Personal, Indoor and Mobile Radio Communications (PIMRC'05)*, vol. 1, pp. 664–668, Berlin, Germany.

Czink N, Cera P, Salo J, Bonek E, Nuutinen J and Ylitalo J 2005b Automatic clustering of MIMO channel parameters using the multi-path component distance measure *Proceedings of Wireless Personal Multimedia Communications WPMC'05*, Aalborg, Denmark.

Czink N, Galdo GD, Yin X and Meklenbrauker C 2006 A novel environment characterisation metric for clustered MIMO channels used to validate a SAGE parameter estimator *Proceedings of the 15th IST Mobile & Wireless Communication Summit, Myconos, Greece.*

Czink N, Tian R, Wyne S, Eriksson G, Zemen T, Ylitalo J, Tufvesson F and Molisch A 2007a Cluster parameters for time-variant MIMO channel models *Antennas and Propagation, 2007. EuCAP 2007. The Second European Conference on*, pp. 1–8.

Czink N, Tian R, Wyne S, Tufvesson F, Nuutinen JP, Ylitalo J, Bonek E and Molisch A 2007b Tracking time-variant cluster parameters in MIMO channel measurements *Second International Conference on Communications and Networking in China (CHINACOM)*, pp. 1147–1151.

Daniels R and Heath R 2007 60 GHz wireless communications: emerging requirements and design recommendations. *IEEE Vehicular Technology Magazine* **2**(3), 41–50.

De Jong Y and Herben M 1999 High-resolution angle-of-arrival measurement of the mobile radio channel. *IEEE Transactions on Antennas and Propagation* **47**(11), 1677–1687.

(ed. Chatfield C) 1989 *The Analysis of Time Series: An Introduction* 4th edn. Chapman and Hall.

Fessler JA and Hero AO 1994 Space-alternating generalized expectation-maximization algorithm. *IEEE Transactions on Signal Processing* **42**(10), 2664–2677.

Fleury BH, Tschudin M, Heddergott R, Dahlhaus D and Pedersen KL 1999 Channel parameter estimation in mobile radio environments using the SAGE algorithm. *IEEE Journal on Selected Areas in Communications* **17**(3), 434–450.

Frenkel L and Feder M 1999 Recursive expectation-maximization algorithms for time-varying parameters with applications to multiple target tracking. *IEEE Transactions on Signal Processing* **47**(2), 306–320.

Friedlander B 1990 A sensitivity analysis of the music algorithm. *Acoustics, Speech, and Signal Processing [see also IEEE Transactions on Signal Processing], IEEE Transactions on* **38**(10), 1740–1751.

Gustafson C, Bolin D and Tufvesson F 2014a Modeling the cluster decay in mm-wave channels *Antennas and Propagation (EuCAP), 2014 8th European Conference on*, pp. 804–808.

Gustafson C, Haneda K, Wyne S and Tufvesson F 2014b On mm-wave multipath clustering and channel modeling. *IEEE Transactions on Antennas and Propagation* **62**(3), 1445–1455.

Gustafson C, Tufvesson F, Wyne S, Haneda K and Molisch A 2011 Directional analysis of measured 60 GHz indoor radio channels using SAGE *IEEE 73rd Vehicular Technology Conference (VTC Spring)*, pp. 1–5.

Hawbaker D and Rappaport T 1990a Indoor wideband radio propagation measurement system at 1.3 GHz and 4.0 GHz *Vehicular Technology Conference, 1990 IEEE 40th*, pp. 626–630.

Hawbaker D and Rappaport T 1990b Indoor wideband radiowave propagation measurements at 1.3 GHz and 4.0 GHz. *Electronics Letters* **26**(21), 1800–1802.

Hur S, Cho YJ, Lee J, Kang NG, Park J and Benn H 2014 Synchronous channel sounder using horn antenna and indoor measurements on 28 GHz *IEEE International Black Sea Conference on Communications and Networking (BlackSeaCom)*, pp. 83–87, Odessa.

IST 2007 WINNER II channel models.

ITU 2008 Guidelines for evaluation of radio interface technologies for IMT-Advanced.

Jämsä T, Kyösti P and Kusume K 2014 Deliverable D1.2 initial channel models based on measurements. Technical report, Project Name: Mobile and Wireless Communications Enablers for the Twenty-twenty Information Society (METIS), Document Number: ICT-317669-METIS/D1.2.

Jäntti TP 1992 The influence of extended sources on the theoretical performance of the MUSIC and ESPRIT methods: narrow-band sources. *Proc. IEEE Int. Conf. Acoust., Speech, Signal Processing*, pp. II–429–II–432.

Jeon K, Hui B, Chang K, Park H and Park Y 2012 SISO polarized flat fading channel modeling for dual-polarized antenna systems *Information Networking (ICOIN), 2012 International Conference on*, pp. 368–373.

Jiang L, Thiele L and Jungnickel V 2007 On the modelling of polarized MIMO channel. *Proc. European Wireless 2007* pp. 1–4.

Joo J, Moon S, Yoon Y and Kim K 2003 Effects of fast frequency hopping multiple access systems due to the frequency and timing offset under Rayleigh fading *IEEE Wireless Communications and Networking, 2003*, vol. 1, pp. 126–131, Spokane, Washington, USA.

Karttunen A, Haneda K, Järveläinen J and Putkonen J 2015 Polarisation characteristics of propagation paths in indoor 70 GHz channels *Proceedings of European Conference on Antenna and Propagation (EuCAP)*, pp. 1–5, Lisbon.

Kaveh M and Barabell AJ 1986 The statistical performance of the MUSIC and the minimum-norm algorithms in resolving plane waves in noise. *IEEE Transactions on Acoustics, Speech, and Signal Processing* **ASSP-34**(2), 331–342.

Kim M, Takada J, Chang Y, Shen J and Oda Y 2015 Large scale characteristics of urban cellular wideband channels at 11 GHz *Proceedings of European Conference on Antenna and Propagation*, pp. 1–4, Lisbon.

Kozlowski S, Szumny R, Kurek K and Modelski J 2008 Statistical modelling of a wideband propagation channel in the factory environment *Wireless Technology, 2008. EuWiT 2008. European Conference on*, pp. 190–193.

Krim H and Proakis J 1994 Smoothed eigenspace-based parameter estimation. *Automatica* **30**(1), 27–38.

Krim H, Forster P and Proakis J 1992 Operator approach to performance analysis of root-music and root-min-norm. *IEEE Transactions on Signal Processing* **40**, 1687–1696.

Lee W and Yeh Y 1972 Polarization diversity system for mobile radio. *Communications, IEEE Transactions on [legacy, pre-1988]* **20**(5), 912–923.

Lee WY and Brandt R 1973 The elevation angle of mobile radio signal arrival. *IEEE Transactions on Vehicular Technology* **22**(4), 110–113.

MacCartney G and Rappaport T 2014 73 GHz millimeter wave propagation measurements for outdoor urban mobile and backhaul communications in New York City *Proceedings of IEEE International Conference on Communications (ICC)*, pp. 4862–4867, Sydney.

MacCartney G, Zhang J, Nie S and Rappaport T 2013 Path loss models for 5G millimeter wave propagation channels in urban microcells *Proceedings of IEEE Global Communications Conference (GLOBECOM)*, pp. 3948–3953, Atlanta.

Marcos S, Marsal A and Benidir M 1994 Performances analysis of the propagator method for source bearing estimation *Proceedings of IEEE International Conference on Acoustics, Speech and Signal Processing (ICASSP)*, vol. IV, pp. 19–22.

Marcos S, Marsal A and Benidir M 1995 The propagator method for source bearing estimation. *Signal Processing* **42**, 121–138.

Martinez-Ingles MT, Molina-Garcia-Pardo JM, Rodriguez JV, Pascual-Garcia J and Juan-Llacer L 2013 Experimental comparison of UWB against mm-wave indoor radio channel characterization *Proceeedings of IEEE Antennas and Propagation Society International Symposium (APSURSI)*, pp. 1946–1947, Orlando.

Moon T 1997 The expectation-maximization algorithm. *IEEE Signal Processing Magazine* **13**(6), 47–60.

Moraitis N and Panagopoulos AD 2015 Millimeter wave channel measurements and modeling for indoor femtocell applications *Proceedings of European Conference on Antenna and Propagation (EuCAP)*, pp. 1–5, Lisbon.

Nie S, MacCartney GR, Sun S and Rappaport TS 2013a 72 GHz millimeter wave indoor measurements for wireless and backhaul communications *Personal Indoor and Mobile Radio Communications (PIMRC), 2013 IEEE 24th International Symposium on*, pp. 2429–2433.

Nie S, MacCartney GR, Sun S and Rappaport TS 2013b 72 GHz millimeter wave indoor measurements for wireless and backhaul communications *Proceedings of IEEE International Symposium on Personal Indoor and Mobile Radio Communications (PIMRC)*, pp. 2429–2433, London.

Nielsen SF 2000 The stochastic EM algorithm: estimation and asymptotic results. *Bernoulli* **6**(3), 381–570.

Okumura Y, Ohmori E, Kawano T and Fukuda K 1968 Field strength and its variability in VHF and UHF land-mobile radio services. *Review of the Electrical Communication Laboratory* **16**(9), 825–837.

Paulraj, A.; Roy RKT 1986 A subspace rotation approach to signal parameter estimation. *Proceedings of the IEEE* **74**, 1044–1046.

Piersanti S, Annoni L and Cassioli D 2012 Millimeter waves channel measurements and path loss models *Proceedings of IEEE International Conference on Communication (ICC)*, pp. 4552–4556, Ottawa.

Rao B and Hari K 1993 Weighted subspace methods and spatial smoothing, analysis and comparison. *IEEE Transactions on Signal Processing* **41**(2), 788–803.

Rao, B.D.; Hari K 1990 Effect of spatial smoothing on the performance of noise subspace methods. *Proceedings of IEEE International Conference on Acoustics, Speech and Signal Processing* **5**, 2687–2690.

Rappaport T and Seidel S 1989 Multipath propagation models for in-building communications *Mobile Radio and Personal Communications, 1989, Fifth International Conference on*, pp. 69–74.

Rappaport T, Seidel S and Takamizawa K 1991 Statistical channel impulse response models for factory and open plan building radio communicate system design. *IEEE Transactions on Communications* **39**(5), 794–807.

Rappaport T, Sun S, Mayzus R, Zhao H, Azar Y, Wang K, Wong G, Schulz J, Samimi M and Gutierrez F 2013 Millimeter wave mobile communications for 5G cellular: It will work!. *IEEE Access* **1**, 335–349.

Richter A 2004 RIMAX – a flexible algorithm for channel parameter estimation from channel sounding measurements. Technical Report TD-04-045, COST 273, Athens, Greece.

Richter A 2005 *Estimation of radio channel parameters: Models and algorithms* PhD thesis Technische Universität Ilmenau, ISBN 3-938843-02-0 Ilmenau, Germany.

Richter A and Thomä RS 2005 Joint maximum likelihood estimation of specular paths and distributed diffuse scattering *Proceedings of the IEEE 61st Vehicular Technology Conference (VTC-Spring)*, vol. 1, pp. 11–15, Stockholm.

Richter A, Hampicke D, Sommerkorn G and Thoma R 2000 Joint estimation of DoD, time-delay, and DoA for high-resolution channel sounding *Proceedings of IEEE 51st Vehicular Technology Conference (VTC-Spring)*, vol. 2, pp. 1045–1049.

Richter A, Landmann M and Thomä RS 2003 Maximum likelihood channel parameter estimation from multidimensional channel sounding measurements *Proceedings of the 57th IEEE Semiannual Vehicular Technology Conference (VTC)*, vol. 2, pp. 1056–1060.

Salameh A and Tayem N 2006 Conjugate MUSIC for non-circular sources *Proceedings of 2006 IEEE International Conference on Acoustics, Speech and Signal Processing (ICASSP)*, vol. 4, pp. IV–IV.

Saleh A and Valenzuela R 1987 A statistical model for indoor multipath propagation channel. *IEEE Journal of Selected Areas in Communications* **5**(2), 128–137.

Samimi M, Wang K, Azar Y, Wong G, Mayzus R, Zhao H, Schulz J, Sun S, Gutierrez F and Rappaport T 2013 28 GHz angle of arrival and angle of departure analysis for outdoor cellular communications using steerable beam antennas in New York City *IEEE 77th Vehicular Technology Conference (VTC Spring)*, pp. 1–6.

Schmid H 1970 A prediction model for multipath propagation of pulse signals at VHF and UHF over irregular terrain. *IEEE Transactions on Antennas and Propagation* **18**(2), 253–258.

Schuster A 1898 On the investigation of hidden periodicities with application to a supposed 26 day period of meteorological phenomena. *Terrestrial Magnetism and Atmospheric Electricity* **3**, 13–41.

Seidel S, Takamizawa K and Rappaport T 1989 Application of second-order statistics for an indoor radio channel model *Vehicular Technology Conference, 1989, IEEE 39th*, vol. 2, pp. 888–892.

Semkin V, Virk U, Karttunen A, Haneda K and Räisänen AV 2015 E-band propagation channel measurements in an urban street canyon *Proceedings of European Conference on Antenna and Propagation*, pp. 1–4, Lisbon.

Smulders PFM 2009 Statistical characterization of 60-GHz indoor radio channels. *IEEE Transactions on Antennas and Propagation* **57**(10), 2820–2829.

Stoica P and Nehorai A 1989 Music, maximum likelihood, and Cramer–Rao bound. *Acoustics, Speech and Signal Processing* **37**, 720–741.

Suzuki H 1977 A statistical model for urban radio propagation channel. *IEEE Transactions on Communication Systems* **25**, 673–680.

Taparugssanagorn A, Alatossava M, Holappa VM and Ylitalo J 2007 Impact of channel sounder phase noise on directional channel estimation by SAGE. *IET Microwaves, Antennas and Propagation* **1**(3), 803–808.

Tayem N and Kwon H 2005 L-shape 2-dim. arrival angle estimation with propagator method. *AP* **53**, 1622–1630.

Turin G, Clapp F, Johnston T, Fine S and Lavry D 1972 A statistical model of urban multipath propagation channel. *IEEE Transactions on Vehicular Technology* **21**, 1–9.

Wang Y, Chen J and Fang W 2001 TST-MUSIC for joint DOA-delay estimation. *IEEE Transactions on Signal Processing* **46**, 721–729.

Weiler RJ, Peter M, Kühne T, Wisotzki M and Keusgen W 2015 Simultaneous millimeter-wave multi-band channel sounding in an urban access scenario *Proceedings of European Conference on Antenna and Propagation*, pp. 1–5, Lisbon.

Wu X, Zhang Y, Wang CX, Goussetis G, el Hadi M. Aggoune and Alwakeel MM 2015 28 GHz indoor channel measurements and modelling in laboratory environment using directional antennas *Proceedings of European Conference on Antenna and Propagation*, pp. 1–5, Lisbon.

Xiao H and Burr A 2008 Reduced-complexity cluster modeling for time-variant wideband mimo channels *Wireless Conference, 2008. EW 2008. 14th European*, pp. 1–5.

Xiao H, Burr A and de Lamare R 2007 Reduced-complexity cluster modelling for the 3GPP channel model *Communications, 2007. ICC '07. IEEE International Conference on*, pp. 4622–4627.

Yang H, Smulders PF and Herben MH 2006 Frequency selectivity of 60-GHz LOS and NLOS indoor radio channels *Proceedings of. 63th IEEE Vehicular Technology Conference (VTC)*, vol. 6, pp. 2727–2731, Melbourne.

Yegani P and McGillem C 1989a A statistical model for line-of-sight (LOS) factory radio channels *Vehicular Technology Conference, 1989, IEEE 39th*, vol. 2, pp. 496–503.

Yegani P and McGillem C 1989b A statistical model for the obstructed factory radio channel *Global Telecommunications Conference, 1989, and Exhibition. Communications Technology for the 1990s and Beyond. GLOBECOM '89., IEEE*, vol. 3, pp. 1351–1355.

Yegani P and McGillem C 1991 A statistical model for the factory radio channel. *IEEE Transactions on Communications* **39**(10), 1445–1454.

Yegani P and McGillem C 1993 FH-MFSK multiple-access communications systems performance in the factory environment. *IEEE Transactions on Vehicular Technology* **42**(2), 148–155.

Yin X, Fu Y, Liang J and Kim MD 2011 Investigation of large- and small-scale fading cross-correlation using propagation graphs *Proceedings of International Symposium on Antennas and Propagation, SB01-1004*, Jeju, Korea.

Yin X, Liang J, Chen J, Park J, Kim M and Chung H 2012a Empirical models of cross-correlation for small-scale fading in co-existing channels *Proceedings of Asia-Pacific Communication Conference 2012*, pp. 327–332, Jeju, Korea.

Yin X, Liang J, Fu Y, Yu J, Zhang Z, Park JJ, Kim MD and Chung HK 2012b Measurement-based stochastic modeling for co-existing propagation channels in cooperative relay scenarios *Future Network Mobile Summit (FutureNetw), 2012*, pp. 1–8.

Yin X, Liu L, Nielsen D, Pedersen T and Fleury B 2007 A SAGE algorithm for estimation of the direction power spectrum of individual path components *Proceedings of Global Telecommunications Conference, 2007. GLOBECOM '07. IEEE*, pp. 3024–3028.

Yin X, Pedersen T, Czink N and Fleury B 2006a Parametric characterization and estimation of bi-azimuth dispersion path components *Proceedings of IEEE 7th Workshop on Signal Processing Advances in Wireless Communications. SPAWC'06.*, pp. 1–6, Rome, Italy.

Yin X, Pedersen T, Czink N and Fleury B 2006b Parametric characterization and estimation of bi-azimuth and delay dispersion of individual path components *Antennas and Propagation, 2006. EuCAP 2006. First European Conference on*, pp. 1–8.

Yin X, Zeng Z, Cheng X and Zhong Z 2012c Empirical modeling of cross-correlation for spatial-polarimetric channels in indoor scenarios *Personal Indoor and Mobile Radio Communications (PIMRC), 2012 IEEE 23st International Symposium on*, pp. 1677–1681.

Yin X, Zhou X, Zhang Z, Kim MD and Chung HK 2012d Parametric modeling of the cross-correlation for large-scale-fading of propagation channels *Vehicular Technology Conference (VTC Spring), 2012 IEEE 75th*, pp. 1–5.

Zhang N, Yin X, Lu SX, Du M and Cai X 2014 Measurement-based angular characterization for 72 GHz propagation channels in indoor environments *Proceedings of IEEE Global Communication Conference (Globecom)international workshop on mobile communications in higher frequency bands (MCHFB)*, pp. 370–376, Austin.

Zhang Y, Brady M and Smith S 2001 Segmentation of brain MR images through a hidden Markov random field model and the expectation-maximization algorithm. *IEEE Transactions on Medical Imaging* **20**(1), 45–57.

2

Characterization of Propagation Channels

In general, any wireless communication system includes three parts: transmitter (Tx), receiver (Rx), and the wireless channel in between to connect them, as shown in Figure 2.1. Unlike the Tx and Rx, which can be designed to give the system a better tradeoff between reliability and efficiency, the wireless channel cannot be engineered. However, reliable knowledge of the propagation channel is the foundation of the design and analysis of any wireless communication system. The various concepts and definitions of the wireless channel are usually confusing for the beginner. This chapter will attempt to provide a unified and conceptually simple explanation of a morass of concepts around wireless channels.

2.1 Three Phenomena in Wireless Channels

Wireless channels are the real environments in which the Tx and Rx operate. Due to the nature of wireless propagation, fading is inevitable in wireless channels. Fading refers to the time variation of the received signal power induced by changes in the transmission medium or path. Generally speaking, fading can be categorized into large-scale fading, consisting of path loss and shadowing, and small-scale fading. Therefore, in total we have three phenomena in wireless channels.

Path loss, P, and shadowing, S, belong to the large-scale fading category, since they are dominant when the mobile station moves over distances of several tens of wavelengths. As shown in Figure 2.2, path loss P is the attenuation in the transmitted signal while propagating from the Tx to the Rx and is observed over a distance of several hundred or thousand wavelengths. Shadowing S is the slow variations of the received signal power over distances of several tens or hundreds of wavelengths due to large terrain features such as buildings and hills. Large-scale fading is very important for system design at the network level. For example, the cell coverage area, outage, and handoffs are influenced by these effects.

On the other hand, small-scale fading appears as a result of multipath propagation. As shown in Figure 2.2, multipath fading, h, refers to fast variations of the received signal power due to

Propagation Channel Characterization, Parameter Estimation and Modelling for Wireless Communications, First Edition.
Xuefeng Yin and Xiang Cheng.

Figure 2.1 A wireless communication system consisting of three parts

the constructive and destructive interference of the multiple signal paths between the Tx and Rx. These variations are observed over a distance of the order of the wavelength. Small-scale fading plays an important role in determining the link-level performance according to bit error rates, average fade durations, and so on. Therefore, as shown in Figure 2.1, to completely characterize wireless channels we can use the following expression

$$g = P \cdot S \cdot h \tag{2.1}$$

where P is the path loss, S is shadowing, and h is the multipath fading. It must be noted that throughout this book we constrain our interests to the investigation and modeling of small-scale fading for different types of channels, for example fixed-to-mobile (F2M) channels, vehicle-to-vehicle (V2V) channels, cooperative MIMO channels, and so on.

2.2 Path Loss and Shadowing

Path loss is the attenuation in the transmitted signal while propagating from the Tx and Rx. This attenuation is caused by the effects such as free-space loss, refraction, diffraction, and reflection. Significant variations in the path loss are observed over distances of several hundred or thousand wavelengths.

The simplest path-loss model corresponds to a propagation in free space; in other words, in an LoS link between the Tx and Rx. In this case, the received signal power can be expressed as [Stüber 2001]

$$P_R = P_T G_T G_R \frac{\lambda^2}{4\pi D^2} \tag{2.2}$$

where P_T is the transmitted power, G_T and G_R are the transmit and receive antenna gains, respectively, λ is the carrier wavelength, and D is the distance between the Tx and Rx. Note that the path-loss exponents (the power of the distance dependence D) are 2 for free-space propagation. Therefore, under free space propagation, the received power decreases as the square of the distance. Equation (2.2) also shows the path-loss dependency on the carrier wavelength λ; the shorter the wavelength, the higher the path loss.

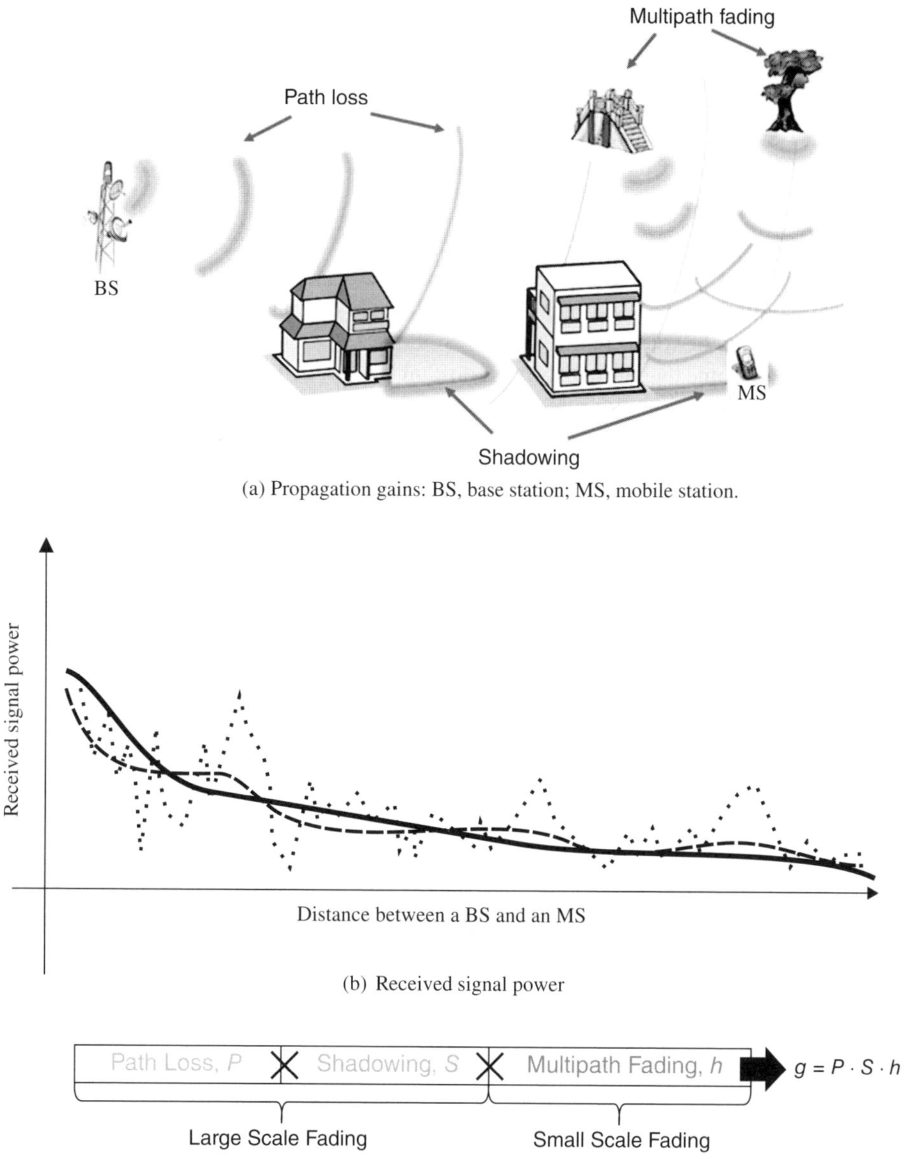

(a) Propagation gains: BS, base station; MS, mobile station.

(b) Received signal power

(c) Splitting the gain to three sources

Figure 2.2 Three phenomena in wireless channels

However, in a real environment, wireless signals seldom experience free-space propagation. Therefore, several different models have been proposed to model path loss in propagation environments such as urban, rural, and indoor areas. Examples of such models are:

- Okumura–Hata [Hata and Nagatsu 1980, Okumura et al. 1968]
- Lee [Lee 2010]
- Walfish–Ikegami [COST 231 1991].

Experiments show that the actual path-loss exponents are around 3–8, suggesting higher attenuation than in free-space propagation conditions. A detailed description of the different path-loss models can be found in the book by Stüber [2001].

These models assume that the path loss is constant at a given distance. However, the presence of obstacles such as buildings and trees leads to random variations of the received power at a given distance. This effect is termed "shadowing" (or shadow fading). Experiments illustrate that shadowing can be modeled as a lognormal random variable, which is consistent with intuition. As shown in Figure 2.2, shadowing is due to the power loss caused by big objects such as high buildings. Therefore, the total power loss – the shadowing – can be calculated by multiplying all power losses caused by big objects. In the log domain, multiplication becomes the addition of every power loss. Based on the central limit theory [Papoulis and Pillai 2002], we know that the shadowing has a normal or Gaussian distribution in the log domain, and thus the shadowing can be modeled as a log-normal distribution.

Therefore, the shadowing distribution is given by [Stüber 2001]:

$$f_{\Omega_p}(x) = \frac{10}{x\sigma_{\Omega_p}\sqrt{2\pi}\ln 10}\exp\left[-\frac{(10\log_{10}x - \mu_{\Omega_p(dBm)})^2}{2\sigma^2_{\Omega_p}}\right] \tag{2.3}$$

where Ω_p denotes the mean-squared envelope level, μ_{Ω_p} designates the area mean expressed in dBm, and σ_{Ω_p} is the standard deviation of the shadowing. Typical values of σ_{Ω_p} range from 5 to 10 dB. Detailed discussions of shadowing can be found in the book by Stüber [2001].

2.3 Multipath Fading

Multipath propagation is when the transmitted signal reaches the Rx by two or more paths. The presence of local scatterers, such as mountains and buildings, often obstructs the direct wave path – the LoS – between the Tx and Rx. Therefore, a non-LoS (NLoS) propagation path will appear between the Tx and Rx. Consequently, the waves must propagate through reflection, diffraction, and scattering. This results in the received waves coming from various directions with different delays. The multiple waves combine vectorially at the receiver antenna (a phenomenon called multipath fading) to produce a composite received signal.

As mentioned above, the presence of local scatterers gives rise to NLoS scenarios, where Rayleigh distribution is the most popular distribution used to describe the fading envelope. Some types of scattering environment have a specular component; in other words, an LoS component or a strong reflected path. These scattering environments are called LoS scenarios, and a Rician distribution is used to describe the fading envelope.

A non-directional channel can be characterized by one of the four system functions, also termed the first set of Bello's functions [Bello and Bengtsson 1963]. These four system functions of non-directional channel are the:

- input delay-spread function (channel impulse response) $h(t, \tau')$
- output Doppler-spread function $H(f_D, f_c)$
- delay Doppler-spread function (spread function) $g(f_D, \tau')$
- time-variant transfer function $G(t, f_c)$,

where t denotes the time, τ' designates the time delay, f_c is the carrier frequency, and f_D is the Doppler shift. The Fourier relationship between the system functions is shown in Figure 2.3.

Whereas for a directional channel, eight system functions [Kattenbach 2002] – an extension of the traditional four-system functions, produced by incorporating direction and space terms – can be used, because the system functions of directional channels include those of non-directional channels as special cases and directional channel description (strictly speaking, double-directional channel description) is very useful for MIMO systems, here we give a brief overview of directional channels and invite interested readers to refer to the literature for more detail [Fleury 2000, Kattenbach 2002]. The eight system functions of directional channels are the:

- time-variant direction-spread impulse response (channel impulse response) $h(t, \tau', \Omega)$
- time- and space-variant impulse response $s(t, \tau', x)$
- direction-Doppler-spread transfer function $H(f_D, f_c, \Omega)$
- Doppler-spread space-variant transfer function $T(f_D, f_c, x)$
- Doppler-direction-spread impulse response (spread function) $g(f_D, \tau', \Omega)$
- Doppler-spread space-variant impulse response $m(f_D, \tau', x)$
- time-variant direction-spread transfer function $M(t, f_c, \Omega)$
- time-space-variant transfer function $G(t, f_c, x)$

where t denotes the time, τ' designates the time delay, f_c and f_D are the carrier frequency and Doppler shift, respectively, x denotes the location of an antenna element in the antenna array in the Tx/Rx, and Ω is the direction of a antenna element in the antenna array in the Tx/Rx including both azimuth angle ϕ and elevation angle θ. The Fourier relationship between the system functions is shown in Figure 2.4. These system functions lay emphasis upon different aspects of the channels. For example, the channel impulse response $h(t, \tau', \Omega)$ focuses on the

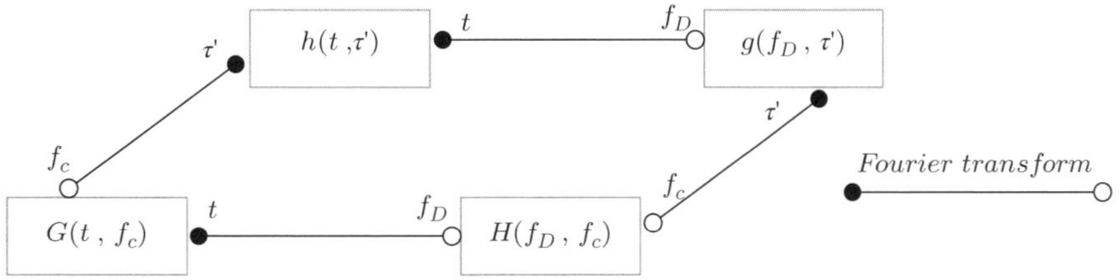

Figure 2.3 Fourier relationship between the system functions of non-directional channels

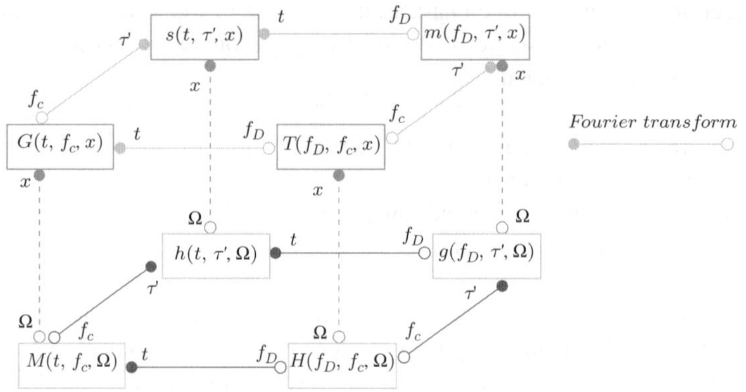

Figure 2.4 Fourier relationship between the system functions of directional channels

description of channels in the time-direction domain, while the time-space-variant transfer function $G(t, f_c, x)$ describes the channels in the frequency-space domain. However, since the channel impulse response $h(t, \tau', \Omega)$ can directly relate the multipath components, it is the most often-used system function and thus will be the one mainly used throughout this book.

Based on a basic knowledge of signals and systems [Oppenheim et al. 2009] and the knowledge of multipath fading introduced above, a wireless system such as the one shown in Figure 2.1 can be represented as a general signal system as shown in Figure 2.5. In this case, the channel impulse response $h(t, \tau', \Omega)$ can be expressed by

$$h(t, \tau', \Omega) = \sum_{l=1}^{L} h_l(t)\delta(\tau' - \tau_l')\delta(\Omega - \Omega_l) \tag{2.4}$$

where L is the total number of resolvable multipath components, $h_l(t)$ is the time-variant complex fading envelope associated with the lth resolvable multipath component arriving with an average time delay τ_l' and an average direction Ω_l. Each time-variant complex fading envelope $h_l(t)$ consists of a number of multipaths and can be expressed as

$$h_l(t) = \sum_{n=1}^{N} c_n(t)e^{-j\phi_n(t)} \tag{2.5}$$

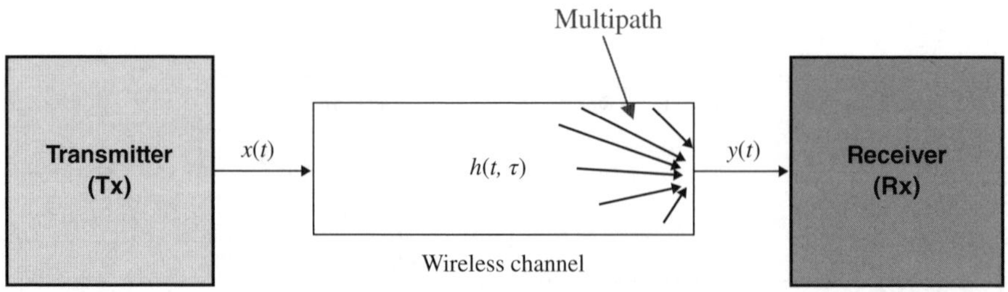

Figure 2.5 A diagram of a signal system represented for a wireless system with multipath fading channels

where N is the number of multipaths, $c_n(t)$ denotes a time-variant amplitude, and $\phi_n(t)$ is the time-variant phase.

Since multipath fading appears only over distances of the order of the wavelength, the fast variations of the received signal power due to multipath fading are mainly caused by the change of phase $\phi_n(t)$. In what follows, we will show how the phase $\phi_n(t)$ is obtained.

Let us start from the most simple scenario, where we consider a fixed Tx and Rx, as shown in Figure 2.6(a). In this case, the complex fading envelope is time-invariant and can be expressed as

$$h_l = \sum_{n=1}^{N} c_n e^{-j\phi_n} \tag{2.6}$$

From Figure 2.6(a), it is clear that the phase ϕ_n is caused by the multipaths and can be shown as

$$\phi_n = 2\pi f_c \frac{d'}{c} \tag{2.7}$$

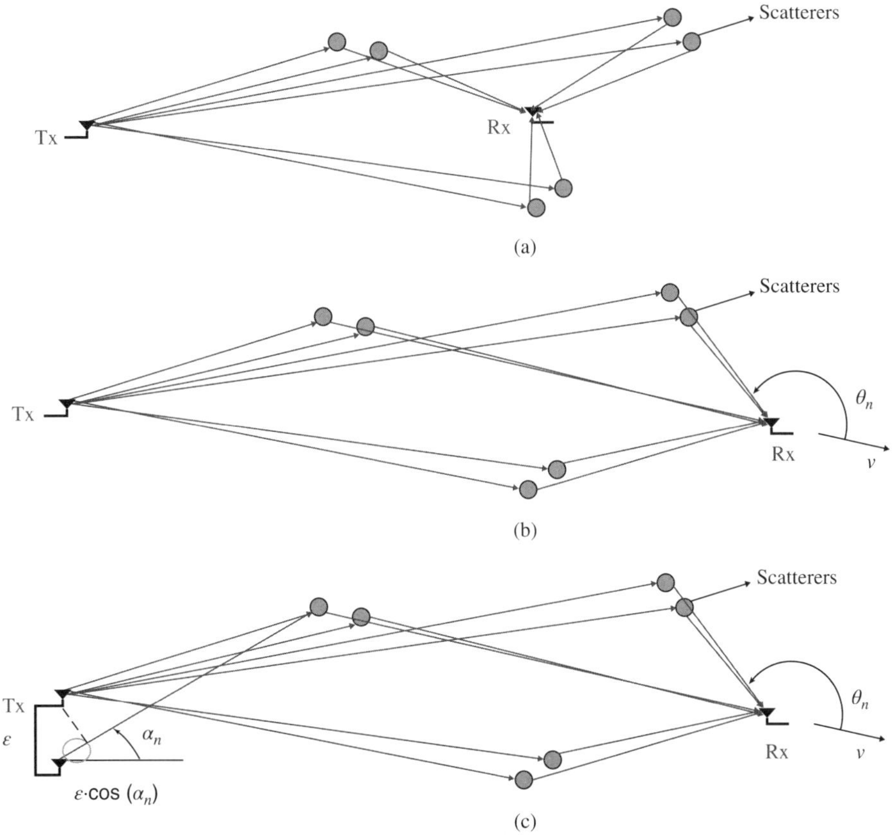

Figure 2.6 Typical wireless communication scenarios: (a) fixed Tx and Rx; (b) moving Rx; (c) multiple antennas

where f_c is the carrier frequency, c is the speed of light, and $d' = d_n - d_{min}$ denotes difference between:

- the travel distance d_n from the Tx to the Rx via the nth scatterer
- the minimum travel distance d_{min} from the Tx to the Rx.

Let us now consider that the Rx is moving and thus the complex fading envelope is time-variant:

$$h_l(t) = \sum_{n=1}^{N} c_n(t) e^{-j\phi_n(t)} \tag{2.8}$$

In this case, as shown in Figure 2.6(b), the travel distance difference consists of two parts: the travel distance difference caused by multipaths and the travel distance difference caused by the motion of the Rx. The phase can then be expressed as

$$\phi_n(t) = 2\pi f_c \frac{d''}{c} = 2\pi f_c \frac{d'_n + \nu t \ \cos(\theta_n)}{c} \tag{2.9}$$

where ν denotes the moving velocity and θ_n is the angle of incidence. If we consider multiple antennas scenarios, as shown in Figure 2.6(c), the time-variant complex fading envelope can be expressed as

$$h_l^{pq}(t) = \sum_{n=1}^{N} c_n^{pq}(t) e^{-j\phi_n^{pq}(t)} \tag{2.10}$$

where the superscript $(\cdot)^{pq}$ means the link between the pth antenna in the Tx and the qth antenna in the Rx. From Figure 2.6(c), the travel distance difference includes three parts: the travel distance difference caused by the multipaths, the motion of the Rx, and the multiple antennas. In this case, the phase can be expressed as

$$\phi_n^{pq}(t) = 2\pi f_c \frac{d'''}{c} = 2\pi f_c \frac{d'' + \epsilon \cos(\alpha_n)}{c} \tag{2.11}$$

where ϵ denotes the antenna-element spacing and α_n is the radiation angle as shown in Figure 2.6(c). Therefore, in total, the phase $\phi_n^{pq}(t)$ in Eq. (2.11) actually includes three parts and can be expressed completely as

$$\phi_n^{pq}(t) = 2\pi f_c \frac{d_n - d_{min}}{c} + 2\pi f_D \cos(\theta_n) + 2\pi \lambda^{-1} \epsilon \cos(\alpha_n) \tag{2.12}$$

where $f_D = f_c \frac{\nu}{c}$ is the maximum Doppler frequency and $\lambda = \frac{f_c}{c}$ is the wavelength. From Eq. (2.12), it is clear that the complete phase shift includes three terms, which can be termed the multipath-induced phase shift, the motion-induced phase shift, and the multiple antenna-induced phase shift, respectively.

2.4 Stochastic Characterization of Multipath Fading

The characterization of the multipath fading channel is essential for understanding and modeling it properly. Due to the huge number of factors that influence the channel, a deterministic characterization is not possible. The only feasible manner to characterize the multipath fading channel is to characterize its statistics.

A full statistical description of the system functions is only given by their multidimensional PDFs, which is practically not feasible. The most important and often-used approximate descriptions are the first-order received envelope and phase PDF of multipath fading and some second-order descriptions of multipath fading, such as the level cross rate (LCR), average fading duration (AFD), and the correlation properties.

2.4.1 *Received Envelope and Phase Distribution*

2.4.1.1 Received Envelope Distribution

In many wireless communication scenarios, the composite received signal consists of a large number of plane waves, as expressed in Eq. (2.5). According to the central limit theory [Papoulis and Pillai 2002], the PDF of complex fading envelope $h_l(t)$ in Eq. (2.5) exhibits the complex Gaussian distribution. In this case, the complex envelope $c(t) = |h_l(t)|$ has a Rayleigh distribution at any time t:

$$f_c(x) = \frac{2x}{\Omega_p} \exp\{-\frac{x^2}{\Omega_p}\} \quad x \geq 0 \tag{2.13}$$

where $\Omega_p = E[c^2]$ is the average envelope power.

Some types of wireless communication environment have a specular or LoS component. In this case, the complex fading envelope can be expressed as

$$h_l(t) = \sqrt{\frac{K}{K+1}} c_{LoS} e^{-j\phi_{LoS}} + \sqrt{\frac{1}{K+1}} \sum_{n=1}^{N} c_n(t) e^{-j\phi_n(t)} \tag{2.14}$$

where K is the Rician factor and defined as the ratio of the specular power s^2 to scattered power Ω_p, and c_{LoS} and ϕ_{LoS} are the amplitude and phase of the LoS component, respectively. The complex envelope $c(t) = |h_l(t)|$ has a Rician distribution at any time t, as shown in following:

$$f_c(x) = \frac{2x}{\Omega_p} \exp\{-\frac{x^2 + s^2}{\Omega_p}\} I_0(\frac{2xs}{\Omega_p}) \quad x \geq 0 \tag{2.15}$$

with $s^2 = h_l^I(t)^2 + h_l^Q(t)^2$, where $h_l^I(t)$ and $h_l^Q(t)$ are the inphase and quadrature components of $h_l(t)$, respectively; in other words, $h_l(t) = h_l^I(t) + h_l^Q(t)$.

Traditionally, the Rayleigh and Rician models have been the most popular for characterizing multipath fading due to their simplicity and acceptable accuracy. Nowadays, some more realistic and flexible models are available, such as Nakagami fading [Stüber 2001] and Weibull fading [Parsons 2000]. By adjusting the parameters of these models, Nakagami and Weibull models can mimic Rician fading, Rayleigh fading, and also worse-than-Rayleigh fading. We invite interested readers to refer to the literature for further details [Parsons 2000, Stüber 2001].

2.4.1.2 Phase Distribution

The phase of the complex fading envelope $h_l(t) = h_l^I(t) + h_l^Q(t)$ can be also expressed as

$$\phi(t) = \tan^{-1}(\frac{h_l^Q(t)}{h_l^I(t)}) \tag{2.16}$$

The phase is normally a non-uniform distribution and has a very complicated form, except for Rayleigh fading, where the phase has a uniform distribution over the interval $[-\pi, \pi]$:

$$f_{\phi(t)}(x) = \frac{1}{2\pi} \quad -\pi \leq x \leq \pi \tag{2.17}$$

2.4.2 *Envelope Level Cross Rate and Average Fade Duration*

Envelope LCR and AFD are two important second-order statistics associated with envelope fading. The LCR, $L_c(r_g)$, is by definition the average number of times per second that the signal envelope, $c(t) = |h_l(t)|$, crosses a specified level r_g with positive/negative slope. Using the traditional PDF-based method [Youssef et al. 2005], the closed-form expression of LCR for Rician fading can be derived as [Stüber 2001]:

$$L_c(r_g) = \sqrt{2\pi(K+1)} f_D \rho e^{-K-(K+1)\rho^2} I_0(2\rho\sqrt{K(K+1)}) \tag{2.18}$$

where $\rho = r_g/\sqrt{\Omega_p}$. For the case of Rayleigh fading ($K = 0$) and isotropic scattering, the above expression can be simplified to

$$L_c(r_g) = \sqrt{2\pi} f_D \rho e^{-\rho^2} \tag{2.19}$$

Here, we would like to highlight that isotropic scattering environments were first mentioned by Clark in 1968, in the context of ideal scenarios in which the scatterers were assumed to be uniformly distributed around the Tx/Rx. Real wireless communication environments normally involve non-isotropic scattering. In this book, we will concentrate on the investigation and modeling of wireless channels in non-isotropic scattering scenarios.

The envelope AFD, $T_{c-}(r_g)$, is the average time over which the signal envelope, $c(t)$, remains below a certain level r_g. Therefore, the envelope AFD is a measure of how long the envelope remains below a specified level. In general, the AFD $T_{c-}(r_g)$ for Rician fading channels is defined by [Pätzold and Laue 1998]:

$$T_{c-}(r_g) = \frac{P_{c-}(r_g)}{L_c(r_g)} = \frac{1 - Q\left(\sqrt{2K}, \sqrt{2(K+1)}r_g\right)}{L_c(r_g)} \tag{2.20}$$

where $P_{c-}(r_g)$ indicates a cumulative distribution function of $c(t)$ with $Q(\cdot,\cdot)$ denoting the generalized Marcum Q function. By setting $K = 0$ in Eq. (2.20), we can have the envelope AFD for the Rayleigh fading case.

2.4.3 *Correlation Functions*

The correlation properties raise the important issue of how multipath fading varies with time, frequency, and distance. Note that the rate of channel variation has a significant impact on several aspects of the communication problem. For the case of non-directional SISO channels, we have the following four correlation functions based on the four system functions introduced in Section 2.3:

$$R_h(t_1, t_2; \tau_1', \tau_2') = E[h^*(t_1, \tau_1')h(t_2, \tau_2')] \tag{2.21}$$

$$R_H(f_{c1}, f_{c2}; f_{D1}, f_{D2}) = E[H^*(f_{c1}, f_{D1})H(f_{c2}, f_{D2})] \tag{2.22}$$
$$R_G(f_{c1}, f_{c2}; t_1, t_2) = E[G^*(f_{c1}, t_1)G(f_{c2}, t_2)] \tag{2.23}$$
$$R_g(\tau_1', \tau_2'; f_{D1}, f_{D2}) = E[g^*(\tau_1', f_{D1})g(\tau_2', f_{D2})]. \tag{2.24}$$

According to the Fourier relationship between Bello's system functions, as shown in Figure 2.3, there is a double Fourier transform relationship between the correlation functions, as shown in Figure 2.7.

In accordance to the aforementioned SISO case, we can define correlation functions of the MIMO system functions; in other words, the directional system functions introduced in Section 2.3. This leads again to eight different correlation functions, in which the angle-resolved correlation functions are listed as follows:

- Doppler-delay and direction correlation function
  $R_g(f_{D1}, f_{D2}; \tau_1', \tau_2'; \Omega_1, \Omega_2;) = E[g^*(f_{D1}, \tau_1', \Omega_1)g(f_{D2}, \tau_2', \Omega_2)]$
- Time-delay and direction correlation function
  $R_h(t_1, t_2; \tau_1', \tau_2'; \Omega_1, \Omega_2;) = E[h^*(t_1, \tau_1', \Omega_1)h(t_2, \tau_2', \Omega_2)]$
- Doppler-frequency and direction correlation function
  $R_H(f_{D1}, f_{D2}; f_{c1}, f_{c2}; \Omega_1, \Omega_2;) = E[H^*(f_{D1}, f_{c1}, \Omega_1)H(f_{D2}, f_{c2}, \Omega_2)]$
- Time-frequency and direction correlation function
  $R_M(t_1, t_2; f_{c1}, f_{c2}; \Omega_1, \Omega_2;) = E[M^*(t_1, f_{c1}, \Omega_1)M(t_2, f_{c2}, \Omega_2)]$.

The antenna-variant correlation functions are listed as:

- Time-delay and space correlation function
  $R_s(t_1, t_2; \tau_1', \tau_2'; x_1, x_2;) = E[s^*(t_1, \tau_1', x_1)s(t_2, \tau_2', x_2)]$
- Doppler-delay and space correlation function
  $R_m(f_{D1}, f_{D2}; \tau_1', \tau_2'; x_1, x_2;) = E[m^*(f_{D1}, \tau_1', x_1)m(f_{D2}, \tau_2', x_2)]$
- Time-frequency and space correlation function
  $R_G(t_1, t_2; f_{c1}, f_{c2}; x_1, x_2;) = E[G^*(t_1, f_{c1}, x_1)G(t_2, f_{c2}, x_2)]$
- Doppler-frequency and space correlation function
  $R_T(f_{D1}, f_{D2}; f_{c1}, f_{c2}; x_1, x_2;) = E[T^*(f_{D1}, f_{c1}, x_1)T(f_{D2}, f_{c2}, x_2)]$.

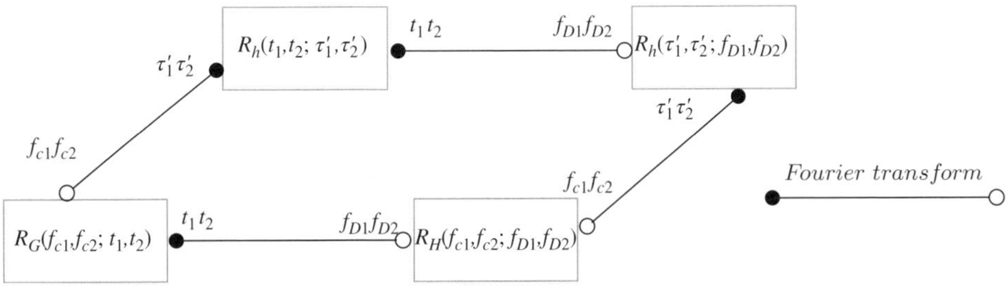

Figure 2.7 Fourier relationship between the correlation functions of the non-directional system functions

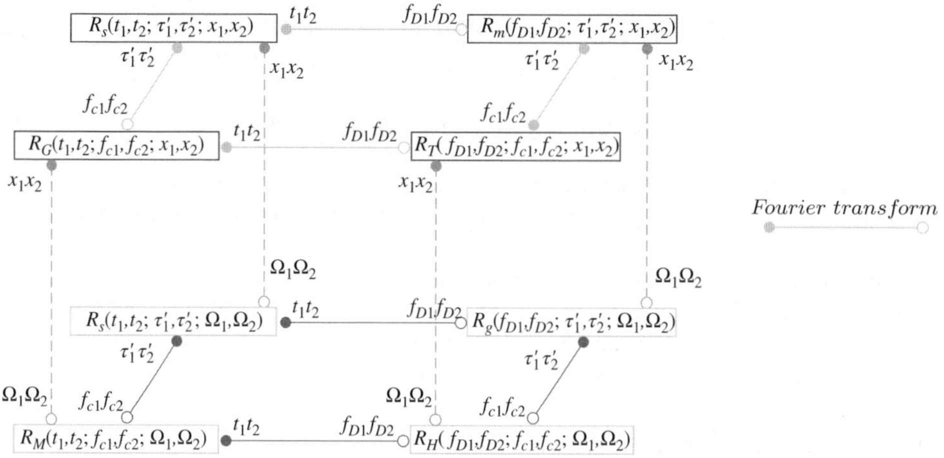

Figure 2.8 Fourier relationship between the correlation functions of the directional system functions

Based on the Fourier relationship between directional system functions, as shown in Figure 2.4, we have a double Fourier transform relationship between the correlation functions, as shown in Figure 2.8.

2.5 Duality of Multipath Fading

Selectivity and dispersion are basic and important properties in radio channels. Moreover, they have a so-called "dual" relationship, which means that they allow one to investigate the channels from different domains [Bello 1964]. Selectivity means that the multipath fading expresses variation behavior in one domain, while dispersion means that the multipath fading expresses spread behavior in the other domain. In what follows, we will show that multipath fading expresses selectivity in the time, frequency, and space domains, while it presents dispersion in the Doppler, delay, and angle/direction domains, as shown in Figure 2.9.

Time selectivity is the property that the channel changes over time due to the motion of the Tx, the Rx, and/or the scatterers. When viewed in the frequency domain, time selectivity appears as Doppler shifts in the transmitted signal, resulting in a broadening of the transmitted signal spectrum. This effect is termed "frequency dispersion". Therefore, time selectivity and frequency-dispersion are one type of selectivity–dispersion duality. Normally, coherence time T_c and its approximate reciprocal, the Doppler spread σ_{f_D}, are used to measure the time selectivity and frequency dispersion, respectively. The coherence time T_c and Doppler spread σ_{f_D} can be calculated by the correlation functions introduced in Section 2.4.3. Their detailed expressions can be found in the literature [Fleury 2000, Kattenbach 2002, Stüber 2001]. In general, the larger the Doppler spread σ_{f_D}, the smaller the coherence time T_c, and thus the faster the variation in the time domain and the broader the spread in the Doppler domain.

When the Doppler spread σ_{f_D} is sufficiently large to result in the coherence time T_c being smaller than the transmitted symbol duration T_s, the channel changes within the transmitted symbol duration T_s. This kind of channel is named a fast-fading channel. On the other hand, if

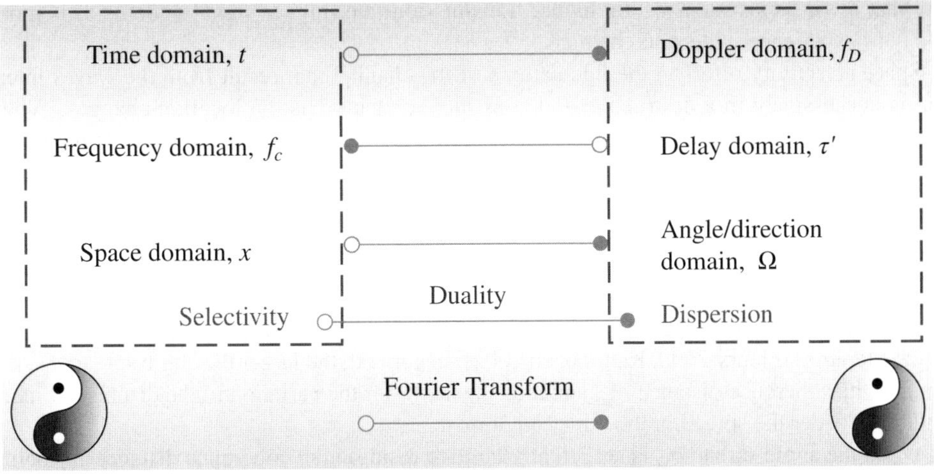

Figure 2.9 Duality relationship between selectivity and dispersion of multipath fading

the Doppler spread σ_{f_D} is sufficiently small to lead to the coherence time T_c being larger than the transmitted frame duration T_f, the channel is approximately constant within the transmitted frame duration T_f. In this case, the channel is termed a slow fading channel. Therefore, we can conclude that a wireless channel with large Doppler spread σ_{f_D} (i.e. smaller coherence time T_c) is more likely to be a fast-fading channel. Moreover, we can say that a wireless system with a higher transmission rate (higher signal wideband) is more likely to experience slow-fading channels.

Frequency selectivity refers to the property that the channel changes with frequency. Multipath components that arrive with different time delays result in this frequency selectivity. From the time domain, multipath components with different time delays lead to the spread of the transmitted signal. This effect is called time dispersion. Therefore, frequency selectivity and time dispersion are another type of selectivity–dispersion duality. To measure the frequency selectivity and time dispersion, we use the coherence bandwidth B_c and its reciprocal, delay spread $\sigma_{\tau'}$, respectively. The coherence bandwidth B_c and delay spread $\sigma_{\tau'}$ can be calculated by the correlation functions introduced in Section 2.4.3. The detailed expressions can be found in the literature [Fleury 2000, Kattenbach 2002, Stüber 2001]. In general, the larger the delay spread $\sigma_{\tau'}$, the smaller the coherence bandwidth B_c, and thus the the faster the variation in the frequency domain and the broader the spread in the delay domain.

When the delay spread $\sigma_{\tau'}$ is larger than the symbol duration T_s, which corresponds to the coherence bandwidth B_c being smaller than the signal bandwidth B_s, the transmitted frequencies experience different amplitude and phase changes, the channel is termed a frequency-selective (or wideband) channel. On the other hand, if the delay spread $\sigma_{\tau'}$ is smaller than the symbol duration T_s – in other words, the coherence bandwidth B_c is larger than the signal bandwidth B_s – the transmitted frequencies undergo approximately identical amplitude and phase changes and the channel is called a frequency flat (narrowband) channel. Therefore, we can conclude that a wireless channel with a larger delay spread $\sigma_{\tau'}$ (i.e. a smaller coherence bandwidth B_c) is more likely to be a wideband channel. Moreover, we can say that a wireless system with a higher transmission rate (higher signal wideband) is more likely to experience wideband channels.

Space selectivity refers to the fluctuations of the channel that result from the waves interfering successively in a destructive and constructive fashion as the location changes. When viewed in the angle domain, these space-variant fluctuations are caused by multipath components with different directions over the location. This effect is termed "direction dispersion". Therefore, space selectivity and direction dispersion are another type of selectivity–dispersion duality. Similar to the two dualities already mentioned, space selectivity and direction dispersion are measured by the coherence distance D_c and its reciprocal, the angle spread σ_Ω, respectively. The coherence distance D_c and angle/direction spread σ_Ω can be calculated by the correlation functions introduced in Section 2.4.3. Their detailed expressions can be found in the literature [Fleury 2000, Kattenbach 2002]. In general, the larger the angle spread σ_Ω, the smaller the coherence distance D_c, and thus the the faster the variation in the distance domain and the broader the spread in the direction domain.

When the angle spread σ_Ω is sufficiently large to result in the coherence distance D_c being smaller than the space separation between any two antenna elements A_s, the antennas are uncorrelated. This kind of channel is named a spatial-uncorrelated channel. Otherwise, if the angle spread σ_Ω is sufficiently small to lead to the coherence space D_c being larger than the space separation between any two antenna elements A_s, the antennas are correlated. In this case, the channel is termed a spatial-correlated channel. Therefore, we can conclude that a wireless channel with a larger angle spread σ_Ω (i.e. a smaller coherence distance D_c) is more likely to be a spatial-uncorrelated channel. Moreover, we can say that a multiple antenna wireless system with larger antenna spacing is more likely to experience spatial-uncorrelated channels.

From the description here, it is clear that the classifications of wireless channels – wideband (frequency-selectivity) channels vs narrowband (frequency-flat) channels, fast-fading channels vs slow-fading channels, and spatial-correlated channels vs spatial-uncorrelated channels – are not purely based on the characteristics of the channels. Instead, they are based on a comparison of the wireless channel characteristics and wireless *system* characteristics. Finally, the duality of multipath fading is summarized in Figure 2.10.

2.6 WSSUS Assumption of Multipath Fading

In the context of wireless radio channels, wide-sense stationary uncorrelated scattering (WSSUS) is usually assumed. WSSUS was originally defined for non-directional (i.e. SISO) channels, with the wide-sense stationary (WSS) property assumed with respect to both time t and frequency f when regarding the properties of a time-variant transfer function $G'(t, f)$ [Bello and Bengtsson 1963, Steele 1992]. In the context of the complex channel impulse response for non-directional channels $h'(t, \tau')$, the WSSUS assumption represents WSS with respect to time t and uncorrelated scattering (US) with respect to the time delay τ'. Therefore, we have the term WSSUS and can completely describe the properties of a WSSUS channel in only the time related domains; that is, the time and delay domains. By applying the WSSUS assumption to the correlation functions of non-directional channels, we can derive corresponding simplified versions. For example, the four-dimensional correlation function of

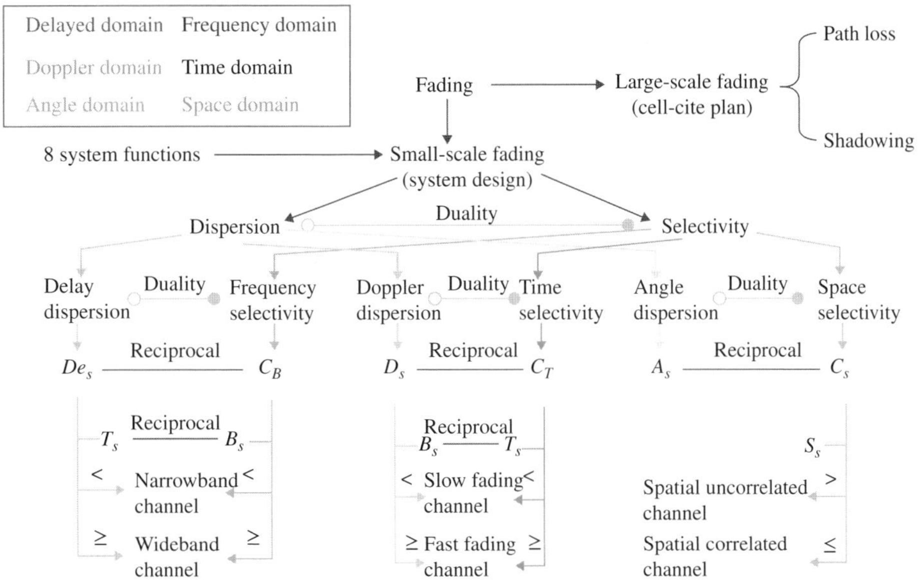

Figure 2.10 Summarization of duality of multipath fading

the channel impulse response simplifies to

$$R_h(t_1, t_2; \tau_1', \tau_2') = P_h(\Delta t, \tau_1')\delta(\tau_2' - \tau_1') \tag{2.25}$$

where $P_h(\Delta t, \tau_1')$ is the cross-power density of $h(t_1, \tau_1')$ and $h(t_1 + \Delta t, \tau_1')$. The resulting correlation functions and the Fourier relationship between them are shown in Figure 2.11.

From the dualities between the time-Doppler relation, the frequency-delay relation, and the distance-direction relation, as described in Section 2.5, the WSSUS concept for non-directional channels can be extended straightforwardly for directional channels [Fleury 2000, Kattenbach

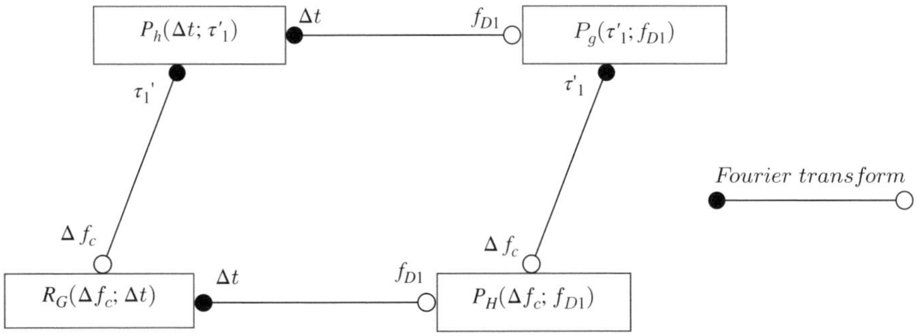

Figure 2.11 Fourier relationship between the correlation functions of the non-directional system functions for WSSUS channels

2002]. In this case, a directional WSSUS channel shows US with respect, not only to the time delay τ' and Doppler shift f_D, but also to the direction Ω. Since US with respect to the time delay τ' is equivalent to the WSS property with respect to the carrier frequency f_c, US with respect to the direction Ω will result in the WSS property with respect to the distance x.

According to the WSSUS case of the SISO channel correlation functions, stationarity of the MIMO correlation functions can be defined. A fully stationary MIMO channel is characterized by a Doppler-delay-direction correlation function

$$R_g(f_{D1}, f_{D2}; \tau_1', \tau_2'; \Omega_1, \Omega_2) = P_g(\Omega_1, \tau_1', f_{D1})\delta(\Omega_2 - \Omega_1)\delta(\tau_2' - \tau_1')\delta(f_{D2} - f_{D1}) \quad (2.26)$$

The corresponding time-frequency-space correlation function therefore only depends on the shift in all dimensions, but no longer on the absolute values

$$R_G(t_1, t_2; f_{c1}, f_{c2}; x_1, x_2;) = P_G(\Delta x, \Delta f_c, \Delta t) \quad (2.27)$$

The relationship between all the correlation functions for WSSUS directional channels is given in Figure 2.12.

WSSUS, in a strict sense, is never fulfilled since it requires the channel statistics to stay constant for an infinite time. However, the WSSUS assumption is fulfilled for many channels over short periods of time or distances of the order of tens of wavelengths [Stüber 2001]. Therefore, to significantly simplify the modelling complexity, the WSSUS assumption is widely used to model many radio channels [Lee 2010, Steele 1992, Stüber 2001]. However, the WSSUS assumption for some newly emerging communication scenarios, such as V2V, is no longer

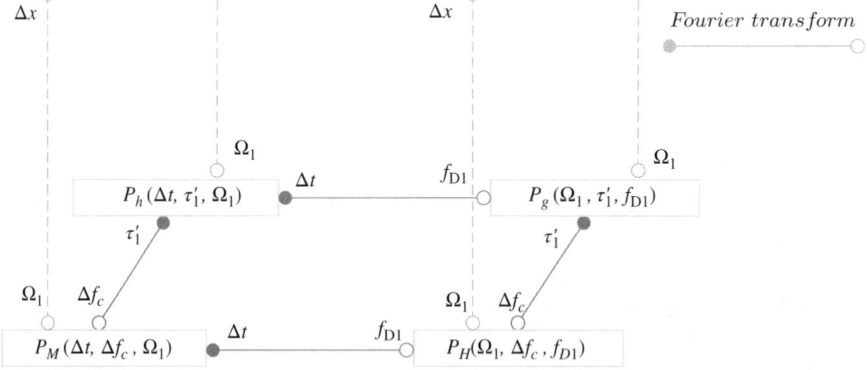

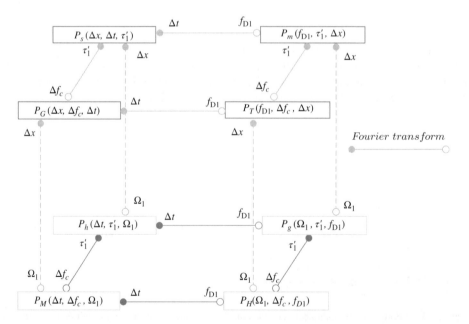

Figure 2.12 Fourier relationship between the correlation functions of the directional system functions for WSSUS channels

valid. Therefore, in these new scenarios, we have to investigate and model the non-stationarity of wireless channels.

2.7 A Review of Propagation Channel Modeling

In the previous sections, we addressed the importance of investigating wireless channels and described their important characteristics. Here, we will talk about the next key issue: how to properly model them. To answer this question, we need to clearly know what the aims of channel modeling are. In general, there are two main aims:

- to investigate and properly model wireless channels for the design, parameter optimization, and testing of wireless communication systems
- to offer up a simple channel model as a platform for the validation of various communication systems.

The first aim means that the main task of channel modeling is to simulate the main influences on a signal transmission so as to create a basis for the development of transmission systems. The second aim is motivated by concerns in system engineering, where simple and easy-to-use channel models with acceptable accuracy are needed.

It is clear that there is a tradeoff between complexity (realization) and accuracy (performance): the first aim emphasizes performance while the second emphasizes practicality. Of course, there is also a tradeoff even in the first aim. To gain a good understanding the channel, we need to investigate its properties as much as we can. On the other hand, tractable mathematical models are needed for the analysis, design, and testing of wireless communication systems.

Nowadays, the two aims have started to merge; that is, more accurate channel models are needed for the validation of communication systems.

As described in Section 2.3, there are eight system functions that can be used to model the multipath fading of wireless channels. Among them, the channel impulse response has been widely chosen for modeling multipath fading of wireless channels. Specifically, the double-directional time-variant complex impulse response of a wireless channel can be modeled as the superposition of L resolvable paths or taps

$$h(t,\tau',\Omega_T,\Omega_R)=\sum_{l=1}^{L}h_l(t)\delta(\tau'-\tau_l'(t))\delta(\Omega_T-\Omega_{T,l}(t))\delta(\Omega_R-\Omega_{R,l}(t)) \tag{2.28}$$

where $\tau_l'(t)$, $\Omega_{T,l}(t)$, and $\Omega_{R,l}(t)$ represent the excess delay, direction of departure (DoD), and direction of arrival (DoA) of the lth resolved path at time t, respectively, $\delta(\cdot)$ denotes Dirac delta function, and $h_l(t)$ denotes the complex fading envelope of the lth ($l=1,\ldots,L$) resolved path and can be expressed as

$$h_l(t)=\sum_{n=1}^{N}a_{l,n}(t)e^{j2\pi f_{D,l,n}(t)t}e^{j\vec{k}(\theta_{T,l,n}(t))\vec{d}_T}e^{j\vec{k}(\theta_{R,l,n}(t))\vec{d}_R} \tag{2.29}$$

From Eq. (2.29), it is clear that each resolvable path $h_l(t)$ also consists of multiple unresolvable subpaths with complex amplitudes represented by $a_{l,n}(t)$ ($n=1,2,\ldots,N$). Here,

$f_{D,l,n}(t) = v(t) f_c \cos \beta_{l,n}(t)/c$ is the Doppler frequency of the nth unresolvable subpath within the lth resolvable path at time t induced by the motion of both the Tx and Rx, $v(t)$ denotes the relative velocity, f_c is the carrier frequency, $\beta_{l,n}(t)$ is the aggregate phase angle of the nth subpath, and c is the speed of light. The terms $e^{j\vec{k}(\theta_{T,l,n}(t))\vec{d}_T}$ and $e^{j\vec{k}(\theta_{R,l,n}(t))\vec{d}_R}$ are the corresponding distance-induced phase shifts, where $\theta_{T,l,n}(t)$ and $\theta_{R,l,n}(t)$ denote the DoD and DoA of the nth subpath within the lth path at time t, respectively, $\vec{d}_T$ and $\vec{d}_R$ are the vectors of the chosen element position measured from an arbitrary but fixed reference points on the corresponding arrays, and $\vec{k}$ is the wave vector so that $\vec{k}(\theta_{T(R),l,n}(t))\vec{d}_{T(R)} = \frac{2\pi}{\lambda}(x \; \cos\varphi_{T(R)}(t) \cos\phi_{T(R)}(t) + y \; \cos\varphi_{T(R)}(t) \sin\phi_{T(R)}(t) + z \; \sin\varphi_{T(R)}(t))$, where $\varphi_{T(R)}$ and $\phi_{T(R)}$ denote elevation and azimuth angles, respectively.

In the next section, we will introduce different modeling approaches, the main aim of which is to accurately and/or efficiently generate a channel impulse response with the desired channel characteristics.

2.7.1 Classification of MIMO Channel Models

Channel models can be classified in many different manners. A potential way of distinguishing the individual models is with regard to the type of channel that is being considered, say wide-band models versus narrowband models, time-variant models versus time-invariant models, 2D propagation environment models versus 3D propagation environment models, stationary models versus non-stationary models, and so on. Another very popular way to distinguish the individual models is based on whether they are obtained from measurement data or not. In this case, we have pure theoretical models versus measurement-based models. Finally, based on whether the models are completely deterministic or stochastic, we have deterministic models versus stochastic models. The most popular manner to classify MIMO channel models is into physical versus analytical models, based on the double-directional radio propagation and the

- Deterministic models
  - Ray-tracing and Stored measurements
- Stochastic models
  - Physical models
    - ∗ Geometry-based models
      - · GBSM<Geometry-Based Stochastic Model> (One-ring, Two-ring Ellipse)
    - ∗ Non-geometrical models
      - · PSM<Parametric Stochastic Model>
(Extension of the Saleh-Valenzueh and Zwick model)
  - Analytical models
    - ∗ Correlated-based models
      - · i.i.d model, Kronecker model and Weichsellberger model
    - ∗ Propagation-motivated models
      - · Finite scatterer model, Maximum entropy model and
Virtual channel representation

Figure 2.13 Classification of MIMO channel models

MIMO channel matrix, respectively. An overview of this classification is shown in Figure 2.13. MIMO channel models are first classified into two main categories: deterministic models and stochastic models.

The deterministic model determines the physical parameters in a fully deterministic manner, for example through ray-tracing and stored measurements. These models often consider some pre-determined and fixed structure, which is used in scatterer placement in an effort to match a specific type of propagation environment or even to represent site-specific obstacles. Simple geometrical optics are then used to track the propagation of the waves through the environment, and the time/space parameters are recorded for each path for use in constructing the MIMO channel matrix.

For stochastic models, the fundamental distinction is between physical and analytical models. Physical channel models characterize an environment on the basis of electromagnetic wave propagation by describing the double-directional multipath propagation [Molisch et al. 2006, Steinbauer et al. 2001] between the location of the Tx array and the location of the Rx array. They explicitly model wave-propagation parameters, such as the complex amplitude, AoD, AoA, and delay of a multipath component. Depending on the chosen complexity, physical models allow for accurate reproduction of radio propagation. Physical models are independent of antenna configurations (antenna pattern, number of antennas, array geometry, polarization, mutual coupling) and system bandwidth. In contrast to physical models, analytical channel models characterize the impulse response (or, equivalently, the transfer function) of the channel between the individual transmit and receive antennas in a mathematical/analytical way without explicitly accounting for wave propagation. The individual impulse responses are subsumed in a MIMO channel matrix. Analytical models are very popular for synthesizing MIMO matrices in the context of system and algorithm development and verification.

2.7.1.1 Physical MIMO Channel Models

Physical models can be classified as geometry-based stochastic models (GBSM) and non-geometric stochastic models (NGSM). The term GBSM refers to the fact that the modeled impulse response is related to the geometrical location of scatterers and other interacting objects. The channel model is derived from the positions of the scatterers, by applying the fundamental laws of specular reflection, diffraction, and scattering of electromagnetic waves. Firstly, the GBSM gives a mathematical function of the channel impulse response obtained from the electromagnetic properties of environment, and then utilizes the proposed stochastic distribution of the scatterers along with geometrical knowledge to get the statistical properties of the parameters, such as the distributions of AoA, AoD, ToA, angle spread and so on, in the given channel response.

The important and often-used GBSMs include the one-ring, two-ring, and elliptical scattering models. A MIMO narrowband channel based on the geometrical one-ring scattering model was first proposed by Shiu et al. [2000] and was further developed by Abdi and Kaveh [2002]. The one-ring model shown in Figure 2.14(a) is appropriate for describing environments in which the BS is elevated and unobstructed, whereas the MS is surrounded by a large number of local scatterers. Such models have been suggested for macrocell-type environments with moderate values of angular spread and large values of delay spread.

The two-ring model is based on the assumption that scatterers surrounding the transmitter and receiver control the AoD and AoA, respectively. Therefore, two rings are drawn, with

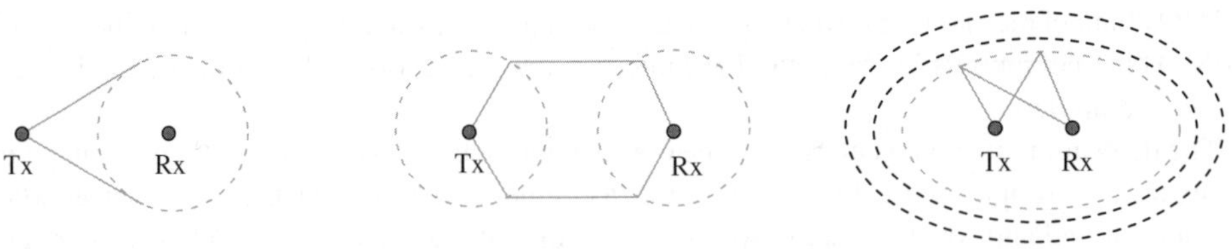

Figure 2.14 The sketches of three important GBSMs: (a) one-ring model; (b) two-ring model; (c) elliptical model

centers at the transmit and receive locations, and the radii of which represent the average distance between each communication node and their respective scatterers. The scatterers are then placed randomly on these rings. Comparison with experimental measurements has revealed that when a wave propagates through this simulated environment, each transmit and receive scatterer participates in the propagation of only one wave (transmit and receive scatterers are randomly paired). The scenario is depicted in Figure 2.14(b). These two-ring models are very simple to generate and provide flexibility in modeling different environments through adaptation of the scattering ring radii and the scatterer distributions along the ring. For example, in something like a forest environment the scatterers might be placed in a uniform distribution around the ring. In contrast, in an indoor environment a few groups of closely spaced scatterers might be used to mimic the clustered multipath behavior frequently observed. From a mathematic point of view, the two-ring model can be considered a generalization of the one-ring model. This means it is easy to get the two-ring model from the one-ring model.

The one-ring and two-ring model are frequency nonselective, which limits their utility to narrowband MIMO systems. In contrast to the one-ring and the two-ring models, the elliptical model is used for the modeling of wideband MIMO channels. The method of elliptical models for choosing scatterer placement is to draw a set of ellipses with varying focal lengths whose foci correspond to the transmit and receive positions, as shown in Figure 2.14(c). Scatterers are then placed according to a predetermined scheme on these ellipses, and only single reflections are considered. In this model, all waves bouncing off scatterers located on the same ellipse will have the same propagation-time delay, leading to the designation of constant delay ellipses. The spacing of the ellipses should be determined according to the arrival time resolution desired from the model, which would typically correspond to the inverse of the frequency bandwidth of the communication signal.

In addition to their relative simplicity, these models have two interesting features. First, once the scattering environment has been realized using an appropriate mechanism, one or both of the communication nodes can move within the environment to simulate mobility. Second, with certain statistical scatterer distributions, simple and closed-form statistical distributions can be found for delay spread, angular spread, and spatial correlation [Abdi and Kaveh 2002, Ertel et al. 1999, Norklit and Vaughan 1998].

In contrast, NGSMs, also named as parametric stochastic models (PSMs), describes and determines physical parameters, such as DoD, DoA, and delay, in a completely stochastic manner by prescribing underlying probability distribution functions without assuming an underlying geometry. Examples are the models in Wallace and Jensen [2001] and Wallace et al. [2003], which are the extensions of the Saleh–Valenzuela model. For a PSM, we first give a mathematical representation of channel response, which includes all of the parameters of the wireless channels. Then, we utilize measurement to get the statistical properties of these parameters, which are substituted into the mathematical channel response to get the final result. In contrast to to GBSMs, PSMs do not take into account the propagation environment. PSMs aim to get the statistical properties of the parameters in a given channel response by using measurement.

2.7.1.2 Analytical MIMO Channel Models

Analytical models can be further subdivided into correlation-based models (CBMs) and propagation-motivated models (PMMs). Correlation-based models characterize the MIMO channel matrix statistically in terms of the correlations between the matrix entries. Popular correlation-based analytical channel models are the Kronecker model [Chizhik et al. 2000, Chuah et al. 1998, Kermoal et al. 2002, Shiu et al. 2000] and the Weichselberger model [Weichselberger et al. 2006]. The first subclass models the channel matrix via propagation parameters. Examples are the finite scatterer model [Burr 2003], the maximum entropy model [Debbah and Muller 2005], and the virtual channel representation [Sayeed 2002].

CBMs treat the MIMO channel using correlations between channel matrix entries. CBMs assume the channel coefficients to be complex Gaussian distributed, and that the first- and second-order moments fully characterize the statistical behavior of the wireless channel. A CBM is based on the second-order statistics of the MIMO channel. The CBM is an analytical model and thus does not take into account the propagation environment as GBSM does. Developers of CBMs aim at obtaining the spatial-temporal correlation function of the channel, which they use to try to give the channel response or some other statistical property of the wireless channel.

PMMs treat the channel matrix by modeling propagation parameters. The PMM can be defined as the combination of CBMs and physical models.

Finally, the relationship between the different MIMO channel models according to categories of accurate versus simple or physical versus analytical is shown in Figure 2.15.

2.7.2 *Classification of V2V Channel Models*

In this section, we will give a brief overview of recent advances in V2V channel models. In terms of the modeling approach, these models can be categorized into geometry-based deterministic models (GBDMs) [Maurer et al. 2008] and stochastic models, and the latter can be further classified as NGSMs [Acosta-Marum and Brunelli 2007, Sen and Matolak 2008] and GBSMs [Akki 1994, Akki and Haber 1986, Karedal et al. 2009, Pätzold et al. 2008, Zajic and Stuber 2008, 2009, Zajic et al. 2009]. Here, we will briefly introduce these various channel models. A more detailed treatment of GBSMs will be given in Chapter 4.

2.7.2.1 Geometry-based Deterministic Models

GBDMs characterize V2V physical channel parameters in a completely deterministic manner. A GBDM based on the ray-tracing method for M2M channels was proposed in Maurer et al. 2008. The approach aims at reproducing the actual physical radio propagation process for a

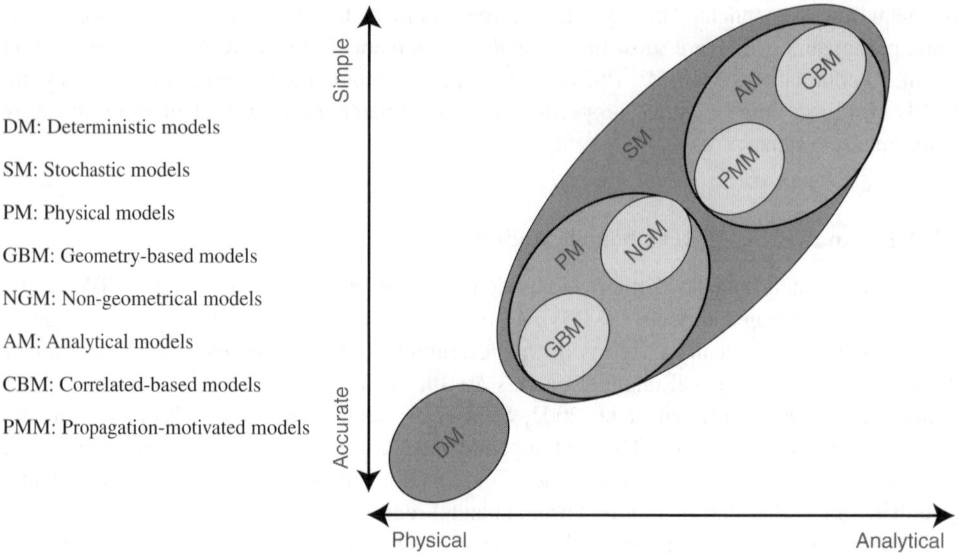

Figure 2.15 Tradeoff in MIMO channel models. DM, deterministic models; SM, stochastic models; PM, physical models; GBM, geometry-based models; NGM, non-geometrical models; AM, analytical models; CBM, correlation-based models; PMM, propagation-motivated models

given environment. As illustrated in Figure 2.16, the representation of the real environment duplication includes two major parts: the modeling of the dynamic road traffic (moving cars, vans, trucks, and so on) and the modeling of the roadside environment (buildings, parked cars, road signs, trees, and so on). Then, an accurate model of wave propagation in this real environment is implemented by generating possible paths (rays) from the Tx to the Rx according to geometric considerations and the rules of geometrical optics. In Maurer et al. [2008], a 3D ray-tracing approach was used in a wave propagation model. The resulting complex impulse response incorporated complete channel information – the statistical non-stationarity of the channel, the impact of vehicular traffic density (VTD) on channel statistics, and the impact of the elevation angle on channel statistics – and thus agreed very well with measurements. However, a GBDM requires a detailed and time-consuming description of site-specific propagation environments and consequently cannot be easily generalized to a wide class of scenarios.

2.7.2.2 Non-geometrical Stochastic Models

A NGSM determines the physical parameters of a V2V channel in a completely stochastic manner without presuming any underlying geometry. The SISO NGSM proposed by Acosta-Marum and Brunelli [2007] is the origin of the V2V channel model standardized by IEEE 802.11p. The complex impulse response of the SISO V2V channels in Acosta-Marum and Brunelli [2007] can be modeled from Eqs (2.28) and (2.29) by removing the terms $\delta(\Omega_T - \Omega_{T,l}(t))$, $\delta(\Omega_R - \Omega_{R,l}(t))$ and the corresponding distance-induced phase shifts

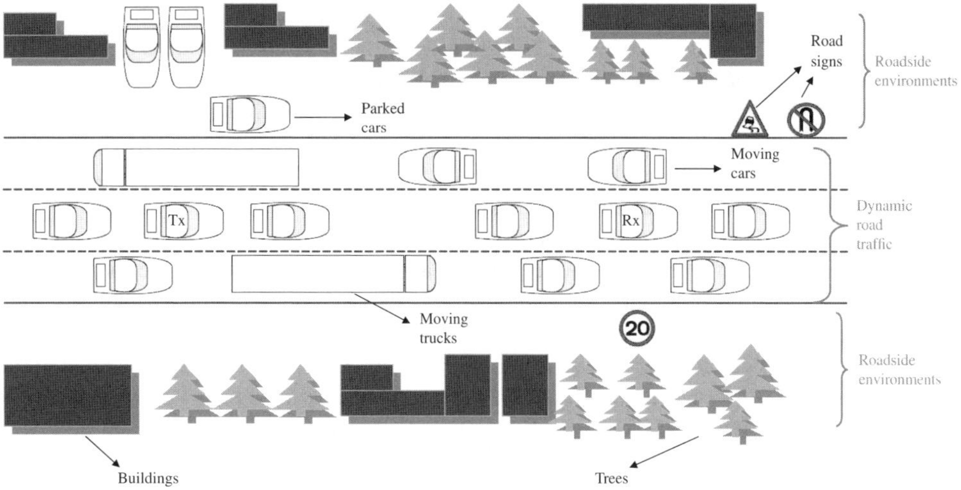

Figure 2.16 A typical V2V environment and the corresponding geometrical description of the GBDM

$e^{j\vec{k}(\theta_{T,l,n}(t))\vec{d}_T}$ and $e^{j\vec{k}(\theta_{R,l,n}(t))\vec{d}_R}$ while the aggregate phase angle $\beta_{l,n}(t)$ is placed in a 2D plane; in other words, it has zero elevation. Based on the tap-delay line structure, this model consists of L taps, with tap amplitude PDF being either Rician or Rayleighan, and thus has the ability to incorporate per-tap channel statistics. Furthermore, each tap contains N unresolvable subpaths that have different types of Doppler spectra: flat, round, classic-3-dB, and classic-6-dB shapes. This allows the synthesis of almost arbitrary Doppler spectra for each tap. However, this NGSM is still based on the WSSUS assumption and has not been used to investigate the impact of VTD on channel statistics.

Recently, a SISO NGSM was proposed by Sen and Matolak [2008]. This took into account the non-stationarity of the channel by modeling multipath component persistence via Markov chains. The impact of VTD on channel statistics was also investigated. The complex impulse response of the SISO V2V channels in the study was derived from the one in Acosta-Marum and Brunelli [2007] by adding an additional term: the persistence process, $z_l(t)$, which accounts for the finite lifetime of the lth resolvable path. The NGSM developed can easily capture the effect of a sudden disappearance of strong multipaths, typically caused by rapid blockage or obstruction by another vehicle or some other obstacle. However, the model did not consider the drift of scatterers into different delay bins (resolvable paths) and therefore the transitional probabilities of the Markov model for the persistence processes may not be accurate. This may reduce the ability of this NGSM to accurately capture the non-stationarity of real V2V channels and this question therefore deserves more investigation.

2.7.2.3 Geometry-based Stochastic Models

A GBSM is derived from a predefined stochastic distribution of effective scatterers by applying the fundamental laws of wave propagation. Such models can be easily adapted to different scenarios by changing the shape of the scattering region and/or the PDF of the location of the scatterers. GBSMs can be further classified as regular-shaped GBSMs (RS-GBSMs) or irregular-shaped GBSMs (IS-GBSMs) depending on whether effective scatterers are placed on regular shapes (one/two-ring, ellipse, and so on) or irregular shapes (randomly). Therefore, the introduced one-ring, two-ring, and elliptical-ring MIMO F2M channel models discussed in Section 2.7.1 actually belongs to the category of RS-GBSMs.

In general, RS-GBSMs are used for the theoretical analysis of channel statistics and theoretical design and comparison of communication systems. Therefore, to preserve the mathematical tractability, RS-GBSMs assume all effective scatterers are located on a regular shape. In contrast, IS-GBSMs aim to reproduce physical reality and thus need to modify the location and properties of the effective scatterers of RS-GBSMs. IS-GBSMs place effective scatterers with specified properties at random locations with certain statistical distributions. The signal contributions of the effective scatterers are determined from a greatly simplified ray-tracing method and the final signal is summed up to obtain the complex impulse response, which can be expressed as in Eqs (2.28) and (2.29). More detailed introduction of GBSMs will be given in Chapter 4.

Bibliography

Abdi A and Kaveh M 2002 A space-time correlation model for multielement antenna systems in mobile fading channels. *IEEE Journal on Selected Areas in Communications* **20**(3), 550–560.

Acosta-Marum G and Brunelli D 2007 Six time- and frequency- selective empirical channel models for vehicular wireless LANs. *IEEE Vehicular Technology Magazine* **2**(4), 4–11.

Akki A 1994 Statistical properties of mobile-to-mobile land communication channels. *IEEE Transactions on Vehicular Technology* **43**(4), 826–831.

Akki A and Haber F 1986 A statistical model of mobile-to-mobile land communication channel. *IEEE Transactions on Vehicular Technology* **35**(1), 2–7.

Bello PA 1964 Time-frequency duality. *IEEE Transactions on Information Theory* **10**(1), 18–33.

Bello PA and Bengtsson M 1963 Characterization of randomly time-invariant linear channels. *IEEE Transactions on Communication Systems* **CS-11**, 360–393.

Burr AG 2003 Capacity bounds and estimates for the finite scatterers MIMO wireless channel. *IEEE Journal on Selected Areas in Communications* **21**(5), 812–818.

Chizhik D, Foschini G and Valenzuela R 2000 Capacities of multi-element transmit and receive antennas: Correlations and keyholes. *Electronics Letters* **36**(13), 1099–1100.

Chuah CN, Kahn J and Tse D 1998 Capacity of multi-antenna array systems in indoor wireless environment *Global Telecommunications Conference, 1998. GLOBECOM 1998. The Bridge to Global Integration. IEEE*, vol. 4, pp. 1894–1899.

Clarke RH 1968 A statistical theory of mobile-radio reception. *Bell System Technical Journal* **47**(6), 957–1000.

COST 231 1991 TD(973)119-REV 2 (WG2): Urban transmission loss models for mobile radio in the 900- and 1,800-MHz bands.

Debbah M and Muller R 2005 MIMO channel modeling and the principle of maximum entropy. *IEEE Transactions on Information Theory* **51**(5), 1667–1690.

Ertel R, Hu Z and Reed J 1999 Antenna array hardware amplitude and phase compensation using baseband antenna array outputs *Vehicular Technology Conference, 1999 IEEE 49th*, vol. 3, pp. 1759–1763.

Fleury B 2000 First- and second-order characterization of direction dispersion and space selectivity in the radio channel. *IEEE Transactions on Information Theory* **46**(6), 2027–2044.

Hata M and Nagatsu T 1980 Mobile location using signal strength measurements in a cellular system. *IEEE Transactions on Vehicular Technology* **29**(2), 245–252.

Karedal J, Tufvesson F, Czink N, Paier A, Dumard C, Zemen T, Mecklenbrauker C and Molisch A 2009 A geometry-based stochastic MIMO model for vehicle-to-vehicle communications. *IEEE Transactions on Wireless Communications* **8**(7), 3646–3657.

Kattenbach R 2002 Statistical modeling of small-scale fading in directional radio channels. *IEEE Journal on Selected Areas in Communications* **20**(3), 584–592.

Kermoal J, Schumacher L, Pedersen K, Mogensen P and Frederiksen F 2002 A stochastic MIMO radio channel model with experimental validation. *IEEE Journal on Selected Areas in Communications* **20**(6), 1211–1226.

Lee W 2010 *Mobile Communications Design Fundamentals* 2nd edn. John Wiley & Sons.

Maurer J, Fügen T, Porebska M, Zwick T and Wisebeck W 2008 A ray-optical channel model for mobile to mobile communications *COST 2100 4th MCM, COST 2100 TD(08) 430*, pp. 6–8, Wroclaw, Poland.

Molisch A, Asplund H, Heddergott R, Steinbauer M and Zwick T 2006 The COST259 directional channel model – Part I: Overview and methodology. *IEEE Transactions on Wireless Communications* **5**(12), 3421–3433.

Norklit O and Vaughan R 1998 Reducing the fading rate with antenna arrays *Global Telecommunications Conference, 1998. GLOBECOM 1998. The Bridge to Global Integration. IEEE*, vol. 6, pp. 3187–3192.

Okumura Y, Ohmori E, Kawano T and Fukuda K 1968 Field strength and its variability in VHF and UHF land-mobile radio services. *Review of the Electrical Communication Laboratory* **16**(9), 825–837.

Oppenheim AV, Willsky AS and Nawab SH 2009 *Signals and Systems* 2nd edn. PHI Learning.

Papoulis A and Pillai SU 2002 *Probability, Random Variables and Stochastic Processes* 4th edn. McGraw-Hill, NJ.

Parsons JD 2000 *The Mobile Radio Propagation Channel* 2nd edn. John Wiley and Sons.

Pätzold M and Laue F 1998 Statistical properties of Jakes' fading channel simulator *Vehicular Technology Conference, 1998. VTC 98. 48th IEEE*, vol. 2, pp. 712–718.

Pätzold M, Hogstad B and Youssef N 2008 Modeling, analysis, and simulation of MIMO mobile-to-mobile fading channels. *IEEE Transactions on Wireless Communications* **7**(2), 510–520.

Sayeed A 2002 Deconstructing multiantenna fading channels. *IEEE Transactions on Signal Processing* **50**(10), 2563–2579.

Sen I and Matolak D 2008 Vehicle-vehicle channel models for the 5-GHz band. *IEEE Transactions on Intelligent Transportation Systems* **9**(2), 235–245.

Shiu DS, Foschini G, Gans M and Kahn J 2000 Fading correlation and its effect on the capacity of multielement antenna systems. *IEEE Transactions on Communications* **48**(3), 502–513.

Steele R 1992 *Mobile Radio Communications*. London UK: Pentech Press.

Steinbauer M, Molisch A and Bonek E 2001 The double-directional radio channel. *IEEE Antennas and Propagation Magazine* **43**(4), 51–63.

Stüber GL 2001 *Principles of Mobile Communications* 2nd edn. Kluwer Academic.

Wallace J and Jensen M 2001 Characteristics of measured 4×4 and 10×10 MIMO wireless channel data at 2.4-GHz *Antennas and Propagation Society International Symposium, 2001. IEEE*, vol. 3, pp. 96–99.

Wallace J, Jensen M, Swindlehurst A and Jeffs B 2003 Experimental characterization of the MIMO wireless channel: data acquisition and analysis. *IEEE Transactions on Wireless Communications* **2**(2), 335–343.

Weichselberger W, Herdin M, Ozcelik H and Bonek E 2006 A stochastic MIMO channel model with joint correlation of both link ends. *IEEE Transactions on Wireless Communications* **5**(1), 90–100.

Youssef N, Wang CX and Pätzold M 2005 A study on the second order statistics of Nakagami-Hoyt mobile fading channels. *IEEE Transactions on Vehicular Technology* **54**(4), 1259–1265.

Zajic A and Stuber G 2008 Three-dimensional modeling, simulation, and capacity analysis of space–time correlated mobile-to-mobile channels. *IEEE Transactions on Vehicular Technology* **57**(4), 2042–2054.

Zajic A and Stuber G 2009 Three-dimensional modeling and simulation of wideband MIMO mobile-to-mobile channels. *IEEE Transactions on Wireless Communications* **8**(3), 1260–1275.

Zajic A, Stuber G, Pratt T and Nguyen S 2009 Wideband MIMO mobile-to-mobile channels: Geometry-based statistical modeling with experimental verification. *IEEE Transactions on Vehicular Technology* **58**(2), 517–534.

3

Generic Channel Models

A received electromagnetic wave with arbitrary wavefront can be viewed as the superposition of multiple plane waves. By using the high-resolution parameter estimation algorithm to process the signals sampled in space, time, and frequency, these plane waves can be resolved by their distinctive direction of arrival (DoA), direction of departure (DoD), delay, Doppler frequency, and polarization complex attenuation coefficients. In cases in which the locations of the transmitter and the receiver are known, propagation paths can be reconstructed based on the parameter estimates.

Figure 3.1 illustrates an experimental example in which the direction parameters of some estimated specular propagation paths are overlaid on the photographs taken from the point of view of the transmitter and the receiver. Reconstructed trajectories of the propagation paths are also depicted by overlaying on the map. It can be seen that paths numbers 1–6 exhibit their bouncing points along the edge of building B3. The interaction between the waves propagating along these paths and the building edge could be diffraction. Paths 9–19, with DoAs and DoDs overlapping with the area where a lot of tree stems and the edges of buildings exist, may be caused by scattering due to the cluttered environment there. Paths 21–23 involve bouncing points on the metal sculptures and the building facade. It is possible that reflections occur on these obstacles when the waves propagate along these paths.

In fact, reflection, diffraction, and scattering are considered to be the three basic propagation mechanisms. They occur depending on the size of the object compared with the wavelength.

Reflection

Reflection occurs when a plane electromagnetic wave encounters an obstacle much larger than the wavelength.

Diffraction

Diffraction happens when the obstacle's size is of the same order as the wavelength. It arises because of the way in which wave propagates. This is described by the Huygens–Fresnel principle and the principle of superposition of waves. The propagation of a wave can be visualized

Propagation Channel Characterization, Parameter Estimation and Modelling for Wireless Communications, First Edition.
Xuefeng Yin and Xiang Cheng.

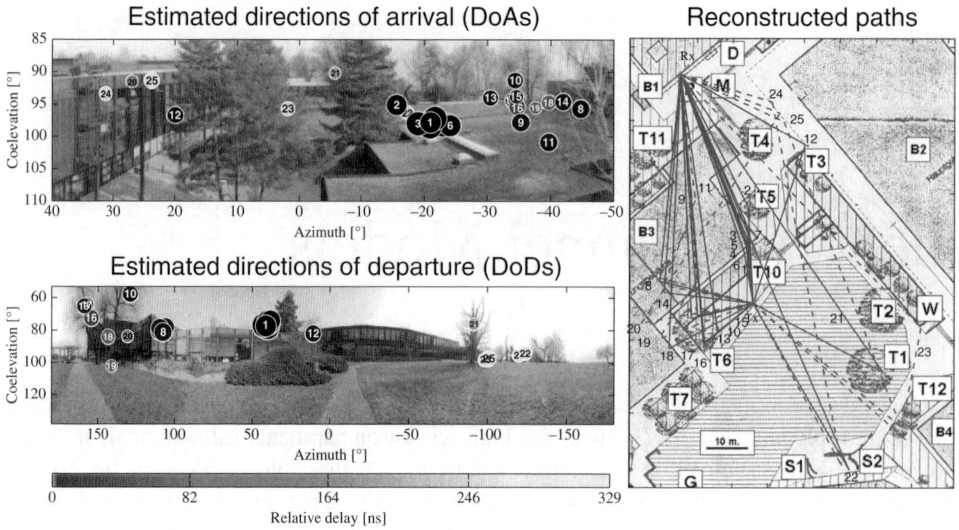

Figure 3.1 Estimated propagation paths in an outdoor environment

by considering every point on a wave front as a point source for a secondary spherical wave. The wave displacement at any subsequent point is the sum of these secondary waves. When waves are added together, their sum is determined by the relative phases as well as the amplitudes of the individual waves so that the summed amplitude of the waves can have any value between zero and the sum of the individual amplitudes. Hence, diffraction patterns usually have a series of maxima and minima.

Scattering

Scattering happens when a plane wave encounters an obstacle much smaller than the wavelength. This obstacle becomes a new source emitting waves in multiple directions. Recently, modeling channel composition based on different propagation mechanisms has been performed for outdoor environments [Medbo et al. 2012]. It was found that the estimated paths considered to be one-bounce paths can be further identified as the specular reflected paths and the diffracted paths. The former paths exhibit less path loss than the latter.

Although it is possible to estimate both the mechanism involved in propagation along a specific path and the path's geometrical and attenuation parameters, in practice, without knowing the exact obstacles in the environment we may use a generic model to describe the composition of the channel. In this chapter, we introduce some generic models that adopt different kinds of parameter to characterize the behavior of individual components in a channel. These models are proposed from different perspectives, such as specular and dispersive, time invariant and variant, as well as deterministic and stochastic path components. They can be used in different contexts for channel estimation. The estimated model parameters are useful for constructing channel models afterwards.

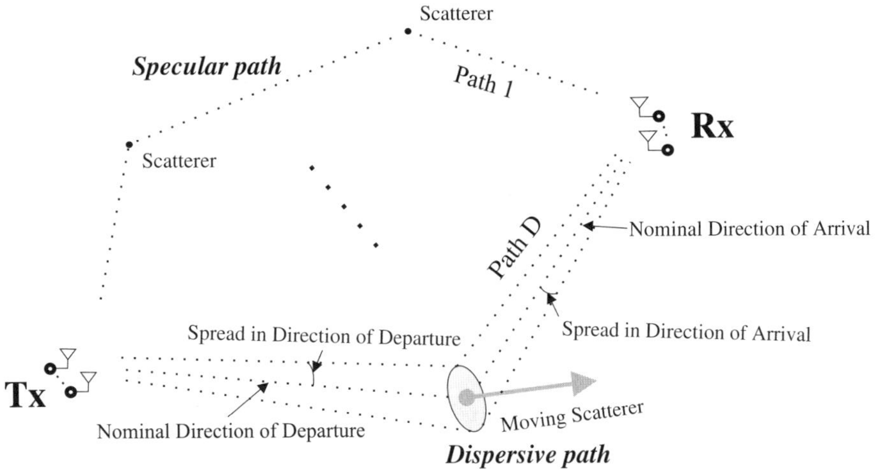

Figure 3.2 Channel dispersion in multiple dimensions

3.1 Channel Spread Function

The propagation of a wave in an environment can be considered to follow different paths. This phenomenon is usually called as multipath propagation. Figure 3.2 shows a channel dispersive in multiple dimensions. These elements may be different in many aspects:

- the geometrical dimensions
  - delay
  - Doppler frequency
  - DoA (i.e. the azimuth and elevation of arrival)
  - DoD (i.e. the azimuth and elevation of departure)
- the fading properties including complex attenuation
- the polarimetric properties indicated by the polarization matrix of a path
- the propagation mechanisms such as reflection, diffraction and scattering.

However, in practice, whether a dispersion dimension should be considered when constructing a generic channel model to be applied for channel parameter estimation depends on the intrinsic resolving ability of the measurement equipment in this dimension.

As illustrated in Figure 3.2, a propagation channel may contain multiple components. For simplicity, we assume that these components are specular path components. A number L of waves are departing from the location of the transmitter (Tx) and impinging in a region surrounding the location of receiver (Rx). For both the Tx and the Rx sites, we assume individual coordinate systems being specified at arbitrary origins $\mathcal{O}_{\rm Tx}$ and $\mathcal{O}_{\rm Rx}$ in the region surrounding the Tx and the Rx respectively.

The location of the Tx and the Rx antennas are determined by two unique vectors $x_{\mathrm{Tx}} \in \mathbb{R}^3$ and $x_{\mathrm{Rx}} \in \mathbb{R}^3$, respectively, where $\mathbb{R}$ denotes the real line. In the case where L components are all specular path components, the output signal of the Rx antenna located at x_{Rx} while the Tx antenna located at x_{Tx} transmits signals can be written as

$$r(x_{\mathrm{Tx}}, x_{\mathrm{Rx}}; t) = \sum_{\ell=1}^{L} \alpha_\ell \exp\{j2\pi\lambda_0^{-1}(\mathbf{\Omega}_{\mathrm{Tx},\ell} \cdot x_{\mathrm{Tx}})\} \exp\{j2\pi\lambda_0^{-1}(\mathbf{\Omega}_{\mathrm{Rx},\ell} \cdot x_{\mathrm{Rx}})\} \times \exp\{j2\pi\nu_\ell t\} s(t - \tau_\ell) \tag{3.1}$$

In the above expression, $s(t)$ denotes the modulating signal at the input of the Tx antenna and λ_0 is the wavelength. The other parameters are, respectively, the complex amplitude α_ℓ, the delay τ_ℓ, the DoA $\mathbf{\Omega}_{\mathrm{Rx},\ell}$, the DoD $\mathbf{\Omega}_{\mathrm{Tx},\ell}$, and the Doppler frequency ν_ℓ of the ℓth impinging wave. In Eq. (3.1), $(\cdot)$ denotes the scalar product. A direction is represented with a unit vector $\mathbf{\Omega}$. For DoA, the direction has its terminal point located at the origin $\mathcal{O}$ of the coordinate system, and its initial point located on a sphere $\mathbb{S}_2$ of unit radius centered at $\mathcal{O}_{\mathrm{Tx}}$ as depicted in Figure 3.3 [Jakes 1974, Watson 1983]. For DoD, the direction has its terminal point located on a sphere $\mathbb{S}_2$ of unit radius centered at $\mathcal{O}_{\mathrm{Rx}}$, and its initial point located at the origin $\mathcal{O}_{\mathrm{Rx}}$. The DoA is uniquely determined by its spherical coordinates $(\phi_{\mathrm{Tx}}, \theta_{\mathrm{Tx}})$ according to

$$\mathbf{\Omega}_{\mathrm{Tx}} = e(\phi_{\mathrm{Tx}}, \theta_{\mathrm{Tx}}) \doteq [\cos(\phi_{\mathrm{Tx}})\sin(\theta_{\mathrm{Tx}}), \sin(\phi_{\mathrm{Tx}})\sin(\theta_{\mathrm{Tx}}), \cos(\theta_{\mathrm{Tx}})]^{\mathrm{T}} \in \mathbb{S}_2 \tag{3.2}$$

where $\phi_{\mathrm{Tx}}, \theta_{\mathrm{Tx}}$ represent the azimuth and the elevation of the DoD respectively. Similarly, the DoA $\mathbf{\Omega}_{\mathrm{Rx}}$ is uniquely determined by its spherical coordinate $(\phi_{\mathrm{Rx}}, \theta_{\mathrm{Rx}})$. Notice that α_ℓ in

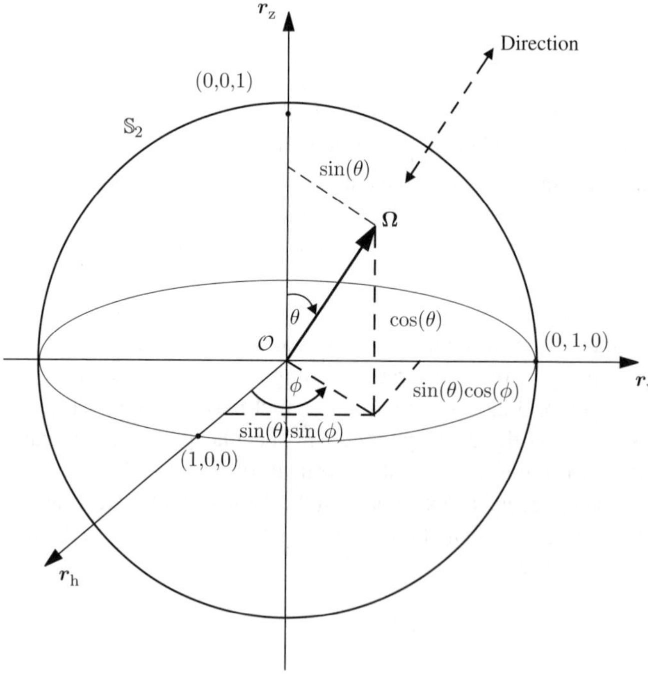

Figure 3.3 Direction of incidence characterized by $\mathbf{\Omega}$

Eq. (3.1) is a function of the complex electric field of the ℓth wave as well as of the Rx antenna response such as, for example, its radiation pattern.

The summation of individual specular path components as represented in Eq. (3.1) can be generalized to the form of integral expression

$$
\begin{aligned}
r(x_{\rm Tx}, x_{\rm Rx}; t) = \int\int\int\int & \exp\{j2\pi\lambda_0^{-1}(\boldsymbol{\Omega}_{\rm Tx}\cdot x_{\rm Tx})\}\exp\{j2\pi\lambda_0^{-1}(\boldsymbol{\Omega}_{\rm Rx}\cdot x_{\rm Rx})\} \\
& \times \exp\{j2\pi\nu t\}s(t-\tau)\, h(\boldsymbol{\Omega}_{\rm Tx}, \boldsymbol{\Omega}_{\rm Rx}, \tau, \nu){\rm d}\boldsymbol{\Omega}_{\rm Tx}{\rm d}\boldsymbol{\Omega}_{\rm Rx}{\rm d}\tau{\rm d}\nu \qquad (3.3)
\end{aligned}
$$

In the special case where the channel can be decomposed into multiple specular path components, the function $h(\boldsymbol{\Omega}_{\rm Tx}, \boldsymbol{\Omega}_{\rm Rx}, \tau, \nu)$ takes the form

$$
h(\boldsymbol{\Omega}_{\rm Tx}, \boldsymbol{\Omega}_{\rm Rx}, \tau, \nu) = \sum_{\ell=1}^{L} \alpha_\ell \delta(\boldsymbol{\Omega}_{\rm Tx} - \boldsymbol{\Omega}_{{\rm Tx},\ell})\delta(\boldsymbol{\Omega}_{\rm Rx} - \boldsymbol{\Omega}_{{\rm Rx},\ell})\delta(\tau-\tau_\ell)\delta(\nu-\nu_\ell) \qquad (3.4)
$$

Function $h(\boldsymbol{\Omega}_{\rm Tx}, \boldsymbol{\Omega}_{\rm Rx}, \tau, \nu)$ is called the channel's bidirection-delay-Doppler spread function. The delay-Doppler spread function has been used to denote the dispersive behavior of a time-variant channel considered as a linear system [Bello and Bengtsson 1963]. It has also been extended to include DoA [Fleury 2000]. Here the spread function is further extended to include bidirectionality; that is to say both DoA *and* DoD.

The spread function can be used to describe the situation of diffuse scattering, i.e. a large amount of impinging waves arising from the diffuse scattering on objects with a large geometrical extent. The integral expression in Eq. (3.3) describes the input–output relationship of the linear system that comprises the Tx antenna, the propagation channel, and the Rx antenna. In our case, this linear system is referred to as the propagation channel. The integral expression in Eq. (3.3) is only an approximation of the real input–output relationship of the radio channel when the range $x_{\rm Tx}$, $x_{\rm Rx}$ and t are small so that the received signal can be considered to be the superposition of infinitesimal elementary plane waves originating from the elementary surfaces of the scattering objects.

The expectation of the spread function is 0:

$$
{\rm E}[h(\boldsymbol{\Omega}_{\rm Tx}, \boldsymbol{\Omega}_{\rm Rx}, \tau, \nu)] = 0 \qquad (3.5)
$$

Here, ${\rm E}[\cdot]$ denotes the expectation of the random element given as an argument. According to Fleury [2000], this property is justified from a physical point of view, because the phase of the impinging waves can be reasonably assumed to be uniformly distributed between 0 and 2π at the carrier frequencies under consideration.

The radio channel is uncorrelated scattering (US) when the bidirection-delay-Doppler spread function is an orthogonal stochastic measure (OSM) [Gihman and Skorohod 1974, Ch. IV]. Similar to the notation introduced in Bello and Bengtsson [1963], the following identity

$$
\begin{aligned}
& {\rm E}[h(\boldsymbol{\Omega}_{\rm Tx}, \boldsymbol{\Omega}_{\rm Rx}, \tau, \nu)h(\boldsymbol{\Omega}'_{\rm Tx}, \boldsymbol{\Omega}'_{\rm Rx}, \tau', \nu')] \\
& \quad = P(\boldsymbol{\Omega}_{\rm Tx}, \boldsymbol{\Omega}_{\rm Rx}, \tau, \nu)\delta(\boldsymbol{\Omega}_{\rm Tx} - \boldsymbol{\Omega}'_{\rm Tx})\delta(\boldsymbol{\Omega}_{\rm Rx} - \boldsymbol{\Omega}'_{\rm Rx})\delta(\tau-\tau')\delta(\nu-\nu') \qquad (3.6)
\end{aligned}
$$

is justified by the uncorrelated scattering assumption and thus the zero-mean process $h(\boldsymbol{\Omega}_{\rm Tx}, \boldsymbol{\Omega}_{\rm Rx}, \tau, \nu)$ is an OSM. In Eq. (3.6)

$$
P(\boldsymbol{\Omega}_{\rm Tx}, \boldsymbol{\Omega}_{\rm Rx}, \tau, \nu) = {\rm E}[|h(\boldsymbol{\Omega}_{\rm Tx}, \boldsymbol{\Omega}_{\rm Rx}, \tau, \nu)|^2] \qquad (3.7)
$$

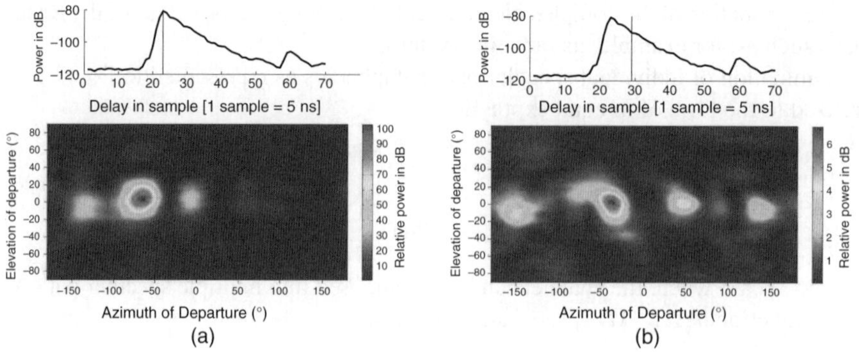

Figure 3.4 Estimated power DoD spectra for the channel response observed at different delays: (a) for $\tau = 115$ ns (23 samples); (b) for $\tau = 145$ ns (29 samples)

is called the bidirection-delay-Doppler power spectrum, which describes how the average impinging power is distributed in the dimensions of the bidirection, delay, and Doppler frequency.

The spread function and the power spectrum of the channel in single dimensions can be calculated by computing the marginal of the multi-dimensional spread function and power spectrum respectively. For example, the spread function in bidirection $h(\mathbf{\Omega}_{\text{Tx}}, \mathbf{\Omega}_{\text{Rx}})$ can be computed as

$$h(\mathbf{\Omega}_{\text{Tx}}, \mathbf{\Omega}_{\text{Rx}}) = \int\int h(\mathbf{\Omega}_{\text{Tx}}, \mathbf{\Omega}_{\text{Rx}}, \tau, \nu)\text{d}\tau\text{d}\nu \tag{3.8}$$

Similarly, the bidirection power spectrum is calculated as

$$P(\mathbf{\Omega}_{\text{Tx}}, \mathbf{\Omega}_{\text{Rx}}) = \int\int P(\mathbf{\Omega}_{\text{Tx}}, \mathbf{\Omega}_{\text{Rx}}, \tau, \nu)\text{d}\tau\text{d}\nu \tag{3.9}$$

Figure 3.4 depicts the experimental results for the estimated power DoD spectrum $P(\mathbf{\Omega}_{\text{Tx}}; \tau = 115\ \text{ns})$ and $P(\mathbf{\Omega}_{\text{Tx}}; \tau = 145\ \text{ns})$ of a channel calculated from the measurement data collected in an indoor environment where the line-of-sight (LoS) path exists. These power-spectrum estimates are calculated using the Bartlett beamforming method [Bartlett 1948]. It can be observed that for $\tau = 115$ ns, where the power delay profile of the channel achieves its maximum, the power DoD spectrum is mostly concentrated on a single DoD, which is very likely to be the DoD of the LoS component; for $\tau = 145$ ns, the channel is more dispersive and concentrated on four dominant DoDs, which have similar elevations of departure, close to the horizontal plane, and which are widely spread in the azimuth.

3.2 Specular-path Model

The specular-path model is based on the assumption that electromagnetic waves propagate in an environment along multiple specular paths. In this section, a bidirection-delay-Doppler-dual-polarization specular path-model is introduced, in which the parameters describing a specular path include its delay, DoA, DoD, Doppler frequency and polarization matrix. Notice that other parameters may be also included, depending on the characteristics of interest. One example would be the time-variability of the path parameters when the time-variant channel is investigated.

Dual polarization is the polarization of an electromagnetic wave, especially transverse electric and magnetic waves, into two orthogonal directions. From the parameter estimation point of view, by considering dual polarization, more samples of the channel observation are obtained, and consequently, the estimation accuracy of the path parameters is higher than when only single-polarization channel is observed. Furthermore, since use of dual polarized antenna arrays has been widely considered as a way to boost the capacity of a MIMO system in advanced wireless communication standards [Deng et al. 2005, Kyritsi and Cox 2002, Pedersen and Mogensen 1999, Vaughan 1990, Yin et al. 2003], modeling the channels in the dual-polarization domain has begun to attract researchers' attention [Acosta-Marum et al. 2010, Degli-Esposti et al. 2007, Hämäläinen et al. 2005, Kwon and Stüber 2010]. It is necessary to design estimation algorithms for extracting the polarization characteristics of multipath components in the propagation channel.

Figure 3.5 depicts propagation when the Rx and Tx antenna arrays have dual-polarized configurations. In the diagram, each of the dual antennas transmits/receives the signal in two polarizations at the same time. This is consistent with the fact that an antenna cannot transmit or receive signals in one polarization only. In the case where it is necessary to differentiate the two polarizations, one of them is called the main polarization, specifying the dominant direction of the signal field pattern. The other is correspondingly referred to the complementary polarization.

In order to describe the wave polarization, a polarization matrix $\boldsymbol{A}_\ell$ is introduced, which is composed of the complex weights for the attenuations along the propagation paths. The signal

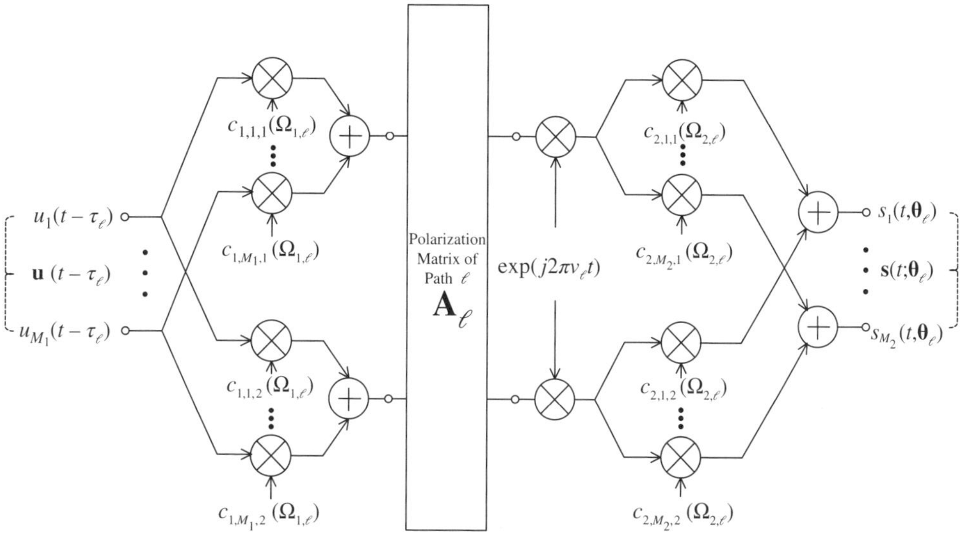

Figure 3.5 Contribution of the ℓth wave to the received signal in a MIMO system incorporating dual-antenna arrays

model describing the contribution of the ℓth wave to the output of the MIMO system reads

$$s(t;\boldsymbol{\theta}_\ell) = \exp(j2\pi\nu_\ell t)\left[\boldsymbol{c}_{2,1}(\boldsymbol{\Omega}_{2,\ell})\ \ \boldsymbol{c}_{2,2}(\boldsymbol{\Omega}_{2,\ell})\right]\begin{bmatrix}\alpha_{\ell,1,1} & \alpha_{\ell,1,2}\\ \alpha_{\ell,2,1} & \alpha_{\ell,2,2}\end{bmatrix} \cdot \left[\boldsymbol{c}_{1,1}(\boldsymbol{\Omega}_{1,\ell})\ \ \boldsymbol{c}_{1,2}(\boldsymbol{\Omega}_{1,\ell})\right]^T \cdot \boldsymbol{u}(t-\tau_\ell) \quad (3.10)$$

where $\boldsymbol{c}_{i,p_i}(\boldsymbol{\Omega})$ denotes the steering vector of the transmitter array ($i = 1$) with a total of M_1 entries, or the receiver array ($i = 2$) with a total of M_2 entries, where M_1 and M_2 are the numbers of antennas in the Tx and Rx respectively. Here p_i, ($p_i = 1, 2$) denotes the polarization index.

Written in matrix form, Eq. (3.10) is reformulated as

$$s(t;\boldsymbol{\theta}_\ell) = \exp(j2\pi\nu_\ell t)\boldsymbol{C}_2(\boldsymbol{\Omega}_{2,\ell})\boldsymbol{A}_\ell\boldsymbol{C}_1^{\mathrm{T}}(\boldsymbol{\Omega}_{1,\ell})\boldsymbol{u}(t-\tau_\ell) \quad (3.11)$$

with

$$\boldsymbol{C}_2(\boldsymbol{\Omega}_{2,\ell}) = \left[\boldsymbol{c}_{2,1}(\boldsymbol{\Omega}_{2,\ell})\ \ \boldsymbol{c}_{2,2}(\boldsymbol{\Omega}_{2,\ell})\right] \quad (3.12)$$

$$\boldsymbol{C}_1(\boldsymbol{\Omega}_{1,\ell}) = \left[\boldsymbol{c}_{1,1}(\boldsymbol{\Omega}_{1,\ell})\ \ \boldsymbol{c}_{1,2}(\boldsymbol{\Omega}_{1,\ell})\right] \quad (3.13)$$

$$\boldsymbol{A}_\ell = \begin{bmatrix}\alpha_{\ell,1,1} & \alpha_{\ell,1,2}\\ \alpha_{\ell,2,1} & \alpha_{\ell,2,2}\end{bmatrix} = [\alpha_{\ell,p_2,p_1}] \quad (3.14)$$

$$\boldsymbol{u}(t) = [u_1(t),\ldots,u_M(t)]^T \quad (3.15)$$

Equation (3.11) can be recast as

$$\begin{aligned} s(t;\boldsymbol{\theta}_\ell) &= \exp(j2\pi\nu_\ell t)\\ &\quad\cdot\{[\alpha_{\ell,1,1}\boldsymbol{c}_{2,1}(\boldsymbol{\Omega}_{2,\ell})\boldsymbol{c}_{1,1}^{\mathrm{T}}(\boldsymbol{\Omega}_{1,\ell}) + \alpha_{\ell,1,2}\boldsymbol{c}_{2,1}(\boldsymbol{\Omega}_{2,\ell})\boldsymbol{c}_{1,2}^{\mathrm{T}}(\boldsymbol{\Omega}_{1,\ell})\\ &\qquad + \alpha_{\ell,2,1}\boldsymbol{c}_{2,2}(\boldsymbol{\Omega}_{2,\ell})\boldsymbol{c}_{1,1}^{\mathrm{T}}(\boldsymbol{\Omega}_{1,\ell}) + \alpha_{\ell,2,2}\boldsymbol{c}_{2,2}(\boldsymbol{\Omega}_{2,\ell})\boldsymbol{c}_{1,2}^{\mathrm{T}}(\boldsymbol{\Omega}_{1,\ell})]\boldsymbol{u}(t-\tau_\ell)\} \quad (3.16)\\ &= \exp(j2\pi\nu_\ell t)\cdot\left(\sum_{p_2=1}^{2}\sum_{p_1=1}^{2}\alpha_{\ell,p_2,p_1}\boldsymbol{c}_{2,p_2}(\boldsymbol{\Omega}_{2,\ell})\boldsymbol{c}_{1,p_1}^{T}(\boldsymbol{\Omega}_{1,\ell})\right)\boldsymbol{u}(t-\tau_\ell) \quad (3.17)\end{aligned}$$

Model for Time-division-multiplexing Channel Sounding

Channel sounding using multiple Tx and Rx antennas can be conducted with a radio-frequency switch that connects a single Tx antenna with the transmit front-end chain, or an Rx antenna with the receiver front-end chain sequentially. We call this kind of sounding technique the time-division-multiplexing (TDM) sounding technique. Examples of measurement equipment that uses TDM sounding are:

- PROPsound [Stucki 2001]
- RUSK [Almers et al. 2005, MEDAV GmbH 2001]
- rBECS [Kim et al. 2012, Park et al. 2012, Yin et al. 2012a,b,c].

We consider a widely used TDM structure, as depicted in Figure 3.6, to construct the signal model.

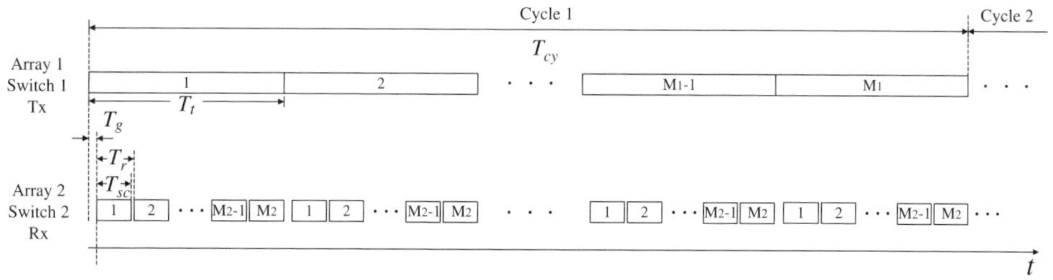

Figure 3.6 Timing structure of sounding and sensing windows

The m_1th antenna element of Array 1 is active during the sounding windows:[1]

$$q_{1,m_1}(t) = \sum_{i=1}^{I} q_{T_t}(t - t_{i,m_1} + T_g), m_1 = 1, \ldots, M_1 \tag{3.18}$$

where i denotes the cycle index and

$$t_{i,m_1} = (i-1)T_{cy} + (m_1 - 1)T_t$$

Here $q_{1,m_1}(t)$ is a real function, with a value of 1 or 0 corresponding to the active or inactive moments of the m_1th window. Let us define the sounding window vector

$$\boldsymbol{q}_1(t) \doteq [q_{1,1}(t), \ldots, q_{1,M_1}(t)]^T$$

The so-called "sensing window"

$$q_{T_{sc}}(t - t_{i,m_1,m_2}), m_2 = 1, \ldots, M_2, m_1 = 1, \ldots, M_1$$

corresponds to the case where

- the m_1th Tx antenna is active
- the m_2th Rx antenna is sensing,

where

$$t_{i,m_2,m_1} = (i-1)T_{cy} + (m_1 - 1)T_t + (m_2 - 1)T_r.$$

The sensing window for the m_2th Rx dual antenna is given by the real function

$$q_{1,m_2}(t) = \sum_{i}^{I} \sum_{m_1=1}^{M_1} q_{T_{sc}}(t - t_{i,m_2,m_1}) \tag{3.19}$$

We can define the sensing window vector

$$\boldsymbol{q}_2(t) \doteq [q_{2,1}(t), \ldots, q_{2,M_2}(t)]^T$$

as well as

$$q_2(t) = \sum_{i=1}^{I} \sum_{m_2=1}^{M_2} \sum_{m_1=1}^{M_1} q_{T_{sc}}(t - t_{i,m_2,m_1}) \tag{3.20}$$

---

[1] If another ordering of polarization sounding/sensing is used, the sounding windows $q_1(t)$ and the sensing windows $q_2(t)$ need merely to be appropriately redefined.

Transmitted Signal

Making use of the sounding window vector $\boldsymbol{q}_1(t)$, we get the explicit transmitted signal $\boldsymbol{u}(t)$ by concatenating the inputs of the M_1 elements of Array 1

$$\boldsymbol{u}(t) = \boldsymbol{q}_1(t)u(t) \tag{3.21}$$

Received Signal

The signal at the output of Switch 2 can be written as

$$Y(t) = \sum_{\ell=1}^{L} \boldsymbol{q}_2^T(t)\boldsymbol{s}(t;\boldsymbol{\theta}_\ell) + \sqrt{\frac{N_o}{2}}q_2(t)W(t) \tag{3.22}$$

with

$$\boldsymbol{s}(t;\boldsymbol{\theta}_\ell) = \exp(j2\pi\nu_\ell t)\boldsymbol{q}_2^{\mathrm{T}}(t)\boldsymbol{C}_2(\boldsymbol{\Omega}_{2,\ell})\boldsymbol{A}_\ell\boldsymbol{C}_1(\boldsymbol{\Omega}_{1,\ell})^{\mathrm{T}}\boldsymbol{q}_1(t-\tau_\ell)u(t-\tau_\ell) \tag{3.23}$$

Implementing Eq. (3.16), we can rewrite as:

$$\begin{aligned} s(t;\boldsymbol{\theta}_\ell) = &\exp(j2\pi\nu_\ell t)\cdot\sum_{p_2=1}^{2}\sum_{p_1=1}^{2}\alpha_{\ell,p_2,p_1}\boldsymbol{q}_2^{\mathrm{T}}(t)\boldsymbol{c}_{2,p_2}(\boldsymbol{\Omega}_{2,\ell})\boldsymbol{c}_{1,p_1}^{\mathrm{T}}(\boldsymbol{\Omega}_{1,\ell})\boldsymbol{q}_1(t) \\ &\cdot u(t-\tau_\ell) \end{aligned} \tag{3.24}$$

Equation (3.24) is the extension of the first equation in Eq. (7) in Fleury et al. 2002 to incorporate polarization. Following the same approach as used in that paper, we define the $M_2 \times M_1$ sounding matrices

$$\boldsymbol{U}(t;\tau_\ell) = \boldsymbol{q}_2(t)\boldsymbol{q}_1(t)^{\mathrm{T}}u(t-\tau_\ell) \tag{3.25}$$

With this definition, Eq. (3.24) can be further written as

$$s(t;\boldsymbol{\theta}_\ell) = \exp(j2\pi\nu_\ell t)\sum_{p_2=1}^{2}\sum_{p_1=1}^{2}\alpha_{\ell,p_2,p_1}\boldsymbol{c}_{2,p_2}^{\mathrm{T}}(\boldsymbol{\Omega}_{2,\ell})\boldsymbol{U}(t;\tau_\ell)\boldsymbol{c}_{1,p_1}(\boldsymbol{\Omega}_{1,\ell}) \tag{3.26}$$

We can also express $s(t;\boldsymbol{\theta}_\ell)$ as

$$s(t;\boldsymbol{\theta}_\ell) = \sum_{p_2=1}^{2}\sum_{p_1=1}^{2} s_{p_2,p_1}(t;\boldsymbol{\theta}_\ell) \tag{3.27}$$

where

$$s_{p_2,p_1}(t;\boldsymbol{\theta}_\ell) \doteq \alpha_{\ell,p_2,p_1}\exp(j2\pi\nu_\ell t)\boldsymbol{c}_{2,p_2}^{\mathrm{T}}(\boldsymbol{\Omega}_{2,\ell})\boldsymbol{U}(t;\tau_\ell)\boldsymbol{c}_{1,p_1}(\boldsymbol{\Omega}_{1,\ell}) \tag{3.28}$$

an expression similar to (7) in Fleury et al. 2002.

3.3 Dispersive-path Model

In the specular path model introduced in Section 3.2, the received signal is assumed to be the superposition of the contributions of multiple waves propagating along uncorrelated paths. Here "uncorrelated paths" refers to the fact that the parameters of these paths have differences larger than the intrinsic resolution of the measurement equipment. However, in real situations the propagation paths could be correlated; in other words, the parameters of these paths might be close to each other, with differences *less* than the intrinsic resolution of the measurement equipment. Such a set of paths is called a "dispersive" path. It is obvious that the conventional specular path model is inappropriate to describe accurately the contributions of dispersive paths.

Recently, dispersive path estimation, also called "slightly spatially distributed source estimation", has received a lot of attention. Different approaches have been put forward, and these can be generally categorized into two classes:

- finding approximate models for the slightly distributed sources, such as the first-order Taylor expansion approximation model of Tan et al. [2003] and the two-ray model of Bengtsson and Ottersten [2000]
- finding high-resolution estimators for estimating the slightly distributed sources or dispersive paths, examples being DSPE [Valaee et al. 1995], DISPARE [Meng et al. 1996, Trump and Ottersten 1996], and spread root-MUSIC, ESPRIT, and MODE [Bengtsson and Ottersten 2000].

In this section, we introduce the generic model of dispersive paths adopted in these publications.

The contribution of multiple slightly distributed sources to the received signal at the output of the mth Rx antenna array can be modeled as

$$s_m(\boldsymbol{\theta}) = \sum_{c=1}^{C} \sum_{\ell=1}^{L_c} \gamma_{c,\ell} \cdot c_m(\theta_{c,\ell}) \tag{3.29}$$

where C is the number of clusters, L_c is the number of multipaths in cth cluster, m is the data index in the frequency and spatial domain, $\boldsymbol{\theta}$ is a parameter vector containing all the unknown parameters in the model, $\gamma_{c,\ell}$ is the path weight of the ℓth path in the cth cluster, and $c_m(\theta_{c,\ell})$ denotes the response, which is given by $c_m(\theta_{c,\ell}) \doteq e^{-j2\pi(m-1)(\Delta s/\lambda)\cos(\theta_{c,\ell})}$. Here Δs is the array-element spacing, $\theta_{c,\ell}$ is the angle between the impinging direction of the ℓth path in the cth cluster and the array axis, and λ is the carrier wavelength.

The complex weight $\gamma_{c,\ell} = \alpha_{c,\ell} e^{j\psi_{c,\ell}}$, where $\alpha_{c,\ell}$ represents a real-valued amplitude and $\psi_{c,\ell}$ denotes the initial phase. If one assumes that the initial phase is fixed as a constant in a stationary time-invariant environment then $s_m(\boldsymbol{\theta})$ is a deterministic signal. However, in some applications it is difficult to ensure the same initial phase in all snapshots. Thus, it is reasonable to assume that the initial phases are independent random variables uniformly distributed on $[-\pi, \pi]$. Correspondingly $\gamma_{c,\ell}$ are random variables as well.

The deviation of the direction of arrival is calculated as $\tilde{\theta}_{c,\ell} = \theta_{c,\ell} - \theta_c$, with θ_c denoting the nominal direction of arrival of all rays. For some applications, $\tilde{\theta}_{c,\ell}$ are assumed to be constant for multiple channel realizations. However, they can be also assumed to be random variables, following an approximately Gaussian distribution $\mathcal{N} \sim (0, \sigma_\theta^2)$.

For reasons of simplicity, we focus on a one-cluster scenario. So the contribution of the cth cluster is

$$s_m(\boldsymbol{\theta}_c) = \sum_{\ell=1}^{L_c} \gamma_{c,\ell} \cdot c_m(\theta_{c,\ell}) \tag{3.30}$$

Assuming that the transmitted signal $u(t)$ is unitary, the received signal originating from the slightly distributed source and additive white Gaussian noise can be given by:

$$\begin{aligned} x_m &= s_m(\boldsymbol{\theta}_c) + w_m \\ &= \sum_{\ell=1}^{L_c} \gamma_{c,\ell} \cdot c_m(\theta_{c,\ell}) + w_m \end{aligned} \tag{3.31}$$

where w is complex circularly symmetric additive white Gaussian noise with variance of σ_w^2. In order to reduce the number of parameters to estimate, it is necessary to propose simple models that effectively approximate the contribution of the slightly distributed source.

It can be shown that the first-order Taylor expansion of a function $a(\phi)$ with parameter ϕ of sight increment $\tilde{\phi}$ from an original value ϕ_o can be written as

$$a(\phi_o + \tilde{\phi}) \approx a(\phi_o) + \tilde{\phi}\frac{\partial a(\phi_o)}{\partial \phi_o}$$

Applying this principle to Eq. (3.30), we obtain an approximate model for slightly distributed sources as

$$\begin{aligned} s_m(\boldsymbol{\theta}_c) &= \sum_{\ell=1}^{L_c} \gamma_{c,\ell} \cdot c_m(\theta_c + \tilde{\theta}_{c,\ell}) \\ &\approx \sum_{\ell=1}^{L_c} \gamma_{c,\ell} \cdot c_m(\theta_c) + \sum_{\ell=1}^{L_c} \gamma_{c,\ell} \cdot \tilde{\theta}_{c,\ell} \cdot \frac{\partial c_m(\theta_c)}{\partial \theta_c} \end{aligned} \tag{3.32}$$

where θ_c is the nominal direction of arrival and $\tilde{\theta}_{c,\ell}$ is the angle spread of the ℓth wave in the cth distributed source. Introducing parameters

$$\gamma_c = \sum_{\ell=1}^{L_c} \gamma_{c,\ell}, \; \psi_c = \sum_{\ell=1}^{L_c} \gamma_{c,\ell} \cdot \tilde{\theta}_{c,\ell}$$

and noticing that when the field pattern of the antenna elements is assumed to be isotropic, the steering vector and its derivative can be written as

$$\begin{aligned} c_m(\theta_c) &= e^{-j2\pi(m-1)(\Delta s/\lambda)\cos(\theta_c)} \\ \frac{\partial c_m(\theta_c)}{\partial \theta_c} &= (j2\pi(m-1)(\Delta s/\lambda)\sin(\theta_c))e^{-j2\pi(m-1)(\Delta s/\lambda)\cos(\theta_c)} \end{aligned} \tag{3.33}$$

Thus Eq. (3.32) is rewritten as

$$s_m(\boldsymbol{\theta}_c) \approx \gamma_c \cdot e^{-j2\pi(m-1)(\Delta s/\lambda)\cos(\theta_c)} + \psi_c \cdot j2\pi(m-1)(\Delta s/\lambda)\sin(\theta_c))e^{-j2\pi(m-1)(\Delta s/\lambda)\cos(\theta_c)}$$

The above expression can be written in matrix notation as

$$\boldsymbol{s}(\boldsymbol{\theta}_c) = \boldsymbol{D}(\bar{\boldsymbol{\theta}}_c)\boldsymbol{\beta}_c \tag{3.34}$$

where $\bar{\boldsymbol{\theta}}_c \doteq [\theta_c]$, $\boldsymbol{\beta}_c$ is a column vector $\boldsymbol{\beta}_c \doteq [\gamma_c, \psi_c]^T$ and $\boldsymbol{D}(\bar{\boldsymbol{\theta}}_c)$ is $2 \times M$ matrix defined as

$$\boldsymbol{D}(\bar{\boldsymbol{\theta}}_c) \doteq [\boldsymbol{a}(\bar{\boldsymbol{\theta}}_c)\ \boldsymbol{b}(\bar{\boldsymbol{\theta}}_c)]$$

with

$$\boldsymbol{a}(\bar{\boldsymbol{\theta}}_c) \doteq [c_m(\theta_c); m = 1, \ldots, M]^T,\ \boldsymbol{b}(\bar{\boldsymbol{\theta}}_c) \doteq \left[\frac{\partial c_m(\theta_c)}{\partial \theta_c}; m = 1, \ldots, M\right]^T$$

Written in matrix notation, the received signal $\boldsymbol{x}$ reads

$$\begin{aligned}\boldsymbol{x} &= \boldsymbol{s}(\boldsymbol{\theta}_c) + \boldsymbol{w} \\ &= \boldsymbol{D}(\bar{\boldsymbol{\theta}}_c)\boldsymbol{\beta}_c + \boldsymbol{w}\end{aligned}$$

The nominal incident angle θ_c is constant across all the snapshots. However, the angle spreads $\tilde{\theta}_{c,\ell}, \ell = 1, \ldots, L_c$ could have different constellations from snapshot to snapshot. We assume that $\tilde{\theta}_{c,\ell}, \ell = 1, \ldots, L_c$ follows the Gaussian distribution with zero mean and variance σ_θ^2, and they are independent of the complex attenuations. When the complex attenuation $\alpha_{c,\ell}, \ell = 1, \ldots, L_c$ is assumed to be a random variable with $\mathcal{N}(0, \frac{1}{L_c}\sigma_{\gamma_c}^2)$, it can be easily shown that the elements in the vector $\boldsymbol{\beta}_c$ are also random variables. Their first and second moments can be computed as:

$$E[\gamma_c] = E\left[\sum_{\ell=1}^{L_c} \gamma_\ell\right] = \sum_{\ell=1}^{L_c} E[\gamma_\ell] = 0$$

$$E[\psi_c] = E\left[\sum_{\ell=1}^{L_c} \gamma_\ell \tilde{\theta}_\ell\right] = \sum_{\ell=1}^{L_c} E[\gamma_\ell]E[\tilde{\theta}_\ell] = 0$$

$$E[\gamma_c \gamma_c^*] = E\left[\sum_{\ell=1}^{L_c} \gamma_\ell \sum_{\ell'=1}^{L_c} \gamma_{\ell'}^*\right] = \sum_{\ell=1}^{L_c} E[\gamma_\ell \gamma_\ell^*] = \sigma_{\gamma_c}^2$$

$$E[\psi_c \psi_c^*] = E\left[\sum_{\ell=1}^{L_c} \gamma_\ell \tilde{\theta}_\ell \sum_{\ell'=1}^{L_c} \gamma_{\ell'}^* \tilde{\theta}_\ell^*\right] = \sum_{\ell=1}^{L_c} E[\gamma_\ell \gamma_\ell^*]E[\tilde{\theta}_\ell \tilde{\theta}_\ell^*] = L_c \frac{1}{L_c} \sigma_{\gamma_c}^2 \sigma_\theta^2 = \sigma_{\gamma_c}^2 \sigma_\theta^2$$

For notational convenience, we use $\sigma_{\psi_c}^2$ to denote $\sigma_{\gamma_c}^2 \sigma_\theta^2$ in what follows. It can be also shown that $E[\gamma_c \psi_c^*] = 0$.

The covariance matrix of the received signal $\boldsymbol{x}$, which consists of the signal of a single slightly distributed source, is calculated as

$$\begin{aligned}\boldsymbol{R} &= E[\boldsymbol{x}\boldsymbol{x}^H] \\ &= \boldsymbol{D}(\bar{\boldsymbol{\theta}}_c)E[\boldsymbol{\beta}_c\boldsymbol{\beta}_c^*]\boldsymbol{D}(\bar{\boldsymbol{\theta}}_c)^H + \sigma_w^2\boldsymbol{I} \\ &= \boldsymbol{c}(\boldsymbol{\theta}_c)\boldsymbol{P}_c\boldsymbol{c}(\boldsymbol{\theta}_c)^H + \sigma_w^2\boldsymbol{I}\end{aligned}$$

where

$$\boldsymbol{P}_c = \begin{bmatrix}\sigma_{\gamma_c}^2 & 0 \\ 0 & \sigma_{\psi_c}^2\end{bmatrix}$$

Another approximate model using the first-order Taylor expansion can be written as

$$\begin{aligned}s_m(\boldsymbol{\theta}_c) &\approx \gamma_c \cdot c_m(\theta_c) + \psi_c \cdot \frac{\partial c_m(\theta_c)}{\partial \theta_c} \\ &\approx \gamma_c \cdot c_m(\theta_c + \frac{\psi_c}{\gamma_c}) \\ &= \gamma_c \cdot e^{-j2\pi(m-1)(\Delta s/\lambda)\cos(\theta_c + j\Delta\theta_c)}\end{aligned} \tag{3.35}$$

where θ_c is redefined to be $\theta_c + \mathcal{R}\{\frac{\psi_c}{\gamma_c}\}$, and $\Delta\theta_c \doteq \mathcal{I}\{\frac{\psi_c}{\gamma_c}\}$ is the angle spread. Written in vector notation we obtain

$$\boldsymbol{s}(\boldsymbol{\theta}_c) = \gamma_c \cdot \boldsymbol{c}(\bar{\boldsymbol{\theta}}_c) \tag{3.36}$$

where $\bar{\boldsymbol{\theta}}_c = [\theta_c, \Delta\theta_c]$. The received signal can be written in vector notation as

$$\boldsymbol{x} = \boldsymbol{c}(\bar{\boldsymbol{\theta}}_c)\gamma_c + \boldsymbol{w}$$

The parameter θ_c and $\Delta\theta_c$ are assumed to be deterministic over realizations. The covariance matrix of the received signal $\boldsymbol{x}$ is written as

$$\begin{aligned}\boldsymbol{R} &= E[\boldsymbol{x}\boldsymbol{x}^H] \\ &= \boldsymbol{c}(\bar{\boldsymbol{\theta}}_c)E[\gamma_c\gamma_c^*]\boldsymbol{c}(\bar{\boldsymbol{\theta}}_c)^H + \sigma_w^2\boldsymbol{I} \\ &= \sigma_{\gamma_c}^2\boldsymbol{c}(\bar{\boldsymbol{\theta}}_c)\boldsymbol{c}(\bar{\boldsymbol{\theta}}_c)^H + \sigma_w^2\boldsymbol{I}\end{aligned}$$

3.4 Time-evolution Model

Recently, the temporal behavior of propagation paths has attracted much attention [Czink et al. 2007, Kwakkernaat and Herben 2007]. In Kwakkernaat and Herben [2007] the time-variant characteristics of path parameters are considered as an additional degree of freedom for path clustering. In Czink et al. [2007], the evolution of the clusters of paths is illustrated using measurement data.

In these studies, the path-evolution characteristics are obtained indirectly from the path parameter estimates computed from individual observation snapshots. The estimation methods used – the SAGE algorithm [Czink et al. 2007] and the Unitary ESPRIT [Kwakkernaat and Herben 2007] – are derived under the assumption that the path parameters in different observation snapshots are independent. This (unrealistic) assumption results in a "loss of information" in the estimation of the path evolution in time. Furthermore, due to model-order mismatch and heuristic settings in these algorithms such as the (usually fixed) dynamic range, a time-variant path may remain undetected in some snapshots. Consequently, a time-variant path can be erroneously taken to be several paths. These effects influence the performance of clustering algorithms and the effectiveness of the channel models based on them. It is therefore of great importance to use appropriate algorithms to estimate the temporal characteristics of paths directly.

In recent years, methods have been proposed to track time-variant paths for MIMO channel sounding [Chung and Böhme 2005, Richter et al. 2006, Salmi et al. 2006]. In Chung and Böhme [2005], recursive expectation-maximization (EM) and recursive space-alternating generalized EM (SAGE)-inspired algorithms have been proposed as ways of tracking the azimuths of arrival (AoAs) of paths. In Richter et al. [2006] and Salmi et al. [2006], the extended Kalman filter (EKF) was used for tracking of the delays, DoAs, DoDs, and complex amplitudes of time-variant paths. Yin et al. [2008a,b] used a sequential Monte-Carlo method – a so-called "particle filter" – to track the parameters of time-variant paths. Differing from the EKF and the recursive EM and SAGE-inspired algorithms, the particle filter can be applied when the observation model is nonlinear.

In these methods, state-space modeling of the time-variant path parameters has been adopted. In this section, the state-space model describing the time-evolving path parameters, and the observation model for the received signal in the Rx of the sounding equipment are introduced. For simplicity of presentation, these models are discussed while considering a single-path scenario. Extension of these models to multiple-path scenarios is straightforward.

State-space Model

We consider a scenario in which the environment consists of time-variant specular paths. The parameters of a path are the delay τ, the azimuth of departure (AoD) ϕ_1, the azimuth of arrival (AoA) ϕ_2, the Doppler frequency ν, the rates of change of these parameters, denoted by $\Delta\tau$, $\Delta\phi_1$, $\Delta\phi_2$, and $\Delta\nu$, respectively, as well as the complex amplitude α. The kth observation of the state vector of a path is defined as

$$\boldsymbol{\Omega}_k = [\boldsymbol{P}_k^{\mathrm{T}}, \alpha_k^{\mathrm{T}}, \Delta_k^{\mathrm{T}}]^{\mathrm{T}} \tag{3.37}$$

where $[\cdot]^{\mathrm{T}}$ denotes the transpose operation, $\boldsymbol{P}_k \doteq [\tau_k, \phi_{1,k}, \phi_{2,k}, \nu_k]^{\mathrm{T}}$ represents the "position" parameter vector, $\boldsymbol{\Delta}_k \doteq [\Delta\tau_k, \Delta\phi_{1,k}, \Delta\phi_{2,k}, \Delta\nu_k]^{\mathrm{T}}$ denotes the "rate-of-change" parameter vector, and $\boldsymbol{\alpha}_k \doteq [|\alpha_k|, \arg(\alpha_k)]^{\mathrm{T}}$ is the amplitude vector with $|\alpha_k|$ and $\arg(\alpha_k)$ representing the magnitude and the argument of α_k respectively. The state vector $\boldsymbol{\Omega}_k$ is modelled as a Markov process:

$$p(\boldsymbol{\Omega}_k|\boldsymbol{\Omega}_{1:k-1}) = p(\boldsymbol{\Omega}_k|\boldsymbol{\Omega}_{k-1}), \quad k \in [1, \ldots, K] \tag{3.38}$$

where $\boldsymbol{\Omega}_{1:k-1} \doteq \{\boldsymbol{\Omega}_1, \ldots, \boldsymbol{\Omega}_{k-1}\}$ is a sequence of state values from the 1st to the $(k-1)$th observation, and K denotes the total number of observations. The transition of $\boldsymbol{\Omega}_k$ with respect to k is modeled as

$$\underbrace{\begin{bmatrix} \boldsymbol{P}_k \\ \alpha_k \\ \Delta_k \end{bmatrix}}_{\boldsymbol{\Omega}_k} = \underbrace{\begin{bmatrix} \boldsymbol{I}_4 & 0_{4\times 2} & T_k \boldsymbol{I}_4 \\ J_k & \boldsymbol{I}_2 & 0_{2\times 4} \\ 0_{4\times 4} & 0_{4\times 2} & \boldsymbol{I}_4 \end{bmatrix}}_{\boldsymbol{F}_k \doteq} \underbrace{\begin{bmatrix} \boldsymbol{P}_{k-1} \\ \alpha_{k-1} \\ \Delta_{k-1} \end{bmatrix}}_{\boldsymbol{\Omega}_{k-1}} + \underbrace{\begin{bmatrix} 0_{4\times 1} \\ \boldsymbol{v}_{\alpha,k} \\ \boldsymbol{v}_{\Delta,k} \end{bmatrix}}_{\boldsymbol{v}_k \doteq} \tag{3.39}$$

where $\boldsymbol{I}_n$ represents the $n \times n$ identity matrix, $\boldsymbol{0}_{b\times c}$ is the all-zero matrix of dimension $b \times c$,

$$\boldsymbol{J}_k = \begin{bmatrix} 0 & 0 & 0 & 0 \\ 0 & 0 & 0 & 2\pi T_k \end{bmatrix},$$

and T_k denotes the interval between the starts of the $(k-1)$th and the kth observation periods. The vector $\boldsymbol{v}_k$ in Eq. (3.39) contains the driving process in the amplitude vector

$$\boldsymbol{v}_{\alpha,k} \doteq [v_{|\alpha|,k}, v_{\arg(\alpha),k}]^{\mathrm{T}} \tag{3.40}$$

and in the rate-of-change parameter vector

$$\boldsymbol{v}_{\Delta,k} \doteq [v_{\Delta\tau,k}, v_{\Delta\phi_1,k}, v_{\Delta\phi_2,k}, v_{\Delta\nu,k}]^{\mathrm{T}}. \tag{3.41}$$

The entries $v_{(\cdot),k}$ in Eqs (3.40) and (3.41) are independent Gaussian random variables $v_{(\cdot),k} \sim \mathcal{N}(0, \sigma^2_{(\cdot)})$.

In this section, we consider the case with $T_k = T$, $k \in [1, \ldots, K]$. For notational brevity, we will drop the subscript k in $\boldsymbol{F}_k$.

Observation Model

In the kth observation period, the discrete-time signals at the output of the m_2th Rx antenna when the m_1th Tx antenna transmits can be written as

$$y_{k,m_1,m_2}(t) = x_{k,m_1,m_2}(t; \boldsymbol{\Omega}_k) + n_{k,m_1,m_2}(t), \quad t \in [t_{k,m_1,m_2}, t_{k,m_1,m_2} + T),$$
$$m_1 = 1, \ldots, M_1, m_2 = 1, \ldots, M_2 \tag{3.42}$$

where t_{k,m_1,m_2} denotes the time instant when the m_2th Rx antenna starts to receive signals while the m_1th Tx antenna transmits, T is the sensing duration of each Rx antenna, and M_1 and M_2 represent the total number of Tx antennas and Rx antennas respectively. The signal contribution $x_{k,m_1,m_2}(t; \boldsymbol{\Omega}_k)$ reads

$$x_{k,m_1,m_2}(t; \boldsymbol{\Omega}_k) = \alpha_k \exp(j 2\pi \nu_k t) c_{1,m_1}(\phi_{k,1}) c_{2,m_2}(\phi_{k,2}) \cdot u(t - \tau_k) \tag{3.43}$$

Here, $c_{1,m_1}(\phi)$ and $c_{2,m_2}(\phi)$ represent, respectively, the response in azimuth of the m_1th Tx antenna, and the response in azimuth of the m_2th Rx antenna, $u(t - \tau_k)$ denotes the transmitted signal delayed by τ_k. The noise $n_{k,m_1,m_2}(t)$ in Eq. (3.42) is a zero-mean Gaussian process with spectrum height σ_n^2. For notational convenience, we use the vector $\boldsymbol{y}_k$ to represent all the samples received in the kth observation period and $\boldsymbol{y}_{1:k} \doteq \{\boldsymbol{y}_1, \boldsymbol{y}_2, \ldots, \boldsymbol{y}_k\}$ to denote a sequence of observations.

3.5 Power Spectral Density Model

In conventional parametric models of the MIMO wideband propagation channel [Czink 2007, Correia 2001, Medbo et al. 2006, Chapter 3], a dispersed component is modeled using a cluster of multiple specular components estimated from measurement data. The cluster parameters are the nominal directions, the delay spread, and the direction spread. The estimates of these parameters are calculated from the parameter estimates of the specular components assigned to this cluster. However, as shown in Bengtsson and Völcker [2001], the extracted dispersion parameters of the specular components do not accurately characterize the true behavior of a dispersed component. This effect is caused by the generic modeling of the received signal in terms of well-separated specular components. In order to construct correct channel models, we need suitable generic parametric models characterizing the dispersive components and to use efficient parameter estimators to get accurate estimates from real measurements.

In recent years, many algorithms have been proposed for the estimation of the dispersive characteristics of individual components in the channel response [Besson and Stoica 1999, Betlehem et al. 2006, Ribeiro et al. 2004, Trump and Ottersten 1996]. These algorithms estimate the power spectral density (PSD) of each component. A component PSD can be irregular in real environments and thus a gross description relying on certain characteristic parameters, such as the center of gravity and spreads of the PSD, is usually adopted. The algorithms involved estimate these parameters by approximating the shape of the normalized component PSD with a certain probability density function (PDF), for example in AoA [Besson and Stoica 1999, Ribeiro et al. 2004, Trump and Ottersten 1996], and in AoA and AoD [Betlehem et al. 2006]. The parameter estimates obtained by using these algorithms depend on the selected PDFs. However, no rationale behind the selection is given in these contributions. Furthermore, the applicability of the PDFs in characterizing the PSD and the performance of the estimation algorithms have not been investigated experimentally with real measurement data.

Maximum-entropy-principle-based Generic PSD Models

It is well-known that the amount of the information conveyed by observations of a specific distribution can be measured via the entropy calculated using the PDF of the distribution [Shannon 1948]. When the power distribution is considered, the amount of information conveyed by the observations of a power distribution can also be measured by using the entropy computed using the power spectral density (PSD). It follows naturally that in order to estimate the parameters of the power distribution, using the PSD that maximizes the entropy enables us to extract the most information from the power distribution. The PSD properties that are considered important are the constraints that it must satisfy. Thus a generic form of the PSD that can be used to estimate these properties based on observations should be derived by maximizing the entropy subject to these constraints. The maximum-entropy (MaxEnt) theorem described by Mardia [1975] gives a general framework for deriving the MaxEnt PDF of a random vector ψ over a space $\mathcal{A}$ subject to certain constraints. Exploiting the analogy between normalized PSDs and PDFs, this MaxEnt theorem can be used to derive an MaxEnt normalized PSD.

The MaxEnt theorem [Jaynes 2003] has been used for the derivation of the PSD characterizing the component power distribution [Yin et al. 2006a,b; 2007a,b]. This rationale assumes that each component has a fixed center of gravity and spread and that no additional information about the PSD exists. The center of gravity and the spreads of a component PSD are described by the first and second moments of the corresponding power distribution. Using the MaxEnt principle, we derive a PSD that has fixed first and second moments, while maximizing the entropy of any other constraint. The estimates of the dispersion parameters obtained by modeling the component PSD with this entropy-maximizing PSD provide the "safest" results, in the sense that they are more accurate than the estimates computed using another form of PSD subject to constraints that are invalid in real situations. Based on this rationale, we derived the MaxEnt PSDs to characterize the component power distribution in AoA and AoD [Yin et al. 2006a], in elevation and azimuth [Yin et al. 2007a,b], and in biazimuth (AoA and AoD) and delay [Yin et al. 2006b]. These entropy-maximizing PDFs actually coincide with, respectively, the bivariate von-Mises–Fisher PDF [Mardia et al. 2003], the Fisher–Bingham-5 (FB_5) PDF [Kent 1982], and the extended von-Mises–Fisher PDF [Mardia et al. 2003]. Preliminary investigations using measurement data demonstrate that these characterizations are applicable in real environments.

In the following, an extension of the work reported by Yin et al. [2007a,b; 2006a,b] is considered. We first define the parameters that characterize dispersion of individual components in six dimensions: the bidirection (i.e. DoD and DoA), delay and Doppler frequency, and then the constraint set is determined for those parameters specified. Finally, we derive the MaxEnt normalized bidirection-delay-Doppler frequency PSD subject to the constraint set.

Constraints for Deriving the PSD

In this section, we consider the dispersion vector

$$\boldsymbol{\psi} \triangleq [\boldsymbol{\Omega}_1, \boldsymbol{\Omega}_2, \tau, \nu]^{\mathrm{T}}$$

with range $\mathcal{A}^* = \mathbb{S}_2 \times \mathbb{S}_2 \times \mathbb{R}_+ \times [\mathbb{R}]$, where

$$\mathbb{S}_2 = \{\boldsymbol{x} \in \mathbb{R}^3 : \| \boldsymbol{x} \| = 1\} \subset \mathbb{R}^3 \tag{3.44}$$

denotes a unit sphere.

We are interested in knowing the centers of gravity, the spreads, and the coupling coefficients of the power spectral density $f(\boldsymbol{\psi})$ in the six dispersion dimensions, $\boldsymbol{\Omega}_1$, $\boldsymbol{\Omega}_2$, τ, and ν. Notice that we also include the coupling coefficients, as the coupling of different dispersion dimensions is an effect that has been recently observed experimentally [Yin et al. 2006a, Fig. 1]. Furthermore, recent studies also show that the capacity and the diversity gains of MIMO systems depend on the coupling of the DoA and DoD dimensions [Betlehem et al. 2006]. Thus, it is necessary to include the coupling coefficients of the PSD. In the following, we define the centers of gravity, the spreads, and the coupling coefficients of the distribution characterized by using the PSD $f(\boldsymbol{\psi})$.

Center of gravity

The centers of gravity of the bidirection-delay-Doppler frequency power distribution are determined by the mean DoD $\boldsymbol{\mu}_{\Omega_1}$, mean DoA $\boldsymbol{\mu}_{\Omega_2}$, mean delay μ_τ, and mean Doppler frequency μ_ν. They are defined as, respectively,

$$\boldsymbol{\mu}_{\boldsymbol{\Omega}_k} = \int_{\mathcal{A}^*} \boldsymbol{\Omega}_k f(\boldsymbol{\psi})\mathrm{d}\boldsymbol{\psi}, k = 1, 2 \tag{3.45}$$

$$\mu_\tau = \int_{\mathcal{A}^*} \tau f(\boldsymbol{\psi})\mathrm{d}\boldsymbol{\psi} \tag{3.46}$$

$$\mu_\nu = \int_{\mathcal{A}^*} \nu f(\boldsymbol{\psi})\mathrm{d}\boldsymbol{\psi} \tag{3.47}$$

Spreads

Following the nomenclature of Fleury [2000], the spreads of $f(\boldsymbol{\psi})$ in DoD, DoA, delay, and Doppler frequency, denoted as σ_{Ω_1}, σ_{Ω_2}, σ_τ, and σ_ν, respectively, can be computed as

$$\sigma_{\Omega_k} = \sqrt{1 - |\boldsymbol{\mu}_{\Omega_k}|^2}, k = 1, 2 \tag{3.48}$$

$$\sigma_\tau = \sqrt{\int_{\mathcal{A}^*} (\tau - \mu_\tau)^2 f(\boldsymbol{\psi})\mathrm{d}\boldsymbol{\psi}} \tag{3.49}$$

$$\sigma_\nu = \sqrt{\int_{\mathcal{A}^*} (\nu - \mu_\nu)^2 f(\boldsymbol{\psi})\mathrm{d}\boldsymbol{\psi}} \tag{3.50}$$

where $|\cdot|$ denotes the Euclidean norm.

Coupling coefficients

In the propagation scenario, the directions $\boldsymbol{\Omega}_1$ and $\boldsymbol{\Omega}_2$ are uniquely determined by the azimuths and elevations of the directions: $\boldsymbol{\Omega}_k \triangleq \boldsymbol{e}(\phi_k, \theta_k), k = 1, 2$. In the case where only azimuths are considered, the coupling coefficient of two directions can be defined as one scalar. However, when both azimuth and elevation are considered we need to introduce a covariance matrix to describe the correlation of the components of directions on three axes of the Cartesian coordinate system. Thus for the case considered, we define the coupling coefficient matrix $\boldsymbol{\rho}_{\Omega_1\Omega_2}$ of two directions $\boldsymbol{\Omega}_1$ and $\boldsymbol{\Omega}_2$ as

$$\rho_{\Omega_1\Omega_2} \triangleq \frac{1}{\sigma_{\Omega_1}\sigma_{\Omega_2}} \int_{\mathcal{A}^*} (\boldsymbol{\Omega}_1 - \mu_{\Omega_1})^{\mathrm{T}} \boldsymbol{R}_1 \boldsymbol{R}_2^{\mathrm{T}} (\boldsymbol{\Omega}_2 - \mu_{\Omega_2}) f(\boldsymbol{\psi})\mathrm{d}\boldsymbol{\psi} \in \mathcal{R}^{3\times 3} \tag{3.51}$$

where $\boldsymbol{R}_k, k = 1, 2$ are two orthonormal matrices. The function of $\boldsymbol{R}_k$ is to align the difference vectors $\boldsymbol{\Omega}_k - \boldsymbol{\mu}_{\Omega_k}$ to the same Cartesian coordinate system. The expression of $\boldsymbol{R}_k$ can be easily defined in the case where directions are determined by azimuth only; in other words with directions' ends located on a unit circle. However, it is non-trivial to write an analytical expression for $\boldsymbol{R}_k$ when the directions' ends are located on a unit sphere. In the following, we provide a method for computing $(\boldsymbol{\Omega}_k - \boldsymbol{\mu}_{\Omega_k})^{\mathrm{T}}\boldsymbol{R}_k$ rather than giving the exact expression of $\boldsymbol{R}_k$. This computation method allows for alignment of $(\boldsymbol{\Omega}_k - \boldsymbol{\mu}_{\Omega_k}), k = 1, 2$ in the same spherical coordinate system as required for computing $\boldsymbol{\rho}_{\Omega_1\Omega_2}$ in Eq. (3.51).

The direction $\boldsymbol{\Omega}$ in the spherical coordinate system can be written as

$$\boldsymbol{\Omega} = \boldsymbol{e}_r \cdot 1 + \boldsymbol{e}_\theta \cdot \theta + \boldsymbol{e}_\phi \cdot \phi \tag{3.52}$$

where $\boldsymbol{e}_r$, $\boldsymbol{e}_\theta$ and $\boldsymbol{e}_\phi$ denote respectively the radius, the co-elevation, and the azimuth axes of the spherical coordinate system. Rotating $\boldsymbol{\Omega}$ to the direction $\boldsymbol{\Omega}' = \boldsymbol{e}(\phi', \theta')$ can be expressed as follows:

$$\boldsymbol{\Omega} \oplus \Delta\boldsymbol{\Omega} = \boldsymbol{\Omega}' \tag{3.53}$$

where $\Delta\boldsymbol{\Omega} = \boldsymbol{e}_r \cdot 0 + \boldsymbol{e}_\theta \cdot (\theta' - \theta) + \boldsymbol{e}_\phi \cdot (\phi' - \phi)$, and $\oplus$ denotes the axis-wise summation. Thus, the operation $(\boldsymbol{\Omega}_1 - \boldsymbol{\mu}_{\Omega_1})^{\mathrm{T}}\boldsymbol{R}_1$ in Eq. (3.51) can be performed using the following steps:

1. Calculate $(\boldsymbol{\Omega}_1 - \boldsymbol{\mu}_{\Omega_1})$ in the spherical coordinate system.
2. Specify a rotational vector $\Delta\boldsymbol{\Omega} = \boldsymbol{e}_r \cdot 0 + \boldsymbol{e}_\theta \cdot (\bar{\theta}_2 - \bar{\theta}_1) + \boldsymbol{e}_\phi \cdot (\bar{\phi}_2 - \bar{\phi}_1)$.
3. Perform the operation $(\boldsymbol{\Omega}_1 - \boldsymbol{\mu}_{\Omega_1}) \oplus \Delta\boldsymbol{\Omega}$.
4. Rewrite the obtained vector in the Cartesian coordinate system.

The same method is used for computing $(\boldsymbol{\Omega}_2 - \boldsymbol{\mu}_{\Omega_2})^{\mathrm{T}}\boldsymbol{R}_2$, with the rational vector $\Delta\boldsymbol{\Omega} = \boldsymbol{e}_r \cdot 0 + \boldsymbol{e}_\theta \cdot (\bar{\theta}_1 - \bar{\theta}_2) + \boldsymbol{e}_\phi \cdot (\bar{\phi}_1 - \bar{\phi}_2)$. Then the calculation of $(\boldsymbol{\Omega}_1 - \boldsymbol{\mu}_{\Omega_1})^{\mathrm{T}}\boldsymbol{R}_1\boldsymbol{R}_2^{\mathrm{T}}(\boldsymbol{\Omega}_2 - \boldsymbol{\mu}_{\Omega_2})$ in the integral of Eq. (3.51) can be calculated as the outer product of the two vectors obtained, represented in the Cartesian coordinate system.

The correlation coefficients between the dispersion in direction $\boldsymbol{\Omega}_k$, $k \in [1, 2]$ and the delay τ, as well as the Doppler frequency ν, are defined as

$$\rho_{\Omega_k\tau} \triangleq \frac{1}{\sigma_{\Omega_k}\sigma_\tau}\int_{\mathcal{A}^*}(\boldsymbol{\Omega}_k - \mu_{\Omega_k})^{\mathrm{T}}\boldsymbol{R}_k(\tau - \mu_\tau)f(\psi)\mathrm{d}\boldsymbol{\psi} \in \mathcal{R}^{1\times 3}, k = 1, 2 \tag{3.54}$$

$$\rho_{\Omega_k\nu} \triangleq \frac{1}{\sigma_{\Omega_k}\sigma_\nu}\int_{\mathcal{A}^*}(\boldsymbol{\Omega}_k - \mu_{\Omega_k})^{\mathrm{T}}\boldsymbol{R}_k(\nu - \mu_\nu)f(\psi)\mathrm{d}\boldsymbol{\psi} \in \mathcal{R}^{1\times 3}, k = 1, 2 \tag{3.55}$$

The coupling coefficient between the dispersion in τ and ν is defined as

$$\rho_{\tau\nu} \triangleq \frac{1}{\sigma_\tau\sigma_\nu}\int_{\mathcal{A}^*}(\tau - \mu_\tau)(\nu - \mu_\nu)f(\psi)\mathrm{d}\boldsymbol{\psi} \in \mathcal{R}, k = 1, 2 \tag{3.56}$$

Notice that we also introduce $(\boldsymbol{\Omega}_k - \boldsymbol{\mu}_{\Omega_k})^{\mathrm{T}}\boldsymbol{R}_k$ in Eqs (3.54)–(3.56) in order to maintain the rotational invariance of $\boldsymbol{\rho}_{\Omega_k\tau}, \boldsymbol{\rho}_{\Omega_k\nu}$. In other words, $\boldsymbol{\rho}_{\Omega_k\tau}$ and $\boldsymbol{\rho}_{\Omega_k\nu}$ do not change when the center of gravity of $f(\boldsymbol{\Omega}_k, \tau, \nu) \triangleq \int_{\mathcal{A}^*} f(\boldsymbol{\psi})\mathrm{d}\boldsymbol{\Omega}_{k'}$ with $k' \in [1, 2]$ and $k' \neq k$ is changed.

It can be straightforward to show that the coefficients in the matrix $\boldsymbol{\rho}_{\Omega_1\Omega_2}$, the vectors $\boldsymbol{\rho}_{\Omega_1\tau}, \boldsymbol{\rho}_{\Omega_1\nu}, \boldsymbol{\rho}_{\Omega_2\tau}, \boldsymbol{\rho}_{\Omega_2\nu}$, and $\rho_{\tau\nu}$, are in the range $[-1, 1]$. They are equal to zero when the spreads in the corresponding dimensions are decoupled; in other words, when the bidirection-delay-Doppler-frequency PSD can be factorized as the product of the marginal PSDs in the considered dimensions. When the spreads in any pair (a, b) of $\boldsymbol{\Omega}_1$, $\boldsymbol{\Omega}_2, \tau$, and ν are linearly dependent – when the marginal PSD $f(a, b)$ is a straight line – the coefficient ρ_{ab} is close to 1 or -1, with its sign determined by the increasing or decreasing slope of the line. Furthermore, these coefficients are rotational invariant. Papers in the literature have put forward coefficients for describing the correlation of two circular variables [Jupp and Mardia 1980] and the correlation of a circular variable and a linear variable [Mardia 1976]. A common drawback of these coefficients is that it is impossible to use them to write analytical expressions of the parameters of the MaxEnt normalized PSD. We postulate that in typical cases of MIMO channel sounding, the newly proposed coupling coefficients can be used in the component MaxEnt normalized PSD as explicit parameters. This analytical parametric representation of the PSD allows derivation of the maximum likelihood estimators of the parameters of interest based on the received signals.

For notational convenience, we use

$$\boldsymbol{\theta} = (\mu_\tau, \mu_\nu, \boldsymbol{\mu}_{\Omega_1}, \boldsymbol{\mu}_{\Omega_2}, \sigma_\tau, \sigma_\nu, \sigma_{\Omega_1}, \sigma_{\Omega_2}, \boldsymbol{\rho}_{\Omega_1\Omega_2}, \boldsymbol{\rho}_{\Omega_1\tau}, \boldsymbol{\rho}_{\Omega_1\nu}, \boldsymbol{\rho}_{\Omega_2\tau}, \boldsymbol{\rho}_{\Omega_2\nu}, \rho_{\tau\nu}) \tag{3.57}$$

to denote the parameters characterizing the bidirection-delay-Doppler frequency PSD $f(\psi)$.

Derivation of the MaxEnt Bidirection-delay-Doppler-frequency PSD

As described at the beginning of this section, we seek a bidirection-delay-Doppler-frequency PSD that maximizes the entropy under the constraint that the parameters in Eq. (3.57) are all fixed. Such a PSD leaves the highest uncertainty about the shape of the PSD of a path component provided the parameters in Eq. (3.57) are known. This means that estimation of the these parameters with any other form of PSD will be better than when using the MaxEnt PSD.

The constraints applied in the derivation of the MaxEnt component PSD are as follows:

Constraint 1: $\int_{\mathcal{A}^*} \boldsymbol{\Omega}_1 f(\psi)\mathrm{d}\psi$ is fixed

Constraint 2: $\int_{\mathcal{A}^*} \boldsymbol{\Omega}_2 f(\psi)\mathrm{d}\psi$ is fixed

Constraint 3: $\int_{\mathcal{A}^*} \tau f(\psi)\mathrm{d}\psi$ is fixed

Constraint 4: $\int_{\mathcal{A}^*} \tau^2 f(\psi)\mathrm{d}\psi$ is fixed

Constraint 5: $\int_{\mathcal{A}^*} \boldsymbol{\Omega}_2^{\mathrm{T}}\boldsymbol{R}_1\boldsymbol{R}_2^{\mathrm{T}}\boldsymbol{\Omega}_1 f(\psi)\mathrm{d}\psi$ is fixed

Constraint 6: $\int_{\mathcal{A}^*} \boldsymbol{\Omega}_1^{\mathrm{T}}\boldsymbol{R}_1\tau f(\psi)\mathrm{d}\psi$ is fixed

Constraint 7: $\int_{\mathcal{A}^*} \boldsymbol{\Omega}_2^{\mathrm{T}}\boldsymbol{R}_2\tau f(\psi)\mathrm{d}\psi$ is fixed

Constraint 8: $\int_{\mathcal{A}^*} \nu f(\psi)\mathrm{d}\psi$ is fixed

Constraint 9: $\int_{\mathcal{A}^*} \nu^2 f(\psi)\mathrm{d}\psi$ is fixed

Constraint 10: $\int_{\mathcal{A}^*} \boldsymbol{\Omega}_1^{\mathrm{T}}\boldsymbol{R}_1\nu f(\psi)\mathrm{d}\psi$ is fixed

Constraint 11: $\int_{\mathcal{A}^*} \boldsymbol{\Omega}_2^{\mathrm{T}}\boldsymbol{R}_2\nu f(\psi)\mathrm{d}\psi$ is fixed

Constraint 12: $\int_{\mathcal{A}^*} \tau\nu f(\psi)\mathrm{d}\psi$ is fixed

(3.58)

Invoking the MaxEnt theorem, the PSD can be calculated to be

$$\begin{aligned} f_{\mathrm{MaxEnt}}(\psi) = \exp\{&b_0 + \boldsymbol{b}_1^{\mathrm{T}}\boldsymbol{\Omega}_1 + \boldsymbol{b}_2^{\mathrm{T}}\boldsymbol{\Omega}_2 + b_3\tau + b_4\tau^2 + \boldsymbol{b}_5\boldsymbol{\Omega}_2^{\mathrm{T}}\boldsymbol{R}_1\boldsymbol{R}_2^{\mathrm{T}}\boldsymbol{\Omega}_1 + \boldsymbol{b}_6\boldsymbol{\Omega}_1^{\mathrm{T}}\boldsymbol{R}_1\tau \\ &+ \boldsymbol{b}_7\boldsymbol{\Omega}_2^{\mathrm{T}}\boldsymbol{R}_2\tau + b_8\nu + b_9\nu^2 + \boldsymbol{b}_{10}\boldsymbol{\Omega}_1^{\mathrm{T}}\boldsymbol{R}_1\nu + \boldsymbol{b}_{11}\boldsymbol{\Omega}_2^{\mathrm{T}}\boldsymbol{R}_2\nu + b_{12}\tau\nu\} \end{aligned} \tag{3.59}$$

where b_0 is the normalization factor such that $\int_{\mathcal{A}^*} f_{\mathrm{MaxEnt}}(\psi)\mathrm{d}\psi = 1$, and $\boldsymbol{b}_1, \boldsymbol{b}_2, b_3, \ldots, b_{12}$ are obtained by applying Constraints $1, 2, \ldots, 12$ respectively.

We use a vector $\boldsymbol{b}$ to represent the parameters in the PSD Eq. (3.59):

$$\boldsymbol{b} = (\boldsymbol{b}_1, \boldsymbol{b}_2, b_3, b_4, \boldsymbol{b}_5, \boldsymbol{b}_6, \boldsymbol{b}_7, b_8, b_9, \boldsymbol{b}_{10}, \boldsymbol{b}_{11}, b_{12}).$$

As $\boldsymbol{b}_1, \boldsymbol{b}_2, \boldsymbol{b}_6, \boldsymbol{b}_7, \boldsymbol{b}_{10}, \boldsymbol{b}_{11} \in \mathbb{R}^3$, $\boldsymbol{b}_5 \in \mathbb{R}^{3\times 3}$, the total number of parameters of the PSD Eq. (3.59) is 30. Notice that these parameters are not completely independent. This is because there is another constraint that has not been considered in the derivation, namely that the directions are all unit vectors. By taking into account this constraint, the 30 parameters can be reduced to 21, similar to the case of the six-variate Gaussian distribution.

Our study shows that it is non-trivial to derive the analytical expression of the parameters in $\boldsymbol{\theta}$ as functions of $\boldsymbol{b}$. In such a case, the elements of $\boldsymbol{\theta}$ need to be calculated according to their definitions using the MaxEnt PSD $f_{\text{MaxEnt}}(\boldsymbol{\psi})$.

In order to obtain the expression of $f_{\text{MaxEnt}}(\boldsymbol{\psi})$ in terms of $\boldsymbol{\theta}$, we consider the case where $f_{\text{MaxEnt}}(\boldsymbol{\psi})$ is highly concentrated. In such a case, the entropy-maximizing PSD can be approximated by the Gaussian PSD with $\boldsymbol{\theta}$ as the parameters. By change of parameters, we may identify the expression of $\boldsymbol{bsb}$ in terms of $\boldsymbol{\theta}$, and then obtain a novel expression of $f_{\text{MaxEnt}}(\boldsymbol{\psi})$ parameterized by $\boldsymbol{\theta}$. Since this newly derived $f_{\text{MaxEnt}}(\boldsymbol{\psi})$ still satisfies the maximum-entropy requirement subject to the constraints defined, the PSD is applicable regardless of the condition that the power distribution is highly concentrated. The benefit of using this new expression of $f_{\text{MaxEnt}}(\boldsymbol{\psi})$ in terms of $\boldsymbol{\theta}$ is that it allows estimation of $\boldsymbol{\theta}$ directly by using, for example, maximum-likelihood-based algorithms based on the generic form of the PSD from the received signals.

Highly Concentrated Direction-delay-Doppler-frequency PSD

The special case in which the bidirection-delay-Doppler-frequency PSD $f_{\text{MaxEnt}}(\boldsymbol{\psi})$ is highly concentrated is now considered. From the definitions of the parameters in $\boldsymbol{\theta}$, it can be shown that in this case, $\boldsymbol{\theta}$ can be approximated with the center of gravity, spreads, and coupling of the spreads of a biazimuth-bielevation-delay-Doppler-frequency power distribution. We use $f(\boldsymbol{\omega})$ to denote the PSD of the biazimuth-bielevation-delay-Doppler-frequency power distribution, where the vector

$$\boldsymbol{\omega} \triangleq [\phi_1, \theta_1, \phi_2, \theta_2, \tau, \nu]^{\text{T}}$$

has the support $\mathcal{C} \triangleq (0, 2\pi] \times [-\pi/2, \pi/2\pi] \times (0, 2\pi] \times [-\pi/2, \pi/2\pi] \times \mathcal{R}_+ \times \mathcal{R}$. The center of gravity

$$\mu_{\boldsymbol{\omega}} = [\mu_{\phi_1}, \mu_{\theta_1}, \mu_{\phi_2}, \mu_{\theta_2}, \mu_\tau, \mu_\nu]^{\text{T}}$$

of the biazimuth-bielevation-delay-Doppler frequency component PSD $f(\boldsymbol{\omega})$ is calculated as

$$\mu_{\boldsymbol{\omega}} = \int_{\mathcal{C}} \boldsymbol{\omega} f(\boldsymbol{\omega}) \text{d}\boldsymbol{\omega} \tag{3.60}$$

The spreads of $f(\boldsymbol{\omega})$ in biazimuth, bielevation, delay, and Doppler frequency are computed as

$$\sigma_{\boldsymbol{\omega}} = \begin{bmatrix}\sigma_{\phi_1} & \sigma_{\theta_1} & \sigma_{\phi_2} & \sigma_{\theta_2} & \sigma_\tau & \sigma_\nu\end{bmatrix}^{\text{T}} = \sqrt{\int_{\mathcal{C}} (\boldsymbol{\omega} - \mu_{\boldsymbol{\omega}})^2 f(\boldsymbol{\omega}) \text{d}\boldsymbol{\omega}} \tag{3.61}$$

The coupling coefficients of the spreads in different dimensions are defined by analogy with the correlation coefficients of two linear random variables as

$$\rho_{ab}^2 = \int_{\mathcal{C}} (a - \mu_a)(b - \mu_b) f(\boldsymbol{\omega}) \text{d}\boldsymbol{\omega}$$

with a and b representing any pair of $\phi_1, \theta_1, \phi_2, \theta_2, \tau$ and ν. For notational convenience, we use

$$\boldsymbol{\eta} = (\mu_{\phi_1}, \mu_{\theta_1}, \mu_{\phi_2}, \mu_{\theta_2}, \mu_\tau, \mu_\nu, \sigma_{\phi_1}, \sigma_{\theta_1}, \sigma_{\phi_2}, \sigma_{\theta_2}, \sigma_\tau, \sigma_\nu,$$
$$\rho_{\phi_1\phi_2}, \rho_{\phi_1\theta_1}, \rho_{\phi_1\theta_2}, \rho_{\phi_1\tau}, \rho_{\phi_1\nu}, \rho_{\phi_2\theta_1}, \rho_{\phi_2\theta_2}, \rho_{\phi_2\tau}, \rho_{\phi_2\nu}, \rho_{\theta_1\theta_2}, \rho_{\theta_1\tau}, \rho_{\theta_1\nu}, \rho_{\theta_2\tau}, \rho_{\theta_2\nu}, \rho_{\tau\nu})$$

to denote the parameters of the bidirection-delay-Doppler frequency PSD $f(\boldsymbol{\omega})$.

The estimates of $\boldsymbol{\theta}$ can therefore be approximated by the estimates of $\boldsymbol{\eta}$. To accurately estimate $\boldsymbol{\eta}$, we derive the biazimuth-bielevation-delay-Doppler frequency PSD that maximizes the entropy subject to the constraint of fixed $\boldsymbol{\eta}$. This PSD can be written as

$$f_{\text{MaxEnt}}(\boldsymbol{\omega}) \propto \exp\left[-\frac{1}{2|\boldsymbol{B}|}\sum_{j=1}^{6}\sum_{i=1}^{6}|\boldsymbol{B}|_{jk}\left(\frac{\boldsymbol{\omega}_j - \boldsymbol{\mu}_{\omega,j}}{\sigma_{\omega,j}}\right)\left(\frac{\boldsymbol{\omega}_k - \boldsymbol{\mu}_{\omega,k}}{\sigma_{\omega,k}}\right)\right] \quad (3.62)$$

where

$$\boldsymbol{\mu}_\omega = \begin{bmatrix}\mu_{\phi_1}\\ \mu_{\theta_1}\\ \mu_{\phi_2}\\ \mu_{\theta_2}\\ \mu_\tau\\ \mu_\nu\end{bmatrix}, \quad \boldsymbol{B} = \begin{bmatrix} 1 & \rho_{\phi_1\theta_1} & \rho_{\phi_1\phi_2} & \rho_{\phi_1\theta_2} & \rho_{\phi_1\tau} & \rho_{\phi_1\nu}\\ \rho_{\theta_1\phi_1} & 1 & \rho_{\theta_1\phi_2} & \rho_{\theta_1\theta_2} & \rho_{\theta_1\tau} & \rho_{\theta_1\nu}\\ \rho_{\phi_2\phi_1} & \rho_{\phi_2\theta_1} & 1 & \rho_{\phi_2\theta_2} & \rho_{\phi_2\tau} & \rho_{\phi_2\nu}\\ \rho_{\theta_2\phi_1} & \rho_{\theta_2\theta_1} & \rho_{\theta_2\phi_2} & 1 & \rho_{\theta_2\tau} & \rho_{\theta_2\nu}\\ \rho_{\tau\phi_1} & \rho_{\tau\theta_1} & \rho_{\tau\phi_2} & \rho_{\tau\theta_2} & 1 & \rho_{\tau\nu}\\ \rho_{\nu\phi_1} & \rho_{\nu\theta_1} & \rho_{\nu\phi_2} & \rho_{\nu\theta_2} & \rho_{\nu\tau} & 1 \end{bmatrix}$$

$|\boldsymbol{B}|$ denotes the determinant of $\boldsymbol{B}$, $|\boldsymbol{B}|_{jk}$ is the cofactor of $\boldsymbol{B}_{jk}$, where $\boldsymbol{B}_{jk}$ represents the (j,k)th entry, $\boldsymbol{\mu}_{\omega,j}$ denotes the jth entry of the column vector $\boldsymbol{\mu}_{\omega,j}$, and $\boldsymbol{\sigma}_{\omega,j}$ represents the jth entry of $\boldsymbol{\sigma}_\omega$.

The subtraction of angular variables, for example $\phi_k - \mu_{\phi_k}$, $\theta_k - \mu_{\theta_k}$, $k = 1, 2$, arising when calculating $\boldsymbol{\omega} - \boldsymbol{\mu}_\omega$ is defined in such a way that the resulting azimuth lies in the range $[-\pi, \pi)$ and elevation lies in the range $[-\pi/2, \pi/2)$. This rationale is applicable for the subtraction of angular variables throughout this book.

Notice that the MaxEnt biazimuth-bielevation-delay-Doppler frequency PSD $f_{\text{MaxEnt}}(\boldsymbol{\omega})$ in Eq. (3.62) has the same form as a truncated Gaussian PDF with the support of $\mathcal{C}$. Strictly speaking the traditional meaning of $\sigma_{\phi_1}, \sigma_{\theta_1}, \sigma_{\phi_2}, \sigma_{\theta_2}, \rho_{\phi_1\theta_1}, \rho_{\phi_1\phi_2}, \rho_{\phi_1\theta_2}, \rho_{\phi_1\tau}, \rho_{\phi_1\nu}, \rho_{\theta_1\phi_2}, \rho_{\theta_1\theta_2}, \rho_{\theta_1\tau}, \rho_{\theta_1\nu}, \rho_{\phi_2\theta_2}, \rho_{\phi_2\tau}, \rho_{\phi_2\nu}, \rho_{\theta_2\tau}, \rho_{\theta_2\nu}$, and $\rho_{\tau\nu}$ as second-order central moments of a six-variate Gaussian distribution no longer applies to Eq. (3.62), due to the fact that the angular ranges are bounded. However, these parameters provide good approximations of these moments when $\sigma_{\phi_1}, \sigma_{\theta_1}, \sigma_{\phi_2}$, and σ_{θ_2} are small.

According to the MaxEnt theorem [Mardia 1976], a PDF that maximizes the entropy subject to certain constraints is unique. As $f_{\text{MaxEnt}}(\boldsymbol{\omega})$ in Eq. (3.59) and $f_{\text{MaxEnt}}(\boldsymbol{\omega})$ in Eq. (3.62) both maximize the entropy subject to similar constraints, it is reasonable to postulate that the approximation

$$f_{\text{MaxEnt}}(\boldsymbol{\psi})|_{\boldsymbol{\psi}=(\boldsymbol{e}(\phi_1,\theta_1),\boldsymbol{e}(\phi_2,\theta_2),\tau,\nu)} \approx f_{\text{MaxEnt}}(\boldsymbol{\omega}) \quad (3.63)$$

holds in the case of highly concentrated PSD. In what follows, this postulation is used to find the expression of $\boldsymbol{b}$ in terms of $\boldsymbol{\theta}$.

We first rewrite $\boldsymbol{b}_1$ and $\boldsymbol{b}_2$ in $f_{\text{MaxEnt}}(\boldsymbol{\psi})$, Eq. (3.59) as

$$\boldsymbol{b}_i = \kappa_i \boldsymbol{e}(\bar{\phi}_k, \bar{\theta}_k), k = 1, 2, \text{with } \kappa_i = |\boldsymbol{b}_i| \quad (3.64)$$

and

$$\boldsymbol{R}_1 \triangleq \boldsymbol{e}_r \cdot 0 + \boldsymbol{e}_\theta \cdot (\bar{\theta}_2 - \bar{\theta}_1) + \boldsymbol{e}_\phi \cdot (\bar{\phi}_2 - \bar{\phi}_1) \quad (3.65)$$

$$\boldsymbol{R}_2 \triangleq \boldsymbol{e}_r \cdot 0 + \boldsymbol{e}_\theta \cdot (\bar{\theta}_1 - \bar{\theta}_2) + \boldsymbol{e}_\phi \cdot (\bar{\phi}_1 - \bar{\phi}_2) \quad (3.66)$$

in the spherical coordinate system. Inserting Eqs (3.64), (3.65), and (3.66) into Eq. (3.59) yields (with slight abuse of notation):

$$\begin{aligned}
&f_{\text{MaxEnt}}(\boldsymbol{\omega};\boldsymbol{b})\\
&= \exp\{b_0 + b_1\cos(\phi_1-\bar{\phi}_1) + b_2\cos(\theta_1-\bar{\theta}_1) + b_3\cos(\phi_2-\bar{\phi}_2) + b_4\cos(\theta_2-\bar{\theta}_2)\\
&\quad + b_5(\tau-\bar{\tau})^2 + b_6(\nu-\bar{\nu})^2 + b_7\cos((\phi_1-\bar{\phi}_1)-(\phi_2-\bar{\phi}_2))\\
&\quad + b_8\cos((\theta_1-\bar{\theta}_1)-(\theta_2-\bar{\theta}_2)) + b_9\cos((\phi_1-\bar{\phi}_1)-(\theta_2-\bar{\theta}_2))\\
&\quad + b_{10}\cos((\theta_1-\bar{\theta}_1)-(\phi_2-\bar{\phi}_2)) + b_{11}\cos((\phi_1-\bar{\phi}_1)-(\theta_1-\bar{\theta}_1))\\
&\quad + b_{12}\cos((\phi_2-\bar{\phi}_2)-(\theta_2-\bar{\theta}_2)) + b_{13}(\tau-\bar{\tau})\sin(\phi_1-\bar{\phi}_1)\\
&\quad + b_{14}(\tau-\bar{\tau})\sin(\theta_1-\bar{\theta}_1) + b_{15}(\tau-\bar{\tau})\sin(\phi_2-\bar{\phi}_2) + b_{16}(\tau-\bar{\tau})\sin(\theta_2-\bar{\theta}_2)\\
&\quad + b_{17}(\nu-\bar{\nu})\sin(\phi_1-\bar{\phi}_1) + b_{18}(\nu-\bar{\nu})\sin(\theta_1-\bar{\theta}_1) + b_{19}(\nu-\bar{\nu})\sin(\phi_2-\bar{\phi}_2)\\
&\quad + b_{20}(\nu-\bar{\nu})\sin(\theta_2-\bar{\theta}_2) + b_{21}(\tau-\bar{\tau})(\nu-\bar{\nu})\}
\end{aligned} \tag{3.67}$$

In the case where the component PSD is highly concentrated, the following approximations hold:

$$\cos(\phi_k-\bar{\phi}_k) \approx 1-(\phi_k-\bar{\phi}_k)^2/2 \tag{3.68}$$

$$\sin(\phi_k-\bar{\phi}_k) \approx \phi_k-\bar{\phi}_k, \tag{3.69}$$

$$\cos(\theta_k-\bar{\theta}_k) \approx 1-(\theta_k-\bar{\theta}_k)^2/2 \tag{3.70}$$

$$\sin(\theta_k-\bar{\theta}_k) \approx \theta_k-\bar{\theta}_k, k=1,2 \tag{3.71}$$

Inserting Eqs (3.68)–(3.71) into Eq. (3.62) and invoking the postulation of Eq. (3.63), we obtain the analytical expressions of the elements of $\boldsymbol{b} = b_1, b_2, \ldots, b_{12}$ in terms of the elements of $\boldsymbol{\eta}$:

$$b_1 = \frac{|\boldsymbol{B}|_{11}}{|\boldsymbol{B}|\sigma_{\phi_1^2}} - (b_7+b_9+b_{11}),\; b_2 = \frac{|\boldsymbol{B}|_{22}}{|\boldsymbol{B}|\sigma_{\theta_1^2}} - (b_8+b_{10}+b_{11}),$$

$$b_3 = \frac{|\boldsymbol{B}|_{33}}{|\boldsymbol{B}|\sigma_{\phi_2^2}} - (b_7+b_{10}+b_{12}) \tag{3.72}$$

$$b_4 = \frac{|\boldsymbol{B}|_{44}}{|\boldsymbol{B}|\sigma_{\theta_2^2}} - (b_8+b_9+b_{12}),\; b_5 = -\frac{|\boldsymbol{B}|_{55}}{2|\boldsymbol{B}|\sigma_\tau^2},\; b_6 = -\frac{|\boldsymbol{B}|_{66}}{2|\boldsymbol{B}|\sigma_\nu^2},\; b_7 = \frac{-|\boldsymbol{B}|_{13}-|\boldsymbol{B}|_{31}}{2|\boldsymbol{B}|\sigma_{\phi_1}\sigma_{\phi_2}} \tag{3.73}$$

$$b_8 = \frac{-|\boldsymbol{B}|_{24}-|\boldsymbol{B}|_{42}}{2|\boldsymbol{B}|\sigma_{\theta_1}\sigma_{\theta_2}},\; b_9 = \frac{-|\boldsymbol{B}|_{14}-|\boldsymbol{B}|_{41}}{2|\boldsymbol{B}|\sigma_{\phi_1}\sigma_{\theta_2}},\; b_{10} = \frac{-|\boldsymbol{B}|_{23}-|\boldsymbol{B}|_{32}}{2|\boldsymbol{B}|\sigma_{\theta_1}\sigma_{\phi_2}},\; b_{11} = \frac{-|\boldsymbol{B}|_{12}-|\boldsymbol{B}|_{21}}{2|\boldsymbol{B}|\sigma_{\phi_1}\sigma_{\theta_1}} \tag{3.74}$$

$$b_{12} = \frac{-|\boldsymbol{B}|_{34}-|\boldsymbol{B}|_{43}}{2|\boldsymbol{B}|\sigma_{\phi_2}\sigma_{\theta_2}},\; b_{13} = \frac{-|\boldsymbol{B}|_{15}-|\boldsymbol{B}|_{51}}{2|\boldsymbol{B}|\sigma_{\phi_1}\sigma_\tau},\; b_{14} = \frac{-|\boldsymbol{B}|_{25}-|\boldsymbol{B}|_{52}}{2|\boldsymbol{B}|\sigma_{\theta_1}\sigma_\tau},\; b_{15} = \frac{-|\boldsymbol{B}|_{35}-|\boldsymbol{B}|_{53}}{2|\boldsymbol{B}|\sigma_{\phi_2}\sigma_\tau} \tag{3.75}$$

$$b_{16} = \frac{-|\boldsymbol{B}|_{45} - |\boldsymbol{B}|_{54}}{2|\boldsymbol{B}|\sigma_{\theta_2}\sigma_\tau},\ b_{17} = \frac{-|\boldsymbol{B}|_{16} - |\boldsymbol{B}|_{61}}{2|\boldsymbol{B}|\sigma_{\phi_1}\sigma_\nu},\ b_{18} = \frac{-|\boldsymbol{B}|_{26} - |\boldsymbol{B}|_{62}}{2|\boldsymbol{B}|\sigma_{\theta_1}\sigma_\nu} \tag{3.76}$$

$$b_{19} = \frac{-|\boldsymbol{B}|_{36} - |\boldsymbol{B}|_{63}}{2|\boldsymbol{B}|\sigma_{\phi_2}\sigma_\nu},\ b_{20} = \frac{-|\boldsymbol{B}|_{46} - |\boldsymbol{B}|_{64}}{2|\boldsymbol{B}|\sigma_{\theta_2}\sigma_\nu},\ b_{21} = \frac{-|\boldsymbol{B}|_{56} - |\boldsymbol{B}|_{65}}{2|\boldsymbol{B}|\sigma_\tau\sigma_\nu} \tag{3.77}$$

Replacing the elements of $\boldsymbol{\eta}$ by their approximations in $\boldsymbol{\theta}$, we can obtain the mapping between $\boldsymbol{b}$ and $\boldsymbol{\theta}$. Using this mapping, $f_{\text{MaxEnt}}(\boldsymbol{\psi};\boldsymbol{b})$ in Eq. (3.67) can be written as $f_{\text{MaxEnt}}(\boldsymbol{\psi};\boldsymbol{\theta})$.

Notice that in the case where the component PSD is highly concentrated, the parameters $\boldsymbol{\theta}$ of $f_{\text{MaxEnt}}(\boldsymbol{\psi};\boldsymbol{\theta})$ represent the parameters of interest; in other words the values of $\boldsymbol{\theta}$ can be considered as estimates of the parameters of interest. When the component PSD is not highly concentrated, the expression $f_{\text{MaxEnt}}(\boldsymbol{\psi};\boldsymbol{\theta})$ is still applicable. However, the values of $\boldsymbol{\theta}$ when the spreads of the PSD are large do not have the meaning of the parameters of interest. In such a case, the parameters of interest $\boldsymbol{\theta}$ have to be calculated based on their definitions using $f_{\text{MaxEnt}}(\boldsymbol{\psi};\boldsymbol{\theta})$.

Model of the Received Signal in MIMO Systems

In this section, we introduce the signal model for MIMO channel sounding and state our assumptions for the properties of dispersion of the propagation channel in bidirection, delay, Doppler frequency, and polarization.

The channel sounder considered has M_1 transmit antennas and M_2 receive antennas. We consider a scenario in which the components of the channel response are dispersed in DoD, DoA, delay, Doppler frequency and polarization. Following the nomenclature of Fleury [2000], we use a unit vector $\boldsymbol{\Omega}$ to characterize a direction. The initial point of the vector is anchored at the origin $\mathcal{O}$ of a coordinate system specified in the region surrounding the antenna array of interest. The end point of $\boldsymbol{\Omega}$ is on a unit sphere $\mathbb{S}_2$ centered at $\mathcal{O}$. In this scenario, a direction is uniquely determined by the spherical coordinates $(\phi, \theta) \in [-\pi, \pi) \times [0, \pi]$ of its end point according to the relation

$$\boldsymbol{\Omega} = [\cos(\phi)\sin(\theta), \sin(\phi)\sin(\theta), \cos(\theta)]^{\mathrm{T}} \tag{3.78}$$

where $[\cdot]^{\mathrm{T}}$ denotes the transpose operator. The angles ϕ and θ are referred to as the azimuth and elevation of the direction respectively. In what follows, $\boldsymbol{\Omega}_i$, $i = 1, 2$ are used to denote the DoD and DoA respectively. The angles (ϕ_i, θ_i), $i = 1, 2$ represent the azimuth-elevation of departure and the azimuth-elevation of arrival respectively. In the scenario considered, the propagation is dispersive in dual polarization; that is, in vertical and horizontal polarization. For the case where a total of D propagation paths exist between the Tx and the Rx, the bidirection-delay-Doppler-dual-polarized spread function $\boldsymbol{H}(\boldsymbol{\Omega}_1, \boldsymbol{\Omega}_2, \tau, \nu)$ of the channel can be written as

$$\boldsymbol{H}(\boldsymbol{\Omega}_1, \boldsymbol{\Omega}_2, \tau, \nu) = \sum_{d=1}^{D} \boldsymbol{H}_d(\boldsymbol{\Omega}_1, \boldsymbol{\Omega}_2, \tau, \nu) \tag{3.79}$$

where

$$\begin{aligned}\boldsymbol{H}_d(\boldsymbol{\Omega}_1, \boldsymbol{\Omega}_2, \tau, \nu) &= [H_{d,p_1,p_2}(\boldsymbol{\Omega}_1, \boldsymbol{\Omega}_2, \tau, \nu); \{p_1, p_2\} \in [1, 2]] \\ &= \begin{bmatrix} H_{d,1,1}(\boldsymbol{\Omega}_1, \boldsymbol{\Omega}_2, \tau, \nu) & H_{d,2,1}(\boldsymbol{\Omega}_1, \boldsymbol{\Omega}_2, \tau, \nu) \\ H_{d,1,2}(\boldsymbol{\Omega}_1, \boldsymbol{\Omega}_2, \tau, \nu) & H_{d,2,2}(\boldsymbol{\Omega}_1, \boldsymbol{\Omega}_2, \tau, \nu) \end{bmatrix}\end{aligned} \tag{3.80}$$

represents the dual-polarization-bidirection-delay-Doppler frequency spread function of the dth component, p_i, $i = 1, 2$ denote polarization status, with $p_i = 1$ being vertical polarization and $p_i = 2$ being horizontal polarization. The baseband representation of the output signal of the $M_1 \times M_2$ MIMO sounding system considered can be written as

$$\begin{aligned} \boldsymbol{Y}(t) = \int_{\mathbb{S}_2} \int_{\mathbb{S}_2} \int_{-\infty}^{+\infty} \int_{-\infty}^{+\infty} \boldsymbol{C}_2(\Omega_2) \boldsymbol{H}(\boldsymbol{\Omega}_1, \boldsymbol{\Omega}_2, \tau, \nu) \boldsymbol{C}_1(\Omega_1)^{\mathrm{T}} u(t-\tau) \\ \times \exp\{j2\pi\nu t\} \mathrm{d}\boldsymbol{\Omega}_1 \mathrm{d}\boldsymbol{\Omega}_2 \mathrm{d}\tau \mathrm{d}\nu + \boldsymbol{W}(t) \end{aligned} \quad (3.81)$$

where $\boldsymbol{Y}(t) \in \mathcal{C}^{M_2 \times M_1}$ is a $M_2 \times M_1$ matrix with the (m_2, m_1)th entry $Y_{m_2,m_1}(t)$ being the output of the m_2th Rx antenna when the m_1th Tx antenna transmits. $\boldsymbol{C}_1(\boldsymbol{\Omega}_1)$ and $\boldsymbol{C}_2(\boldsymbol{\Omega}_2)$ in Eq. (3.81) represent the dual-polarized array response of the Tx and Rx respectively. They are written as

$$\boldsymbol{C}_i(\Omega_i) = [\boldsymbol{c}_{i,1}(\boldsymbol{\Omega}_1), \boldsymbol{c}_{i,2}(\boldsymbol{\Omega}_1)] \in \mathcal{C}^{M_i \times 2}, i = 1, 2 \quad (3.82)$$

where $\boldsymbol{c}_{i,p_i}(\boldsymbol{\Omega}_i)$, $p_i = 1, 2$ denote the array response in vertical and horizontal polarization, which can be written as

$$\boldsymbol{c}_{i,p_i}(\boldsymbol{\Omega}_i) = [c_{i,1,p_i}(\boldsymbol{\Omega}_i), \ldots, c_{i,M_i,p_i}(\boldsymbol{\Omega}_i)]^{\mathrm{T}} \quad (3.83)$$

with $c_{i,M_i,p_i}(\boldsymbol{\Omega})$ representing the response of an individual antenna with polarization p_i.

In Eq. (3.81), $u(t)$ is the complex baseband representation of the transmitted signal, and $\boldsymbol{W}(t) \in \mathbb{C}^{M_2 \times M_1}$ is a matrix containing i.i.d. circularly symmetric white Gaussian noise components. The spectral height for the entries of $\boldsymbol{W}(t)$ equals N.

Inserting Eq. (3.79) into Eq. (3.81), we can rewrite $\boldsymbol{Y}(t)$ as

$$\boldsymbol{Y}(t) = \sum_{d=1}^{D} \boldsymbol{S}_d(t) + \boldsymbol{W}(t) \quad (3.84)$$

with $\boldsymbol{S}_d(t)$ being the contribution of the dth component:

$$\begin{aligned} \boldsymbol{S}_d(t) = \int_{\mathbb{S}_2} \int_{\mathbb{S}_2} \int_{-\infty}^{+\infty} \int_{-\infty}^{+\infty} \boldsymbol{C}_2(\Omega_2) \boldsymbol{H}_d(\boldsymbol{\Omega}_1, \boldsymbol{\Omega}_2, \tau, \nu) \boldsymbol{C}_1(\Omega_1)^{\mathrm{T}} u(t-\tau) \\ \times \exp\{j2\pi\nu t\} \mathrm{d}\boldsymbol{\Omega}_1 \mathrm{d}\boldsymbol{\Omega}_2 \mathrm{d}\tau \mathrm{d}\nu \end{aligned} \quad (3.85)$$

Stochastic Properties of the Channel

We assume that the dual-polarized component spread functions $H_{d,p_1,p_2}(\boldsymbol{\Omega}_1, \boldsymbol{\Omega}_2, \tau, \nu)$, $d \in \{1, \ldots, D\}$ are uncorrelated complex (zero-mean) orthogonal stochastic measures:

$$\begin{aligned} &\mathrm{E}[H_{d,p_1,p_2}(\boldsymbol{\Omega}_1, \boldsymbol{\Omega}_2, \tau, \nu) H_{d',p_1',p_2'}(\boldsymbol{\Omega}_1', \boldsymbol{\Omega}_2', \tau', \nu')^*] \\ &= P_{d,p_1,p_2}(\boldsymbol{\Omega}_1, \boldsymbol{\Omega}_2, \tau, \nu) \delta_{dd'} \delta_{p_1 p_1'} \delta_{p_2 p_2'} \delta(\boldsymbol{\Omega}_1 - \boldsymbol{\Omega}_1') \delta(\boldsymbol{\Omega}_2 - \boldsymbol{\Omega}_2') \delta(\tau - \tau') \delta(\nu - \nu') \end{aligned} \quad (3.86)$$

where $(\cdot)^*$ denotes complex conjugation, $\delta_{..}$ and $\delta(\cdot)$ represent the Kronecker delta and the Dirac delta function respectively, and $P_{d,p_1,p_2}(\boldsymbol{\Omega}_1, \boldsymbol{\Omega}_2, \tau, \nu)$ represents the power spectrum of the dth component with Tx polarization p_1 and Rx polarization p_2, which can be calculated as

$$P_{d,p_1,p_2}(\boldsymbol{\Omega}_1, \boldsymbol{\Omega}_2, \tau, \nu) = \mathrm{E}[|H_{d,p_1,p_2}(\boldsymbol{\Omega}_1, \boldsymbol{\Omega}_2, \tau, \nu)|^2] \tag{3.87}$$

The polarized component PS $P_{d,p_1,p_2}(\boldsymbol{\Omega}_1, \boldsymbol{\Omega}_2, \tau, \nu)$ can be further written as

$$P_{d,p_1,p_2}(\boldsymbol{\Omega}_1, \boldsymbol{\Omega}_2, \tau, \nu) = P_{d,p_1,p_2} \cdot f_{d,p_1,p_2}(\boldsymbol{\Omega}_1, \boldsymbol{\Omega}_2, \tau, \nu) \tag{3.88}$$

where P_{d,p_1,p_2} represents the average power of the dth (p_1, p_2)-polarized component, and $f_{d,p_1,p_2}(\boldsymbol{\Omega}_1, \boldsymbol{\Omega}_2, \tau, \nu)$ denotes the PSD of the dth component at polarizations (p_1, p_2).

It can be easily shown from the property defined in Eq. (3.86) that

$$\begin{aligned} &\mathrm{E}[\boldsymbol{H}_d(\boldsymbol{\Omega}_1, \boldsymbol{\Omega}_2, \tau, \nu) \odot \boldsymbol{H}_{d'}(\boldsymbol{\Omega}_1', \boldsymbol{\Omega}_2', \tau', \nu')^*] \\ &\quad = \boldsymbol{P}_d(\boldsymbol{\Omega}_1, \boldsymbol{\Omega}_2, \tau, \nu)\delta_{dd'}\delta(\boldsymbol{\Omega}_1 - \boldsymbol{\Omega}_1')\delta(\boldsymbol{\Omega}_2 - \boldsymbol{\Omega}_2')\delta(\tau - \tau')\delta(\nu - \nu') \end{aligned} \tag{3.89}$$

where $\odot$ is the element-wise product, and $\boldsymbol{P}_d(\boldsymbol{\Omega}_1, \boldsymbol{\Omega}_2, \tau, \nu)$ is called the "dual-polarized component PS matrix":

$$\begin{aligned} \boldsymbol{P}_d(\boldsymbol{\Omega}_1, \boldsymbol{\Omega}_2, \tau, \nu) &= \mathrm{E}[\boldsymbol{H}_d(\boldsymbol{\Omega}_1, \boldsymbol{\Omega}_2, \tau, \nu) \odot \boldsymbol{H}_d(\boldsymbol{\Omega}_1, \boldsymbol{\Omega}_2, \tau, \nu)^*] \\ &= \begin{bmatrix} P_{d,1,1}(\boldsymbol{\Omega}_1, \boldsymbol{\Omega}_2, \tau, \nu) & P_{d,1,2}(\boldsymbol{\Omega}_1, \boldsymbol{\Omega}_2, \tau, \nu) \\ P_{d,2,1}(\boldsymbol{\Omega}_1, \boldsymbol{\Omega}_2, \tau, \nu) & P_{d,2,2}(\boldsymbol{\Omega}_1, \boldsymbol{\Omega}_2, \tau, \nu) \end{bmatrix} \end{aligned} \tag{3.90}$$

We can further rewrite $\boldsymbol{P}_d(\boldsymbol{\Omega}_1, \boldsymbol{\Omega}_2, \tau, \nu)$ as

$$\boldsymbol{P}_d(\boldsymbol{\Omega}_1, \boldsymbol{\Omega}_2, \tau, \nu) = \boldsymbol{P}_d \odot \boldsymbol{f}_d(\boldsymbol{\Omega}_1, \boldsymbol{\Omega}_2, \tau, \nu) \tag{3.91}$$

where

$$\boldsymbol{P}_d = \begin{bmatrix} P_{d,1,1} & P_{d,1,2} \\ P_{d,2,1} & P_{d,2,2} \end{bmatrix} \text{ and } \boldsymbol{f}_d(\boldsymbol{\Omega}_1, \boldsymbol{\Omega}_2, \tau, \nu) = \begin{bmatrix} f_{d,1,1}(\boldsymbol{\Omega}_1, \boldsymbol{\Omega}_2, \tau, \nu) & f_{d,1,2}(\boldsymbol{\Omega}_1, \boldsymbol{\Omega}_2, \tau, \nu) \\ f_{d,2,1}(\boldsymbol{\Omega}_1, \boldsymbol{\Omega}_2, \tau, \nu) & f_{d,2,2}(\boldsymbol{\Omega}_1, \boldsymbol{\Omega}_2, \tau, \nu) \end{bmatrix}$$

represent respectively the average power matrix and the dual-polarized PSD matrix of the dth component.

Invoking Eq. (3.79) and Eq. (3.86), it can be seen that the dual-polarized spread function matrix $\boldsymbol{H}(\boldsymbol{\Omega}_1, \boldsymbol{\Omega}_2, \tau, \nu)$ of the channel is also an orthogonal stochastic measure:

$$\begin{aligned} &\mathrm{E}[\boldsymbol{H}(\boldsymbol{\Omega}_1, \boldsymbol{\Omega}_2, \tau, \nu) \odot \boldsymbol{H}(\boldsymbol{\Omega}_1', \boldsymbol{\Omega}_2', \tau', \nu')] \\ &\quad = \boldsymbol{P}(\boldsymbol{\Omega}_1, \boldsymbol{\Omega}_2, \tau, \nu)\delta(\boldsymbol{\Omega}_1 - \boldsymbol{\Omega}_1')\delta(\boldsymbol{\Omega}_2 - \boldsymbol{\Omega}_2')\delta(\tau - \tau')\delta(\nu - \nu') \end{aligned} \tag{3.92}$$

where $\boldsymbol{P}(\boldsymbol{\Omega}_1, \boldsymbol{\Omega}_2, \tau, \nu)$ is the dual-polarized bidirection-delay-Doppler PS matrix of the channel considered:

$$\begin{aligned} \boldsymbol{P}(\boldsymbol{\Omega}_1, \boldsymbol{\Omega}_2, \tau, \nu) &= \mathrm{E}[\boldsymbol{H}(\boldsymbol{\Omega}_1, \boldsymbol{\Omega}_2, \tau, \nu) \odot \boldsymbol{H}(\boldsymbol{\Omega}_1, \boldsymbol{\Omega}_2, \tau, \nu)^*] \\ &= \sum_{d=1}^{D} \boldsymbol{P}_d(\boldsymbol{\Omega}_1, \boldsymbol{\Omega}_2, \tau, \nu) \end{aligned} \tag{3.93}$$

We assume that the power spectral densities of the dth component with different combinations of polarizations are identical:

$$f_{d,p_1,p_2}(\boldsymbol{\Omega}_1, \boldsymbol{\Omega}_2, \tau, \nu) = f_d(\boldsymbol{\Omega}_1, \boldsymbol{\Omega}_2, \tau, \nu), \text{ for } p_1 = 1, 2, p_2 = 1, 2 \tag{3.94}$$

Under this assumption, Eq. (3.91) can be simplified to

$$\boldsymbol{P}_d(\boldsymbol{\Omega}_1, \boldsymbol{\Omega}_2, \tau, \nu) = \boldsymbol{P}_d \cdot f_d(\boldsymbol{\Omega}_1, \boldsymbol{\Omega}_2, \tau, \nu) \tag{3.95}$$

It is of interest to extract the PS matrix $\boldsymbol{P}(\boldsymbol{\Omega}_1, \boldsymbol{\Omega}_2, \tau, \nu)$ of a propagation channel from measurement data. One approach is to use parameter estimation methods to estimate the parameters modeling the PSs of individual components. Thus it is necessary to find a suitable generic parametric representation of the component PSs.

One feasible way is to assume that the component PSD $f_d(\boldsymbol{\Omega}_1, \boldsymbol{\Omega}_2, \tau, \nu)$ in Eq. (3.95) can be approximated as

$$f_d(\boldsymbol{\Omega}_1, \boldsymbol{\Omega}_2, \tau, \nu) \approx f_{\text{MaxEnt}}(\boldsymbol{\Omega}_1, \boldsymbol{\Omega}_2, \tau, \nu; \boldsymbol{\theta}_d) \tag{3.96}$$

where $\boldsymbol{\theta}_d$ denotes the component-specific parameters

$$\begin{aligned}\theta_d \triangleq [&\mu_{\phi_{1,d}}, \mu_{\theta_{1,d}}, \mu_{\phi_{2,d}}, \mu_{\theta_{2,d}}, \mu_{\tau_d}, \mu_{\nu_d}, \sigma_{\phi_{1,d}}, \sigma_{\theta_{1,d}}, \sigma_{\phi_{2,d}}, \sigma_{\theta_{2,d}}, \sigma_{\tau_d}, \sigma_{\nu_d},\\ &\rho_{\phi_1\tau,d}, \rho_{\phi_1\theta_1,d}, \rho_{\phi_1\phi_2,d}, \rho_{\phi_1\theta_2,d}, \rho_{\phi_1\tau,d}, \rho_{\phi_1\nu,d}, \rho_{\theta 1\phi_2,d}, \rho_{\theta_1\theta_2,d}, \rho_{\theta_1\tau,d},\\ &\rho_{\theta_1\nu,d}, \rho_{\phi_2\theta_2,d}, \rho_{\phi_2\tau,d}, \rho_{\phi_2\nu,d}, \rho_{\tau\nu,d}]\end{aligned} \tag{3.97}$$

In the scenario in which the channel response consists of D components, the power spectrum $P(\boldsymbol{\Omega}_1, \boldsymbol{\Omega}_2, \tau, \nu)$ of the channel is parameterized by

$$\boldsymbol{\Theta} = (\boldsymbol{\Theta}_1, \boldsymbol{\Theta}_2, \ldots, \boldsymbol{\Theta}_D) \tag{3.98}$$

where $\boldsymbol{\Theta}_d = (P_d, \boldsymbol{\theta}_d)$ denotes the parameters of the component d. Using this parametric approach, the estimation of $P(\boldsymbol{\Omega}_1, \boldsymbol{\Omega}_2, \tau, \nu)$ is then equivalent to the estimation of the parameters $\boldsymbol{\Theta}$.

3.6 Model for Keyhole Channel

The keyhole phenomenon [Chizhik et al. 2000, Shin and Lee 2003] reveals the existence of a rank-1 channel, which is the worst case for MIMO transmission. This effect is also observable when the propagation paths between the Tx and the Rx involve the same one-dimensional diffraction [Chizhik et al. 2002]. Experimental evidence of the existence of the keyhole has been presented by Almers et al. [2003]. In the same paper, the keyhole phenomenon was generalized to the pinhole scenario, which results in low-rank channels. The pinhole effect is realistic and observable in many scenarios, such as diffraction by building edges and corners, waveguiding in street canyons and corridors, as well as propagation above roofs. In general, the keyhole and pinhole effects usually occur when a propagation channel can be decomposed into two separable subchannels and the propagation connecting the two subchannels has a small number of degrees of freedom. Examples of this scenario are outdoor-to-indoor propagation through windows, and propagation through two environments connected by a hallway or a tunnel.

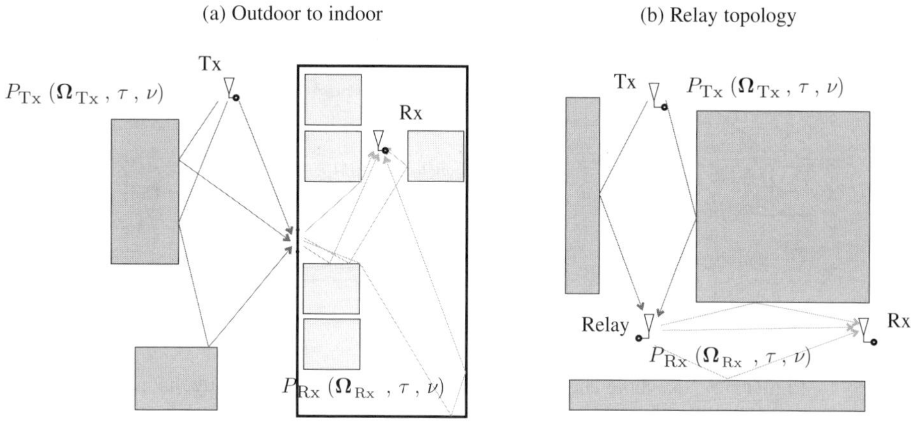

Figure 3.7 Two propagation scenarios in which the keyhole or pinhole effect may occur

The keyhole effect can also be used to describe the propagation in a relay network where a repeater or a relay station receives signals from an environment and transmits the signals to another isolated area. Figure 3.7 shows two examples of propagation where a keyhole channel can be formed.

The conventional keyhole model focuses on the impact of the keyhole on the statistics of the channel matrix. In Gesbert et al. [2002], the keyhole channel is described using the MIMO correlated fading channel. The entries of a keyhole channel transfer matrix are modeled as an i.i.d. double-Rayleigh distribution. In Karasawa et al. [2007], the distributions of the eigenvalues of the keyhole and multi-keyhole channel matrix are analyzed. In Levin and Loyka [2008], the distribution of the outage capacity of correlated keyhole MIMO channels is derived. These models are useful for studying the impact of a keyhole channel analytically. However, they are insufficient for describing the space-time-frequency characteristics of wideband keyhole channels.

In this section, a generic model is outlined for characterization of the dispersion of the keyhole propagation channel in delay, DoD, DoA, and Doppler frequency. The marginal dispersive behavior of the keyhole channel and the analytical expressions for its transfer matrix and correlation matrix are presented. These results give the insights into the propagation mechanism and characteristics of the keyhole channel. Using measurement data collected with a wideband MIMO channel sounder, the existence of keyhole effect is verified and the applicability of the proposed model is assessed. As relay-based topologies will be widely used in future wireless communication systems, ray-based investigations of wave propagation in keyhole or pinhole scenarios are of great importance.

Generic Models for Channel Spread Functions in Keyhole Scenarios

Let us consider the scenario depicted in Figure 3.8. The spread function of the propagation channel between the Tx and the keyhole and the channel between the keyhole and the Rx site are denoted respectively by $h_{\mathrm{Tx}}(\mathbf{\Omega}_{\mathrm{Tx}}, \tau, \nu)$ and $h_{\mathrm{Rx}}(\mathbf{\Omega}_{\mathrm{Rx}}, \tau, \nu)$. Here, $\mathbf{\Omega}_{\mathrm{Tx}}$ and $\mathbf{\Omega}_{\mathrm{Rx}}$ represent,

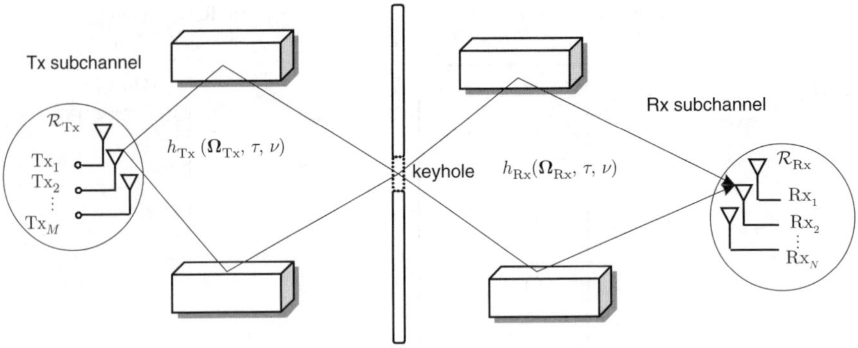

Figure 3.8 Keyhole channel that can be split into Tx and Rx subchannels

respectively, the DoD and the DoA, τ is the delay, and ν denotes the Doppler frequency. We assume that the spread functions $h_{\rm Tx}(\mathbf{\Omega}_{\rm Tx}, \tau, \nu)$ and $h_{\rm Rx}(\mathbf{\Omega}_{\rm Rx}, \tau, \nu)$ are two OSMs:

$$\mathrm{E}[h_{\rm Tx}(\mathbf{\Omega}_{\rm Tx}, \tau, \nu) h_{\rm Rx}(\mathbf{\Omega}_{\rm Rx}, \tau, \nu)^*] = 0 \tag{3.99}$$

We use $h(\mathbf{\Omega}_{\rm Tx}, \mathbf{\Omega}_{\rm Rx}, \tau, \nu)$ to denote the spread function of the propagation channel between the Tx and the Rx. The model proposed for $h(\mathbf{\Omega}_{\rm Tx}, \mathbf{\Omega}_{\rm Rx}, \tau, \nu)$ is written in terms of $h_{\rm Tx}(\mathbf{\Omega}_{\rm Tx}, \tau, \nu)$ and $h_{\rm Rx}(\mathbf{\Omega}_{\rm Rx}, \tau, \nu)$ as

$$h(\mathbf{\Omega}_{\rm Tx}, \mathbf{\Omega}_{\rm Rx}, \tau, \nu) = \alpha_{\rm k} \cdot h_{\rm Tx}(\mathbf{\Omega}_{\rm Tx}, \tau, \nu) * h_{\rm Rx}(\mathbf{\Omega}_{\rm Rx}, \tau, \nu) \tag{3.100}$$

where $\alpha_{\rm k}$ denotes the complex attenuation caused by the scattering effect of the keyhole, and $*$ represents the convolution operation. The convolution is applied because the propagation channel between the Tx and the keyhole and the channel between the keyhole and the Rx are independent. In what follows, the expressions of the marginal and conditional dispersive behavior of the keyhole channel in delay, Doppler frequency, DoA, DoD, and bidirection are discussed.

Delay Spread Function

The delay spread function $h(\tau)$ of the keyhole channel can be calculated from the bi-direction delay and Doppler frequency spread function $h(\mathbf{\Omega}_{\rm Tx}, \mathbf{\Omega}_{\rm Rx}, \tau, \nu)$ as

$$h(\tau) = \int_{-\infty}^{+\infty} \int_{\mathbb{S}_2} \int_{\mathbb{S}_2} h(\mathbf{\Omega}_{\rm Tx}, \mathbf{\Omega}_{\rm Rx}, \tau, \nu) \mathrm{d}\mathbf{\Omega}_{\rm Tx} \mathrm{d}\mathbf{\Omega}_{\rm Rx} \mathrm{d}\nu \tag{3.101}$$

From Eq. (3.100) it can be shown that

$$h(\tau) = \alpha_{\rm k} \cdot h_{\rm Tx}(\tau) * h_{\rm Rx}(\tau) \tag{3.102}$$

where

$$h_{\mathrm{Tx}}(\tau) = \int_{-\infty}^{+\infty}\int_{\mathbb{S}_2} h_{\mathrm{Tx}}(\mathbf{\Omega}_{\mathrm{Tx}},\tau,\nu)\mathrm{d}\mathbf{\Omega}_{\mathrm{Tx}}\mathrm{d}\nu \quad \text{and} \tag{3.103}$$

$$h_{\mathrm{Rx}}(\tau) = \int_{-\infty}^{+\infty}\int_{\mathbb{S}_2} h_{\mathrm{Rx}}(\mathbf{\Omega}_{\mathrm{Rx}},\tau,\nu)\mathrm{d}\mathbf{\Omega}_{\mathrm{Rx}}\mathrm{d}\nu \tag{3.104}$$

are the delay spread function of the Tx subchannel and the Rx subchannel, respectively.

Doppler Frequency Spread Function

Similar to $h(\tau)$, the Doppler frequency spread function $h(\nu)$ of the keyhole channel can be calculated from $h(\mathbf{\Omega}_{\mathrm{Tx}},\mathbf{\Omega}_{\mathrm{Rx}},\tau,\nu)$ as

$$\begin{aligned} h(\nu) &= \int_{-\infty}^{+\infty}\int_{\mathbb{S}_2}\int_{\mathbb{S}_2} h(\mathbf{\Omega}_{\mathrm{Tx}},\mathbf{\Omega}_{\mathrm{Rx}},\tau,\nu)\mathrm{d}\mathbf{\Omega}_{\mathrm{Tx}}\mathrm{d}\mathbf{\Omega}_{\mathrm{Rx}}\mathrm{d}\tau \\ &= \alpha_{\mathrm{k}}\cdot h_{\mathrm{Tx}}(\nu) * h_{\mathrm{Rx}}(\nu) \end{aligned} \tag{3.105}$$

where

$$h_{\mathrm{Tx}}(\nu) = \int_{0}^{+\infty}\int_{\mathbb{S}_2} h_{\mathrm{Tx}}(\mathbf{\Omega}_{\mathrm{Tx}},\tau,\nu)\mathrm{d}\mathbf{\Omega}_{\mathrm{Tx}}\mathrm{d}\tau \quad \text{and} \tag{3.106}$$

$$h_{\mathrm{Rx}}(\nu) = \int_{0}^{+\infty}\int_{\mathbb{S}_2} h_{\mathrm{Rx}}(\mathbf{\Omega}_{\mathrm{Rx}},\tau,\nu)\mathrm{d}\mathbf{\Omega}_{\mathrm{Rx}}\mathrm{d}\tau \tag{3.107}$$

are the Doppler frequency spread function of the Tx subchannel and the Rx subchannel, respectively.

Bidirection Spread Function

The bidirection spread function $h(\mathbf{\Omega}_{\mathrm{Tx}},\mathbf{\Omega}_{\mathrm{Rx}})$ of the keyhole channel is calculated as

$$h(\mathbf{\Omega}_{\mathrm{Tx}},\mathbf{\Omega}_{\mathrm{Rx}}) = \int_{-\infty}^{+\infty}\int_{-\infty}^{+\infty} h(\mathbf{\Omega}_{\mathrm{Tx}},\mathbf{\Omega}_{\mathrm{Rx}},\tau,\nu)\mathrm{d}\tau\mathrm{d}\nu \tag{3.108}$$

Inserting Eq. (3.100) into Eq. (3.108) yields

$$h(\mathbf{\Omega}_{\mathrm{Tx}},\mathbf{\Omega}_{\mathrm{Rx}}) = h_{\mathrm{Tx}}(\mathbf{\Omega}_{\mathrm{Tx}})\cdot\alpha_{\mathrm{k}}\cdot h_{\mathrm{Rx}}(\mathbf{\Omega}_{\mathrm{Rx}}) \tag{3.109}$$

where

$$h_{\mathrm{a}}(\mathbf{\Omega}_{\mathrm{a}}) = \int_{-\infty}^{+\infty}\int_{-\infty}^{+\infty} h_{\mathrm{a}}(\mathbf{\Omega}_{\mathrm{a}},\tau,\nu)\mathrm{d}\tau\mathrm{d}\nu$$

with a substituted by Tx and Rx representing, respectively, the DoD spread function of the Tx subchannel and the DoA spread function of the Rx subchannel.

The DoA spread function of the keyhole channel under the condition that the DoD is specified, say $\mathbf{\Omega}_{\mathrm{Tx}} = \mathbf{\Omega}'_{\mathrm{Tx}}$, is calculated as

$$h(\mathbf{\Omega}_{\mathrm{Tx}}, \mathbf{\Omega}_{\mathrm{Rx}}; \mathbf{\Omega}_{\mathrm{Tx}} = \mathbf{\Omega}'_{\mathrm{Tx}}) = \alpha_{\mathrm{k}} \cdot \alpha_{\mathbf{\Omega}'_{\mathrm{Tx}}} \cdot h_{\mathrm{Rx}}(\mathbf{\Omega}_{\mathrm{Rx}}) \tag{3.110}$$

with

$$\alpha_{\mathbf{\Omega}'_{\mathrm{Tx}}} = \int_{-\infty}^{+\infty}\int_{-\infty}^{+\infty} h_{\mathrm{Tx}}(\mathbf{\Omega}_{\mathrm{Tx}}, \tau, \nu; \mathbf{\Omega}_{\mathrm{Tx}} = \mathbf{\Omega}'_{\mathrm{Tx}})\mathrm{d}\tau\mathrm{d}\nu$$

Eq. (3.110) indicates that the paths in the keyhole channel with a specific DoD exhibit the same dispersive constellation in DoA as the other paths with other DoDs.

DoD/DoA Spread Function

The DoD spread function $h(\mathbf{\Omega}_{\mathrm{Tx}})$ and the marginal DoA spread function $h(\mathbf{\Omega}_{\mathrm{Rx}})$ of the keyhole channel can be calculated as

$$h(\mathbf{\Omega}_{\mathrm{Tx}}) = \alpha_{\mathrm{k}} \cdot \alpha_{\mathrm{Rx}} \cdot h_{\mathrm{Tx}}(\mathbf{\Omega}_{\mathrm{Tx}}) \tag{3.111}$$

$$h(\mathbf{\Omega}_{\mathrm{Rx}}) = \alpha_{\mathrm{k}} \cdot \alpha_{\mathrm{Tx}} \cdot h_{\mathrm{Rx}}(\mathbf{\Omega}_{\mathrm{Rx}}) \tag{3.112}$$

respectively, where

$$\alpha_{\mathrm{a}} = \int_{\mathbb{S}_2}\int_{-\infty}^{+\infty}\int_{-\infty}^{+\infty} h_{\mathrm{a}}(\mathbf{\Omega}_{\mathrm{a}}, \tau, \nu)\mathrm{d}\tau\mathrm{d}\nu\mathrm{d}\mathbf{\Omega}_{\mathrm{a}}$$

with a substituted by Tx and Rx.

The Transfer and Correlation Matrices of a Keyhole Channel

The transfer matrix $\boldsymbol{H}$ of the MIMO radio channel can be calculated as

$$\boldsymbol{H} = \int_{-\infty}^{+\infty}\int_{-\infty}^{+\infty}\int_{\mathbb{S}_2}\int_{\mathbb{S}_2} \boldsymbol{c}_2(\mathbf{\Omega}_{\mathrm{Rx}})\boldsymbol{c}_1(\mathbf{\Omega}_{\mathrm{Tx}})^{\mathrm{T}} r(\tau, \nu) * h(\mathbf{\Omega}_{\mathrm{Tx}}, \mathbf{\Omega}_{\mathrm{Rx}}, \tau, \nu)\mathrm{d}\mathbf{\Omega}_{\mathrm{Tx}}\mathrm{d}\mathbf{\Omega}_{\mathrm{Rx}}\mathrm{d}\tau\mathrm{d}\nu \tag{3.113}$$

where $\boldsymbol{c}_1(\cdot)$ and $\boldsymbol{c}_2(\cdot)$ denote respectively the response of the Tx and Rx antenna array, and

$$r(\tau, \nu) = \int_{-\infty}^{+\infty} s(t)s^{*(t-\tau)} \exp\{-j2\pi\nu t\}\mathrm{d}t \tag{3.114}$$

represents the autocorrelation function of the transmitted signal $s(t)$ with respect to delay and Doppler frequency. We use $\boldsymbol{h} = \mathrm{vec}(\boldsymbol{H})$ to represent a vector that contains the entries of $\boldsymbol{H}$. It is easy to show that

$$\boldsymbol{h} = \int_{-\infty}^{+\infty}\int_{-\infty}^{+\infty}\int_{\mathbb{S}_2}\int_{\mathbb{S}_2} \boldsymbol{c}_2(\mathbf{\Omega}_{\mathrm{Rx}}) \otimes \boldsymbol{c}_1(\mathbf{\Omega}_{\mathrm{Tx}}) r(\tau, \nu) * h(\mathbf{\Omega}_{\mathrm{Tx}}, \mathbf{\Omega}_{\mathrm{Rx}}, \tau, \nu)\mathrm{d}\mathbf{\Omega}_{\mathrm{Tx}}\mathrm{d}\mathbf{\Omega}_{\mathrm{Rx}}\mathrm{d}\tau\mathrm{d}\nu \tag{3.115}$$

where $\otimes$ represents the Kronecker product operation. The correlation matrix of $\boldsymbol{h}$ is then calculated as

$$\begin{aligned}\boldsymbol{\Sigma}_{\boldsymbol{h}} &= \mathrm{E}[\boldsymbol{h}\boldsymbol{h}^{\mathrm{H}}] \\ &= \int_{-\infty}^{+\infty}\int_{-\infty}^{+\infty}\int_{\mathbb{S}_2}\int_{\mathbb{S}_2}[\boldsymbol{c}_2(\boldsymbol{\Omega}_{\mathrm{Rx}})\boldsymbol{c}_2(\boldsymbol{\Omega}_{\mathrm{Rx}})^{\mathrm{H}}]\otimes[\boldsymbol{c}_1(\boldsymbol{\Omega}_{\mathrm{Tx}})\boldsymbol{c}_1(\boldsymbol{\Omega}_{\mathrm{Tx}})^{\mathrm{H}}]\cdot[r(\tau,\nu)*r^*(\tau,\nu)] \\ &\quad * P(\boldsymbol{\Omega}_{\mathrm{Tx}},\boldsymbol{\Omega}_{\mathrm{Rx}},\tau,\nu)\mathrm{d}\boldsymbol{\Omega}_{\mathrm{Tx}}\mathrm{d}\boldsymbol{\Omega}_{\mathrm{Rx}}\mathrm{d}\tau\mathrm{d}\nu \end{aligned} \tag{3.116}$$

where

$$P(\boldsymbol{\Omega}_{\mathrm{Tx}},\boldsymbol{\Omega}_{\mathrm{Rx}},\tau,\nu) = \mathrm{E}[h(\boldsymbol{\Omega}_{\mathrm{Tx}},\boldsymbol{\Omega}_{\mathrm{Rx}},\tau,\nu)h^*(\boldsymbol{\Omega}_{\mathrm{Tx}},\boldsymbol{\Omega}_{\mathrm{Rx}},\tau,\nu)] \tag{3.117}$$

denotes the power spectrum of the keyhole channel. Applying the assumption that the spread functions of the Tx subchannel and the Rx subchannel are OSMs; that is:

$$\mathrm{E}[h_{\mathrm{Tx}}(\boldsymbol{\Omega}_{\mathrm{Tx}},\tau,\nu)h_{\mathrm{Rx}}(\boldsymbol{\Omega}_{\mathrm{Rx}},\tau,\nu)^*] = 0 \tag{3.118}$$

we obtain for the bi-directional delay and Doppler frequency power spectrum

$$P(\boldsymbol{\Omega}_{\mathrm{Tx}},\boldsymbol{\Omega}_{\mathrm{Rx}},\tau,\nu) = P_{\mathrm{Tx}}(\boldsymbol{\Omega}_{\mathrm{Tx}},\tau,\nu) * P_{\mathrm{Rx}}(\boldsymbol{\Omega}_{\mathrm{Rx}},\tau,\nu) \tag{3.119}$$

where $P_{\mathrm{a}}(\boldsymbol{\Omega}_{\mathrm{a}},\tau,\nu) = \mathrm{E}[h_{\mathrm{a}}(\boldsymbol{\Omega}_{\mathrm{a}},\tau,\nu)h_{\mathrm{a}}^*(\boldsymbol{\Omega}_{\mathrm{a}},\tau,\nu)]$ with a replaceable by Tx and Rx to represent, respectively, the power spectrum of the Tx subchannel and of the Rx subchannel. Equation (3.119) indicates that the power spectrum of the keyhole propagation channel can be computed by the convolution of the power spectrum of two separable subchannels.

In Yin et al. 2009, the proposed generic models for keyhole channels were evaluated by using measurement data. The results reported demonstrated that channel characteristics observed from measurements are consistent with the behavior predicted with generic models.

Bibliography

Acosta-Marum G, Walkenhorst BT and Baxley RJ 2010 An empirical doubly-selective dual-polarization vehicular MIMO channel model *Vehicular Technology Conference (VTC 2010-Spring), 2010 IEEE 71st*, pp. 1–5.

Almers P, Tufvesson F and Molisch A 2003 Measurement of keyhole effect in a wireless multiple-input multiple-output (MIMO) channel. *IEEE Communications Letters* **7**(8), 373–375.

Almers P, Wyne S, Tufvesson F and Molisch A 2005 Effect of random walk phase noise on MIMO measurements *Vehicular Technology Conference, 2005. VTC 2005-Spring. 2005 IEEE 61st*, vol. 1, pp. 141–145.

Bartlett M 1948 Smoothing periodograms from time series with continuous spectra. *Nature* **161**, 686–687.

Bello PA and Bengtsson M 1963 Characterization of randomly time-invariant linear channels. *IEEE Transactions on Communication Systems* **CS-11**, 360–393.

Bengtsson M and Ottersten B 2000 Low-complexity estimators for distributed sources. *IEEE Transactions on Signal Processing* **48**(8), 2185–2194.

Bengtsson M and Völcker B 2001 On the estimation of azimuth distributions and azimuth spectra *Proceedings of the 54th IEEE Vehicular Technology Conference (VTC2001-Fall)*, vol. 3, pp. 1612–1615, Atlantic City, USA.

Besson O and Stoica P 1999 Decoupled estimation of DoA and angular spread for spatially distributed sources. *IEEE Transactions on Signal Processing* **49**, 1872–1882.

Betlehem T, Abhayapala TD and Lamahewa TA 2006 Space-time MIMO channel modelling using angular power distributions *Proceedings of the 7th Australian Communications Theory Workshop*, pp. 165–170, Perth, Australia.

Chizhik D, Foschini G and Valenzuela R 2000 Capacities of multi-element transmit and receive antennas: Correlations and keyholes. *Electronics Letters* **36**(13), 1099–1100.

Chizhik D, Foschini G, Gans M and Valenzuela R 2002 Keyholes, correlations, and capacities of multielement transmit and receive antennas. *IEEE Transactions on Wireless Communications* **1**(2), 361–368.

Chung P and Böhme JF 2005 Recursive EM and SAGE-inspired algorithms with application to DOA estimation. *IEEE Transactions on Signal Processing* **53**(8), 2664–2677.

Czink N 2007 *The random-cluster model – a stochastic MIMO channel model for broadband wireless communication systems of the 3rd generation and beyond* PhD thesis Technische Universitat Wien, Vienna, Austria, FTW Dissertation Series.

Czink N, Tian R, Wyne S, Eriksson G, Tufvesson F, Zemen T, Nuutinen J, Ylitalo J, Bonek E and Molisch A 2007 Tracking time-variant cluster parameters in MIMO channel measurements: Algorithm and results *Proceedings of International Conference on Communications and Networking in China (ChinaCOM)*, pp. 1147–1151, Shanghai, China.

Degli-Esposti V, Kolmonen VM, Vitucci E, Fuschini F and Vainikainen P 2007 Analysis and ray tracing modelling of co- and cross-polarization radio propagation in urban environment. *Antennas and Propagation, 2007. EuCAP 2007. The Second European Conference on* pp. 1–4.

Deng Y, Burr A and White G 2005 Performance of MIMO systems with combined polarization multiplexing and transmit diversity *Proceedings of the 61st Vehicular Technology Conference (VTC 2005-Spring)*, vol. 2, pp. 869–873.

(ed. Correia L) 2001 *Wireless Flexible Personalised Communication (COST 259: European Co-Operation in Mobile Radio Research)*. John Wiley & Sons.

Fleury B 2000 First- and second-order characterization of direction dispersion and space selectivity in the radio channel. *IEEE Transactions on Information Theory* **46**(6), 2027–2044.

Fleury BH, Jourdan P and Stucki A 2002 High-resolution channel parameter estimation for MIMO applications using the SAGE algorithm *Proceedings of International Zurich Seminar on Broadband Communications*, vol. 30, pp. 1–9.

Gesbert D, Bolcskei H, Gore D and Paulraj A 2002 Outdoor MIMO wireless channels: models and performance prediction. *IEEE Transactions on Communications* **50**(12), 1926–1934.

Gihman II and Skorohod AV 1974 *The Theory of Stochastic Processes*. Springer.

Hämäläinen J, Nuutinen JP, Wichman R, Ylitalo J and Jämsä T 2005 Analysis and measurements for indoor polarization MIMO in 5.25 GHz band. *Proc. IEEE Vehicular Technology Conference, VTC 2002 Spring* pp. 252–256.

Jakes WC 1974 *Microwave Mobile Communications*. IEEE Press.

Jaynes E 2003 *Probability Theory*. Cambridge University Press.

Jupp PE and Mardia KV 1980 A general correlation coefficient for directional data and related regression problems. *Biometrika* **67**, 163–173.

Karasawa Y, Tsuruta M and Taniguchi T 2007 Multi-keyhole model for MIMO radio-relay systems. *Antennas and Propagation, 2007. EuCAP 2007. The Second European Conference on* pp. 1–6.

Kent JT 1982 The Fisher–Bingham distribution on the sphere. *Journal of the Royal Statistical Society, Series B (Methodological)* **44**, 71–80.

Kim MD, Park JJ, Chung HK and Yin X 2012 Cross-correlation characteristics of multi-link channel based on channel measurements at 3.7 GHz *Advanced Communication Technology (ICACT), 2012 14th International Conference on*, pp. 351–355.

Kwakkernaat M and Herben M 2007 Analysis of clustered multipath estimates in physically nonstationary radio channels *Proceedings of the 17th IEEE International Symposium on Personal, Indoor and Mobile Radio Communications (PIMRC)*, pp. 1–5, Athens, Greece.

Kwon SC and Stüber G 2010 3-D geometry-based statistical modeling of cross-polarization discrimination in wireless communication channels *Vehicular Technology Conference (VTC 2010-Spring), 2010 IEEE 71st*, pp. 1–5.

Kyritsi P and Cox D 2002 Effect of element polarization on the capacity of a MIMO system *Wireless Communications and Networking Conference, WCNC2002*, vol. 2, pp. 892–896.

Levin G and Loyka S 2008 On the outage capacity distribution of correlated keyhole MIMO channels. *IEEE Transactions on Information Theory* **54**(7), 3232–3245.

Mardia K 1976 Linear-circular correlation coefficients and rhythmometry. *Biometrika* **63**(2), 403–405.

Mardia K, Kent J and Bibby J 2003 *Multivariate Analysis*. Academic Press.

Mardia KV 1975 Statistics of directional data. *Journal of the Royal Statistical Society. Series B (Methodological)* **37**, 349–393.

MEDAV GmbH 2001 *Manual W701W1.096: RUSK MIMO: Broadband vector channel sounder for MIMO channels*.

Medbo J, Asplund H, Berg JE and Jalden N 2012 Directional channel characteristics in elevation and azimuth at an urban macrocell base station *The 6th European Conference on Antennas and Propagation (EUCAP)*, pp. 428–432.

Medbo J, Riback M and Berg J 2006 Validation of 3GPP spatial channel model including WINNER wideband extension using measurements *Proceedings of IEEE 64th Vehicular Technology Conference, (VTC2006-Fall)*, pp. 1–5, Montréal, Canada.

Meng Y, Stoica P and Wong K 1996 Estimation of the directions of arrival of spatially dispersed signals in array processing. *IEE Proceedings Radar, Sonar and Navigation* **143**(1), 1–9.

Park JJ, Kim MD, Kwon HK, Chung HK, Yin X and Fu Y 2012 Measurement-based stochastic cross-correlation models of a multilink channel in cooperative communication environments. *ETRI Journal on Wired and Wireless Telecommunication Technologies* **34**(6), 858–868.

Pedersen K and Mogensen P 1999 Simulation of dual-polarized propagation environments for adaptive antennas *Vehicular Technology Conference, 1999. VTC 1999 - Fall. IEEE VTS 50th*, vol. 1, pp. 62–66.

Ribeiro CB, Ollila E and Koivunen V 2004 Stochastic maximum likelihood method for propagation parameter estimation *Proceedings of the 15th IEEE International Symposium on Personal, Indoor and Mobile Radio Communications (PIMRC'06)*, vol. 3, pp. 1839–1843, Helsinki, Finland.

Richter A, Salmi J and Koivunen V 2006 An algorithm for estimation and tracking of distributed diffuse scattering in mobile radio channels *Proceedings of the 7th IEEE International Workshop on Signal Processing Advances for Wireless Communications (SPAWC'06)*, pp. 1–5.

Salmi J, Richter A and Koivunen V 2006 Enhanced tracking of radio propagation path parameters using state-space modeling *Proceedings of the 2006 European Signal Processing Conference (EUSIPCO)*, pp. 1–5, Florence, Italy.

Shannon C 1948 A mathematical theory of communication. *The Bell System Technical Journal* **27**, 379–423, 623–656.

Shin H and Lee JH 2003 Capacity of multiple-antenna fading channels: spatial fading correlation, double scattering, and keyhole. *IEEE Transactions on Information Theory* **49**(10), 2636–2647.

Stucki A 2001 PropSound system specifications document: Concept and specifications. Technical report, Elektrobit AG, Switzerland.

Tan C, Beach M and Nix A 2003 Enhanced-SAGE algorithm for use in distributed-source environments. *Electronics Letters* **39**(8), 697–698.

Trump T and Ottersten B 1996 Estimation of nominal direction of arrival and angular spread using an array of sensors. *Signal Processing* **50**, 57–69.

Valaee S, Champagne B and Kabal P 1995 Parametric localization of distributed sources. *IEEE Transactions on Signal Processing* **43**, 2144–2153.

Vaughan R 1990 Polarization diversity in mobile communications. *IEEE Transactions on Vehicular Technology* **39**(3), 177–186.

Watson GS 1983 *Statistics on Spheres*. Wiley.

Yin X, Fleury BH, Jourdan P and Stucki A 2003 Polarization estimation of individual propagation paths using the SAGE algorithm *Proceedings of the IEEE International Symposium on Personal, Indoor and Mobile Radio Communications (PIMRC)*, pp. 1795–1799, Beijing, China.

Yin X, Liang J, Chen J, Park J, Kim M and Chung H 2012a Empirical models of cross-correlation for small-scale fading in co-existing channels *Proceedings of Asia-Pacific Communication Conference 2012*, pp. 327–332, Jeju, Korea.

Yin X, Liang J, Fu Y, Yu J, Zhang Z, Park JJ, Kim MD and Chung HK 2012b Measurement-based stochastic modeling for co-existing propagation channels in cooperative relay scenarios *Future Network Mobile Summit (FutureNetw), 2012*, pp. 1–8.

Yin X, Liang J, Fu Y, Zhang Z, Park JJ, Kim MD and Chung HK 2012c Measurement-based stochastic models for the cross-correlation of multi-link small-scale fading in cooperative relay environments *Antennas and Propagation (EUCAP), 2012 6th European Conference on*, pp. 1–5.

Yin X, Liu L, Nielsen D, Czink N and Fleury BH 2007a Characterization of the azimuth-elevation power spectrum of individual path components *Proceedings of the International ITG/IEEE Workshop on Smart Antennas (WSA 2007)*, pp. 1–5, Vienna, Austria.

Yin X, Liu L, Nielsen D, Pedersen T and Fleury B 2007b A SAGE algorithm for estimation of the direction power spectrum of individual path components *Proceedings of Global Telecommunications Conference, 2007. GLOBECOM '07. IEEE*, pp. 3024–3028.

Yin X, Pedersen T, Czink N and Fleury B 2006a Parametric characterization and estimation of bi-azimuth dispersion path components *Proceedings of IEEE 7th Workshop on Signal Processing Advances in Wireless Communications. SPAWC'06.*, pp. 1–6, Rome, Italy.

Yin X, Pedersen T, Czink N and Fleury B 2006b Parametric characterization and estimation of bi-azimuth and delay dispersion of individual path components *Antennas and Propagation, 2006. EuCAP 2006. First European Conference on*, pp. 1–8.

Yin X, Pedersen T, Steinbock G, Kirkelund G, Blattnig P, Jaquier A and Fleury B 2008a Tracking of the multi-dimensional parameters of a target signal using particle filtering *Radar Conference, 2008. RADAR '08. IEEE*, pp. 1–6.

Yin X, Steinbock G, Kirkelund G, Pedersen T, Blattnig P, Jaquier A and Fleury B 2008b Tracking of time-variant radio propagation paths using particle filtering *Communications, 2008. ICC '08. IEEE International Conference on*, pp. 920–924.

Yin X, Zhou Y and Liu F 2009 A generic wideband channel model for keyhole propagation scenarios and experimental evaluation *Communications and Networking in China, 2009. ChinaCOM 2009. Fourth International Conference on*, pp. 1–5.

4

Geometry-based Stochastic Channel Modeling

Geometry-based stochastic modeling approach is one of the most important and popular modeling approaches due to its flexibility: it can be either very simple and thus useful for theoretical investigation of channels, or relatively complicated and useful for simulating real channels. No matter what the aim is – simple theoretical investigations of channels or complete reproduction of real channels – the approach deals with scatterers and thus can capture the essential of the channels. Chapter 4 will introduce the geometry-based stochastic modeling approach in more detail and explain how it is used to model and study real channels.

4.1 General Modeling Procedure

As described in Chapter 2, the geometry-based stochastic modeling approach belongs to the so-called "scattering modeling" category, which also includes geometry-based deterministic modeling. Compared with its deterministic equivalent, which needs a detailed description of the real communication environment, geometry-based stochastic modeling approach is more simple and general and thus has been widely used. The general modeling procedure involved is summarized in Figure 4.1, and consists of the following steps.

1. Basic setting of communication environment: this includes the position and/or moving direction and velocity of the Tx/Rx, as well as the classification of effective scatterers, into, for example, moving scatterers and static scatterers.
2. Scatterer placement: place scatterers in the predefined scattering region based on a PDF. As mentioned in Chapter 2, the approach can be categorized as regular- or irregular-shaped depending on the distribution of scatterers in this region – regular (e.g. one/two-ring, ellipse, and so on) or irregular (random).
3. Parameterization: in this step, there are two ways to parameterize scatterers. The first considers a finite number of scatterers and assigns fading properties to each based on measurement data. The second assumes an infinite number of scatterers and thus the channel characteristics can be determined only by their PDF, and therefore without

Propagation Channel Characterization, Parameter Estimation and Modelling for Wireless Communications, First Edition.
Xuefeng Yin and Xiang Cheng.

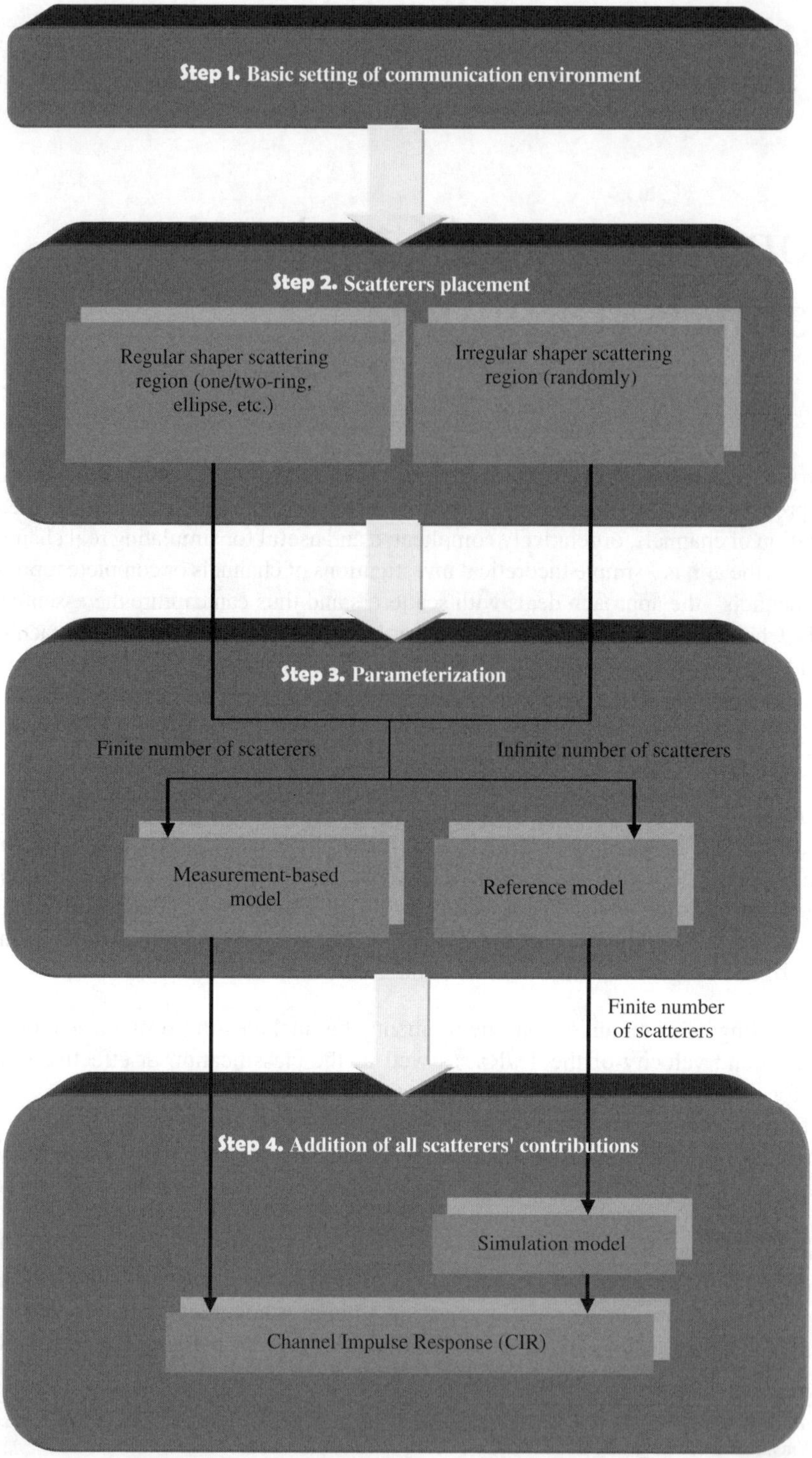

Figure 4.1 A general modeling procedure for the geometry-based stochastic modeling approach

assigning fading properties to each. In this case, the channel model obtained cannot be implemented in practice and is referred to as the reference model. This is useful for the theoretical analyses of the channel characteristics.

4. Addition of all scatterers' contributions: sum at the received side of all these scatterers' contributions to obtain the channel impulse response. Note that since the reference model has an infinite number of scatterers, the corresponding simulation model, which has a finite number of scatterers and thus is realizable in practice, should be obtained first in this step.

4.2 Regular-shaped Geometry-based Stochastic Models

As described in Chapter 2, a channel model obtained via geometry-based stochastic modeling is termed a GBSM. As explained in Section 4.1, a GBSM is derived from a predefined stochastic distribution of effective scatterers by applying the fundamental laws of wave propagation. Such models can be easily adapted to different scenarios by changing the shape of the scattering region and/or the PDF of the location of the scatterers. From Section 4.1, we know that GBSMs can be further classified as regular-shaped GBSMs (RS-GBSMs) or irregular-shaped GBSMs (IS-GBSMs) depending on whether the effective scatterers are placed on regular shapes (for example one/two-ring, ellipse, and so on) or irregularly (randomly). In this section, RS-GBSMs will be described in more detail.

In general, RS-GBSMs are used for the theoretical analysis of channel statistics and theoretical design and comparison of communication systems. This requires a good deal of mathematical tractability, and this is why RS-GBSMs assume that all effective scatterers are located on a regular shape. In the following, we will first introduce a very well-known RS-GBSM for conventional cellular systems and then described new RS-GBSMs that have been developed for vehicle-to-vehicle (V2V) communication systems.

4.2.1 RS-GBSMs for Conventional Cellular Communication Systems

The one-ring narrowband MIMO fixed-to-mobile (F2M) RS-GBSM is very common for use in conventional cellular macrocell scenarios. It was first proposed by Chen et al. [2000] and further developed by Abdi and Kaveh [2002]. The one-ring model has been widely used for narrowband MIMO F2M channels in macrocell scenarios (see Figure 4.2), due to its close agreement with the measured data [Abdi et al. 2002] and its mathematical tractability.

Let us consider a one-ring narrowband MIMO RS-GBSM in the scenario shown in Figure 4.3, where the effective scatterers are located on a ring surrounding the mobile station (MS) with radius R. Here the effective scatterers – a terminology first proposed in Lee's model [Liberti and Rappaport 1999] – are used to represent the effect of many scatterers with similar spatial locations. The BS and MS have M_T and M_R omnidirectional antenna elements in the horizontal plane, respectively. Without loss of generality, we consider uniform linear antenna arrays with $M_T = M_R = 2$ (a 2×2 MIMO channel). The antenna element spacings at the BS and MS are designated by δ_T and δ_R, respectively. It is usually assumed that the radius R is much smaller than D, the distance between the BS and MS. Furthermore, it is assumed that both R and D are much larger than the antenna element spacings δ_T and δ_R; that is, $D \gg R \gg \max\{\delta_T, \delta_R\}$. The multi-element antenna tilt angles are denoted by β_T and β_R. The MS moves with a speed in the direction determined by the angle of

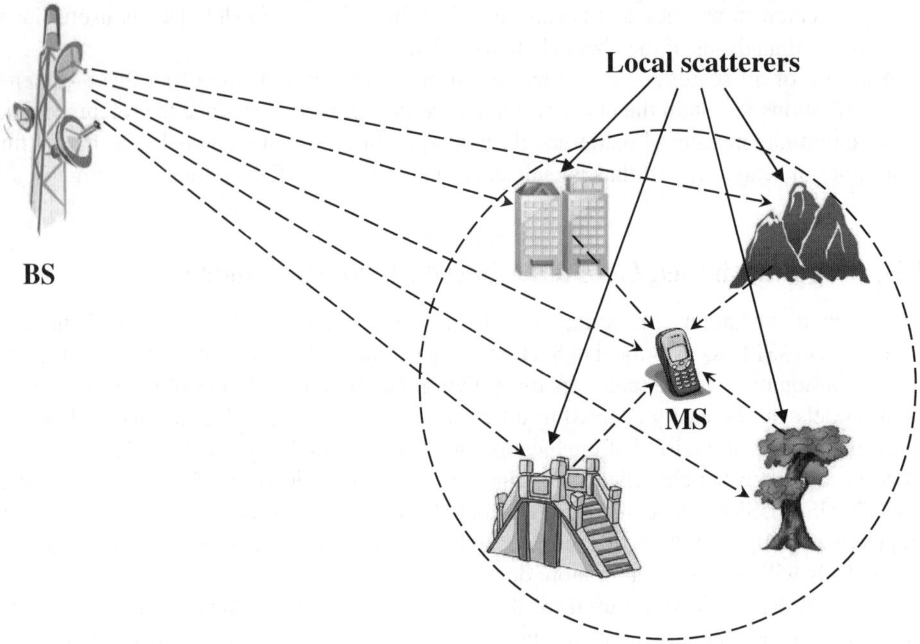

Figure 4.2 A typical F2M cellular propagation environment for macro-cell scenarios

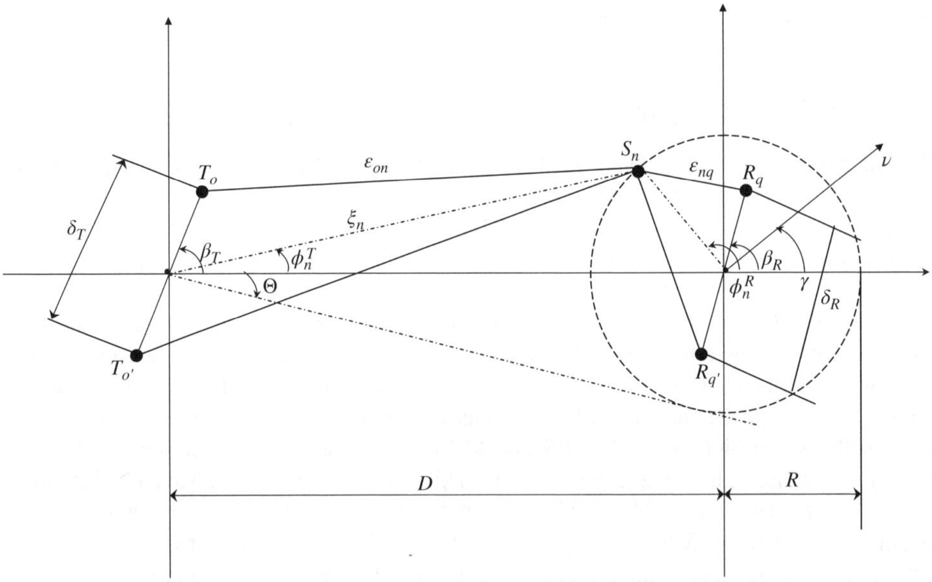

Figure 4.3 Geometrical configuration of a narrowband one-ring channel model with local scatterers around the mobile user

motion γ. The angle spread seen at the BS is denoted by Θ, which is related to R and D by $\Theta \approx \arctan\,(R/D) \approx R/D$.

Based on knowledge introduced in Chapter 2, the MIMO fading channel can be described by a matrix $\mathbf{H}(t) = [h_{oq}(t)]_{M_R \times M_T}$ of size $M_R \times M_T$. Without a LoS component, the subchannel complex fading envelope between the oth ($o = 1,...,M_T$) BS and the qth ($q = 1,...,M_R$) MS at the carrier frequency f_c can be expressed as

$$h_{oq}(t) = \lim_{N\to\infty} \frac{1}{\sqrt{N}} \sum_{n=1}^{N} \exp\ \left\{ j[\psi_n - 2\pi f_c \tau_{oq,n} + 2\pi f_D t \cos(\phi_n^R - \gamma)] \right\} \tag{4.1}$$

with $\tau_{oq,n} = (\varepsilon_{on} + \varepsilon_{nq})/c$, where $\tau_{oq,n}$ is the travel time of the wave through the link $T_o - S_n - R_q$ scattered by the nth scatterer S_n and c is the speed of light. The AoA of the wave travelling from the nth scatterer towards the MS is denoted by ϕ_n^R, while ε_{on} and ε_{nq} can be expressed as a function of ϕ_n^R as

$$\varepsilon_{on} \approx \xi_n - \delta_T \left[\cos(\beta_T) + \Theta \sin(\beta_T) \sin(\phi_n^R)\right]/2 \tag{4.2a}$$

$$\varepsilon_{no} \approx R - \delta_R \cos\left(\phi_n^R - \beta_R\right)/2 \tag{4.2b}$$

where $\xi_n \approx D + R\cos(\phi_n^R)$. The phases ψ_n are independent and identically distributed (i.i.d.) random variables with uniform distributions over $[0, 2\pi)$, f_D is the maximum Doppler frequency, and N is the number of independent effective scatterers S_n around the MS.

Since we assume that the number of effective scatterers in one effective cluster in this reference model tends to infinity, as shown in Eq. (4.1), the discrete AoA ϕ_n^R can be replaced by the continuous expressions ϕ^R. In the literature, many different scatterer distributions have been proposed to characterize the AoA ϕ^R. Examples of these PDFs include:

- uniform [Salz and Winters 1994]
- Gaussian [Adachi et al. 1986]
- wrapped Gaussian [Schumacher et al. 2002]
- cardioid [Byers and Takawira 2004].

In this chapter, the von Mises PDF [Abdi et al. 2002] is used. This can approximate all the PDFs in the list above. The von Mises PDF is defined as

$$f(\phi) \triangleq \exp\left[k\cos(\phi - \mu)\right]/2\pi I_0(k) \tag{4.3}$$

where $\phi \in [-\pi, \pi)$, $I_0(\cdot)$ is the zeroth-order modified Bessel function of the first kind, $\mu \in [-\pi, \pi)$ accounts for the mean value of the angle ϕ, and k ($k \geq 0$) is a real-valued parameter that controls the angle spread of the angle ϕ. For $k = 0$ (isotropic scattering), the von Mises PDF reduces to the uniform distribution, while for $k > 0$ (non-isotropic scattering), the von Mises PDF approximates different distributions based on the values of k [Abdi and Kaveh 2002].

The one-ring model just described is a narrowband MIMO F2M cellular channel model. To meet the enormous demand for high-speed communications, wideband MIMO cellular systems have been suggested in many communication standards, leading to an increasing requirement for wideband MIMO F2M channel models. However, the one-ring model, with its assumption that effective scatterers are located on a single ring, is overly simplistic and thus unrealistic for modeling wideband channels [Latinovic et al. 2003]. How to properly extend the narrowband one-ring model to wideband applications is still an open problem. In Chapter 9, we will address this open problem and give one possible solution.

4.2.2 RS-GBSMs for V2V Communication Systems

Unlike the rich history of RS-GBSMs for cellular systems, the development of RS-GBSMs for V2V communication systems is still in its infancy. Akki and Haber were the first to propose a 2D two-ring RS-GBSM with with the consideration of only double-bounced rays for narrowband isotropic scattering SISO V2V Rayleigh fading channels in macrocell scenarios [Akki 1994, Akki and Haber 1986]. Pätzold et al. [2008] presented a two-ring RS-GBSM considering only double-bounce rays for non-isotropic scattering MIMO V2V Rayleigh fading channels in macrocell scenarios. Zajic and Stuber [2008a] proposed a general 2D two-ring RS-GBSM with both single- and double-bounced rays for non-isotropic scattering MIMO V2V Ricean channels in both macrocell and microcell scenarios. The 2D two-ring narrowband model in Zajic and Stuber 2008a was further extended to a 3D two-cylinder narrowband model by the same team [Zajic and Stuber 2008b] and to a 3D two-concentric-cylinder wideband model [Zajic and Stuber 2008a, 2009].

Based on the real V2V environment that was shown in Figure 2.16, Figure 4.4 shows a geometrical description of the 3D two-concentric-cylinder wideband model, which has LoS, single-, and double-bounced rays. To matching real V2V environments, Zajic and Stuber [2008a, 2009] divided the complex impulse response into three parts:

- the LoS component
- the single-bounced rays generated from the effective scatterers located on either of the two cylinders
- the double-bounced rays produced from the effective scatterers located on both cylinders

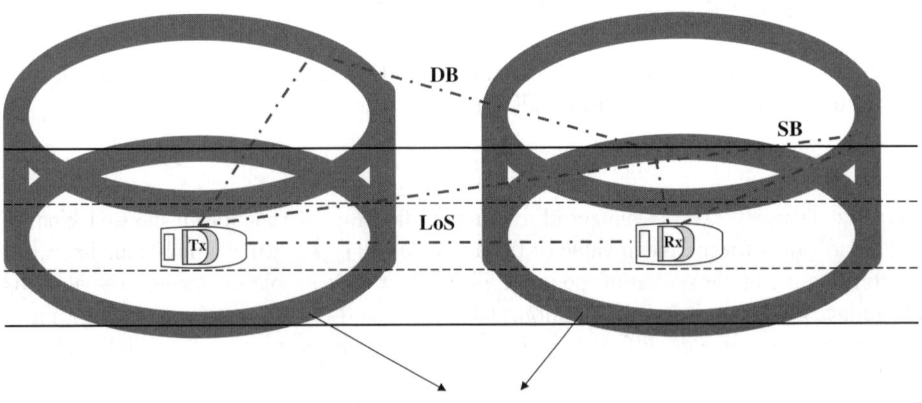

Figure 4.4 Geometrical description for an RS-GBSM of the typical V2V environment of Figure 2.16: SB, single-bounced; DB, double-bounced

This is illustrated in Figure 4.4, and can be expressed mathematically as

$$h_{pq}(t,\tau) = h_{pq}^{LoS}(t,\tau) + h_{pq}^{LoS}(t,\tau) + h_{pq}^{SBR}(t,\tau) + h_{pq}^{DB}(t,\tau) \tag{4.4}$$

As described in Chapter 2, there are, in total, eight system functions. The choice of which system function to use for investigating a channel is mainly based on purpose of the analysis: the selected system function should make the analysis easier. Accordingly, the authors in Zajic and Stuber [2008a, 2009] used the time-variant transfer function instead of the channel impulse response. As addressed in Chapter 2, the time-variant transfer function is the Fourier transform of the channel impulse response and can be written as

$$T_{pq}(t,t) = \mathcal{F}\{h_{pq}(t,\tau)\} = T_{pq}^{LoS}(t,f) + T_{pq}^{LoS}(t,f) + T_{pq}^{SBR}(t,f) + T_{pq}^{DB}(t,f) \tag{4.5}$$

However, all of the RS-GBSMs discussed here cannot be used to study the impact of vehicular traffic density (VTD) on channel statistics or to conduct investigations of per-tap channel statistics in wideband cases. Furthermore, an RS-GBSM cannot be used to study non-stationarity because of the static nature of the geometry in RS-GBSMs.

4.3 Irregular-shaped Geometry-based Stochastic Models

Unlike RS-GBSMs, IS-GBSMs are used with the aim of reproducing physical reality and thus need to incorporate modifications to the location and properties of the effective scatterers of RS-GBSMs. IS-GBSMs place effective scatterers with specified properties at random locations, but in specified statistical distributions. The signal contributions of the effective scatterers are determined via a greatly-simplified ray-tracing method and the components signals are summed to obtain the complex impulse response. In order to provide better agreement with the measurement results presented by Paier et al. [2009], Karedal et al. [2009] further divided the impulse response into four parts:

- the LoS component, which may contain more than just the true LOS signal (e.g. ground reflections)
- discrete components from reflections of mobile scatterers (e.g. moving cars)
- discrete components from reflections of significant (strong) static scatterers (e.g. building and road signs located on the roadside)
- diffuse components from reflections of weak static scatterers located on the roadside.

Their scheme is depicted in Figure 4.5. Therefore, IS-GBSMs are actually greatly simplified versions of the GBDMs introduced in Chapter 2, and can be made suitable for a wide variety of V2V scenarios by properly adjusting the statistical distributions of the locations of the effective scatterers. With the ray-tracing approach, the IS-GBSM of Karedal et al. [2009] can easily handle the non-stationarity of V2V channels by prescribing the motion of the Tx, Rx, and mobile scatterers. Therefore, the complex channel impulse response can be expressed as [Karedal et al. 2009]:

$$h(t,\tau) = h_{LoS}(t,\tau) + \sum_{p=1}^{P} h_{MD}(t,\tau_p) + \sum_{q=1}^{Q} h_{SD}(t,\tau_q) + \sum_{r=1}^{R} h_{DI}(t,\tau_r) \tag{4.6}$$

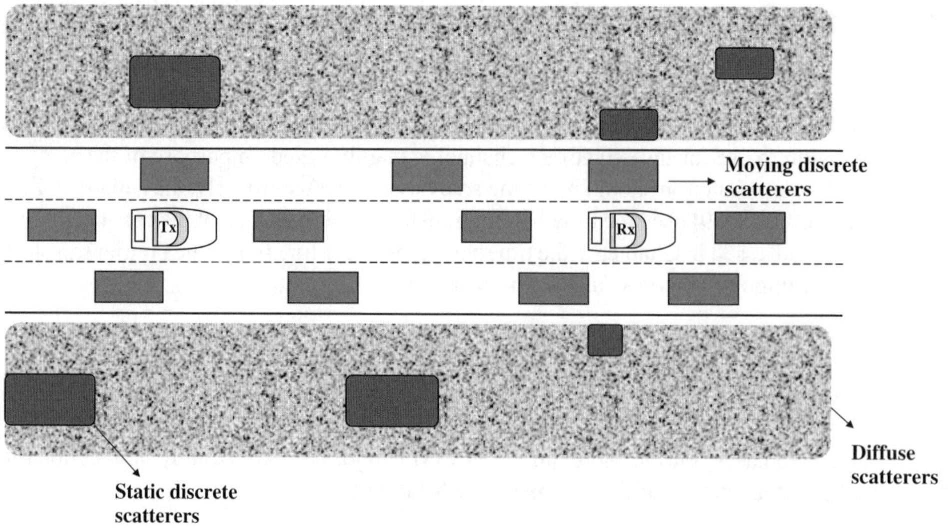

Figure 4.5 Geometrical description for an IS-GBSM of the typical V2V environment of Figure 2.16

where $h_{LoS}(t, \tau)$ is the LoS component, $\sum_{p=1}^{P} h_{MD}(t, \tau_p)$ are the discrete components stemming from reflections off mobile scatterers (MD), with P being the number of mobile discrete scatterers, $\sum_{q=1}^{Q} h_{SD}(t, \tau_q)$ are the discrete components stemming from reflections off static scatterers (SD), with Q being the number of mobile static scatterers, and $\sum_{r=1}^{R} h_{DI}(t, \tau_r)$ are the diffuse components (DI) with R being the number of diffuse scatterers. Note that only single-bounced rays are considered in this IS-GBSM due to the fairly low VTD of the measurements in Paier et al. [2009]. For a high-VTD environment, it is possible that double-bounced rays should be considered as well.

It is worth noting that compared with the NGSM introduced in Chapter 2 [Sen and Matolak 2008], the IS-GBSM of Karedal et al. [2009] can easily handle the drift of scatterers into different delay bins but with relatively higher complexity. Finally, the newly important V2V channel models that have been described in Chapters 2 and 4 are summarized and classified in Table 4.1.

4.4 Simulation Models

As described in Section 4.1, by assuming an infinite number of scatterers, GBSMs can be easily obtained but cannot be implemented in practice due to the resulting infinite complexity. In this case, these GBSMs are called "reference models". Therefore, corresponding simulation models, which have a finite complexity and thereby are realizable in practice, are necessary in practical simulations and performance evaluations of wireless communication systems. Note that the RS-GBSMs introduced in Section 4.2 are actually reference models since they assume an infinite number of effective scatterers, as shown in Eqs (4.1) and (4.4), where the number of effective scatterers N tends to infinity. As mentioned by Stüber [2001], a reference model can be used for theoretical analysis and design of a wireless communication system,

Table 4.1 Important V2V channel models

Channel model	Antenna and FS	Stationarity	Impact of VTD	Per-tap CS	Scatterer region/ distribution	Scattering assumptions	Applicable scenarios
Maurer et al. 2008 GBDM	MIMO wideband	non-stationary	yes	no	3D non-isotropic (deterministic)	SB+MB	Site-specific
Acosta-Marum and Ingram 2007 NGSM	SISO wideband	stationary	no	yes	2D non-isotropic (N/A)	N/A	Micro, pico
Sen and Matolak 2008 NGSM	SISO wideband	non-stationary	yes	yes	2D non-isotropic (N/A)	N/A	Micro, pico
Akki and Haber 1986 RS-GBSM	SISO narrowband	stationary	no	no	2D isotropic (two-ring)	DB	Macro
Pätzold et al. 2008 RS-GBSM	MIMO narrowband	stationary	no	no	2D non-isotropic (two-ring)	DB	Macro, micro
Zajic and Stuber 2008a RS-GBSM	MIMO narrowband	stationary	no	no	2D non-isotropic (two-ring)	SB+DB	Macro, micro
Zajic and Stuber 2008b RS-GBSM	MIMO narrowband	stationary	no	no	3D non-isotropic (two-cylinder)	SB+DB	Macro, micro
Zajic and Stuber 2008a, 2009 RS-GBSM	MIMO wideband	stationary	no	no	3D non-isotropic (two concentric-cylinder)	SB+DB	Macro, micro
Karedal et al. 2009 IS-GBSM	MIMO wideband	non-stationary	yes	no	2D non-isotropic (randomly)	SB	Micro, pico

FS, frequency-selectivity; CS, channel statistics; SB, single-bounced; MB, multiple-bounced; DB, double-bounced; Macro, macrocell; Micro, microcell; Pico, picocell; N/A, not-applicable.

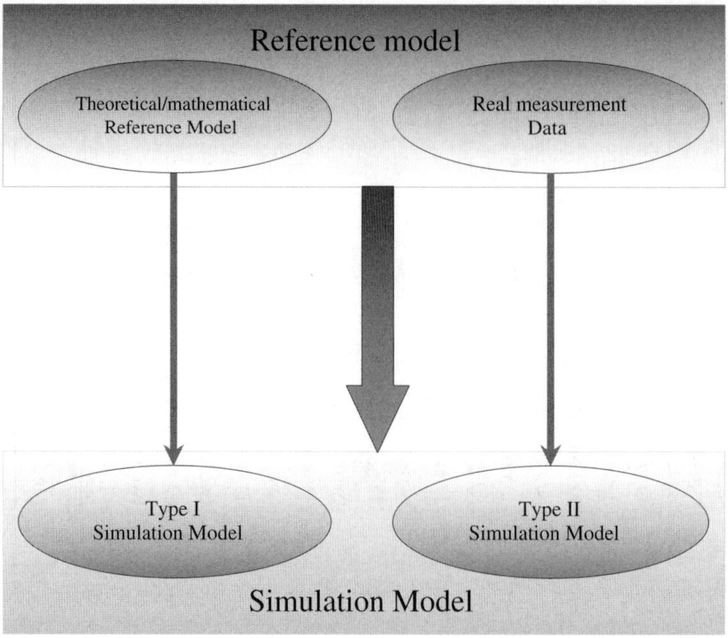

Figure 4.6 Relationship between the reference model and simulation model

and is also a starting point for the design of a realizable simulation model of reasonable complexity – one with a finite number of effective scatterers. Therefore, the development of a simulation model aims for reasonable complexity while representing the desired statistical properties of the reference model as faithfully as possible.

Before introducing different simulation models, it is worth emphasizing that the GBDMs and NGSMs described in Chapter 2 and the IS-GBSMs introduced in Section 4.3 can be categorized as the other type of simulation model, since they have finite complexity; the number of scatterers is finite, and thus the model can be directly implemented in practice. In this sense, the reference model is the real measurement data. Figure 4.6 shows the relationship between the reference and simulation models. This book concentrates on the Type I simulation model. Therefore, for simplicity, the mention of a reference model and a simulation model refers to a theoretical/mathematical reference model and a Type I simulation model, respectively.

There are several different methods for simulating fading channels. The most accepted approaches are filter methods [Fechtel and Feng 1993, Verdin and Tozer 1993, Wang and Cox 2002, Young and Beaulieu 1998, 2000] and SoS methods [Clarke 1968, Jakes 1994, Patel et al. 2005b, Pätzold 2002, Pop and Beaulieu 2001, Wang et al. 2008, Wang and Zoubir 2007, Zajic and Stuber 2006, Zheng and Xiao 2002, 2003].

4.4.1 Filter Simulation Models

The filter method has been widely used and is shown in Figure 4.7. Two uncorrelated Gaussian random processes with zero mean and the same variances are passed through identical lowpass filters to limit and reshape the spectrum. A Gaussian random process is completely

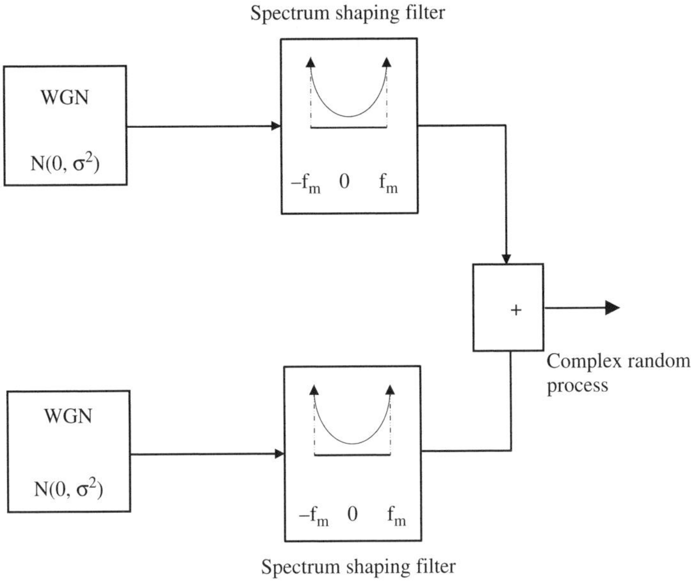

Figure 4.7 Filter method of channel waveform generation

characterized by its mean value and color, which can be described either by the power spectral density (PSD) or, alternatively, by the autocorrelation function [Caver 2002, Hanzo et al. 2000, Pätzold 2002, Stüber 2001]. The purpose of filtering is to give the simulated spectrum the maximum possible resemblance to the theoretical one [Clarke 1968]. These filters only limit and shape the spectrum but do not alter the PDF of the random process. However, they reduce the variances of the processes. The filter transfer function is the square root of the desired spectrum shape. The theoretical spectrum is U-shaped, with limits equal to the maximum positive and negative Doppler shifts, which are equal to ν/λ, where ν is the velocity of the mobile unit and λ is the wavelength of the signal. It is practically not possible to make the spectrum the same shape as the theoretical one, but it can be approximated by controlling the characteristics of the shaping filter. The cut-off frequencies of the filters are equal to the maximum Doppler shift. They can be implemented in the analog as well as in the digital domain. Digital implementation is flexible and gives a very close approximation to the desired U-shaped spectrum. In practical digital implementations, the filters operate at a lower sampling frequency. In order to bring the sampling rate up to the value required by the signal representation, the spectrum shaping filters are followed by interpolators. The interpolation involves upsampling and lowpass filtering. It is efficient to perform interpolation in stages. Therefore, an overall interpolation factor is usually split into a number of cascaded stages to avoid large numbers of filter coefficients.

Filter methods are designed to filter Gaussian noise through appropriate filters to generate a channel waveform with the desired Doppler PSD: U-shaped, flat, and so on. The main limitation of this approach is that only waveforms having rational Doppler PSD can be produced exactly. However, as mentioned by Stüber [2001], it is non-rational Doppler PSDs that are normally encountered in practice. In order to approximate a waveform with a non-rational

Doppler PSD, the models have to include high-order filters, which leads to this approach being significantly complicated and time-consuming. Moreover, the Doppler PSD obtained is not band-limited, because it is difficult to implement the filters with sharp stop-bands in practice. Although the filter method has these limitations, it has been widely accepted as the starting point of the investigation of simulation models.

4.4.2 *Sum-of-sinusoids Simulation Models*

By considering that received signals are the sum, at the receiver side, of all scatterers' contributions, SoS methods generate the channel waveform by superimposing a finite number of appropriate sinusoids, as shown in Figure 4.8. The channel waveform can be expressed as

$$h(t) = \sum_{n=1}^{N} c_n \cos(2\pi f_n t + \theta_n) \tag{4.7}$$

where c_n, f_n, and θ_n are the gains, frequencies, and phases of a SoS simulation model. In contrast to filter simulation models, SoS simulation models have low complexity and produce channel waveforms with high accuracy and a perfectly band-limited Doppler PSD. Furthermore, it is easy to extend the SoS models to develop space-time (ST) correlated simulators for MIMO systems. Therefore, SoS methods are the main simulation method used throughout this book.

An SoS simulation model can be either deterministic or stochastic in terms of the underlying parameters (gains, frequencies, and phases) [Wang et al. 2008]. For a deterministic model, all the parameters are fixed for all simulation trials. In contrast, a stochastic model has at least one random parameter (gains, frequencies, or phases), and this varies for each simulation trial. Therefore, the relevant statistical properties of a stochastic model vary for each simulation trial but converge to the desired ones when averaged over a sufficient number of trials. It is worth noting that a stochastic model with only phases as random variables is actually an ergodic

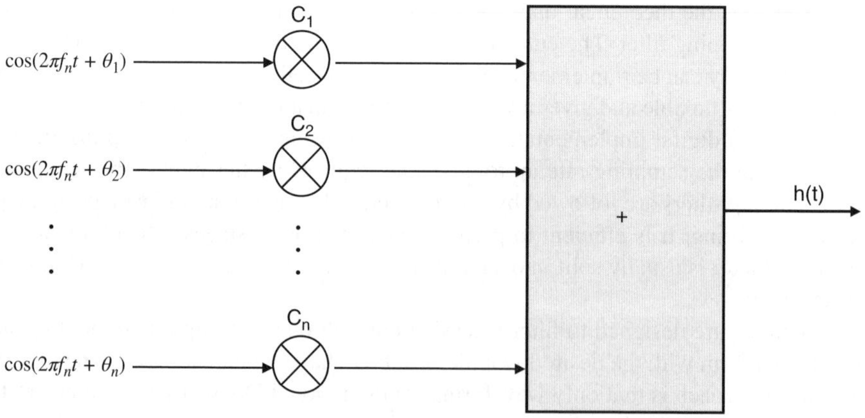

Figure 4.8 SoS method of channel waveform generation

process. Due to the ergodicity, such a stochastic simulation model needs only one simulation trial to converge to the desired statistical properties.

4.4.2.1 SoS Simulation Models for Cellular Channels

Many approaches have been suggested for SoS simulation of SISO F2M Rayleigh fading channels. Jakes [1994] was among the first to propose a deterministic simulation model for SISO F2M Rayleigh fading channels. However, his model does not satisfy most of the statistical properties of Rayleigh fading channels [Pätzold and Laue 1998] and it is not wide-sense stationary [Pop and Beaulieu 2001]. Therefore, various modifications of Jakes' model have been proposed in the literature [Pätzold 2002, Pop and Beaulieu 2001, Wang et al. 2008]. Pop and Beaulieu [2001] proposed a new deterministic simulation model to solve the non-stationarity of Jakes' model, but it could not satisfy most of the other statistical properties of Rayleigh fading channels. Pätzold and his co-workers developed several deterministic simulation models with different parameter computation methods [Pätzold et al. 1994, 1996a, b, 1998]. The method of equal area (MEA) [Pätzold et al. 1994, 1996a] is characterized by the fact that the area under the Doppler PSD between two neighboring discrete frequencies are equiareal. This parameter computation method presents acceptable performance with low complexity. A quasi-optimal procedure is the method of exact Doppler spread (MEDS) [Pätzold et al. 1996b, 1998]. This method outperforms the MEA and even exhibits very similar performance to the optimization method (i.e. the L_p-norm method) [Pätzold et al. 1996b, 1998]. Compared to the MEA and MEDS, the numerical complexity of the optimization method is comparatively high, so that simulation of such isotropic scattering Rayleigh fading channels is often not worth the effort [Pätzold 2002]. More recently, the drawback of the MEDS in generating multiple uncorrelated Rayleigh fading waveforms has been resolved by Wang and his co-workers [Wang et al. 2008, 2009].

In order to satisfy more statistical properties and/or match the desired properties over longer time delays, Zheng and Xiao have proposed several new stochastic simulation models [Xiao et al. 2006, Zheng and Xiao 2002, 2003]. By allowing all three parameters (gains, frequencies, and phases) to be random variables, Zheng and Xiao's model gets statistical properties similar to those of Rayleigh fading channels. Since the models are no longer ergodic processes, their statistical properties vary for each simulation trial, but they converge to the desired properties over a sufficient number of simulation trials (normally 50 to 100). A detailed comparison of the statistical properties of Zheng and Xiao's models is presented in Patel et al. 2005a.

However, all these deterministic and stochastic simulation models are limited to isotropic scattering SISO F2M Rayleigh fading channels; simulation models for MIMO F2M channels under a more realistic scenario of non-isotropic scattering are scarce in the current literature. However, Pätzold and Hogstad [2004] have proposed a narrowband MIMO F2M deterministic simulation for macrocell scenarios. Hogstad et al. [2005] developed new deterministic simulation models for both narrowband and wideband MIMO F2M channels of microcell scenarios. Up to now, only one wideband MIMO F2M deterministic simulation model has been proposed in for macrocell scenarios [Pätzold and Hogstad 2006]. However, since the reference model on which it was based has several drawbacks, their simulation cannot capture the statistical properties of real wideband MIMO F2M channels for macrocell scenarios. This problem will be addressed in Chapter 9.

4.4.2.2 SoS Simulation Models for V2V channels

Several methods for the simulation of V2V channels have been proposed. Patel et al. [2005b] were the first to propose new SoS simulation models for SISO V2V Rayleigh fading channels. They first modified the MEDS model proposed for F2M channels by Pätzold et al. [1996b, 1998] and put forward a new deterministic simulation model. However, this matches the desired statistical properties of the reference model only for a small range of normalised time delays. Therefore, the authors also modified the stochastic model developed for F2M channels by Zheng and Xiao [2002]. This second stochastic model can match the desired statistical properties over a large range of normalised time delays, but at the expense of simulation complexity (it needs 50 simulation trials). The stochastic model in Patel et al. [2005b] was further improved by Zajic and Stuber [2006], by choosing orthogonal functions for the in-phase and quadrature components of the complex fading envelope. More recently, Wang et al. [2009] extended the Rayleigh V2V stochastic simulation model in Patel et al. [2005b] to include a LoS component; that is, for Ricean fading channels.

However, it is worth noting that all these simulation models are limited to application in isotropic scattering environments. So far, only one stochastic simulation model [Zheng 2006] has been proposed for the simulation of non-isotropic scattering V2V Rayleigh fading channels. However, this model has notable difficulty in reproducing the desired statistical properties of the reference model and a comparatively high computational complexity, a problem which is discussed in detail later in this section. Furthermore, accurate deterministic simulation models for non-isotropic scattering V2V Rayleigh fading channels are not available in the current literature.

As for the simulation of MIMO V2V channels, Pätzold et al. [2005] proposed a new deterministic SoS simulation model under the condition of isotropic scattering. This model was further improved and extended to include a LoS component by Zajic and Stuber [2008a]. To simulate MIMO V2V channels under a more realistic scenario of non-isotropic scattering, Pätzold et al. [2008] modified the narrowband deterministic simulation model in their earlier paper [Pätzold et. al. 2005] for isotropic scattering environments and proposed a new parameter computation method. This method is termed modified MEA (MMEA) since it is originated from MEA [Pätzold et al. 1994, 1996a] for isotropic scattering F2M channels. Zajic and Stuber [2008b] proposed new deterministic and stochastic simulation models for non-isotropic scattering narrowband MIMO V2V channels. More recently, Zajic and Stuber [2009] developed a new wideband deterministic simulation model for MIMO V2V channels under non-isotropic scattering conditions. However, all the previously reported non-isotropic scattering MIMO V2V deterministic simulation models have weaknesses in reproducing properly the statistical properties of the reference model. Moreover, stochastic simulation models for non-isotropic scattering wideband MIMO V2V channels are surprisingly unavailable in the current literature. Open issues about MIMO V2V simulation models will be addressed in Chapter 9.

4.5 Simulation Models for Non-isotropic Scattering Narrowband SISO V2V Rayleigh Fading Channels

As described above, most available simulation models for narrowband SISO V2V channels are limited to application in isotropic scattering environments. Simulation models for V2V Rayleigh fading channels under the more realistic scenario of non-isotropic scattering are scarce in the current literature. So far, only one stochastic SoS simulation model has been proposed for the simulation of non-isotropic scattering V2V Rayleigh fading channels [Zheng 2006]. However, this model only considered the symmetrical property of the distributions of the AoA and AoD and was based on the acceptance rejection algorithm (ARA). This leads to it having a difficulty in reproducing the desired statistical properties of the reference model and a comparatively high computational complexity. Moreover, no deterministic simulation models for non-isotropic scattering V2V Rayleigh fading channels are available in the current literature.

To fill the above gap, by taking the traditional two-ring V2V RS-GBSM [Patel et al. 2005b] as a reference model, a new V2V deterministic SoS-based simulation model [Cheng et al. 2011] is now introduced. By allowing at least one parameter (frequencies and/or gains) to be a random variable, this deterministic model can be further modified to be a stochastic model. It is worth noting that the simulation models introduced incorporate the PDFs of the AoA and AoD, and thus can approximate the desired statistical properties of the reference model for any non-isotropic scattering V2V Rayleigh fading channel. Moreover, compared with the ARA stochastic model of Zheng [2006], the stochastic model introduced gives a better approximation of the properties of the reference model with an even smaller number of harmonic functions.

4.5.1 A Two-ring SISO V2V Reference Model

Figure 4.9 shows the geometry of a two-ring SISO V2V RS-GBSM [Patel et al. 2005b]. There are two rings of effective scatterers: one around the Tx and the other around the Rx. Based on this model and taking into account the direction of movement of the Tx and Rx, we can

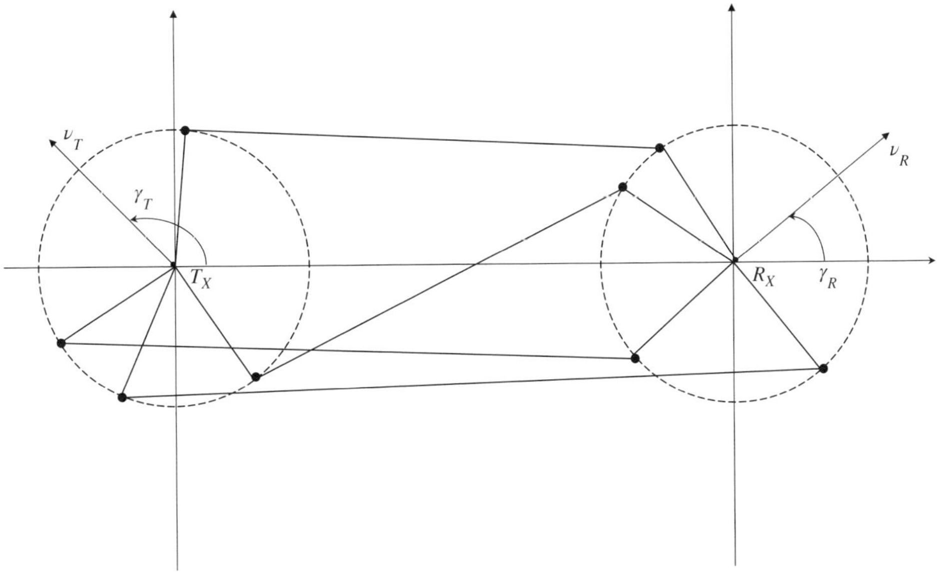

Figure 4.9 Geometrical two-ring model for SISO V2V channels

express the complex fading envelope of our reference model under a narrowband non-isotropic scattering V2V Rayleigh fading assumption as

$$\begin{aligned} h(t) &= h_i(t) + jh_q(t) \\ &= \lim_{N,M\to\infty} \frac{1}{\sqrt{NM}} \sum_{n,m=1}^{N,M} e^{j\psi_{nm}} e^{j\left[2\pi f_{T_{\max}} t \cos(\phi_T^m - \gamma_T) + 2\pi f_{R_{\max}} t \cos(\phi_R^n - \gamma_R)\right]} \end{aligned} \tag{4.8}$$

where $h_i(t)$ and $h_q(t)$ are in-phase and quadrature components of the complex fading envelope $h(t)$, respectively, $j = \sqrt{-1}$, N and M are the number of harmonic functions representing effective scatterers (propagation paths) located on the ring around the Rx and Tx, respectively, and $f_{T_{\max}}$ and $f_{R_{\max}}$ are the maximum Doppler frequency due to the motion of the Rx and Tx, respectively. The Tx and Rx move in directions determined by the angles of motion γ_T and γ_R, respectively. The random AoA and AoD of the nth path are denoted by ϕ_R^n and ϕ_T^m, respectively, and ψ_{nm} is the random phase uniformly distributed on $[-\pi, \pi)$. It is assumed that ϕ_R^n, ϕ_T^m, and ψ_{nm} are mutually independent random variables.

Since the number N and M of effective scatterers in the reference model $h(t)$ tends to infinity, the discrete expressions of the AoA $\phi_R^{(n)}$ and AoD $\phi_T^{(m)}$ can be replaced by the continuous expressions ϕ_R and ϕ_T, respectively. Note that since $h(t)$ describes a non-isotropic scattering V2V Rayleigh fading channel, the AoA ϕ_R and AoD ϕ_T exhibit nonuniform distributions. To characterize the AoA and AoD, we use the von Mises PDF given in Eq. (10.7). Applying the von Mises PDF to the AoA ϕ_R and AoD ϕ_T, we obtain $f(\phi_R) \triangleq \exp[k_R \times \cos(\phi_R - \mu_R)]/[2\pi I_0(k_R)]$ and $f(\phi_T) \triangleq \exp[k_T \cos(\phi_T - \mu_T)]/[2\pi I_0(k_T)]$, respectively.

The autocorrelation function (ACF) of the reference model $h(t)$ can be expressed as

$$\rho_{hh}(\tau) = \mathbf{E}[h(t)h^*(t-\tau)] = \frac{1}{I_0(k_T)I_0(k_R)} I_0\left[(A_T^2 + B_T^2)^{1/2}\right] I_0\left[(A_R^2 + B_R^2)^{1/2}\right] \tag{4.9}$$

with

$$A_T = k_T \cos(\mu_T) + j2\pi\tau f_{T_{\max}} \cos(\gamma_T) \tag{4.10a}$$

$$B_T = k_T \sin(\mu_T) + j2\pi\tau f_{T_{\max}} \sin(\gamma_T) \tag{4.10b}$$

$$A_R = k_R \cos(\mu_R) + j2\pi\tau f_{R_{\max}} \cos(\gamma_R) \tag{4.10c}$$

$$B_R = k_R \sin(\mu_R) + j2\pi\tau f_{R_{\max}} \sin(\gamma_R) \tag{4.10d}$$

where τ is the time separation, $(\cdot)^*$ denotes the complex conjugate operation, and $\mathbf{E}[\cdot]$ is the statistical expectation operator.

4.5.2 SoS Simulation Models

Based on the reference model developed in the Section 4.5.1, we now introduce the corresponding deterministic and stochastic SoS simulation models.

4.5.2.1 Deterministic Simulation Model

Based on the two-ring reference model, we can now design a deterministic simulation model that needs only one simulation trial to obtain the desired statistical properties:

$$\tilde{h}(t) = \tilde{h}_i(t) + j\tilde{h}_q(t) \tag{4.11}$$

$$\tilde{h}_i(t) = \frac{1}{\sqrt{N_i M_i}} \sum_{n_i, m_i = 1}^{N_i, M_i} \cos\left[\tilde{\psi}_{n_i m_i} + 2\pi f_{T_{\max}} t \cos(\tilde{\phi}_T^{m_i} - \gamma_T) + 2\pi f_{R_{max}} t \cos(\tilde{\phi}_R^{n_i} - \gamma_R)\right] \tag{4.12}$$

$$\tilde{h}_q(t) = \frac{1}{\sqrt{N_q M_q}} \sum_{n_q, m_q = 1}^{N_q, M_q} \sin\left[\tilde{\psi}_{n_q m_q} + 2\pi f_{T_{max}} t \cos(\tilde{\phi}_T^{m_q} - \gamma_T) + 2\pi f_{R_{max}} t \cos(\tilde{\phi}_R^{n_q} - \gamma_R)\right] \tag{4.13}$$

where $\tilde{h}_i(t)$ and $\tilde{h}_q(t)$ are the in-phase and quadrature components of the complex fading envelope $\tilde{h}(t)$, respectively, $N_{i/q}$ is the number of effective scatterers located on the ring around the RX, $M_{i/q}$ is the number of effective scatterers located on the ring around the Tx, the AoA $\tilde{\phi}_R^{n_{i/q}}$ and the AoD $\tilde{\phi}_T^{m_{i/q}}$ are discrete realisations of the random variables ϕ_R and ϕ_T, respectively, and the phases $\tilde{\psi}_{n_{i/q} m_{i/q}}$ are the single outcomes of the random phases ψ_{nm} in Eq. (4.8). It is assumed that $\tilde{\phi}_R^{n_{i/q}}$, $\tilde{\phi}_T^{m_{i/q}}$, and $\tilde{\psi}_{n_{i/q} m_{i/q}}$ are mutually independent. Note that the AoA $\tilde{\phi}_R^{(n_{i/q})}$ and the AoD $\tilde{\phi}_T^{(m_{i/q})}$ remain constant for different simulation trials due to the deterministic nature of the proposed simulation model.

From the deterministic simulation model outlined above, it is obvious that the key issue in designing a V2V deterministic simulation model is to find the sets of AoAs $\{\tilde{\phi}_R^{n_{i/q}}\}_{n_{i/q}=1}^{N_{i/q}}$ and AoDs $\{\tilde{\phi}_T^{m_{i/q}}\}_{m_{i/q}=1}^{M_{i/q}}$ that make the simulation model reproduce the desired statistical properties of the reference model as faithfully as possible with reasonable complexity; that is, with a finite number of $N_{i/q}$ and $M_{i/q}$. Under the condition of a non-isotropic scattering environment, the PDFs of the AoA ϕ_R and AoD ϕ_T should be used to design the sets of AoAs $\{\tilde{\phi}_R^{n_{i/q}}\}_{n_{i/q}=1}^{N_{i/q}}$ and AoDs $\{\tilde{\phi}_T^{m_{i/q}}\}_{m_{i/q}=1}^{M_{i/q}}$ that guarantee the uniqueness of the sine and cosine functions related to the AoA $\phi_R^{(n)}$ and AoD $\phi_T^{(n)}$ in the reference model in Eq. (4.8). This means that the sets of AoAs and AoDs of the simulation model $\tilde{h}(t)$ in Eq. (4.11) should meet the following conditions:

- $\cos(\tilde{\phi}_R^{n_{i/q}} - \gamma_R) \neq \cos(\tilde{\phi}_R^{n'_{i/q}} - \gamma_R)$, $n_{i/q} \neq n'_{i/q}$
- $\cos(\tilde{\phi}_T^{m_{i/q}} - \gamma_T) \neq \cos(\tilde{\phi}_T^{m'_{i/q}} - \gamma_T)$, $m_{i/q} \neq m'_{i/q}$

Via extensive investigation of the PDFs of the AoA ϕ_R and AoD ϕ_T, we find that unlike isotropic scattering V2V environments [Patel et al. 2005b], it is difficult to obtain sets of AoAs and AoDs that meet these two conditions for all non-isotropic scattering V2V environments. Therefore, we divide the non-isotropic scattering V2V environment into three categories in terms of the mean AoA μ_R and mean AoD μ_T (i.e., the PDFs of the AoA and AoD), and the angles of motion γ_R and γ_T, and then design the sets of AoAs and AoDs separately for these three cases: 1)

- *Case I*: the main transmitted and received powers come from the same direction as or opposite direction to the movements of the Tx and Rx, respectively: $|\mu_T - \gamma_T| = |\mu_R - \gamma_R| = 0^\circ$ or π.
- *Case II*: the main transmitted and received powers come from directions that are perpendicular to those of the movements of the Tx and Rx, respectively: $|\mu_T - \gamma_T| = |\mu_R - \gamma_R| = 90^\circ$.
- *Case III*: different from *Cases I* and *II*.

For *Case I*, the in-phase component $h_i(t)$ and quadrature component $h_q(t)$ of the reference model $h(t)$ in Eq. (4.8) are correlated. Therefore, to meet this correlation, we set $N_i = N_q = N$ and $M_i = M_q = M$ and thereby the AoA $\tilde{\phi}_R^{n_{i/q}}$ and AoD $\tilde{\phi}_T^{m_{i/q}}$ can be replaced by the $\tilde{\phi}_R^n$ and $\tilde{\phi}_T^m$, respectively. Inspired by the modified MMEA of de Leon and Patzold [2007] for non-isotropic scattering F2M channels, we design the AoA and AOD of our model as

$$\frac{n-1/4}{N} = \int_{\tilde{\phi}_R^{n-1}}^{\tilde{\phi}_R^n} f\,(\tilde{\phi}_R^n) d\tilde{\phi}_R^n, \quad \tilde{\phi}_R^n \in [-\pi, \pi) \qquad n = 1, 2, \ldots, N \tag{4.14a}$$

$$\frac{m-1/4}{M} = \int_{\tilde{\phi}_T^{m-1}}^{\tilde{\phi}_T^m} f\,(\tilde{\phi}_T^m) d\tilde{\phi}_T^m, \quad \tilde{\phi}_T^m \in [-\pi, \pi) \qquad m = 1, 2, \ldots, M \tag{4.14b}$$

where $\tilde{\phi}_R^0 = -\pi$ and $\tilde{\phi}_T^0 = -\pi$, $f(\tilde{\phi}_R^n)$ and $f(\tilde{\phi}_T^m)$ denote the PDFs of the AoA $\tilde{\phi}_R^n$ and the AoD $\tilde{\phi}_T^m$, respectively. The cumulative distribution functions of the AoA $\tilde{\phi}_R$ and the AoD $\tilde{\phi}_T$ are defined as $F_R(x) = \int_{-\infty}^{x} f(\tilde{\phi}_R^n) d\tilde{\phi}_R^n$ and $F_T(x) = \int_{-\infty}^{x} f(\tilde{\phi}_T^n) d\tilde{\phi}_T^n$, respectively. If the inverse functions $F_R^{-1}(\cdot)$ of $F_R(\cdot)$ and $F_T^{-1}(\cdot)$ of $F_T(\cdot)$ exist, Eqs (4.14a) and (4.14b) become

$$\tilde{\phi}_R^n = F_R^{-1}\left(\frac{n-1/4}{N}\right) \tag{4.15a}$$

$$\tilde{\phi}_T^m = F_T^{-1}\left(\frac{m-1/4}{M}\right) \tag{4.15b}$$

Note that the value of $\frac{1}{4}$ used in Eqs (4.15a) and (4.15b) guarantees that the designed sets of AoAs and AoDs can meet the aforementioned two conditions ($\tilde{\phi}_R^n \neq -\tilde{\phi}_R^{(n')} + 2\gamma_R$, $n \neq n'$ and $\tilde{\phi}_T^m \neq -\tilde{\phi}_T^{m'} + 2\gamma_T$, $m \neq m'$) for *Case I*. The proof of the above statement is omitted here since it is easily obtained by following the procedure provided by de Leon and Patzold [2007]. The performance of this design will be validated in the Section 4.5.2.3 on Numerical Results and Analysis (see p. 98).

In *Case II*, the cross-correlation between the in-phase component $h_i(t)$ and quadrature component $h_q(t)$ of the complex fading envelope $h(t)$ in Eq. (4.8) is equal to zero. Therefore, by setting $N_i \neq N_q$ and $M_i \neq M_q$ we can directly use the expression of the simulation model itself to guarantee the cross-correlation is equal to zero, rather than through the design of the

AoAs and AoDs. This makes for a more efficient use of the number of harmonic functions and thus results in better performance of the model. Following a similar parameter computation method to that in *Case I*, we design the AoA and AoD of this model as

$$\tilde{\phi}_R^{n_{i/q}} = F_R^{-1}\left(\frac{n_{i/q} - 1/2}{N_{i/q}}\right), \quad \tilde{\phi}_R^{n_{i/q}} \in [-\pi, \pi) \quad n_{i/q} = 1, 2,..., N_{i/q} \tag{4.16a}$$

$$\tilde{\phi}_T^{m_{i/q}} = F_T^{-1}\left(\frac{m_{i/q} - 1/2}{M_{i/q}}\right), \quad \tilde{\phi}_T^{m_{i/q}} \in [-\pi, \pi) \quad m_{i/q} = 1, 2,..., M_{i/q} \tag{4.16b}$$

Note that a value of $\frac{1}{2}$ is applied in Eqs (4.16a) and (4.16b) instead of the value of $\frac{1}{4}$ in Eqs (4.15a) and (4.15b). The reason is that unlike *Case I*, for *Case II* it is difficult to design sets of AoAs and AoDs that meet the two conditions: it is difficult to find one value that can guarantee the sets of AoAs and AoDs designed give the best possible approximation to the two conditions. However, based on simulations using the modified MEDS by Patel et al. [2005b, Eqs (37), (38)] for the simulation of isotropic scattering V2V channels, we found that with the value of $\frac{1}{2}$, the simulation model performs better than with a value of $\frac{1}{4}$. The performance of this design will be validated in the subsection on Numerical Results and Analysis (see p. 98).

For *Case III*, since the in-phase and quadrature components of the reference model $h(t)$ in Eq. (4.8) are correlated (similar to *Case I*) we set $N_i = N_q = N$ and $M_i = M_q = M$ as well and thus the AoA $\tilde{\phi}_R^{n_{i/q}}$ and AoD $\tilde{\phi}_T^{m_{i/q}}$ can be replaced by the $\tilde{\phi}_R^n$ and $\tilde{\phi}_T^m$, respectively. Following a similar parameter computation method as in *Case I*, we design the AoA and AoD of this model as

$$\tilde{\phi}_R^n = F_R^{-1}\left(\frac{n - 1/2}{N}\right), \quad \tilde{\phi}_R^n \in [-\pi, \pi) \quad n = 1, 2,..., N \tag{4.17a}$$

$$\tilde{\phi}_T^m = F_T^{-1}\left(\frac{m - 1/2}{M}\right), \quad \tilde{\phi}_T^m \in [-\pi, \pi) \quad m = 1, 2,..., M \tag{4.17b}$$

Note that a value of $\frac{1}{2}$ is used in Eqs (4.17a) and (4.17b) for the same reason as in *Case II*.

The correlation properties of the deterministic simulation model proposed must be analysed by using a time average rather than a statistical average. The time-average ACF of the proposed simulation model $\tilde{h}(t)$ is defined as

$$\tilde{\rho}_{\tilde{h}\tilde{h}}(\tau) = \left\langle \tilde{h}(t)\tilde{h}^*(t - \tau)\right\rangle \tag{4.18}$$

where $\langle \cdot \rangle$ denotes the time-average operator. Substituting Eq. (4.11) into Eq. (4.18), we have

$$\tilde{\rho}_{\tilde{h}\tilde{h}}(\tau) = 2\tilde{\rho}_{\tilde{h}_i\tilde{h}_i}(\tau) - 2j\tilde{\rho}_{\tilde{h}_i\tilde{h}_q}(\tau) \tag{4.19}$$

where

$$\tilde{\rho}_{\tilde{h}_i\tilde{h}_i}(\tau) = \frac{1}{2N_iM_i}\sum_{n_i,m_i=1}^{N_i,M_i} \cos\left[2\pi f_{T_{max}}\tau\cos(\tilde{\phi}_T^{m_i} - \gamma_T) + 2\pi f_{R_{max}}\tau\cos(\tilde{\phi}_R^{n_i} - \gamma_R)\right] \tag{4.20a}$$

$$\tilde{\rho}_{\tilde{h}_i\tilde{h}_q}(\tau) = \begin{cases} -\dfrac{1}{2NM} \sum\limits_{n,m=1}^{N,M} \sin[2\pi f_{T_{max}} \tau \cos(\tilde{\phi}_T^m - \gamma_T) + 2\pi f_{R_{max}} \tau \cos(\tilde{\phi}_R^n - \gamma_R)], \\ \qquad N_i = N_q = N \text{and } M_i = M_q = M \ (Cases\ I \text{ and } III) \\ 0, \qquad N_i \neq N_q \text{ and } M_i \neq M_q \ (Case\ II) \end{cases} \tag{4.20b}$$

In Appendix 4-A, we give a brief outline of the derivations of Eqs (4.20a) and (4.20b). From Eq. (4.20b), and based on the corresponding derivation in Appendix 5-A, it is clear that by setting $N_i \neq N_q$ and $M_i \neq M_q$, the cross-correlation between the in-phase component $\tilde{h}_i(t)$ and quadrature component $\tilde{h}_q(t)$ of the proposed simulation model $\tilde{h}(t)$ is equal to zero no matter how the sets of AoAs and AoDs are designed. When N (N_i) and M (M_i) tend to infinity, it is straightforward that the time-average ACF in Eq. (4.19) matches the ensemble average ACF in Eq. (4.9). This allows us to conclude that for $\{N(N_i), M(M_i)\} \to \infty$, the deterministic simulation model introduced here can represent the correlation properties of the reference model.

4.5.2.2 Stochastic Simulation Model

The deterministic model outlined above can be further modified to become a stochastic simulation model by allowing both the phases and frequencies to be random variables. Unlike the deterministic model, the properties of the stochastic model vary for each simulation trial, but will converge to the desired ones when averaged over a sufficient number of simulation trials. A hat is used to distinguish this model from the deterministic one, thus

$$\hat{h}(t) = \hat{h}_i(t) + j\hat{h}_q(t) \tag{4.21}$$

$$\hat{h}_i(t) = \frac{1}{\sqrt{N_i M_i}} \sum_{n_i, m_i=1}^{N_i, M_i} \cos\left[\hat{\psi}_{n_i m_i} + 2\pi f_{T_{max}} t \cos(\hat{\phi}_T^{m_i} - \gamma_T) + 2\pi f_{R_{max}} t \cos(\hat{\phi}_R^{n_i} - \gamma_R)\right] \tag{4.22}$$

$$\hat{h}_q(t) = \frac{1}{\sqrt{N_q M_q}} \sum_{n_q, m_q=1}^{N_q, M_q} \sin\left[\hat{\psi}_{n_q m_q} + 2\pi f_{T_{max}} t \cos(\hat{\phi}_T^{m_q} - \gamma_T) + 2\pi f_{R_{max}} t \cos(\hat{\phi}_R^{n_q} - \gamma_R)\right] \tag{4.23}$$

where $\hat{h}_i(t)$ and $\hat{h}_q(t)$ are in-phase and quadrature components of the complex fading envelope $\hat{h}(t)$, respectively, $\hat{\phi}_R^{n_{i/q}}$ and $\hat{\phi}_T^{m_{i/q}}$ denote the AoA and AoD of this stochastic simulation model, respectively, and the phases $\hat{\psi}_{n_{i/q} m_{i/q}}$ are random variables uniformly distributed on the interval $[-\pi, \pi)$. The parameters $\hat{\phi}_R^{n_{i/q}}$, $\hat{\phi}_T^{m_{i/q}}$, and $\hat{\psi}_{n_{i/q} m_{i/q}}$ are independent of each other. Note that unlike the AoA and AoD in a deterministic simulation model, the AoA $\hat{\phi}_R^{n_{i/q}}$ and AoD $\hat{\phi}_T^{m_{i/q}}$ are random variables and thus vary for different simulation trials.

The fundamental issue for the design of the sets of AoAs $\{\hat{\phi}_R^{n_{i/q}}\}_{n_{i/q}=1}^{N_{i/q}}$ and AoDs $\{\hat{\phi}_T^{m_{i/q}}\}_{m_{i/q}=1}^{M_{i/q}}$ is how to incorporate a random term into the AoA and AoD. Here, to deal with this fundamental issue, we apply the method proposed by Zheng and Xiao [2002] for the simulation of isotropic scattering F2M channels. According to the comparative analysis of Patel et al. [2005a], we can conclude that the smaller (but sufficient)the range on which the AoA $\hat{\phi}_R^{n_{i/q}}$ and AoD $\hat{\phi}_T^{m_{i/q}}$ are designed, the better the performance of the stochastic model. Based on the extensive investigation of the PDFs of the AoA ϕ_R and AoD ϕ_T, we find that unlike isotropic scattering V2V environments [Patel et al. 2005b], the appropriate range on which the AoA and AoD are designed varies for different non-isotropic scattering V2V environments. Therefore, similar to the deterministic model described above, we design the sets of AoAs and AoDs of this stochastic model separately for the following three cases:

- *Case I*: the main transmitted and received powers come from the same direction as the movements of the Tx and Rx; that is along the x-axis: $\mu_T = \gamma_T = \mu_R = \gamma_R = 0°$ or π.
- *Case II*: the main transmitted and received powers come from directions that are perpendicular to those of the movements of the Tx and Rx, respectively: $|\mu_T - \gamma_T| = |\mu_R - \gamma_R| = 90°$.
- *Case III*: any circumstances different from *Cases I* and *II*.

For *Case I*, for the same reason as in the design of our deterministic model, we have $N_i = N_q = N$ and $M_i = M_q = M$ and thus the AoA $\hat{\phi}_R^{n_{i/q}}$ and AoD $\hat{\phi}_T^{m_{i/q}}$ can be replaced by the $\hat{\phi}_R^n$ and $\hat{\phi}_T^m$, respectively. In this case, the PDFs of the AoA and AoD are symmetric with respect to the origin and so the appropriate ranges for the design of both AoA and AoD are from 0 to π because the rest of the range is redundant because the range $[-\pi, 0)$ provides us with the same infomation as $[0, \pi)$. Inspired by the method of Zheng and Xiao [2002] for isotropic scattering F2M channels, we can now design the AoA and AoD of this model as

$$\hat{\phi}_R^n = F_R^{-1}\left(\frac{n - 1/2 + \theta_R}{N}\right), \quad \hat{\phi}_R^n \in [0, \pi) \quad n = 1, 2, ..., N \tag{4.24a}$$

$$\hat{\phi}_T^m = F_T^{-1}\left(\frac{m - 1/2 + \theta_T}{M}\right), \quad \hat{\phi}_T^m \in [0, \pi) \quad m = 1, 2, ..., M \tag{4.24b}$$

where θ_R and θ_T are random variables uniformly distributed on the interval $\left[-\frac{1}{2}, \frac{1}{2}\right)$ and are independent to each other. It is worth stressing that the interval $\left[-\frac{1}{2}, \frac{1}{2}\right)$ and the constant value of $\frac{1}{2}$ are chosen to guarantee that the design of the AoA and AoD is based on the desired range (here, $[0, \pi)$). This indicates that any interval and the corresponding constant value can be chosen only if they can fulfill the aforementioned guarantee (for example, the interval is $[0, 1)$ and the constant value is 1). Note that the introduction of the random terms θ_R and θ_T in the AoA $\hat{\phi}_R^n$ and AoD $\hat{\phi}_T^m$, respectively, leads to the AoA and AoD being random variables and thus varying for different simulation trials.

For *Case II*, analogous to the situation for the deterministic model for *Case II*, we impose $N_i \neq N_q$ and $M_i \neq M_q$ in this stochastic model to guarantee that the cross-correlation between the in-phase component $\hat{h}_i(t)$ and quadrature component $\hat{h}_q(t)$ of the proposed stochastic model $\hat{h}(t)$ for each simulation trial is equal to zero. In this case, since the PDFs of the AoA and AoD are asymmetric with respect to the origin, the full range (from $-\pi$ to π) is needed for the design of the AoA and AoD. Therefore, we can express the AoA and AoD as

$$\hat{\phi}_R^{n_{i/q}} = F_R^{-1}\left(\frac{n_{i/q} - 1/2 + \theta_R}{N_{i/q}}\right), \ \hat{\phi}_R^{n_{i/q}} \in [-\pi, \pi) \ n_{i/q} = 1, 2, ..., N_{i/q} \quad (4.25a)$$

$$\hat{\phi}_T^{m_{i/q}} = F_T^{-1}\left(\frac{m_{i/q} - 1/2 + \theta_T}{M_{i/q}}\right), \ \hat{\phi}_T^{m_{i/q}} \in [-\pi, \pi) \ m_{i/q} = 1, 2, ..., M_{i/q} \quad (4.25b)$$

For *Case III*, for the same reason as in *Case I*, we have $N_i = N_q = N$ and $M_i = M_q = M$ and thus the AoA $\hat{\phi}_R^{n_{i/q}}$ and AoD $\hat{\phi}_T^{m_{i/q}}$ can be replaced by the $\hat{\phi}_R^n$ and $\hat{\phi}_T^m$, respectively. In this case, analogous to *Case II*, we know that the full range from $-\pi$ to π is necessary for the design of the AoA and AoD. Therefore, the AoA and AoD can be designed as

$$\hat{\phi}_R^n = F_R^{-1}\left(\frac{n - 1/2 + \theta_R}{N}\right), \quad \hat{\phi}_R^n \in [-\pi, \pi) \quad n = 1, 2, ..., N \quad (4.26a)$$

$$\hat{\phi}_T^m = F_T^{-1}\left(\frac{m - 1/2 + \theta_T}{M}\right), \quad \hat{\phi}_T^m \in [-\pi, \pi) \quad m = 1, 2, ..., M \quad (4.26b)$$

Unlike the deterministic simulation model above, the time-ACF of the stochastic simulation model should be computed according to $\hat{\rho}_{\hat{h}\hat{h}}(\tau) = \mathbf{E}[\hat{h}(t)\hat{h}^*(t - \tau)]$. It can be shown that the stochastic model exhibits the correlation properties of the reference model irrespective of the values of $N_{i/q}$ and $M_{i/q}$; in other words for any $N_{i/q}$ and $M_{i/q}$. Appendix 4-B outlines the derivation of the ACF $\hat{\rho}_{\hat{h}\hat{h}}(\tau)$ for the model $\hat{h}(t)$.

It is worth noting that this stochastic model performs better and has lower complexity than the ARA model of Zheng [2006]. Zheng did not give a detailed explanation of how to generate the AoAs and AoDs for his model using the ARA, so it is impossible to reproduce it. Therefore, demonstrate the comparative performance, in Figure 4.10 we compare the ACF of the real part of the ARA model obtained from Figure 5 in Zheng's paper with the equivalent from the model outlined here. For a fair comparison, the same parameters are used for both models, namely $f_{T_{max}} = 100$ Hz, $f_{R_{max}} = 50$ Hz, $\mu_T = \pi/4$, $\mu_R = -\pi/4$, $k_T = k_R = 3$, $\gamma_T = \gamma_R = 0°$, and the number of simulation trials $N_{stat} = 10$. Note that the number of harmonic functions used in the ARA model is $N_{ARA} = 144$, while in the new model it is $N_i = M_i = 10$. From Figure 4.10, it is obvious that our model outperforms the ARA model with an even smaller number of harmonic functions: $N_i \times M_i = 100 < N_{ARA}$.

4.5.2.3 Numerical Results and Analysis

In this section, we first validate our deterministic model using the squared error between the correlation properties of the simulation model and those of the reference model. Then validation of the stochastic model is performed by comparing the time-average properties of a single simulation trial from the desired ensemble-average. Performance is also evaluated

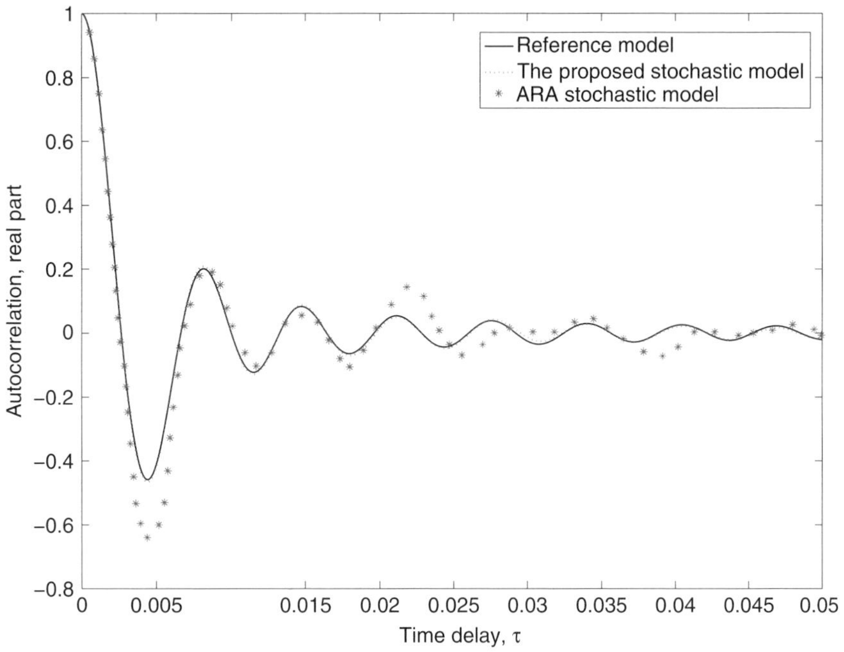

Figure 4.10 Comparison between the introduced stochastic model and the ARA stochastic model

by comparing the correlation properties of the simulation models with those of the reference model. Unless otherwise specified, all the results presented here are obtained using:

- $f_{T_{max}} = f_{R_{max}} = 100$ Hz
- normalized sampling period $f_{T_{max}} T_s = 0.005$ (where T_s is the sampling period)
- For the deterministic model
  - $N_i = M_i = N_q = M_q = 20$ for *Cases I* and *III*
  - $N_i = M_i = 20$, $N_q = M_q = 21$ for *Case II*
- For the stochastic model
  - $N_i = M_i = N_q = M_q = 10$ for *Cases I* and *III*
  - $N_i = M_i = 10$, $N_q = M_q = 11$ for *Case II*

To validate the deterministic model, in Figure 4.11 we compare the difference in the ACF $\tilde{\rho}_{\tilde{h}\tilde{h}}(\tau)$ from the desired $\rho_{hh}(\tau)$ using the squared error $|\tilde{\rho}_{\tilde{h}\tilde{h}}(\tau) - \rho_{hh}(\tau)|^2$ for different non-isotropic V2V scenarios. Similarly, to validate the stochastic model, Figure 4.12 compares the time-averaged ACF of a single simulation trial $\breve{\rho}_{hh}(\tau)$ to the desired ACF $\rho_{hh}(\tau)$ as $\mathbf{E}[|\breve{\rho}_{hh}(\tau) - \rho_{hh}(\tau)|^2]$ for different non-isotropic V2V scenarios. As pointed out by Patel et al. [2005a], this provides a measure of the utility of the stochastic model in simulating the desired channel waveform using finite harmonic functions $N_{i/q}$ and $M_{i/q}$.

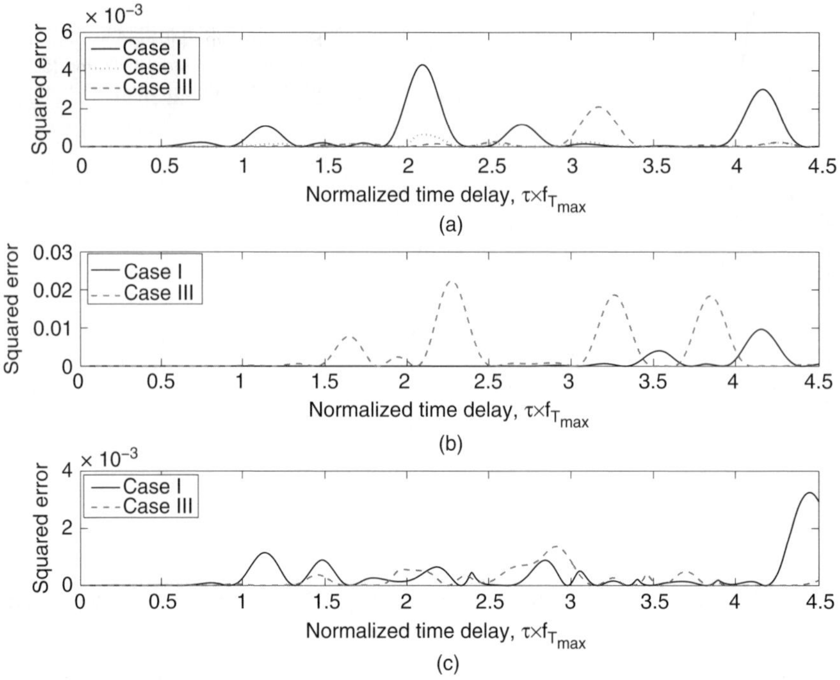

Figure 4.11 Squared error in the CF of the introduced deterministic simulation model with $k = 1$ for different non-isotropic scattering V2V Rayleigh fading channels: (a) $\mu_T = \mu_R = 110°$ and $\gamma_T = \gamma_R = 20°$; (b) $\mu_T = \mu_R = \gamma_T = \gamma_R = 0°$; (c) $\mu_T = 30°$, $\mu_R = 160°$, $\gamma_T = 10°$, and $\gamma_R = 20°$

The results in Figure 4.12 are obtained by averaging over 10^4 simulation trials for each value of time delay τ. Note that for the sake of the readability of figures, the difference of the introduced models for *Case II* is only shown in Figures 4.11(a) and 4.12(a) since it is extremely large for other cases. In addition, for longer time delays, the deviation of the simulation model for all cases become extremely large due to an insufficient number of harmonic functions. To maintain readability of the figures, we have removed the longer time delays and only presented shorter ones, which are those usually of interest for communication systems [Patel et al. 2005b].

From Figures 4.11 and 4.12, it is clear that due to the impact of non-isotropic scattering, no parameter computation method consistently outperforms others for all non-isotropic V2V scenarios. This, therefore validates the utility of models that include three different sets of model parameters rather than only one.

To evaluate the performance of the simulation models, in Figures 4.13–4.15 we give a comparison of the ACF of the reference model and the one of the introduced simulation models for various values of k_T, k_R, μ_T, and μ_R. The results obtained for the stochastic model are averaged over $N_{stat} = 10$ simulation trials. It is obvious that the deterministic model provides a fairly good approximation of the ACF of the reference model in a shorter normalized time-delay range, while the stochastic model gives a much better approximation even with a smaller number of harmonic functions $N_{i/q}$ and $M_{i/q}$. The stochastic model has higher

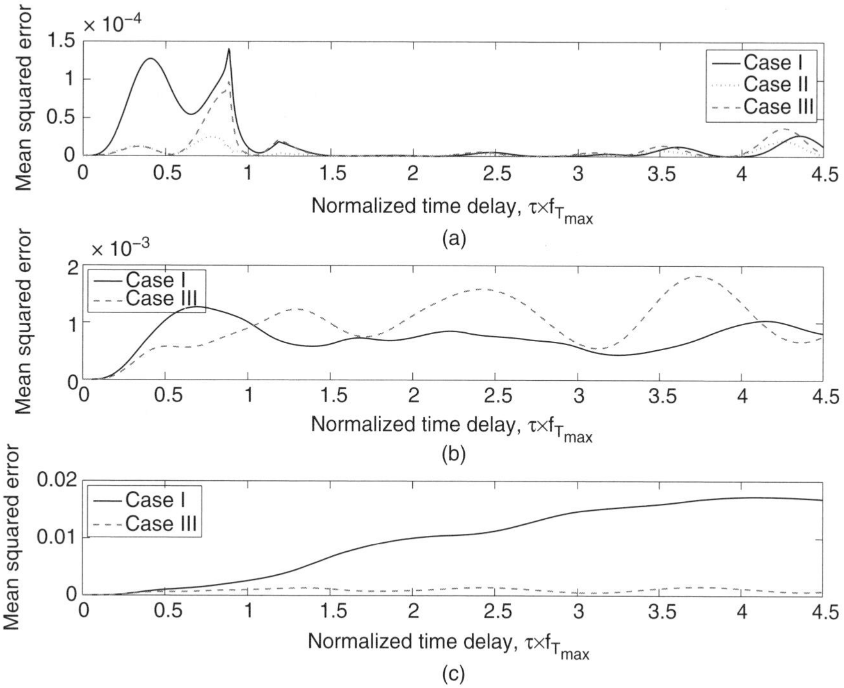

Figure 4.12 Mean squared error in the CF of the introduced stochastic simulation model with $k = 5$ for different non-isotropic scattering V2V Rayleigh fading channels: (a) $\mu_T = \mu_R = 110°$ and $\gamma_T = \gamma_R = 20°$; (b) $\mu_T = \mu_R = \gamma_T = \gamma_R = 0°$; (c) $\mu_T = 20°$, $\mu_R = 10°$, $\gamma_T = 10°$, and $\gamma_R = 20°$

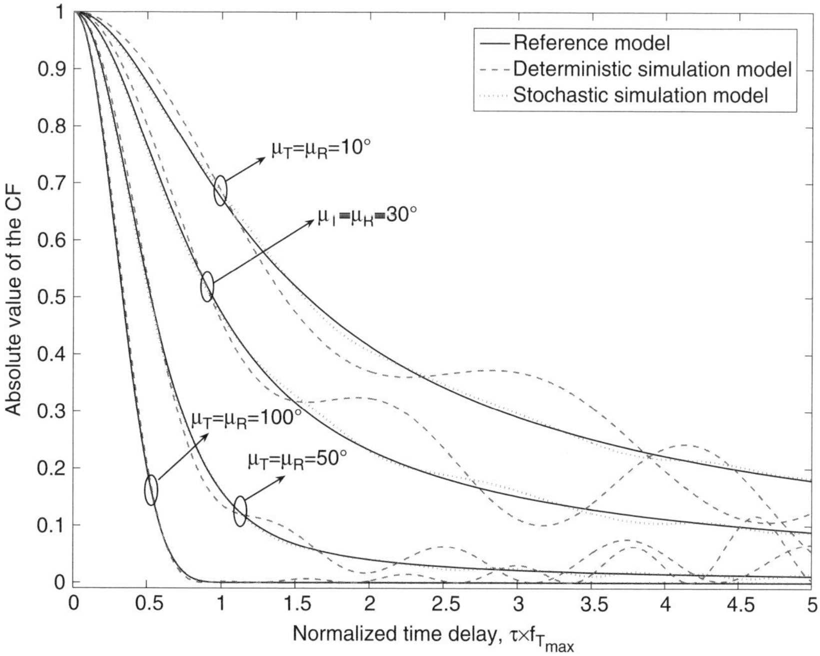

Figure 4.13 Comparison of the CF of the reference model and that of the simulation models with $k_T = k_R = 6$ and $\gamma_T = \gamma_R = 10°$ for various values of the mean AoD μ_T and mean AoA μ_R

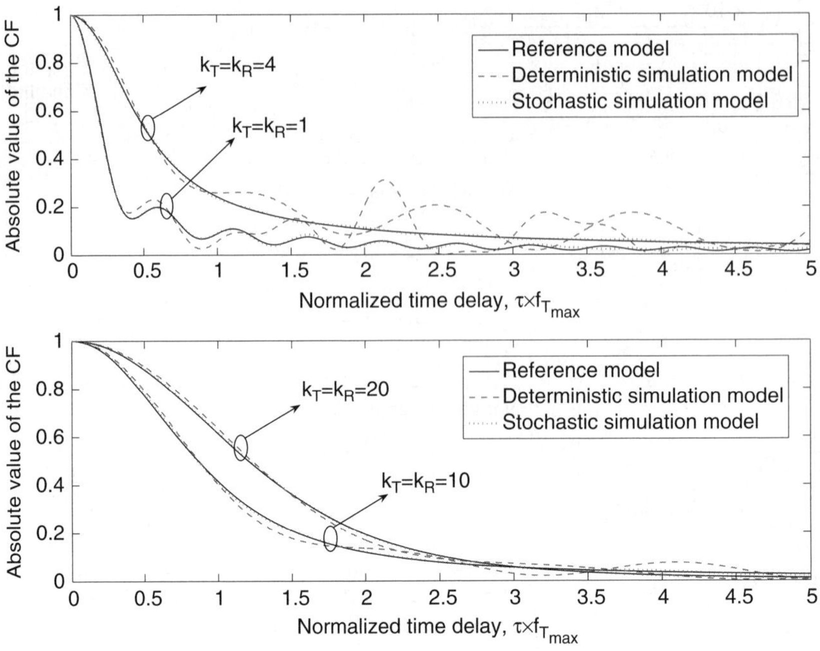

Figure 4.14 Comparison of the CF of the reference model and those of the simulation models with $\mu_T = \mu_R = 60°$ and $\gamma_T = \gamma_R = 30°$ for various values of k_T and k_R

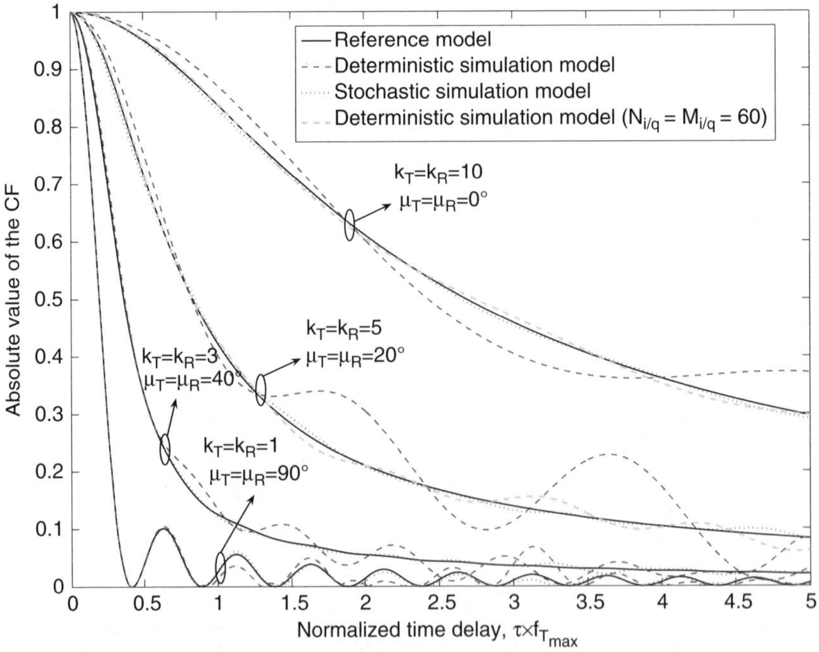

Figure 4.15 Comparison of the CF of the reference model and those of the simulation models with $\gamma_T = \gamma_R = 0°$ for various values of k_T, k_R, the mean AoD μ_T, and mean AoA μ_R

complexity than the deterministic model, since it requires several simulation trials to achieve the desired properties. Notice that the quality of the approximation between the ACFs of our deterministic model and the reference model can be improved by increasing the values of $N_{i/q}$ and $M_{i/q}$, while the quality of the approximation between the ACF of the stochastic model and the one of the reference model can be improved by increasing the values of $N_{i/q}$ and $M_{i/q}$, and/or the value of N_{stat}. More interestingly, from Figure 4.13, we observe that the increase in the values of $|\mu_T - \gamma_T|$ and $|\mu_R - \gamma_R|$ decreases the difficulty the simulation model has in approximating the correlation properties of the reference model. Similarly, Figure 4.14 shows that the difficulty the simulation model has in approximating the correlation properties of the reference model increases with the increase in the values of k_T and k_R. More importantly, Figures 4.13–4.15 show that when the values of $|\mu_T - \gamma_T|$ and $|\mu_R - \gamma_R|$ are small and/or the values of k_T and k_R are large, our deterministic simulation model cannot even approximate the correlation properties of the reference model for short time delays. In such situations, to obtain an acceptable approximation between the correlation properties of the deterministic model and the ones of the reference model, a large number of harmonic functions (here $N_i = M_i = N_q = M_q = 60$, as depicted in Figure 4.15) is necessary. This allows us to conclude that the stochastic simulation model is more suitable than the deterministic model for such non-isotropic scattering V2V scenarios, where the differences between the angles of motion γ_T and γ_R, and the mean AoA μ_T and AoD μ_R are small (that is, the values of $|\mu_T - \gamma_T|$ and $|\mu_R - \gamma_R|$ are small) and/or the received powers are more concentrated in one direction (that is, the values of k_T and k_R are large).

Bibliography

Abdi A and Kaveh M 2002 A space-time correlation model for multielement antenna systems in mobile fading channels. *IEEE Journal on Selected Areas in Communications* **20**(3), 550–560.

Abdi A, Barger J and Kaveh M 2002 A parametric model for the distribution of the angle of arrival and the associated correlation function and power spectrum at the mobile station. *IEEE Transactions on Vehicular Technology* **51**(3), 425–434.

Acosta-Marum G and Ingram M 2007 Six time- and frequency-selective empirical channel models for vehicular wireless LANs *Vehicular Technology Conference, 2007. VTC-2007 Fall. 2007 IEEE 66th*, pp. 2134–2138.

Adachi F, Feeney MT and Parsons JD 1986 Cross-correlation between the envelopes of 900 MHz signals received at a mobile radio base station site. *IEE Proceedings – Communications, Radar & Signal Processing* **133**, 506–512.

Akki A 1994 Statistical properties of mobile-to-mobile land communication channels. *IEEE Transactions on Vehicular Technology* **43**(4), 826–831.

Akki A and Haber F 1986 A statistical model of mobile-to-mobile land communication channel. *IEEE Transactions on Vehicular Technology* **35**(1), 2–7.

Byers G and Takawira F 2004 Spatially and temporally correlated MIMO channels: modeling and capacity analysis. *IEEE Transactions on Vehicular Technology* **53**(3), 634–643.

Caver JK 2002 *Mobile Channel Characteristics*. Kluwer Academic.

Chen TA, Fitz M, Kuo WY, Zoltowski M and Grimm J 2000 A space-time model for frequency nonselective Rayleigh fading channels with applications to space-time modems. *IEEE Journal on Selected Areas in Communications* **18**(7), 1175–1190.

Cheng X, Wang CX, Laurenson DI, Salous S and Vasilakos AV 2011 New deterministic and stochastic simulation models for non-isotropic scattering mobile-to-mobile Rayleigh fading channels. *Wireless Communications and Mobile Computing* **11**(7), 829–842.

Clarke RH 1968 A statistical theory of mobile-radio reception. *Bell System Technical Journal* **47**(6), 957–1000.

de Leon CGD and Patzold M 2007 Sum-of-sinusoids-based simulation of flat fading wireless propagation channels under non-isotropic scattering conditions *Global Telecommunications Conference, 2007. GLOBECOM '07. IEEE*, pp. 3842–3846.

Fechtel S and Feng WY 1993 A novel approach to modeling and efficient simulation of frequency-selective fading radio channels. *IEEE Journal on Selected Areas in Communications* **11**(3), 422–431.

Hanzo L, Webb W and Keller T 2000 *Single and Multi-carrier Quadrature Amplitude Modulation*. West Sussex, England: John Wiley & Sons.

Hogstad BO, Pätzold M, Chopra A, Kim D and Yeom KB 2005 A wideband MIMO channel simulation model based on the geometrical elliptical scattering model *Proc. IEEE WWRF'05*, pp. 1–6, Paris, France.

Jakes WC 1994 *Microwave Mobile Communications* 2nd edn. Wiley-IEEE Press.

Karedal J, Tufvesson F, Czink N, Paier A, Dumard C, Zemen T, Mecklenbrauker C and Molisch A 2009 A geometry-based stochastic MIMO model for vehicle-to-vehicle communications. *IEEE Transactions on Wireless Communications* **8**(7), 3646–3657.

Latinovic Z, Abdi A and Bar-Ness Y 2003 A wideband space-time model for MIMO mobile fading channels *Wireless Communications and Networking, 2003*, vol. 1, pp. 338–342.

Liberti JC and Rappaport TS 1999 *Smart Antennas for Wireless Communication: IS-95 and Third Generation CDMA Applications*. Prentice-Hall.

Maurer J, Fügen T, Porebska M, Zwick T and Wisebeck W 2008 A ray-optical channel model for mobile to mobile communications *COST 2100 4th MCM, COST 2100 TD(08) 430*, pp. 6–8, Wroclaw, Poland.

Paier A, Karedal J, Czink N, Dumard CC, Zemen T, Tufvesson F, Molisch AF and Mecklenbräuker CF 2009 Characterization of vehicle-to-vehicle radio channels from measurements at 5.2 GHz. *Wireless Personal Communications* **50**(1), 19–32.

Patel C, Stuber G and Pratt T 2005a Comparative analysis of statistical models for the simulation of Rayleigh faded cellular channels. *IEEE Transactions on Communications* **53**(6), 1017–1026.

Patel C, Stuber S and Pratt T 2005b Simulation of Rayleigh-faded mobile-to-mobile communication channels. *IEEE Transactions on Communications* **53**(10), 1773.

Pätzold M 2002 *Mobile Fading Channels*. John Wiley and Sons.

Pätzold M and Hogstad B 2006 A wideband space-time MIMO channel simulator based on the geometrical one-ring model *IEEE 64th Vehicular Technology Conference, 2006. VTC-2006 Fall*, pp. 1–6.

Pätzold M and Hogstad J 2004 A space-time channel simulator for MIMO channels based on the geometrical one-ring scattering model *Vehicular Technology Conference, 2004. VTC2004-Fall. 2004 IEEE 60th*, vol. 1, pp. 144–149 Vol. 1.

Pätzold M and Laue F 1998 Statistical properties of Jakes' fading channel simulator *Vehicular Technology Conference, 1998. VTC 98. 48th IEEE*, vol. 2, pp. 712–718.

Pätzold M, Hogstad B and Youssef N 2008 Modeling, analysis, and simulation of MIMO mobile-to-mobile fading channels. *IEEE Transactions on Wireless Communications* **7**(2), 510–520.

Pätzold M, Hogstad B, Youssef N and Kim D 2005 A MIMO mobile-to-mobile channel model: Part I – the reference model *Personal, Indoor and Mobile Radio Communications, 2005. PIMRC 2005. IEEE 16th International Symposium on*, vol. 1, pp. 573–578.

Pätzold M, Killat U and Laue F 1994 A deterministic model for a shadowed Rayleigh land mobile radio channel *Personal, Indoor and Mobile Radio Communications, 1994. Wireless Networks - Catching the Mobile Future., 5th IEEE International Symposium on*, vol. 4, pp. 1202–1210.

Pätzold M, Killat U and Laue F 1996a A deterministic digital simulation model for Suzuki processes with application to a shadowed Rayleigh land mobile radio channel. *IEEE Transactions on Vehicular Technology* **45**(2), 318–331.

Pätzold M, Killat U, Laue F and Li Y 1996b A new and optimal method for the derivation of deterministic simulation models for mobile radio channels *Vehicular Technology Conference, 1996. "Mobile Technology for the Human Race", IEEE 46th*, vol. 3, pp. 1423–1427.

Pätzold M, Killat U, Laue F and Li Y 1998 On the statistical properties of deterministic simulation models for mobile fading channels. *IEEE Transactions on Vehicular Technology* **47**(1), 254–269.

Pop M and Beaulieu N 2001 Limitations of sum-of-sinusoids fading channel simulators. *IEEE Transactions on Communications* **49**(4), 699–708.

Salz J and Winters J 1994 Effect of fading correlation on adaptive arrays in digital mobile radio. *IEEE Transactions on Vehicular Technology* **43**(4), 1049–1057.

Schumacher L, Pedersen K and Mogensen P 2002 From antenna spacings to theoretical capacities – guidelines for simulating MIMO systems *Personal, Indoor and Mobile Radio Communications, 2002. The 13th IEEE International Symposium on*, vol. 2, pp. 587–592.

Sen I and Matolak D 2008 Vehicle-vehicle channel models for the 5-GHz band. *IEEE Transactions on Intelligent Transportation Systems* **9**(2), 235–245.

Stüber GL 2001 *Principles of Mobile Communications* 2nd edn. Kluwer Academic.

Verdin D and Tozer T 1993 Generating a fading process for the simulation of land-mobile radio communications. *Electronics Letters* **29**(23), 2011–2012.

Wang C, Yuan D, hwa Chen H and Xu W 2008 An improved deterministic SoS channel simulator for multiple uncorrelated Rayleigh fading channels. *IEEE Transactions on Wireless Communications* **7**(9), 3307–3311.

Wang LC, Liu WC and Cheng YH 2009 Statistical analysis of a mobile-to-mobile Rician fading channel model. *IEEE Transactions on Vehicular Technology* **58**(1), 32–38.

Wang R and Cox D 2002 Double mobility mitigates fading in ad hoc wireless networks. *IEEE Antennas and Propagation Society International Symposium* **2**, 306–309.

Wang Y and Zoubir A 2007 Some new techniques of localization of spatially distributed source *Signals, Systems and Computers, 2007. ACSSC 2007*, pp. 1807–1811.

Xiao CS, Zheng YR and Beaulieu N 2006 Novel sum-of-sinusoids simulators for ricean fading channels. *IEEE Transactions on Wireless Communications* **5**(12), 3667–3679.

Young D and Beaulieu N 1998 A quantitative evaluation of generation methods for correlated Rayleigh random variates *Global Telecommunications Conference, 1998. GLOBECOM 1998. The Bridge to Global Integration. IEEE*, vol. 6, pp. 3332–3337.

Young D and Beaulieu N 2000 The generation of correlated Rayleigh random variates by inverse discrete Fourier transform. *IEEE Transactions on Communications* **48**(7), 1114–1127.

Zajic A and Stuber G 2006 A new simulation model for mobile-to-mobile Rayleigh fading channels *Wireless Communications and Networking Conference, 2006. WCNC 2006. IEEE*, vol. 3, pp. 1266–1270.

Zajic A and Stuber G 2008a Space-time correlated mobile to mobile channels: Modelling and simulation. *IEEE Transactions on Vehicular Technology* **57**(2), 715–726.

Zajic A and Stuber G 2008b Three-dimensional modeling, simulation, and capacity analysis of space–time correlated mobile-to-mobile channels. *IEEE Transactions on Vehicular Technology* **57**(4), 2042–2054.

Zajic A and Stuber G 2009 Three-dimensional modeling and simulation of wideband MIMO mobile-to-mobile channels. *IEEE Transactions on Wireless Communications* **8**(3), 1260–1275.

Zheng Y 2006 A non-isotropic model for mobile-to-mobile fading channel simulations *Military Communications Conference, 2006. MILCOM 2006. IEEE*, pp. 1–7.

Zheng Y and Xiao C 2002 Improved models for the generation of multiple uncorrelated Rayleigh fading waveforms. *IEEE Communications Letters* **6**(6), 256–258.

Zheng YR and Xiao C 2003 Simulation models with correct statistical properties for Rayleigh fading channels. *IEEE Transactions on Communications* **51**(6), 920–928.

5

Channel Measurements

Measurements of radio propagation channels provide understanding of and insights into the characteristics of radio channels in different environments. The channel characteristics of interest include their large-scale properties, such as path loss, shadowing, and multipath fading, and their small-scale characteristics, represented by the multi-dimensional dispersion of a channel, in for example the direction of arrival (DoA), direction of departure (DoD), Doppler frequency, delay, and polarizations [Bonek et al. 2006, Fleury et al. 2003; 2002b, Fuhl et al. 1997, Karedal et al. 2004, Steinbauer et al. 2000; 2001], and the co- and cross-polarization characteristics [Fleury et al. 2003, Oestges 2005, Yin et al. 2003a]. Channel characteristics obtained from measurements in many scenarios are the basis of empirical stochastic channel models for communication system design and algorithm-performance optimization. Therefore, it is highly important to maintain the accuracy of channel impulse responses and parameter estimates extracted from the measurement data. The system response of measurement equipment and the influence of specific measurement methods should be de-embedded from the observed channel properties.

The efficiency of channel measurements can be influenced by many factors: the antenna characteristics [Sommerkorn et al. 2012], erroneous calibration of the system responses, including the antenna-array responses [Käske et al. 2009], and phase noise [Pedersen et al. 2008a,b]. Nowadays, for millimeter-wave channel measurements, the direction-scan-sounding technique is widely used; it has significant flexibility, such as in the number of scan steps and the selection of antennas with different half-power beamwidth. The scanning strategies adopted can have a significant impact on the channel characterization.

In this chapter, we may not be able to cover all aspects of the impact of measurements on channel characterization. However, as examples, we analyse the impact of imperfections in system calibration such as phase noise, the influence of using specific time-division-multiplexing (TDM) switching modes in the measurements, and the influence of using highly directional antennas. Some of the results have been already published [Taparugssanagorn et al. 2007a, Taparugssanagorn and Ylitalo 2005, Taparugssanagorn et al. 2007b, Yin et al. 2012].

Propagation Channel Characterization, Parameter Estimation and Modelling for Wireless Communications, First Edition. Xuefeng Yin and Xiang Cheng.

5.1 Channel-sounding Equipment/System

Channel measurements are usually performed with a channel sounder, which can be either constructed using the off-the-shelf devices such as a vector network analyzer, oscilloscopes, spectrum analyser, or specifically designed and packaged as a whole. Regardless of the composition of the channel sounder, its design and construction must consider many factors, including the purpose of the measurement campaign, the channel characteristics of interest, and the environments in which measurements will be conducted. Here we focus on wideband channel measurements that require joint estimation of channel characteristics in multiple dimensions, including the frequency, time, space, and polarization domains.

The channel characteristics in the spatial domains, typically around the transmitter and around the receiver, are usually represented by the distribution of channel components in the so-called bi-directional domains: DoD and DoA [Steinbauer et al. 2001]. In order to estimate the bi-directional parameters of propagation paths, a channel sounder needs to be equipped with either antenna arrays in the Tx and the Rx [Krim and Viberg 1996, Steinbauer et al. 2000, Wallace et al. 2003] or a highly directional antenna with its attitude controlled by a multi-axis platform in the Tx and in the Rx [MacCartney and Rappaport 2014]. The former solution with an antenna array has been widely used to construct measurement-based SCM and SCME models, such as:

- the WINNER spatial channel models [Kyösti et al. 2007]
- the IMT-Advanced models [ITU 2009]
- the COST2100 MIMO models [Liu et al. 2012].

The other scheme, using single antennas, has recently been used for channel modeling for 5G wireless communication systems [Rappaport et al. 2013, Xu et al. 1999; 2000, Yin et al. 2016].

For channel sounding with multi-element spatial arrays, the underlying antenna array can be a virtual array generated by stepping a single antenna across specific grids in space [Czink et al. 2005, Steinbauer et al. 2000]. The Tx transmits the sounding signals and the Rx receives the signals distorted by a dispersive propagation channel. The received signal is processed, and the baseband data or the impulse responses of the channels are recorded in a storage device attached to the Rx. Channel parameter estimation needs to be performed based on the data, either in real-time or off-line, depending on the complexity of the underlying estimation methods.

The sounding signals used are usually known to the Rx. They can be various kinds of wideband waveform, such as a phase-shift keying signal modulated by a pseudo-noise sequence [Kattenbach and Weitzel 2000, Stucki 2001, Zetik et al. 2003], multiple-frequency periodic signals with low crest factor [MEDAV GmbH 2001], or a chirp signal [Salous et al. 2002].

Channel sounders equipped with physical multiple-element antenna arrays in the Tx and Rx usually perform a measurement in two manners:

- switched channel sounding [Stucki 2001]
- parallel channel sounding [Salous et al. 2002, Zetik et al. 2003].

A switched channel sounder is equipped with a high-speed radio-frequency (RF) switch at both the Tx and Rx. These switches connect the RF front end to individual elements in the

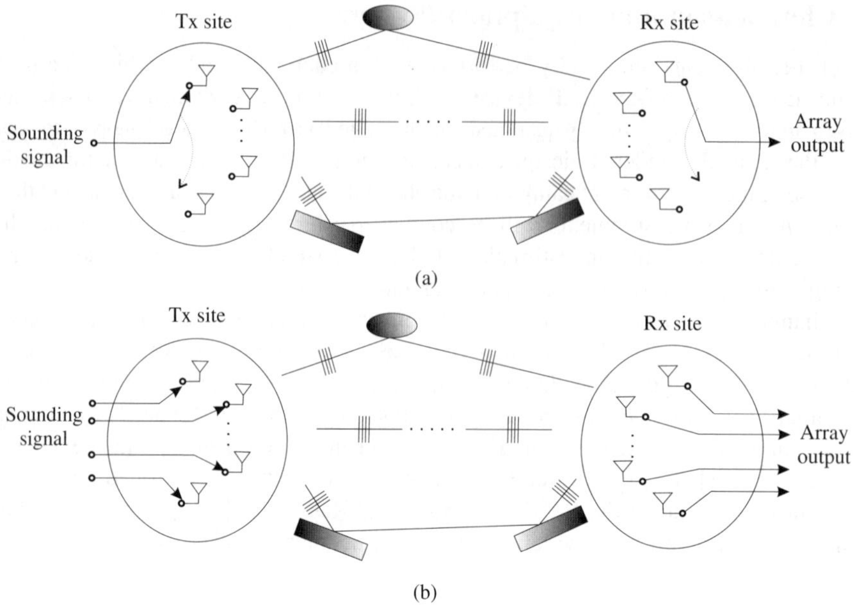

Figure 5.1 MIMO channel sounding systems with (a) a switched architecture and (b) a parallel architecture

antenna arrays. Figure 5.1(a) outlines a switched MIMO channel-sounding system. Using this technique lowers the costs and also the requirement to record in real time.

In the parallel sounding system, the multiple Tx antennas transmit signals simultaneously and the multiple Rx antennas receive signals simultaneously. To separate the signals, the signals transmitted from different Tx antennas should have appropriate autocorrelation and cross-correlation properties. Figure 5.1(b) depicts a parallel MIMO channel-sounding system. Note that the switched and parallel sounding techniques can be used in combination in a channel sounder.

According to Salous et al. [2002] and Zetik et al. [2003], a major disadvantage of switched sounding is the trade-off between the delay estimation range and the highest absolute Doppler frequency. This idea is based on the commonly held belief that a switched sounding system can only estimate Doppler frequencies with absolute values up to half the inverse of the cycle interval. Here, the cycle interval refers to the time between the starting instants of two consecutive measurement cycles. A measurement cycle is the period used to switch each pair of Tx–Rx elements once. It is thought that, in order to estimate high absolute Doppler frequencies, the cycle interval has to be small, so the sensing interval within which an Rx antenna is activated decreases, and thus the estimation range of delay decreases too. However, it has been shown that, with an appropriate switching strategy, the absolute Doppler frequency that a switched sounding system can estimate is up to half the inverse of the *sensing* interval [Pedersen et al. 2008a; 2004, Yin et al. 2003b]. Thus, the trade-off does not exist at all. These studies are going to be elaborated in Section 5.5.

Calibration of the channel-sounding systems is a prerequisite for accurate estimation of the channel parameters. Moreover, distortions inherent to the sounding equipment can heavily influence the accuracy of the estimation; examples include phase noise, clock drifting in synchronization, and so on. In Section 5.3 the impact of phase noise on channel characterization is described, and some preliminary approaches to its mitigation are outlined.

The dimensions of the channel parameters that a sounder combined with a channel estimation technique can resolve is highly dependent on the characteristics of the antenna arrays: their geometrical structure, the radiation patterns of the array elements, and the polarizations. The geometry of the arrays can be linear, circular, planar, cubic, spherical, and so on. The polarizations of the array can be vertical, horizontal, or $\pm 45^\circ$. The radiation pattern can be uniform among all elements, as in the virtual array case, or distinct for different elements. Basically, only when the array has an aperture in certain dimensions can the characteristics of the channel in that dimension be resolved. In Section 5.4, we outline the use of directional antennas as elements in the array, and the impact on channel parameter estimation.

5.2 Post-processing of Measurement Data

In this section, we briefly review some high-resolution methods used to process channel-measurement data. The conventional methods based on the periodogram are not covered because they are easily found in other signal processing textbooks.

Most of high-resolution estimation methods based on the single specular path model were proposed in the final decades of the last century. These methods include:

- subspace-based approaches
  - the multiple signal classification (MUSIC) algorithm [Schmidt 1986]
  - estimation of signal parameters via rotational invariance techniques (ESPRIT) [Roy and Kailath 1989]
- approximations of the maximum-likelihood (ML) methods:
  - the expectation-maximization (EM) algorithm [Moon 1997]
  - the space-alternating generalized expectation-maximization (SAGE) algorithm [Fessler and Hero 1994, Fleury et al. 1999]
  - the RiMAX algorithm [Richter 2004, Richter et al. 2003].

These methods are applied to extract the parameters of the specular path components in multiple dimensions [Fleury et al. 2002c, Heneda et al. 2005, Steinbauer et al. 2001, Zwick et al. 2004].

According to Krim and Viberg [1996], these estimation methods can be categorized into two groups: spectral-based approaches and parametric approaches. Methods belonging to the former group estimate the channel parameters via determination of maxima (or minima) of spectrum-like functions of the dispersion parameters. These techniques are usually computationally attractive as the maxima-searching can be performed in one dimension for all paths. Techniques belonging to the latter category estimate the parameters of an underlying parametric model that characterizes the effect of the propagation channel on the transmitted signal from the signals at the array output. These techniques exhibit better estimation accuracy and higher resolution than spectral-based techniques. However, the computational complexity is usually high due to the multi-dimensional searching required. However, parametric approaches are applicable in cases in which the multipath signals are coherent.[1] In such a case, spectral-based approaches usually fail or perform badly. Both types of algorithm will be introduced in Chapter 6.

5.3 Impact of Phase Noise and Possible Solutions

Most available channel sounders for measuring MIMO spatial channels adopt the time-division-multiplexing mode. Examples of these channel sounders are PROPSound, the RUSK sounder, and the sounder designed by the Center of Radio Communications, Canada. The propagation channels between any pair of Tx and Rx antennas are measured sequentially in timeslots. During the measurements, two switches are applied in the Tx and Rx to connect the RF transceiving chains to the specific Tx and Rx antennas in the dedicated timeslots. As indicated by Taparugssanagorn et al. [2007a], the switching rate should be selected to satisfy two criteria:

- all subchannels between the Tx antennas and the Rx antennas are measured within the channel coherence time
- enough samples of the channels are collected to give adequate resolution for parameter estimation.

In the time-division multiplex (TDM) sounding scheme, phase noise can be generated in: the local oscillators at the Tx and the Rx, the switches at the Tx and the Rx, and in the antennas due to calibration errors of the antenna responses. The latter errors are due to low antenna gain, which is due to the directional characteristics of the antennas, the coupling among them, and the impact of scatterers appearing in the near field of the antenna array.

Phase noise generated by local oscillators in the Tx and the Rx can impact MIMO channel capacity estimation [Baum and Bölcskei 2004, Pedersen et al. 2008b, Taparugssanagorn and Ylitalo 2005] and channel parameter estimation [Taparugssanagorn et al. 2007a,b]. Long-term, slowly varying phase noise can be modeled as the zero-mean non-stationary infinite power Wiener process [Almers et al. 2005]. Usually, long-term varying phase noise can be corrected or mitigated by adding synchronization devices, such as the Rubidium Clock, in the Tx and the Rx, and so it is not necessary to consider the long-term varying phase noise in the MIMO channel sounder. Short-term varying phase noise, over a time period much smaller than 1 s, usually has a significant impact on high-resolution parameter estimation and so its behaviour will be our focus. In the following subsections, we will describe the impact of phase noise on estimation, and the techniques proposed to mitigate it.

Short-term phase noise, observed in less than 1 second, may be modeled as an ARIMA process [Taparugssanagorn et al. 2007a]. Normally, for a MIMO channel sounder, a measurement cycle is usually less than 1 second. For instance, for the measurement campaigns of Czink [2007, Ch. 9], a 50×32 MIMO matrix has been adopted for 5.25 GHz. A subchannel is sounded within a period of 510 μs. Thus, for one cycle of measurement, the time is about 8.42 ms. In the case where four cycles are combined and the data is processed as one snapshot, the time duration is 67 ms. So for this scenario, the properties of phase noise with periods less than 100 ms are of interest.

[1] The signals are called coherent when the signal covariance matrix is singular [Stoica and Moses 1997, p. 240].

The Allan variance is applied to characterize the time-domain statistical behavior, which is calculated as

$$\sigma_y^2(\tau) = \mathrm{E}\left[\frac{(\bar{y}_{k+1} - \bar{y}_k)^2}{2}\right] \tag{5.1}$$

with

$$\begin{aligned}\bar{y}_k &= \frac{1}{\tau}\int_{t_k}^{t_k+\tau} y(t)\mathrm{d}t \\ &= \frac{\phi(t_k+\tau) - \phi(t_k)}{2\pi f_c \tau}\end{aligned} \tag{5.2}$$

where $y(t)$ represents the instantaneous normalized frequency deviation from the carrier frequency f_c, which is computed as

$$y(t) = \frac{1}{2\pi f_c} \cdot \frac{\mathrm{d}\phi(t)}{\mathrm{d}t} \tag{5.3}$$

with $\phi(t)$ denoting the instantaneous phase variation. Assuming that the sampling rate of the phase $\frac{1}{T}$, the samples of the Allan variance at $\tau = mT$ can be estimated as

$$\hat{\sigma_y^2}(mT) = \frac{1}{2(N-2m)(2\pi f_c mT)^2}\sum_{i=1}^{N-2m}(\phi(t_{i+2m}) - 2\phi(t_{i+m}) + \phi(t_i))^2 \tag{5.4}$$

with $m = 1, \ldots, \frac{N-1}{2}$ and N denoting an odd number referring to the total number of samples of the phase.

An example of the measurement of phase noise can be seen in the paper by Taparugssanagorn et al. [2007a], which is briefly described below. The measurement scheme illustrated in Figure 5.2 is used to measure the phase noise of a single-input, single-output channel sounder, where an RF cable connects the Tx and the Rx with a 50-dB fixed attenuator, and the Tx and Rx are each equipped with individual clocks operating in the same way as in real field measurements. The Allan deviation $\hat{\sigma_y}(mT)$ is computed, as illustrated in Figure 5.3, where both the sample Allan variance and its asymptotic characteristics computed from the measured phase-noise sequence are depicted. The curves computed based on the proposed models are also illustrated in Figure 5.3. It was found that an ARMA model with model parameters computed from the sample Allan variance can be used to describe the behavior of the Allan variance. Furthermore, the short-term phase-noise component predominates within the range of $\tau \in [0{,}200\ \mu\mathrm{s}]$; for $\tau > 200\ \mu\mathrm{s}$, the Allan variance of the phase noise is identical with that obtained from white phase noise, as suggested in [IRCC, 1986]; for $\tau > 1$ s, the phase noise can be described using random-walk models.

5.3.1 Mitigating Phase Noise: the Sliding Window

The "sliding window" method proposed by Taparugssanagorn et al. [2007b] for mitigating the impact of phase noise on parameter estimation performance considers multiple consecutive snapshots of measurement data as an observation of the same channel. For the TDM-based sounding system considered, this sliding-window solution is extended to the spatial domain

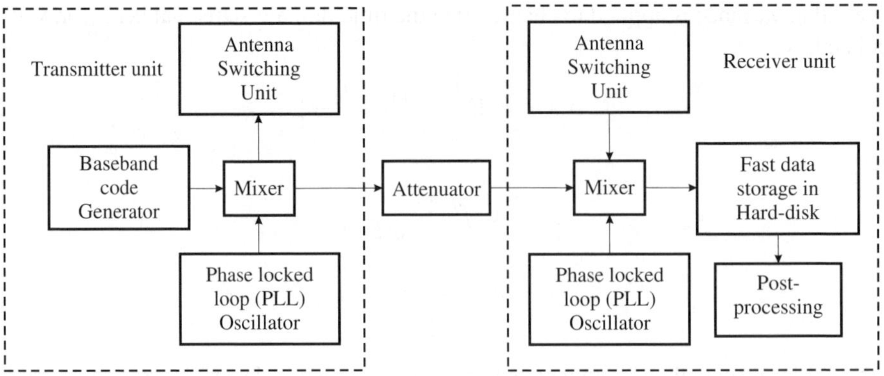

Figure 5.2 The diagram of a measurement of phase noise, as conducted by Taparugssanagorn et al. [2007a]

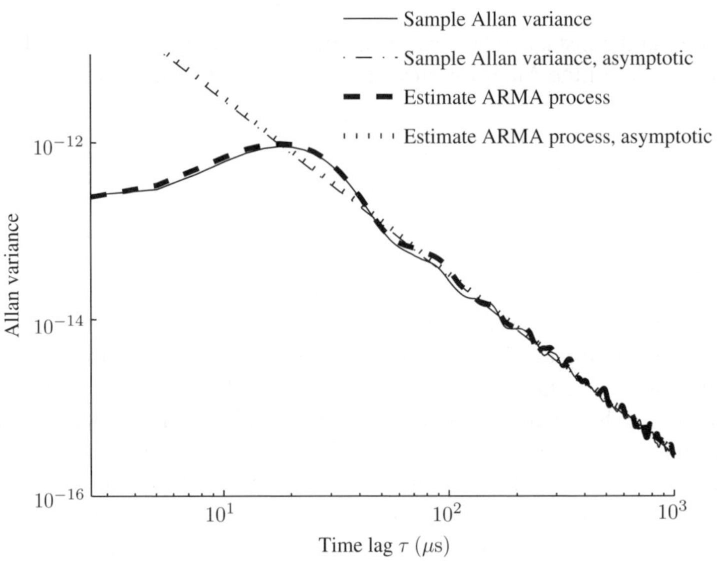

Figure 5.3 Allan deviation

by considering more antennas. In general, the sliding window function is used to calculate the average over multiple channel observations, such that the impact of the phase noise can be mitigated to certain degree. The figures in the paper by Taparugssanagorn et al. [2007b] depict a comparison of the estimation results obtained using the SAGE algorithm, with and without the sliding-window solution, over 20 cycles of data. Different MIMO configurations, such as 50 × 32, 8 × 8, and 4 × 4 are considered. As an example, Figure 5.4 illustrates an environment created in an anechoic chamber where MIMO channel measurements were conducted. The results for 50 × 32 MIMO are shown in Figure 5.5. It can be observed that by using the sliding-window method, it is possible to reduce the probability of generation of artifact estimates.

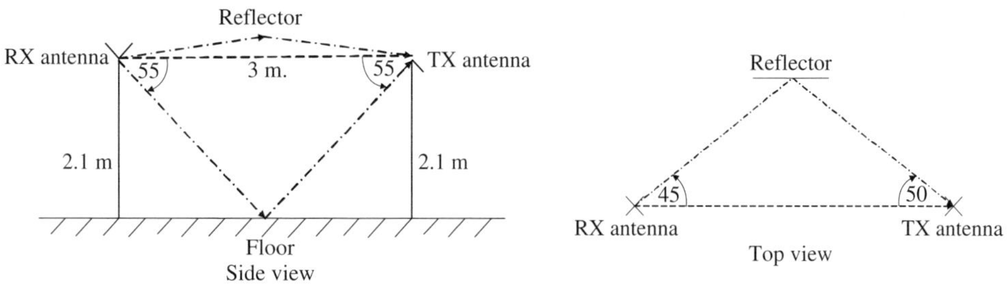

Figure 5.4 Scheme for investigating the impact of phase noise

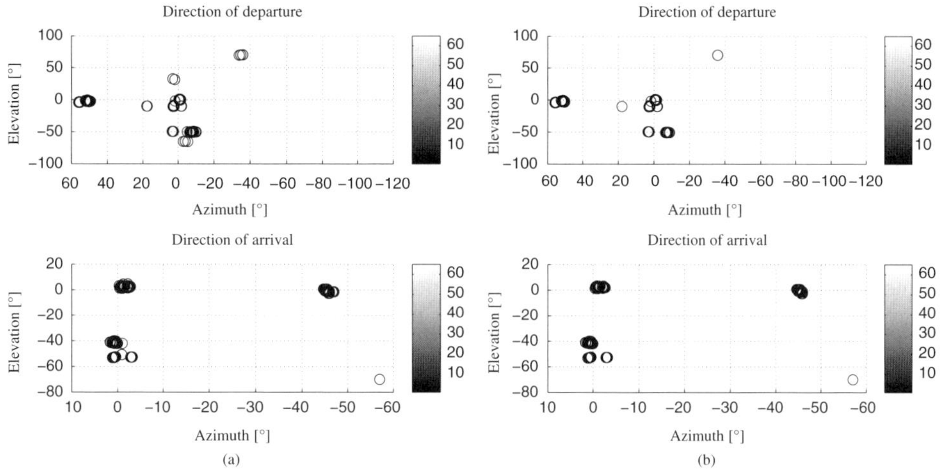

Figure 5.5 The DoA and DoD estimates of multiple propagation paths using 50 × 32 MIMO-TDM sounder: (a) without use of sliding window and (b) applying the sliding window

5.3.2 Mitigating Phase Noise: Whitening and the SAGE algorithm

Another method of combating phase noise is to rederive the high-resolution parameter estimation algorithm, incorporating a whitening function based on the known or calibrated covariance matrix of the phase noise. This approach was introduced by Taparugssanagorn et al. [2007b], with the original specular-path SAGE algorithm of Fleury et al. [1999] being modified with the signal model that was described in Section 4.2. Below is a brief introduction of derivation procedure and the SAGE algorithm obtained.

Let us consider the case in which the TDM sounding scheme is implemented. The original signal model without phase noise can be written as

$$\boldsymbol{Y}(t) = \sum_{\ell=1}^{L} \boldsymbol{S}(t; \boldsymbol{\theta}_\ell) + \boldsymbol{W}(t) \tag{5.5}$$

where vector $\boldsymbol{\theta}_\ell = [\boldsymbol{\Omega}_{1,\ell}, \boldsymbol{\Omega}_{2,\ell}, \tau_\ell, \nu_\ell, \alpha_\ell]$ represents the parameters of the ℓth propagation path, and the signal component $\boldsymbol{S}(t; \boldsymbol{\theta}_\ell)$ is assumed to contain the observations from a total of I measurement cycles:

$$\boldsymbol{S}(t;\boldsymbol{\theta}_\ell) = \begin{bmatrix} s(t_{1,1,1}+t;\boldsymbol{\theta}_\ell) & s(t_{2,1,1}+t;\boldsymbol{\theta}_\ell) & \dots & s(t_{I,1,1}+t;\boldsymbol{\theta}_\ell) \\ s(t_{1,2,1}+t;\boldsymbol{\theta}_\ell) & s(t_{2,2,1}+t;\boldsymbol{\theta}_\ell) & \dots & s(t_{I,2,1}+t;\boldsymbol{\theta}_\ell) \\ \vdots & \vdots & \vdots & \vdots \\ s(t_{1,N,1}+t;\boldsymbol{\theta}_\ell) & s(t_{2,N,1}+t;\boldsymbol{\theta}_\ell) & \dots & s(t_{I,N,1}+t;\boldsymbol{\theta}_\ell) \\ s(t_{1,1,2}+t;\boldsymbol{\theta}_\ell) & s(t_{2,1,2}+t;\boldsymbol{\theta}_\ell) & \dots & s(t_{I,1,2}+t;\boldsymbol{\theta}_\ell) \\ \vdots & \vdots & \vdots & \vdots \\ s(t_{1,N,M}+t;\boldsymbol{\theta}_\ell) & s(t_{2,N,M}+t;\boldsymbol{\theta}_\ell) & \dots & s(t_{I,N,M}+t;\boldsymbol{\theta}_\ell) \end{bmatrix}, t \in [0, T_\mathrm{r}] \tag{5.6}$$

The time sequence $t_{i,n,m}$ calculated as

$$t_{i,n,m} = \left(i - \frac{I+1}{2}\right) T_\mathrm{cy} + \left(n - \frac{N+1}{2}\right) T_\mathrm{r} + \left(m - \frac{M+1}{2}\right) T_\mathrm{t} \tag{5.7}$$

with $i \in [1, \dots, I]$, $n \in [1, \dots, N]$ and $m \in [1, \dots, M]$ refers to the beginning of the interval when the mth Tx antenna transmits and the nth Rx antenna receives in the ith cycle. Here, T_cy, T_r, and T_t represent the intervals between two consecutive cycles (within one burst), the switch interval between two receive antennas, and the switch interval between two transmit antennas, respectively.

The phase noise can be formulated as a matrix $\boldsymbol{\Psi}(t)$ in a format similar to $\boldsymbol{S}(t;\boldsymbol{\theta}_\ell)$:

$$\boldsymbol{\Psi}(t) = \begin{bmatrix} \exp\{j\varphi(t_{1,1,1}+t)\} & \exp\{j\varphi(t_{2,1,1}+t)\} & \dots & \exp\{j\varphi(t_{I,1,1}+t)\} \\ \exp\{j\varphi(t_{1,2,1}+t)\} & \exp\{j\varphi(t_{2,2,1}+t)\} & \dots & \exp\{j\varphi(t_{I,2,1}+t)\} \\ \vdots & \vdots & \vdots & \vdots \\ \exp\{j\varphi(t_{1,N,1}+t)\} & \exp\{j\varphi(t_{2,N,1}+t)\} & \dots & \exp\{j\varphi(t_{I,N,1}+t)\} \\ \exp\{j\varphi(t_{1,1,2}+t)\} & \exp\{j\varphi(t_{2,1,2}+t)\} & \dots & \exp\{j\varphi(t_{I,1,2}+t)\} \\ \vdots & \vdots & \vdots & \vdots \\ \exp\{j\varphi(t_{1,N,M}+t)\} & \exp\{j\varphi(t_{2,N,M}+t)\} & \dots & \exp\{j\varphi(t_{I,N,M}+t)\} \end{bmatrix},$$

$$t \in [0, T_\mathrm{r}]. \tag{5.8}$$

Including the phase-noise components into Eq. (5.5), we obtain the model of signals $\boldsymbol{Y}_\mathrm{p}(t)$ influenced by the phase noise as:

$$\boldsymbol{Y}_\mathrm{p}(t) = \sum_{\ell=1}^{L} \boldsymbol{S}(t;\boldsymbol{\theta}_\ell) \odot \Psi(t) + \boldsymbol{W}(t) \tag{5.9}$$

where $\odot$ denotes the Hadamard product. For convenience in the next steps in the derivation, we rewrite Eq. (5.9) in a vectorized representation as

$$\boldsymbol{y}_\mathrm{p}(t) = \sum_{\ell=1}^{L} \boldsymbol{s}(t;\boldsymbol{\theta}_\ell) \odot \boldsymbol{\psi}(t) + \boldsymbol{w}(t) \tag{5.10}$$

with $\boldsymbol{y}_\mathrm{p}(t) = \mathrm{vec}[\boldsymbol{Y}_\mathrm{p}(t)]$. The vectorization is to rearrange a matrix into a column vector by concatenating all columns of the matrix.

It has been shown by the IRCC [1986] that when the delay lag is larger than 200 μs, the phase noise φ behaves as a white Gaussian random variable. Thus for the case where the interval between two consecutive subchannels is about 510 μs, as in the PROPSound case, the phase noise observed can be considered to be Gaussian distributed. It can be further calculated from Figure 5.3 that for delay lags larger than 5×10^2 μs, the deviation in degree of the phase noise is $\sqrt{2\pi 5 \cdot 10^9 \times 100 \times 10^{-6} \times 10^{-7.5}} \times 180/\pi = 0.4051^\circ$. For such small deviations, it is reasonable to use the Taylor series expansion to approximate $\boldsymbol{\psi}(t)$:

$$\boldsymbol{\psi}(t) \approx \exp\{j\varphi_0\} \times (1 + j\Delta\varphi(t) + \dots) \tag{5.11}$$

where

$$\Delta\varphi(t) = [\varphi(t_{1,1,1} + t)\ \varphi(t_{1,2,1} + t) \dots \varphi(t_{I,N,M} + t)]^T - \varphi(t_{1,1,1} + t) \tag{5.12}$$

denotes the deviation of the phase noise from its average. The term $\exp\{j\varphi_0\}$ can be included into the complex attenuation α_ℓ, $\ell = 1, \dots, I$.

By dropping the high-order Taylor-series component, we obtain

$$\boldsymbol{y}_{\mathrm{p}}(t) \approx \sum_{\ell=1}^{L} \boldsymbol{s}(t; \boldsymbol{\theta}_\ell) \odot [1 + j\Delta\varphi(t)] + \boldsymbol{w}(t) \tag{5.13}$$

$$= \sum_{\ell=1}^{L} \boldsymbol{s}(t; \boldsymbol{\theta}_\ell) + \boldsymbol{s}(t; \boldsymbol{\theta}_\ell) \odot j\Delta\varphi(t) + \boldsymbol{w}(t) \tag{5.14}$$

A realistic assumption is made that during one subchannel the phase noise does not change significantly. The variable t is then dropped from $\Delta\varphi(t)$. The elements of $\Delta\varphi$ are independent Gaussian random variables; that is, $\Delta\varphi \sim \mathcal{CN}(0, \boldsymbol{R}_\varphi)$, with $\boldsymbol{R}_\varphi$ being the covariance matrix of $\Delta\varphi$. It is easy to show that $\boldsymbol{y}(t)$ is Gaussian distributed:

$$\boldsymbol{y}_{\mathrm{p}}(t) \sim \mathcal{CN}\left(\sum_{\ell=1}^{L} \boldsymbol{s}(t; \boldsymbol{\theta}_\ell), \sum_{\ell=1}^{L} \sum_{\ell'=1}^{L} \boldsymbol{s}(t; \boldsymbol{\theta}_\ell) \boldsymbol{s}(t; \boldsymbol{\theta}_\ell)^{\mathrm{H}} \cdot \boldsymbol{R}_\varphi + \sigma_w^2 \boldsymbol{I} \right) \tag{5.15}$$

The above signal model can be then used to derive estimators that jointly extract the path parameters $\boldsymbol{\theta}_\ell$ and the entries of $\boldsymbol{R}_\varphi$ from $\boldsymbol{y}(t)$.

Taparugssanagorn et al. [2007b] derived a SAGE algorithm for estimating $\boldsymbol{\theta}_\ell$ under the assumption that $\boldsymbol{R}_\varphi$ is known. The admissible hidden data is defined to be the contribution of individual propagation paths:

$$\boldsymbol{x}_\ell(t) = \boldsymbol{s}(t; \boldsymbol{\theta}_\ell) \odot [1 + j\boldsymbol{\psi}] + \boldsymbol{w}'(t) \tag{5.16}$$

where $\boldsymbol{w}'(t) \in \mathcal{C}^{NM \times I}$ denotes standard white Gaussian noise with component variance $\sigma_{w'}^2 = \beta_\ell \sigma_w^2$, with β_ℓ being nonnegative and satisfying the equality $\sum_{\ell=1}^{L} \beta_\ell = 1$.

In real measurements, $\boldsymbol{x}_\ell(t)$ and $\boldsymbol{y}(t)$ are discrete. For notational convenience, we use $\boldsymbol{y}$ to represent $\boldsymbol{y} = [\boldsymbol{y}(t); t = t_1, \dots, t_D]$, with D being the total number of samples in the delay domain when sounding a subchannnel. Similarly, $\boldsymbol{x}_\ell = [\boldsymbol{x}_\ell(t); t = t_1, \dots, t_d]$ and $\boldsymbol{s}(\boldsymbol{\theta}_\ell) = [\boldsymbol{s}(t; \boldsymbol{\theta}_\ell); t = t_1, \dots, t_d]$.

In the expectation step of the ith iteration, an objective function $Q(\boldsymbol{\theta}_\ell; \boldsymbol{y}, \boldsymbol{R}_\varphi, \hat{\boldsymbol{\theta}}^{[n-1]})$, defined as

$$Q(\boldsymbol{\theta}_\ell; \boldsymbol{y}, \boldsymbol{R}_\varphi, \hat{\boldsymbol{\theta}}^{[n-1]}) = \int \log p(\boldsymbol{x}_\ell | \boldsymbol{\theta}_\ell, \boldsymbol{R}_\varphi) f(\boldsymbol{x}_\ell | \boldsymbol{Y} = \boldsymbol{y}, \boldsymbol{R}_\varphi, \hat{\boldsymbol{\theta}}^{[n-1]}) \mathrm{d}\boldsymbol{x}_\ell \tag{5.17}$$

is calculated, where the probability density function $p(\boldsymbol{x}_\ell|\boldsymbol{\theta}_\ell, \boldsymbol{R}_\varphi)$ reads

$$p(\boldsymbol{x}_\ell|\boldsymbol{\theta}_\ell, \boldsymbol{R}_\varphi) = \frac{1}{\pi^{NM}|\Sigma_{\boldsymbol{x}_\ell}|} \exp\{-(\boldsymbol{x}_\ell - \boldsymbol{s}(\boldsymbol{\theta}_\ell))^{\mathrm{H}} \Sigma_{\boldsymbol{x}_\ell}^{-1} (\boldsymbol{x}_\ell - \boldsymbol{s}(\boldsymbol{\theta}_\ell))\} \tag{5.18}$$

with $|\cdot|$ being the determinant of the matrix given as argument, and $\Sigma_{\boldsymbol{x}_\ell}$ is the covariance matrix of $\boldsymbol{x}_\ell$

$$\Sigma_{\boldsymbol{x}_\ell} = \boldsymbol{s}(\boldsymbol{\theta}_\ell)\boldsymbol{s}(\boldsymbol{\theta}_\ell)^{\mathrm{H}} \odot \boldsymbol{R}_\varphi + \sigma_w^2 \boldsymbol{I} \tag{5.19}$$

It can be shown that

$$\begin{aligned} &Q(\boldsymbol{\theta}_\ell; \boldsymbol{y}, \boldsymbol{R}_\varphi, \hat{\boldsymbol{\theta}}^{[i-1]}) = \\ &- \log(\pi^{NMD}|\Sigma_{\boldsymbol{X}_\ell}(\boldsymbol{\theta}_\ell)|) - \mathrm{tr}\{\Sigma_{\boldsymbol{X}_\ell}(\boldsymbol{\theta}_\ell)^{-1}(\Sigma_{\boldsymbol{X}_\ell}(\hat{\boldsymbol{\theta}}_\ell^{[i-1]}) - \Sigma_{\boldsymbol{X}_\ell \boldsymbol{Y}}(\hat{\boldsymbol{\theta}}^{[i-1]})\Sigma_{\boldsymbol{Y}}(\hat{\boldsymbol{\theta}}^{[i-1]})^{-1} \\ &\times \Sigma_{\boldsymbol{X}_\ell \boldsymbol{Y}}(\hat{\boldsymbol{\theta}}^{[i-1]})^{\mathrm{H}} + \mu_{\boldsymbol{X}_\ell|\boldsymbol{Y}}(\hat{\boldsymbol{\theta}}^{[i-1]})\mu_{\boldsymbol{X}_\ell|\boldsymbol{Y}}(\hat{\boldsymbol{\theta}}^{[i-1]})^{\mathrm{H}})\} + 2\mathrm{re}\{\boldsymbol{s}(\boldsymbol{\theta}_\ell)^{\mathrm{H}}\Sigma_{\boldsymbol{X}_\ell}(\boldsymbol{\theta}_\ell)^{-1}\mu_{\boldsymbol{X}_\ell|\boldsymbol{Y}}(\hat{\boldsymbol{\theta}}^{[i-1]})\} \\ &- \boldsymbol{s}(\boldsymbol{\theta}_\ell)^{\mathrm{H}}\Sigma_{\boldsymbol{X}_\ell}(\boldsymbol{\theta}_\ell)^{-1}\boldsymbol{s}(\boldsymbol{\theta}_\ell) \end{aligned} \tag{5.20}$$

where

$$\Sigma_{\boldsymbol{X}_\ell \boldsymbol{Y}}(\hat{\boldsymbol{\theta}}^{[i-1]}) = \left[\boldsymbol{s}(\hat{\theta}_\ell^{[i-1]}) \sum_{\ell=1}^{L} \boldsymbol{s}(\hat{\theta}_\ell^{[i-1]})^{\mathrm{H}}\right] \odot \boldsymbol{R}_\varphi + \beta_\ell \sigma_w^2 \boldsymbol{I} \tag{5.21}$$

$$\Sigma_{\boldsymbol{Y}}(\hat{\boldsymbol{\theta}}^{[i-1]}) = \left[\sum_{\ell=1}^{L} \boldsymbol{s}(\hat{\theta}_\ell^{[i-1]}) \sum_{\ell=1}^{L} \boldsymbol{s}(\hat{\theta}_\ell^{[i-1]})^{\mathrm{H}}\right] \odot \boldsymbol{R}_\varphi + \sigma_w^2 \boldsymbol{I} \tag{5.22}$$

$$\mu_{\boldsymbol{X}_\ell|\boldsymbol{Y}}(\hat{\boldsymbol{\theta}}^{[i-1]}) = \boldsymbol{s}(\boldsymbol{\theta}_\ell) + \Sigma_{\boldsymbol{X}_\ell \boldsymbol{Y}}(\hat{\boldsymbol{\theta}}^{[i-1]})\Sigma_{\boldsymbol{Y}}(\hat{\boldsymbol{\theta}}^{[i-1]})^{-1}\left(\boldsymbol{y} - \sum_{\ell=1}^{L} \boldsymbol{s}(\hat{\boldsymbol{\theta}}_\ell^{[i-1]})\right) \tag{5.23}$$

The estimate of $\boldsymbol{\theta}_\ell$ can be updated in the maximization step by solving the optimization problem

$$\hat{\boldsymbol{\theta}}_\ell^{[i]} = \arg\max_{\boldsymbol{\theta}_\ell}\{Q(\boldsymbol{\theta}_\ell; \boldsymbol{y}, \boldsymbol{R}_\varphi, \hat{\boldsymbol{\theta}}^{[n-1]})\} \tag{5.24}$$

The multi-dimensional optimization problem in Eq. (5.24) can be solved by element-wise maximization of the objective function as described by Fleury et al. [1999].

The performance of the derived SAGE algorithm has been evaluated by simulations in a single-path scenario by Taparugssanagorn et al. [2007b], who obtained the covariance matrix of the phase noise Σ_φ using realistic measurement data. They compared the graphs of three objective functions [Taparugssanagorn et al. 2007b, Fig. 3]:

- the standard SAGE algorithm processing no-phase-noise data
- the standard SAGE algorithm with phase-noise data
- the modified SAGE algorithm with phase-noise data.

The curvature of the three graphs at the location of correct estimates was compared and it was shown that when used to process phase-noise-distorted data, the modified SAGE algorithm exhibits a similar objective function as when the standard SAGE processes clean data. This demonstrates that the modified SAGE algorithm has a whitening function, mitigating the impact of phase noise on propagation-path parameter estimation. The comparison of the root mean-square estimation error (RMSEE) curves obtained by Monte-Carlo simulations of different SAGE algorithms were also illustrated in the same paper [Taparugssanagorn et al. 2007b, Fig. 4]. The results show that the RMSEEs for the path parameters – Doppler frequency, azimuth of departure, and azimuth of arrival – are all reduced when the modified SAGE algorithm is applied.

5.4 Directional Radiation Patterns

Directional antennas or arrays are used for interference suppression among multiple sectors in a single cell and coverage improvement for hotspot areas in many wireless communication systems. Recently, the use of high-resolution instantaneous channel characteristics for optimizing the performance of communication system has become more viable, as advanced wideband communication systems equipped with antenna arrays now allow estimation of channels in multiple dimensions, including delay, Doppler frequency, and direction. However, communication systems are not designed with the purpose of measuring channel properties. For example, a directional antenna or arrays with the composite array gain confined in a small range, as illustrated in Figure 5.6, which was generated using a 4 × 1 patch-antenna array in a node-B of a Long-term Evolution system, is often used in channel measurement and parameter estimation.

Channel parameter estimation with directional antenna arrays has not been thoroughly studied. A general principle is that direction estimation can only be performed for the region covered by the main beam of the array [Sanudin et al. 2011a]. How to determine a valid region for parameter estimation is still an open question. Sanudin et al. [2011b] proposed the "Capon-alike" algorithm for DoA estimation using directional antenna arrays. This algorithm works under an assumption that there is no coupling between antennas. Practical experience tell us that without specification of the restricted estimation range, high-resolution parameter estimation algorithms, such as the SAGE algorithm, return spurious or "ghost" paths, which do not exist in reality. Some of these spurious paths have complex attenuations with unrealistically large magnitudes. It is evident that these obviously incorrect estimates will significantly alter the properties and statistics that are extracted for adapting communication modes or for channel modeling.

In this section, the impact of the antenna-array gains on the estimates of channel parameters is analyzed. A practical method of choosing a direction estimation range for the original SAGE algorithm was proposed by Yin et al. [2012], and will now be outlined. This method is applied where directional multi-antenna arrays are used for channel measurement. The direction estimation range is determined by specifying appropriate values for the dynamic range of array gain (DRAG) of directional arrays. Simulation and measurement results show that when the DRAG is properly selected, the estimation errors remain at acceptable levels and spurious paths – estimated paths without real counterparts – are negligible.

As an example, we consider a narrowband 1 × N single-input, multiple-output (SIMO) scenario in which the propagation paths between the Tx and Rx have different directions of

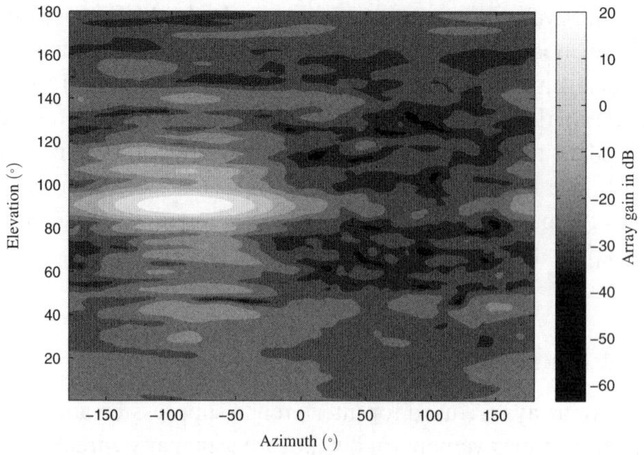

Figure 5.6 The azimuth-elevation pattern of array gain generated by a 4 × 1 patch-antenna array in a node-B of a Long-Term-Evolution (LTE) system

arrival $\boldsymbol{\Omega}$. The narrowband representation of the impulse responses $\boldsymbol{h} \in \mathcal{C}^N$ of this SIMO channel is written as

$$\boldsymbol{h} = \sum_{\ell=1}^{L} \alpha_\ell \boldsymbol{c}(\boldsymbol{\Omega}_\ell) + \boldsymbol{w} \tag{5.25}$$

where ℓ is the index of specular propagation paths, L represents the number of paths, α_ℓ and $\boldsymbol{\Omega}_\ell$ are, respectively, the complex attenuation and the DoA of the ℓth path, $\boldsymbol{w}$ represents standard white Gaussian noise with a spectral height of N_o. The DoA $\boldsymbol{\Omega}_\ell$ is a unit vector uniquely determined by the azimuth of arrival (AoA) $\phi \in [-\pi, \pi]$ and elevation of arrival $\theta \in [0, \pi]$ as $\boldsymbol{\Omega} = [\sin(\theta)\cos(\phi), \sin(\theta)\sin(\phi), \cos(\theta)]$. In Eq. (5.25), $\boldsymbol{c}(\boldsymbol{\Omega})$ represents the array response at a given DoA. The parameters of interest for estimation in the generic channel model Eq. (5.25) are

$$\boldsymbol{\Theta} = (\alpha_1, \boldsymbol{\Omega}_1, \alpha_2, \boldsymbol{\Omega}_2, \ldots, \alpha_N, \boldsymbol{\Omega}_N) \tag{5.26}$$

The SAGE algorithm can be applied as an approximation of the ML method with tractable complexity, via updating parameter estimates $\hat{\boldsymbol{\Theta}}$ element-wise iteratively. For the case considered here, the admissible hidden data space can be selected as the contribution of individual paths to the overall impulse response $\boldsymbol{h}$:

$$\boldsymbol{x}_\ell = \alpha_\ell \boldsymbol{c}(\boldsymbol{\Omega}_\ell) + \boldsymbol{w}, \ell = 1, \ldots, L \tag{5.27}$$

The unknown parameters $\boldsymbol{\Theta}$ are split into multiple subsets for individual paths: $\boldsymbol{\theta}_\ell = [\alpha_\ell, \boldsymbol{\Omega}_\ell]$. In each iteration of the algorithm, the log-likelihood function of the parameters in one subset, say $\boldsymbol{\theta}_\ell$, is calculated as

$$\Lambda(\boldsymbol{\theta}_\ell) = \log p[\boldsymbol{\theta}_\ell | \boldsymbol{h}, \hat{\boldsymbol{\Theta}}^{[i]}] \tag{5.28}$$

where $p(\boldsymbol{\theta})$ denotes the likelihood function of $\boldsymbol{\theta}$ and $\hat{\boldsymbol{\Theta}}^{[i]}$ represents the parameter estimates updated in the ith iteration. Applying the assumption that the noise component $\boldsymbol{w}$ is white

Gaussian, it can be shown that

$$\Lambda(\boldsymbol{\theta}_\ell) = -N\log(2\pi\sigma_w) - \frac{1}{2N\sigma_w^2} \| \hat{\boldsymbol{x}}_\ell^{[i+1]} - \alpha_\ell \boldsymbol{c}(\boldsymbol{\Omega}_\ell) \|^2$$

where

$$\begin{aligned}\hat{\boldsymbol{x}}^{[i+1]} &= \mathrm{E}[\boldsymbol{x}_\ell | \boldsymbol{h}, \hat{\boldsymbol{\Theta}}^{[i]}] \\ &= \boldsymbol{h} - \sum_{\ell'=1,\ell'\neq\ell}^{L} \hat{\alpha}_{\ell'}^{[i]} \boldsymbol{c}(\hat{\boldsymbol{\Omega}}_{\ell'}^{[i]})\end{aligned} \tag{5.29}$$

is an estimate of $\boldsymbol{x}_\ell$ given $\boldsymbol{h}$ and $\hat{\boldsymbol{\Theta}}^{[i]}$.

After some manipulations, it can be shown that the estimate $\hat{\boldsymbol{\Omega}}_\ell^{[i+1]}$ and $\hat{\alpha}_\ell^{[n+1]}$ are calculated respectively as

$$\hat{\boldsymbol{\Omega}}_\ell^{[i+1]} = \arg\max_{\boldsymbol{\Omega}} L(\boldsymbol{\Omega}), \text{ with } L(\boldsymbol{\Omega}) = \frac{(\hat{\boldsymbol{x}}_\ell^{[i+1]})^{\mathrm{H}} \boldsymbol{c}(\boldsymbol{\Omega}) \boldsymbol{c}(\boldsymbol{\Omega})^{\mathrm{H}} \hat{\boldsymbol{x}}_\ell^{[i+1]}}{\boldsymbol{c}(\boldsymbol{\Omega})^{\mathrm{H}} \boldsymbol{c}(\boldsymbol{\Omega})}, \text{ and} \tag{5.30}$$

$$\hat{\alpha}_\ell^{[n+1]} = \frac{\boldsymbol{c}(\hat{\boldsymbol{\Omega}}_\ell^{[i+1]})^{\mathrm{H}} \hat{\boldsymbol{x}}_\ell^{[i+1]}}{\boldsymbol{c}(\hat{\boldsymbol{\Omega}}_\ell^{[i+1]})^{\mathrm{H}} \boldsymbol{c}(\hat{\boldsymbol{\Omega}}_\ell^{[i+1]})} \tag{5.31}$$

where $(\cdot)^{\mathrm{H}}$ is the Hermitian operation. Now we continue by showing how the directional array response influences parameter estimation using SAGE. In real measurements, the exact antenna responses, denoted by $\boldsymbol{c}_t(\boldsymbol{\Omega})$, may be different from the measured response:

$$\boldsymbol{c}_t(\boldsymbol{\Omega}) = \boldsymbol{c}(\boldsymbol{\Omega}) + \Delta(\boldsymbol{\Omega}) \tag{5.32}$$

where $\Delta(\boldsymbol{\Omega})$ represents the calibration error: the deviation between the exact response and the response measured in an anechoic chamber. Inserting $\hat{x}_\ell^{[i+1]} = \alpha_\ell \boldsymbol{c}_t(\boldsymbol{\Omega}_\ell)$ into Eqs (5.30) and (5.31) yields

$$L(\boldsymbol{\Omega}) = |\alpha_\ell|^2 |\boldsymbol{c}(\boldsymbol{\Omega}_\ell)^{\mathrm{H}} \tilde{\boldsymbol{c}}(\boldsymbol{\Omega}) + \Delta(\boldsymbol{\Omega}_\ell)^{\mathrm{H}} \tilde{\boldsymbol{c}}(\boldsymbol{\Omega})|^2, \tag{5.33}$$

$$\hat{\alpha}_\ell = \alpha_\ell \cdot \left(\frac{\boldsymbol{c}(\boldsymbol{\Omega}_\ell)^{\mathrm{H}} \tilde{\boldsymbol{c}}(\hat{\boldsymbol{\Omega}}_\ell)}{\| \boldsymbol{c}(\hat{\boldsymbol{\Omega}}_\ell) \|} + \frac{\Delta(\boldsymbol{\Omega}_\ell)^{\mathrm{H}} \tilde{\boldsymbol{c}}(\hat{\boldsymbol{\Omega}}_\ell)}{\| \boldsymbol{c}(\hat{\boldsymbol{\Omega}}_\ell) \|} \right) \tag{5.34}$$

with $\tilde{\boldsymbol{c}}(\boldsymbol{\Omega}) = \boldsymbol{c}(\boldsymbol{\Omega}) / \| \boldsymbol{c}(\boldsymbol{\Omega}) \|$. Due to the existence of nonzero $\Delta(\boldsymbol{\Omega}_\ell)^{\mathrm{H}} \tilde{\boldsymbol{c}}(\boldsymbol{\Omega})$, $\hat{\boldsymbol{\Omega}}_\ell$ becomes inconsistent with $\boldsymbol{\Omega}_\ell$, and, additionally, $\hat{\alpha}_\ell$ may be an overestimate since the value in the parentheses on the right-hand side of Eq. (5.34) may be very high in cases of small $\| \boldsymbol{c}(\hat{\boldsymbol{\Omega}}_\ell) \|$. These estimation errors can propagate in the iterations of the SAGE algorithm. The parameter estimates $\hat{\boldsymbol{\Theta}}$ then become inapplicable for channel modeling.

In order to avoid the errors introduced by directional antenna arrays, it is necessary to specify a direction estimation range (DER) within which the SAGE algorithm returns reliable estimates. The setting of DER prohibits the SAGE algorithm from estimating any paths in the directional region where $\| \boldsymbol{c}(\boldsymbol{\Omega}) \|$ is small. The exact shape of the array gain in azimuth and elevation needs to be taken into account when determining the DER. Because of this, the dynamic range of array gain (DRAG), denoted by γ_{a}, is introduced to represent the array gain in decibels compared to the maximum of the array gain. In the following, a discussion of how to specify the DRAG is provided.

It is obvious that $\gamma_{\rm a}$ should be within the range of $[-\gamma_{\max}, 0]$, with $\gamma_{\rm a,max}$ computed as

$$\gamma_{\rm a,max} = 10\log_{10}\frac{\max(\| \boldsymbol{c}(\boldsymbol{\Omega}) \|; \boldsymbol{\Omega} \in \mathcal{S}^3)}{\min(\| \boldsymbol{c}(\boldsymbol{\Omega}) \|; \boldsymbol{\Omega} \in \mathcal{S}^3)} \tag{5.35}$$

with $\mathcal{S}^3$ being a unit sphere. Empirically, an appropriate selection of $\gamma_{\rm a}$ is jointly determined by the SNR per path $\gamma_{{\rm p},\ell} = 10\ \log_{10}\frac{P_\ell}{\sigma_w^2}$ and the intrinsic dynamic range $\gamma_{\rm i}$, which is determined by the receiver sensitivity and the thermal noise in the sounding equipment. To guarantee that the parameter estimates returned by the SAGE algorithm are valid, the inequality below should be satisfied for individual paths

$$\gamma_{\rm a} + \gamma_{{\rm p},\ell} \geq \gamma_{\rm i}, \ell \in [1, \ldots, L] \tag{5.36}$$

The lower bound of $\gamma_{\rm a}$ can then be determined as

$$\gamma_{\rm a} \geq \gamma_{\rm i} - \gamma_{\rm p} \tag{5.37}$$

with $\gamma_{\rm p} = \max\{\gamma_{{\rm p},\ell}; \ell \in [1, \ldots, L]\}$. An empirical way of obtaining $\gamma_{\rm p}$ is to estimate a single path by using the SAGE algorithm and then to use the estimate of the path power to compute $\gamma_{\rm p}$.

Based on the above analysis, the procedure for determining the DRAG and DER consists of three steps:

1. For the given measurement data, estimate a single path and get $\gamma_{\rm p}$.
2. Specify $\gamma_{\rm i}$, which depends on the specification of the sounding equipment.
3. Compute the lower bound of $\gamma_{\rm a}$ from Eq. (5.37) and determine DER as the region enclosed by a contour at the height of the DRAG on the directional array gain pattern.

Monte-Carlo simulations are conducted to investigate the impact of different DRAG settings on the SAGE algorithm performance. Received signals are generated for a 1×4 SIMO system with an omnidirectional Tx antenna and a realistic 1×4 patch-antenna array in the Rx. Figure 5.6 depicts the array gain of the Rx computed from the responses of the antennas when measured in an anechoic chamber. It can be observed that the Rx array exhibits significant gain within azimuths ranging from $-150°$ to $-50°$ and elevations from 80° to 100°.

Simulations are performed with respect to γ_p and γ_a. In each snapshot, the specified $\gamma_{\rm p}$ is applied to generate the white Gaussian noise components, and ten propagation paths are created, with their azimuths and elevations of arrival within the range determined from a specified $\gamma_{\rm a}$. Then the SAGE algorithm is applied to extracting the parameters of the ten paths by selecting an $\gamma_{\rm a}$ that fulfills the inequality $\gamma_{\rm a} > \gamma_{\rm p}$. The simulation results shown here were obtained for $\gamma_{\rm a} \in [-12, \ldots, -3]$ dB.

To evaluate whether the SAGE estimation result is consistent with the true values, the environment characterization metric $\boldsymbol{C}$ is applied, which is calculated as [Czink et al. 2006]:

$$\boldsymbol{C} = \left(\sum_{\ell=1}^{L} P_\ell\right)^{-1} \sum_{\ell=1}^{L} P_\ell(\boldsymbol{\Omega}_\ell - \bar{\boldsymbol{\Omega}})(\boldsymbol{\Omega}_\ell - \bar{\boldsymbol{\Omega}})^{\rm T} \tag{5.38}$$

with $P_\ell = |\alpha_\ell|^2$, $(\cdot)^{\mathrm{T}}$ denoting the transpose operation and

$$\bar{\boldsymbol{\Omega}}_\ell = \left(\sum_{\ell=1}^{L} P_\ell\right)^{-1} \sum_{\ell=1}^{L} P_\ell \boldsymbol{\Omega}_\ell.$$

Using $\boldsymbol{C}_{\mathrm{tru}}$ and $\boldsymbol{C}_{\mathrm{est}}$ to represent the ECMs for the true channel and the estimated channel respectively, a measure ζ of the deviation between the two channels is defined as

$$\zeta = 10 \log_{10} \frac{\det(\boldsymbol{C}_{\mathrm{est}})}{\det(\boldsymbol{C}_{\mathrm{tru}})} \tag{5.39}$$

where $\det(\cdot)$ denotes the determinant of the matrix given as an argument.

The determinant of a matrix is the product of its eigenvalues. The eigenvalues of $\boldsymbol{C}$ in the case considered here represent the spreads of the channel in the multi-dimensional parameter space. Therefore, when a negative ζ is obtained, the spread of the estimated channel is less than the spread of the true channel. For positive ζ, the spread of estimated channel is larger than that of the true channel.

Figure 5.7 depicts the simulation results for ζ versus $\gamma_{\mathrm{a}} \in [-12, \ldots, -3]$ dB and $\gamma_{\mathrm{p}} \in [30, 40, \ldots, 70]$ dB. For each combination of γ_{a} and γ_{p}, 200 snapshots are performed to obtain an average ζ. It can be seen from Figure 5.7 that $\zeta \approx 0$ dB when γ_{p} is above 50 dB, indicating that the estimated channel is almost identical to the synthetic channel. When γ_{p} is less than 50 dB, ζ becomes negative, with its absolute value increasing when γ_{p} decreases. This shows that the effect of spurious paths is more significant when the SNR decreases and when directional antenna arrays are used. It can also been seen from Figure 5.7 that for a fixed γ_{p}, the absolute value of ζ increases when $|\gamma_{\mathrm{a}}|$ increases. Specifically, for $\gamma_{\mathrm{a}} > -6$ dB, we observe a deviation $|\zeta| < 6$ dB, which we consider to be acceptable from an empirical point of view. It is interesting to see from these simulation results that ζ usually takes negative

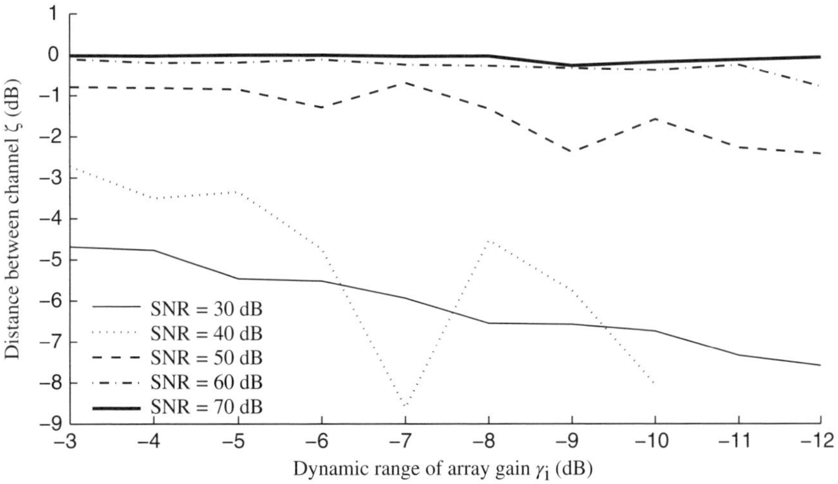

Figure 5.7 The difference between synthetic and estimated channels, denoted by ζ, versus the dynamic range of array gain

values, implying that the estimated channel appears more concentrated than the true channel in the parameter space. Our conjecture is that this is due to the fact that when a spurious path is obtained, its magnitude is usually large in such a way that the channel power spectrum becomes more concentrated around the location of the spurious path.

The measurement data were collected in 2005 in a campaign jointly conducted by Elektrobit and the Technology University of Vienna. The campaign used the wideband MIMO sounder, PROPSound, in a building of Oulu University. Figure 5.8(a) depicts the 50-element antenna array used in the Tx and Rx. Figure 5.8(b) illustrates the indices of the antennas in the array. Notice that the 18 antennas from No. 19 to 36 were not used in the Rx during the measurements. Thus, in total, $50 \times 32 = 1600$ spatial subchannels are measured in one measurement cycle.

We select one cycle of data for the experimental evaluation considered here. The data processing is conducted for three scenarios. In the first scenario, all 50×32 subchannels are considered; in the second and third scenarios, only 50×10 subchannels are considered. The indices of the 10 antennas selected in the Rx array are marked with underscored bold fonts in Figure 5.8(b). Figures 5.8(c) and 5.8(d) illustrate the array gain for the 32-element antenna

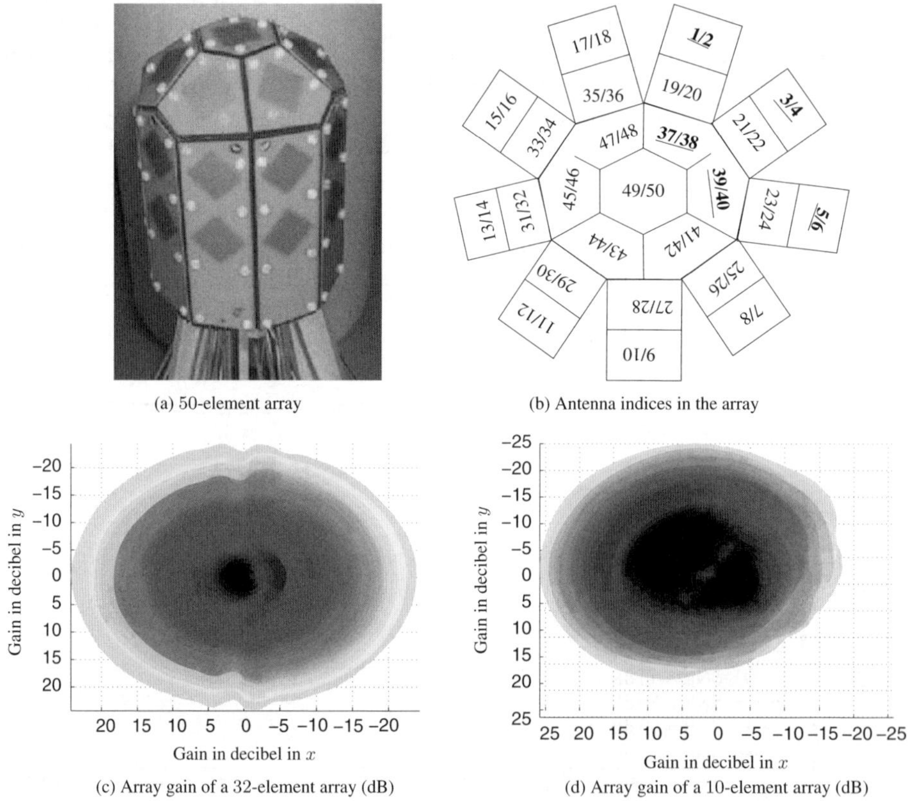

(a) 50-element array

(b) Antenna indices in the array

(c) Array gain of a 32-element array (dB)

(d) Array gain of a 10-element array (dB)

Figure 5.8 Elektrobit/Technology University of Vienna measurement campaign: (a) antenna array used in measurements; (b) indices of the array element; array gain for (c) the omnidirectional and (d) directional Rx array

array and for the 10-element antenna sub-array respectively. It can be seen that when 32 antennas are considered, the Rx array's gain is uniform in azimuth within the range $[0°, 360°)$; when only 10 antennas are considered, the array gain exhibits certain directions. The color in Figure 5.8(c) and 5.8(d) only indicates the shape of the array gain, and does not code the array gain.

Figure 5.9 depicts the average of the power-delay profiles (PDPs) of the 1600 subchannels. Since the 32-element antenna array is almost omnidirectional, the PDP shown in Figure 5.9 is actually obtained with constant array gain irrespective of the impinging directions. Thus we consider the maximum SNR identified from Figure 5.9 to be the value for γ_p; that is $\gamma_\text{p} = 30$ dB. Thus the lower bound of the DRAG γ_a is $\gamma_\text{i} - 30$ dB, with $\gamma_i = 20$ dB [Czink 2007].

The SAGE algorithm is applied to estimate the parameters – DoD, DoA, delay, and complex attenuation – of 10 paths by using 50×32 or 50×10 subchannels. In the latter case, the DRAG γ_a within the range $[-2, -9]$ dB is considered for the Rx array, which imposes corresponding restrictions on the AoA and elevation of arrival (EoA) estimation range.

Figure 5.10(a)–(f) illustrates the estimated DoA, DoD, delay and magnitude of the complex attenuation of 10 paths for three cases:

1. 50×32 subchannels
2. 50×10 subchannels and $\gamma_\text{a} = -8$ dB
3. 50×10 subchannels and $\gamma_\text{a} = -9$ dB.

The gray scale of the spots indicates the delay estimates in sample intervals. The size of the spots codes the magnitude of the complex attenuation in decibels. We can observe from Figure 5.10(a) that there are two groups of paths separated in DoA when all 32 Rx antennas are considered. However, when only 10 Rx antennas are used, the group located in the AoA range $[160°, 170°]$ becomes unobservable (Figure 5.10(c)). It can be also observed from Figure 5.10(c) that:

- the spot observed around $0°$ AoA in the first case in Figure 5.10(a) also appears in Figure 5.10(c) for the second case
- a new group of paths with lower magnitude are estimated within the selected AoA range of $[-164°, 92°]$ and EoA range of $[-62°, 90°]$.

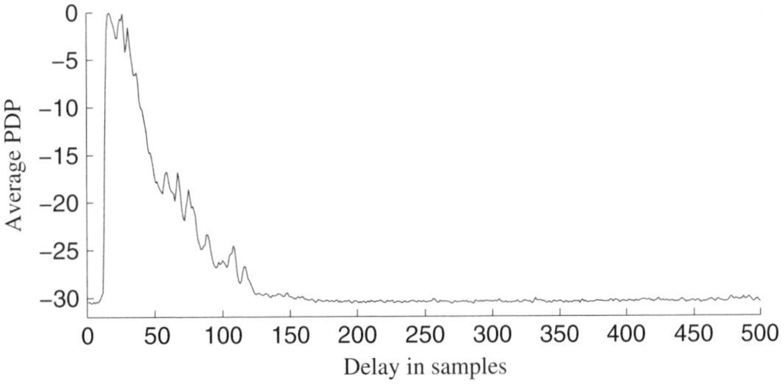

Figure 5.9 Average PDP of channels observed in a cycle considered

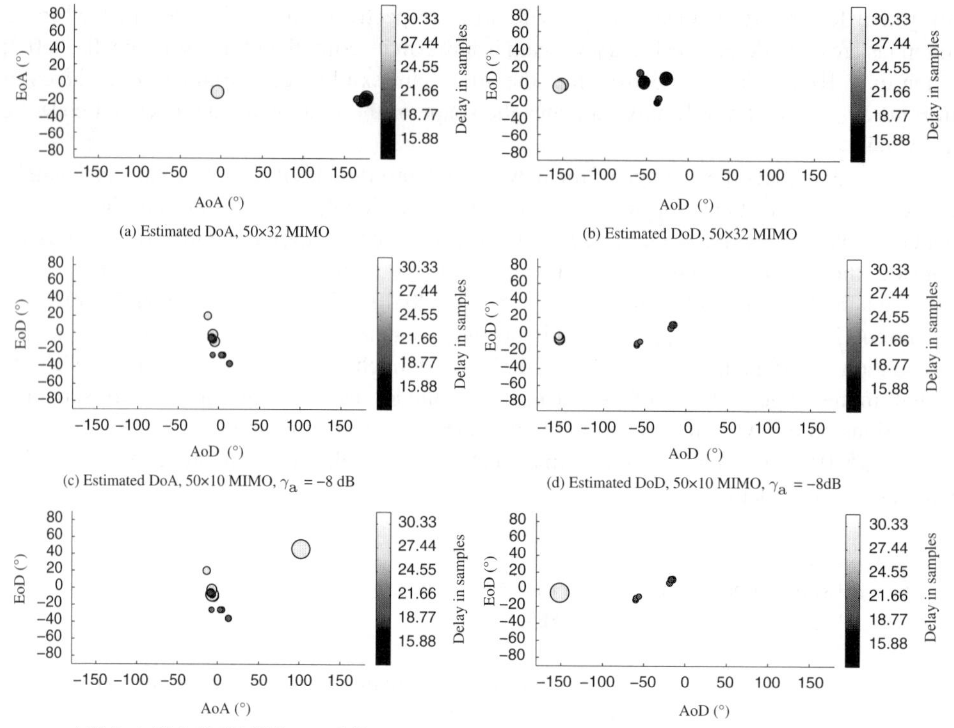

(a) Estimated DoA, 50x32 MIMO (b) Estimated DoD, 50x32 MIMO (c) Estimated DoA, 50x10 MIMO, γ_a = −8 dB (d) Estimated DoD, 50x10 MIMO, γ_a = −8dB (e) Estimated DoA, 50x10 MIMO, γ_a = −9 dB (e) Estimated DoD, 50x10 MIMO, γ_a = −9 dB

Figure 5.10 Path-parameter estimates obtained using the SAGE algorithm from the measurement data with the following settings: (a) and (b), data collected for 50×32 subchannels; (c) and (d), data collected for 50×10 subchannels, and $\gamma_a = -8$ dB; (e) and (f), data collected for 50×10 subchannels, and $\gamma_a = -9$ dB. AoA, azimuth of arrival; AoD, azimuth of departure, EoA, elevation of departure; EoD, elevation of departure

These observations show that when the AoA estimation is restricted to within the range where $\gamma_a > -8$ dB holds, the true paths existing in this region can still be estimated, and new paths with low magnitude can be found as well. However, from Figure 5.10(e) we observe that for $\gamma_a = -9$ dB, which leads to an AoA range of $[-178°, 102°]$ and an EoA range of $[-64°, 90°]$, a spurious path of very large magnitude is found. These results demonstrate the necessity of defining the DRAG properly for parameter estimation when directional arrays are used in measurements.

5.5 Switching-mode Selection

As described in Section 5.1, MIMO chnanel sounders commonly operate with the switched architecture. The sounding signal is fed successively at the ports of the array elements at the Tx, and while any one of these elements transmits, the ports of the antenna elements at the Rx are sensed successively. Some new terminology is adopted here to describe a TDM mode: an *element pair* is a pair containing an element of the Tx array in the first position and an element of the Rx array in the second position; a *measurement cycle* denotes the process where all element pairs are switched once; a *cycle interval* is the period separating the beginning of two consecutive measurement cycles; *the switching interval* refers to the separation between the beginning of two consecutive sensing periods within one measurement cycle; *the cycle rate* and *the switching rate* are the inverses of the cycle interval and the switching interval respectively. The ratio of the switching rate to the cycle rate is at least equal to the product of the element numbers of the two arrays.

It was traditionally believed that the maximum absolute Doppler frequency that can be estimated using the TDM-MIMO sounding technique is half the cycle rate. Therefore, by keeping the switching rate unchanged, large element numbers in the arrays result in a low cycle rate and consequently lead to a small Doppler frequency estimation range (DFER).

In this section, we show that the maximum absolute Doppler frequency that can be estimated using the TDM-MIMO sounding technique actually equals half the *switching* rate. This enlarged DFER is independent of the element numbers of the arrays. This conclusion is drawn under the condition that the parameters are known in advance, except for the Doppler frequency. When the parameters are unknown and need to be estimated, the extension of the DFER may result in ambiguity in the estimation of the Doppler frequency and directions (DoD and DoA). This occurs when the *switching mode* – the temporal order in which the array elements are switched – is chosen inappropriately.

We also analyse the impact of the switching modes of the arrays in TDM-MIMO sounding on the joint estimation of Doppler frequency and directions using the SAGE algorithm. A general principle for optimizing the switching mode is introduced.[2]

5.5.1 Switching-mode for Channel Sounding

Figure 5.11 depicts the time structure used in the sounding of a propagation channel performed in TDM mode. The term SW represents "switch". To characterize the switching mode of a switched array, we first need to define a (spatial) indexing of the array elements, which is then kept fixed. The natural element indexing for a uniform linear array is according to the element

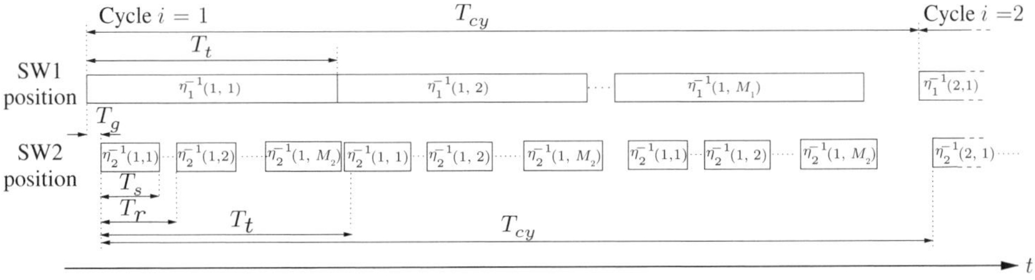

Figure 5.11 The considered TDM measurement mode

[2] Most of the results described in this section were reported by Troels Pedersen [Pedersen and Pedersen 2004, Pedersen et al. 2008a, 2004], the main architect of the work related to switching-mode optimization. For a detailed treatment of switching-mode optimization, readers are referred to the work of Pedersen et al. [2008a].

spatial ordering, starting at one end. Similarly, the natural element indexing of a uniform planar array is determined first by the order of the element row and then by the element order inside its row. The switching mode of an array during one cycle is entirely defined by a permutation of the element indices. Let $\eta_k(i,\cdot)$ denote (the permutation describing) the switching mode of Array k during the ith cycle. Referring to Figure 5.11, the beginning of the interval when the element pair (m_1, m_2) is switched in the ith cycle is

$$t_{i,m_1,m_2} \doteq \left(i - \frac{I+1}{2}\right) T_{\rm cy} + \left(\eta_1(i,m_1) - \frac{M_1+1}{2}\right) T_t + \left(\eta_2(i,m_2) - \frac{M_2+1}{2}\right) T_r. \tag{5.40}$$

Clearly, $\eta_k(i, m_k)$ is the time index of the interval during which the m_kth element of Array k is switched during the ith cycle ($m_k = 1, \ldots, M_k$). Hence, $\eta_k(i,\cdot)$ maps a spatial index onto a time index. The inverse mapping $\eta_k^{-1}(i,\cdot)$ (shown in Figure 5.11) determines the temporal order in which the elements of Array k are sequentially switched in the ith cycle. Notice that the switching mode of Array 2 does not depend on which element of Array 1 is active during each cycle; in other words $\eta_2(i,\cdot)$ does not depend on m_1. For notational convenience we identify the permutation $\eta_k(i,\cdot)$ with the vector

$$\boldsymbol{\eta}_k(i) = [\eta_k(i, m_k), m_k = 1, \ldots, M_k]$$

If $\boldsymbol{\eta}_k(i) = \boldsymbol{\eta}_k, i = 1, \ldots, I$, the switching mode is termed "cycle-independent". The identity switching mode $\boldsymbol{\eta}_k = [1, \ldots, M_k]$ switches the elements of Array k in their spatial order (see Figure 5.5.1).

The scalar signal at the output of SW2 can be described as

$$Y(t) = \sum_{\ell=1}^{L} s(t; \boldsymbol{\theta}_\ell) + \sqrt{\frac{N_0}{2}} q_2(t) W(t) \tag{5.41}$$

For simplicity, we replace the polarization matrix $\boldsymbol{A}_\ell$ of Path ℓ by a scalar α_ℓ that represents the weight of the propagation path. In this case, the contribution of the wave propagating along the ℓth path in the received signal reads

$$s(t; \boldsymbol{\theta}_\ell) = \alpha_\ell \exp\{j2\pi\nu_\ell t\} \boldsymbol{c}_2(\boldsymbol{\Omega}_{2,\ell})^{\rm T} \boldsymbol{U}(t; \tau_\ell) \boldsymbol{c}_1(\boldsymbol{\Omega}_{1,\ell})\}$$

5.5.2 Estimation of Doppler Frequency

At each iteration of the SAGE algorithm, the parameter estimates of the ℓth path are updated in the M-step of the algorithm. This step computes the argument maximizing an objective function $|z(\overline{\boldsymbol{\theta}}_\ell; \hat{x}_\ell)|$, where $\overline{\boldsymbol{\theta}}_\ell \doteq [\boldsymbol{\Omega}_{1,\ell}, \boldsymbol{\Omega}_{2,\ell}, \tau_\ell, \nu_\ell]$ and $|\cdot|$ denotes the norm of the scalar or the vector given as an argument. Notice that the objective function coincides with the ML estimate of $\overline{\boldsymbol{\theta}}_\ell$ in a one-path scenario, in which case $\hat{x}_\ell(t) = y(t)$. In the case where the polarization matrix is replaced by a scalar, the function $z(\overline{\boldsymbol{\theta}}_\ell; \hat{x}_\ell)$ can be calculated to be

$$z(\overline{\boldsymbol{\theta}}_\ell; \hat{x}_\ell) \doteq \tilde{\boldsymbol{c}}_2(\boldsymbol{\Omega}_{2,\ell})^H \boldsymbol{X}_\ell(\tau_\ell, \nu_\ell; \hat{x}_\ell) \tilde{\boldsymbol{c}}_1(\boldsymbol{\Omega}_{1,\ell})^* \tag{5.42}$$

with $(\cdot)^*$ representing the complex conjugate, and

$$\tilde{\boldsymbol{c}}_k(\boldsymbol{\Omega}) \doteq |\boldsymbol{c}_k(\boldsymbol{\Omega})|^{-1} \boldsymbol{c}_k(\boldsymbol{\Omega})$$

being the normalized response of Array k. The entries of the $(M_2 \times M_1)$ matrix $\boldsymbol{X}(\tau_\ell, \nu_\ell; \hat{x}_\ell)$ read

$$X_{\ell,m_2,m_1}(\tau_\ell, \nu_\ell; \hat{x}_\ell) = \sum_{i=1}^{I} \left[\exp\{-j2\pi\nu_\ell t_{i,m_1,m_2}\} \cdot \int_0^{T_{\text{sc}}} u^*(t-\tau_\ell) \exp\{-j2\pi\nu_\ell t\} \hat{x}_\ell(t + t_{i,m_1,m_2})\, \mathrm{d}t \right], \tag{5.43}$$

$m_k = 1, \ldots, M_k$, $k = 1, 2$. In Eq. (5.43) $\hat{x}_\ell(t) = y(t) - \sum_{\ell'=1,\ell'\neq\ell}^{L} s(t; \hat{\boldsymbol{\theta}}_{\ell'})$, with $\hat{\boldsymbol{\theta}}_{\ell'}$ denoting the current estimate of $\boldsymbol{\theta}_{\ell'}$, is an estimate of the admissible hidden data X_ℓ.

We next analyze the behavior of the objective function versus the Doppler frequency, the DoD and DoA. For simplicity, we first make the following four assumptions:

1. The antenna elements are isotropic.
2. The phase change due to the Doppler frequency within T_s is neglected; the term $\exp\{-j2\pi\nu_\ell t\}$ in Eq. (5.43) is set equal to 1. As shown in Appendix A.1 and Section 5.5.6, this effect can be easily included in the model and its impact on the performance of the Doppler frequency estimate proves to be negligible.
3. The remaining interference contributed by the waves ℓ', $\ell' \in \{1, \ldots, L\}/\{\ell\}$ in the estimate $\hat{x}_\ell(t)$ computed in the E-step[3] of Path ℓ is negligible:

$$\hat{x}_\ell(t) = s(t; \boldsymbol{\theta}_\ell) + \sqrt{\frac{N_0}{2}} q_2(t) W(t)$$

   Under this assumption, the M-step of Path ℓ is derived using an equivalent signal model where only Path ℓ is present. If we further focus attention on one particular path, which without loss of generality is selected to be Path 1, then Eq. (5.41) with $L = 1$ is the equivalent signal model for the derivation of the M-step of Path 1. In this case, $\hat{x}_1(t) = y(t)$ and the ML estimate of $\overline{\boldsymbol{\theta}}_1$ is computed in the M-step. For notational convenience we now drop the indexing for the parameters of Path 1.
4. As the focus is on the estimation of the Doppler frequency, DoD, and DoA, we further assume that the SAGE algorithm has perfectly estimated the delay of Path 1 or has knowledge of it. As a result $z(\overline{\boldsymbol{\theta}}; y)$ reduces to a function of $\boldsymbol{\Omega}_1$, $\boldsymbol{\Omega}_2$, and ν according to

$$z(\nu, \boldsymbol{\Omega}_1, \boldsymbol{\Omega}_2; y) = \sum_{i=1}^{I} \sum_{m_2=1}^{M_2} \sum_{m_1=1}^{M_1} \tilde{c}_{1,m_1}(\boldsymbol{\Omega}_1)^* \tilde{c}_{2,m_2}(\boldsymbol{\Omega}_2)^* \cdot \exp\{-j2\pi\nu t_{i,m_1,m_2}\} \times \int_0^{T_{\text{sc}}} u(t-\tau')^* y(t + t_{i,m_1,m_2}) \mathrm{d}t \tag{5.44}$$

   The notation $(\cdot)'$ designates the true value of the parameter given as an argument.

[3] In the E-step, the expectation of the admissible hidden data is calculated under the condition that all parameter estimates are known. It can be shown that the estimate $\hat{x}_\ell(t)$ is obtained by canceling the interferences given by the other paths.

In order to analyze the behavior of the objective function with respect to Doppler frequency, we consider the scenario in which all the parameters in θ but ν and α are known. Based on this assumption, $z(\nu, \mathbf{\Omega}_1, \mathbf{\Omega}_2; y)$ in Eq. (5.44) reduces to a function of the Doppler frequency only, given by

$$z(\nu; y) = \sum_{i=1}^{I} \sum_{m_2=1}^{M_2} \sum_{m_1=1}^{M_1} \tilde{c}_{1,m_1}(\mathbf{\Omega}_1')^* \tilde{c}_{2,m_2}(\mathbf{\Omega}_2')^* \exp\{-j2\pi\nu t_{i,m_2,m_1}\} \times \int_0^{T_{\rm sc}} u(t-\tau')^* y(t + t_{i,m_2,m_1}) \mathrm{d}t. \quad (5.45)$$

The Doppler frequency estimate $\hat{\nu}$ is obtained by maximizing $|z(\nu; y)|$ with respect to ν. Assuming that $c_{k,m_k}(\mathbf{\Omega}) \doteq c_k(\mathbf{\Omega})$, $m_k = 1, \ldots, M_k$, $k = 1, 2$, we can recast $|z(\nu; y)|$ as

$$|z(\nu; y)| = a \cdot \left|\left[\sum_{i=1}^{I} \sum_{m_2=1}^{M_2} \sum_{m_1=1}^{M_1} \frac{\exp\{j2\pi(\nu' - \nu)t_{i,m_2,m_1}\}}{IM_2M_1} + V(\nu)\right]\right| \quad (5.46)$$

where $a \doteq I\sqrt{M_2M_1}PT_{\rm sc}|\alpha|\|c_1(\mathbf{\Omega}_1')\|c_2(\mathbf{\Omega}_2')|$. In the expression of a, P denotes the power of $u(t)$. The noise term $V(\nu)$ can be calculated as (see the derivation in Appendix A.2):

$$V(\nu) = \frac{1}{\sqrt{IM_2M_1\gamma_{\rm o}}} \sum_{i=1}^{I} \sum_{m_2=1}^{M_2} \sum_{m_1=1}^{M_1} W'_{i,m_2,m_1} \exp\{-j2\pi\nu t_{i,m_2,m_1}\}$$

where $W'_{i,m_2,m_1}, i = 1, \ldots, I, m_k = 1, \ldots, M_k, k = 1, 2$ are independent complex circularly symmetric random variables with unit variance, and

$$\gamma_{\rm o} \doteq \frac{IM_2M_1P|\alpha|^2|c_1(\mathbf{\Omega}_1')|^2|c_2(\mathbf{\Omega}_2')|^2}{(N_0/T_{\rm sc})}$$

represents the signal to noise ratio (SNR) in $z(\nu; y)$.

Replacing t_{i,m_2,m_1} with its explicit form Eq. (5.40) in Eq. (5.46) yields

$$|z(\nu; y)| = a \cdot |F(\nu - \nu') + V(\nu)| \quad (5.47)$$

where

$$F(\nu) = F_{\rm cy}(\nu)F_{\rm t}(\nu)F_{\rm r}(\nu) \quad (5.48)$$

with

$$F_{\rm cy}(\nu) \doteq \frac{\sin(\pi\nu IT_{\rm cy})}{I\sin(\pi\nu T_{\rm cy})}, F_{\rm t}(\nu) \doteq \frac{\sin(\pi\nu M_1T_{\rm t})}{M_1\sin(\pi\nu T_{\rm t})}, F_{\rm r}(\nu) \doteq \frac{\sin(\pi\nu M_2T_{\rm r})}{M_2\sin(\pi\nu T_{\rm r})} \quad (5.49)$$

As an example, the graphs of $|F_{\rm cy}(\nu)|$, $|F_{\rm t}(\nu)|$, $|F_{\rm r}(\nu)|$, and $|F(\nu)|$ are shown in Figure 5.12 with the parameter settings given in Table 5.1.

It can be observed that $|F(\nu)|$ exhibits a period identical to that of $|F_{\rm r}(\nu)|$, namely $1/T_{\rm r}$. Since the objective function Eq. (5.47) is also periodic with the same period and coincides with $|F(\nu)|$ in the absence of noise, the valid estimation range for the Doppler frequency using

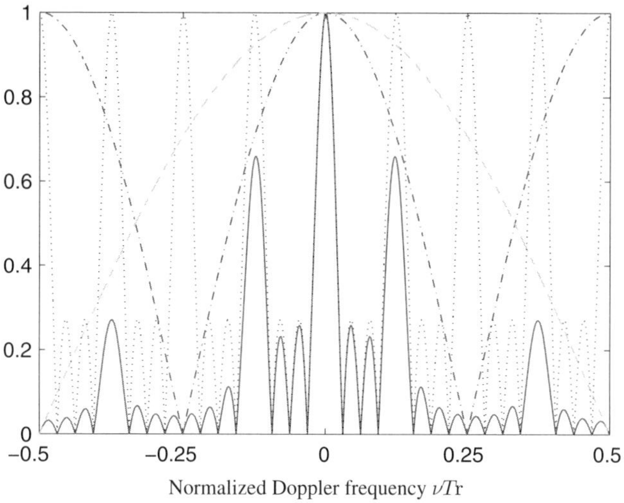

Figure 5.12 The graphs of $|F_{cy}(\nu)|$ (dotted), $|F_t(\nu)|$ (dash-dotted), $|F_r(\nu)|$ (dashed) and $|F(\nu)|$ (solid) for the parameter settings reported in Table 5.1

Table 5.1 Parameter settings used for simulations

I	M_1	M_2	T_{cy} (ms)	T_t (μs)	T_r (μs)	ν (Hz)
4	2	2	3.264	816	408	0

Eq. (5.47) equals $[-1/(2T_r), +1/(2T_r)]$; in other words, the maximum Doppler frequency that can be estimated is half the switching rate.

5.5.3 Ambiguity in Parameter Estimation

The analysis in Section 5.5.2 is carried out under the assumption that the parameters are known except for the Doppler frequency ν and the complex amplitude α. This assumption is not always true in real applications.

In this section, we analyze the objective function in the case in which the directions $\mathbf{\Omega}_1$ and $\mathbf{\Omega}_2$ are unknown and to be estimated jointly with the Doppler frequency. Under this assumption and by dropping a constant term in the objective function Eq. (5.44) and normalizing it by $(IM_1M_2)^{-1}$, Eq. (5.44) can be cast as

$$z(\nu, \mathbf{\Omega}_1, \mathbf{\Omega}_2; y) = \sum_{i=1}^{I} R_i(\check{\nu}) S_i(\check{\mathbf{\Omega}}_1, \check{\nu}) T_i(\check{\mathbf{\Omega}}_2, \check{\nu}) + V(\nu, \mathbf{\Omega}_1, \mathbf{\Omega}_2) \tag{5.50}$$

with the notational convention $\breve{(\cdot)} \doteq (\cdot)' - (\cdot)$. Moreover,

$$R_i(\breve{\nu}) \doteq \frac{1}{I}\exp\{j2\pi\breve{\nu}\left(i - \frac{I+1}{2}\right)T_{\text{cy}}\}$$

$$S_i(\breve{\mathbf{\Omega}}_1, \breve{\nu}) \doteq \frac{1}{M_1}\sum_{m_1=1}^{M_1} \exp\left\{j2\pi\frac{\breve{\mathbf{\Omega}}_1^{\text{T}}\boldsymbol{r}_{1,m_1}}{\lambda} + j2\pi\breve{\nu}\left(\eta_1(i,m_1) - \frac{M_1+1}{2}\right)T_t\right\}$$

$$T_i(\breve{\mathbf{\Omega}}_2, \breve{\nu}) \doteq \frac{1}{M_2}\sum_{m_2=1}^{M_2} \exp\left\{j2\pi\frac{\breve{\mathbf{\Omega}}_2^{\text{T}}\boldsymbol{r}_{2,m_2}}{\lambda} + j2\pi\breve{\nu}\left(\eta_2(i,m_2) - \frac{M_2+1}{2}\right)T_r\right\} \quad (5.51)$$

The noise term $V(\nu, \mathbf{\Omega}_1, \mathbf{\Omega}_2)$ can be derived analogously to $V(\nu)$ in Appendix A.2. Notice that the expressions in the arguments of the exponential terms in the summands of $S_i(\breve{\mathbf{\Omega}}_1, \breve{\nu})$ and $T_i(\breve{\mathbf{\Omega}}_2, \breve{\nu})$ reveal respectively a coupling depending on $\eta_1(i, \cdot)$ in the estimation of the DoD and the Doppler frequency and a coupling depending on $\eta_2(i, \cdot)$ in the estimation of the DoA and the Doppler frequency.

5.5.4 Case Study: TDM-SIMO Channel Sounding with a Uniform Linear Array

In this section we investigate the above mentioned coupling between the estimation of DoA and the Doppler frequency. To keep the discussion simple we restrict the attention to a special case where Array 1 consists of one element ($M_1 = 1$) and Array 2 is uniform and linear. In this case, the DoD cannot be estimated and Eq. (5.50) reduces to

$$z(\nu, \mathbf{\Omega}_2; y) = \sum_{i=1}^{I} R_i(\breve{\nu})T_i(\breve{\mathbf{\Omega}}_2, \breve{\nu}) + V(\nu, \mathbf{\Omega}_2) \quad (5.52)$$

We investigate the behavior of the absolute value of Eq. (5.52) in the noiseless case ($V(\nu, \mathbf{\Omega}_2) = 0$). Array 2 consists of M_2 equidistant isotropic elements with locations

$$\boldsymbol{r}_{2,m_2} = \left[\frac{m_2\lambda}{2}, 0, 0\right]^{\text{T}}, m_2 = 1, \ldots, M_2$$

The inner products arising in the response of this array are calculated as

$$\mathbf{\Omega}_2^{\text{T}}\boldsymbol{r}_{2,m_2} = \omega\frac{m_2\lambda_0}{2}, m_2 = 1, \ldots, M_2$$

where $\omega \doteq \cos(\phi_2)\sin(\theta_2)$. The parameter ω can be interpreted as a spatial frequency. It can be also written as $\omega = \cos(\psi)$ where ψ is the angle between the impinging direction and the array axis. This angle is the only characteristic of the incident direction that can be uniquely determined with a linear array.

The absolute value of Eq. (5.52) reads in this case

$$|z(\nu, \mathbf{\Omega}_2; y)| = |z(\breve{\nu}, \breve{\omega}; y)| \quad (5.53)$$

If the switching mode is cycle-independent, the right-hand expression in Eq. (5.53) factorizes according to

$$|z(\breve{\nu},\breve{\omega};y)| = |G(\breve{\nu})| \cdot |T(\breve{\omega},\breve{\nu})| \tag{5.54}$$

where

$$G(\breve{\nu}) \doteq \frac{\sin(\pi\breve{\nu}IT_{\text{cy}})}{I\sin(\pi\breve{\nu}T_{\text{cy}})}$$

$$T(\breve{\omega},\breve{\nu}) \doteq \frac{1}{M_2}\sum_{m_2=1}^{M_2} \exp\left\{jm_2\pi\breve{\omega} + j2\pi\breve{\nu}\left[\eta_2(m_2) - \frac{M_2+1}{2}\right]T_r\right\}$$

We investigate the impact of different switching modes on Eq. (5.54) for the settings of the TDM-SIMO system and the one-wave scenario specified in Table 5.2. The wave is incident perpendicular to the array axis and its Doppler frequency is 0 Hz. Notice that from Eq. (5.53) the behavior of the objective function only depends on the Doppler frequency deviation from the true Doppler frequency so that the choice of the latter within the range $(-\frac{1}{2T_r}, \frac{1}{2T_r}]$ is irrelevant. Figure 5.13(a)–(c) depicts the graphs of, respectively, $|G(\breve{\nu})|$, $|T(\breve{\omega},\breve{\nu})|$, and $|z(\breve{\nu},\breve{\omega};y)|$ in Eq. (5.54), when the conventionally used identity switching mode is applied. Notice that the range of $\breve{\nu}$ is $(-\frac{1}{2T_r}, \frac{1}{2T_r}] = (-200{,}200]$ Hz.

Clearly, the period of $|G(\breve{\nu})|$ is $\frac{1}{T_{\text{cy}}} = 50$ Hz. The loci of the pairs $(\breve{\nu},\breve{\omega})$ where $|T(\breve{\omega},\breve{\nu})|$ equals its maximum value $(= 1)$ is the line $\breve{\omega} = \breve{\nu}T_r$. As can be observed in Fig. 5.13(c) the product of these two functions $|z(\breve{\nu},\breve{\omega};y)|$ exhibits multiple maxima along the above line, separated by $\frac{1}{T_{\text{cy}}}$ in $\breve{\nu}$. These multiple maxima cause an ambiguity in the joint ML estimation of the Doppler frequency and DoA when the DFER is selected equal to $(-\frac{1}{2T_r}, \frac{1}{2T_r}]$. Notice that $|z(\breve{\nu},\breve{\omega};y)|$ exhibits one unique maximum if $\breve{\nu} \in (-\frac{1}{2T_{\text{cy}}}, \frac{1}{2T_{\text{cy}}}]$. Thus, if this switching mode is used, the DFER has to be restricted to the above interval in order to avoid the ambiguity problem.

Figure 5.13(d) and (e) shows the graphs of $|z(\breve{\nu},\breve{\omega};y)|$ for the cycle-independent switching mode $\boldsymbol{\eta}_2 = [4,2,1,8,5,7,3,6]$ and a cycle-dependent randomly selected switching mode, respectively. With this selection of switching modes, $|z(\breve{\nu},\breve{\omega};y)|$ exhibits a unique maximum and therefore the ambiguity problem does not occur. One can still see clearly the impact of the periodic behavior of $|G(\breve{\nu})|$ on the objective function, depicted in Figure 5.13(d) as sidelobe stripes at the loci of the maxima of $|G(\breve{\nu})|$ when the switching mode is cycle-independent. As exemplified by Figure 5.13(e), this pattern vanishes completely when using a cycle-dependent switching mode. Furthermore, the sidelobes of the third depicted objective function have much lower magnitudes than those of the second objective function.

This study shows that in the worst case (using the identity switching mode), the operational DFER is $(-\frac{1}{2T_{\text{cy}}}, \frac{1}{2T_{\text{cy}}}]$. By appropriately selecting the switching mode the DFER can be

Table 5.2 Case study: Settings of the TDM-SIMO system and parameters of the incident wave

I	M_1	M_2	R	T_{cy} (s)	ν' (Hz)	ω'
8	1	8	1	0.02	0	0

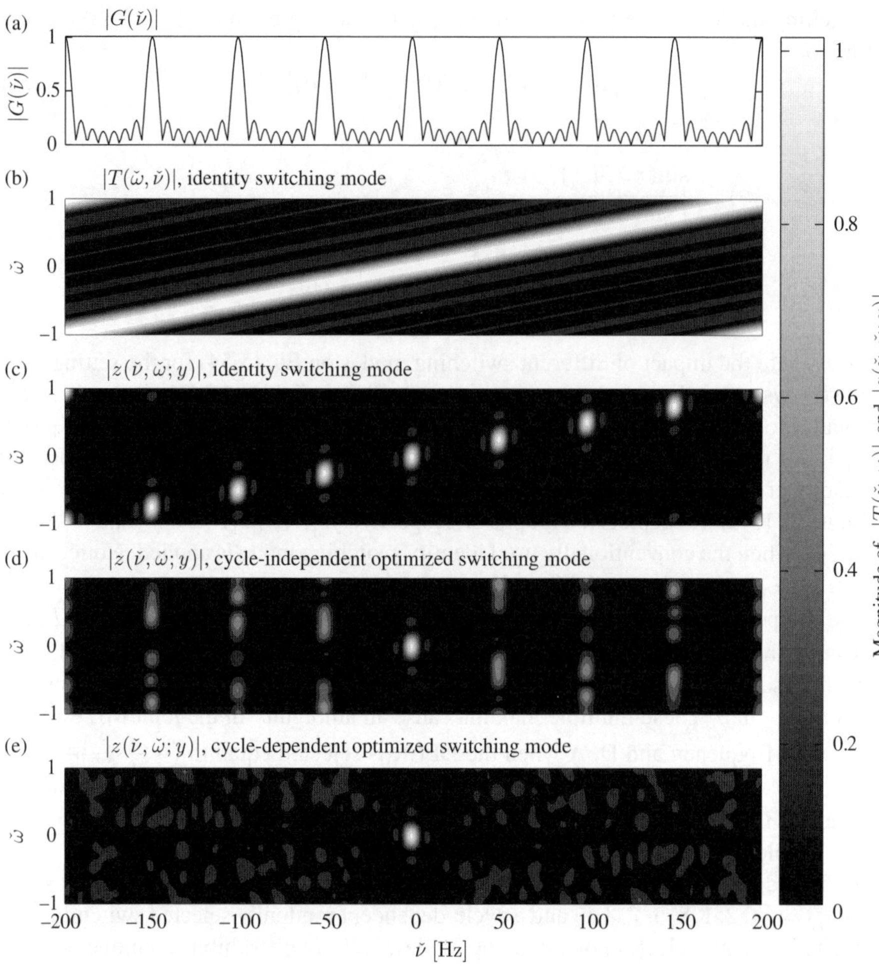

Figure 5.13 Objective functions for the joint Doppler frequency and DoA ML estimates in the case study. The following switching modes are selected: (c) $\boldsymbol{\eta}_2 = [1, 2, \ldots, 8]$; (d) $\boldsymbol{\eta}_2(i) = [4, 2, 1, 8, 5, 7, 3, 6]$; (e) a randomly selected cycle-dependent switching mode. (a) and (b) depict the factors of the objective function (see Eq. (5.54)) for the identity switching mode

extended to $(-\frac{1}{2T_r}, \frac{1}{2T_r}]$; that is, by a factor $M_2 = 8$ in this case study or in general by $M_1 M_2 R$. Furthermore, Figure 5.13(c)–(e) makes it evident that the switching modes significantly affect the magnitudes of the sidelobes of the objective function. This impact is investigated in more detail in Section 5.5.6.

5.5.5 *Switching-mode Optimization*

In this subsection we first derive a necessary and sufficient condition for a cycle-independent switching mode to lead to an objective function exhibiting multiple maxima. Then we show that modulo-type switching modes (and among them the identity switching mode) cause the ambiguity problem when the cycle repetition rate R is integer. Finally, we introduce a principle for selection of switching modes that reduce the variances of the parameter estimators and enhance performance robustness against noise.

The function $z(\breve{\nu}, \breve{\omega}; y)$ in Eq. (5.53) is of the form

$$z(\breve{\nu}, \breve{\omega}; y) = \frac{1}{IM_2} \sum_{i=1}^{I} \sum_{m_2=1}^{M_2} \exp\{j\Phi_{i,m_2}\} \tag{5.55}$$

where

$$\Phi_{i,m_2} \doteq 2\pi\breve{\nu}\left(i - \frac{I+1}{2}\right)T_{\text{cy}} + 2\pi\breve{\nu}\left(\eta_2(i,m_2) - \frac{M_2+1}{2}\right)T_r + \pi\breve{\omega}m_2$$

When $\breve{\omega} = 0$ and $\breve{\nu} = 0$, $|z(\breve{\nu}, \breve{\omega}; y)|$ equals its maximum value of 1. However, a necessary and sufficient condition for $|z(\breve{\nu}, \breve{\omega}; y)| = 1$ to hold is that all the phases in the double sum are congruent modulo 2π. This will be the case if, and only if,

$$\begin{aligned} &\Phi_{i,m_2} - \Phi_{i+1,m_2} \equiv 0 \pmod{2\pi} \\ &\qquad m_2 = 1, \ldots, M_2, i = 1, \ldots, I-1 \end{aligned} \tag{5.56}$$

and

$$\begin{aligned} &\Phi_{i,m_2} - \Phi_{i,m_2+1} \equiv 0 \pmod{2\pi} \\ &\qquad m_2 = 1, \ldots, M_2 - 1, i = 1, \ldots, I \end{aligned} \tag{5.57}$$

Hence $|z(\breve{\nu}, \breve{\omega}; y)|$ exhibits multiple maxima if, and only if, the system of equations defined by Eqs (5.56) and (5.57) has one or more non-trivial solutions $(\breve{\nu}, \breve{\omega}) \in (-\frac{1}{2T_r}, \frac{1}{2T_r}] \times [\omega' - 1, \omega' + 1]$. The trivial solution is $(\breve{\nu}, \breve{\omega}) = (0, 0)$.

We now focus on cycle-independent switching modes. In this case $\eta_2(i, m_2) - \eta_2(i+1, m_2) = 0$ and Eq. (5.56) reduces to $\breve{\nu}T_{\text{cy}} = K$ for $K \in \mathbb{Z} \cap (-\frac{RM_2}{2}, \frac{RM_2}{2}]$, where $\mathbb{Z}$ is the set of integers. Inserting this identity into Eq. (5.57) yields

$$K \cdot \frac{\dot{\eta}_2(m_2)}{RM_2} \equiv \frac{\breve{\omega}}{2} \pmod{1}, m_2 = 1, \ldots, M_2 - 1 \tag{5.58}$$

where $\dot{\eta}_2(m_2) \doteq \eta_2(m_2) - \eta_2(m_2 + 1)$. Hence, provided the switching mode is cycle-independent, a necessary and sufficient condition for the ambiguity problem to occur is that the equation system of Eq. (5.58) has at least one non-trivial solution $(K, \breve{\omega}) \in (\mathbb{Z} \cap (-\frac{RM_2}{2}, \frac{RM_2}{2}]) \times [\omega' - 1, \omega' + 1]$.

A modulo-type switching mode fulfills the congruence

$$(\eta_2(m_2) - 1) \equiv Jm_2 + K \pmod{M_2}$$

for some $J, K \in \mathbb{Z}$ with J and M_2 being relatively prime. As an example, the commonly used identity switching mode $\boldsymbol{\eta}_2 = [1, 2, \ldots, M_2]$ is a modulo-type switching mode with $J = 1$ and $K = 0$. For any modulo-type switching mode, $\{\dot{\eta}_2(m_2); m_2 = 1, \ldots, M_2 - 1\} = \{J, J - M_2\}$. Hence Eq. (5.58) consists of two different congruences. Elimination of $\breve{\omega}$ yields $K =$

RK', with K' taking any value in $\mathbb{Z} \cap (-\frac{M_2}{2}, +\frac{M_2}{2}]$. When $R \in \mathbb{Z}$, the non-trivial solutions for K are the $RM_2 - 1$ values in $\mathbb{Z} \cap (-\frac{RM_2}{2}, \frac{RM_2}{2}] \setminus \{0\}$. This result is in accordance with the eight maxima (corresponding to the seven non-trivial solutions plus the trivial solution) that can be observed in Figure 5.13(c).

5.5.6 *Simulation Studies*

In this subsection Monte-Carlo simulation results are reported for the assessment of the SAGE algorithm performance in estimation of Doppler frequency beyond half the cycle rate. The impact of neglecting phase changes caused by Doppler frequency on the performance of the estimation is also demonstrated. Finally, we show the dependency of the performance on the normalized sidelobe level of the objective function for specified switching modes.

5.5.6.1 Estimation of Doppler Frequency

The RMSEE of $\hat{\nu}_\ell$ that results when the SAGE algorithm is used has been assessed by means of Monte Carlo simulations in a one-wave scenario in two cases:

1. All wave parameters but the Doppler frequency and the complex weight are assumed to be known, which corresponds to the ideal condition, for which the objective function $z(\nu; y)$ was derived on Section 5.5.2;
2. All wave parameters are unknown and are estimated.

The parameters of the TDM mode and of the simulated wave are given in Table 5.3. In the table, K and N_s denote the spreading factor of the pseudo-noise sequence used as the sounding signal and the number of samples per chip duration, respectively. A single antenna is employed in the Tx and an antenna array is used in the Rx. The array consists of a conformal subarray of eight dual-polarized isotropic antennas uniformly spaced on a cylinder together with a uniform planar 2×2 subarray of the same elements placed on the top of the cylinder. All antenna elements are omnidirectional.

Figure 5.14 depicts the simulation results obtained for the two cases. The corresponding Cramér-Rao lower bounds (CRLBs) for ν as calculated by Fleury et al. [1999] according to

$$\text{CRLB} = \frac{1}{\gamma_o} \frac{3}{2\pi^2 R^2 M_2^2 M_1^2 T_r^2 (I^2 - 1)} \tag{5.59}$$

are also shown, as a dotted line.

As shown in Figure 5.14, all the simulation curves exhibit the same behavior: when γ_o is larger than a certain threshold, the RMSEE is close to the corresponding CRLB; When γ_o is below the threshold, the RMSEE increases dramatically. Since the simulated Doppler frequency is equal to 400 Hz which is much larger than half the cycle rate, i.e. 10.2 Hz, the coincidence between the simulation curves and the CRLB when γ_o is beyond the threshold indicates that the SAGE algorithm combined with the TDM mode can achieve maximum performance in estimating the Doppler frequency beyond the range of $[-1/(2T_{cy}), +1/(2T_{cy})]$. Moreover, the RMSEE in Case (2), when $R = 400$, is observed to be very close to that obtained in Case (1).

Table 5.3 Parameter settings used for simulations

Parameter	Value
I	4
M_1	1
M_2	24
K	2
N_s	255
T_{cy}	49 ms
T_t	123 μs
T_r	5.1 μs
α_ℓ	0.2
$\phi_{2,\ell}$	45°
$\theta_{2,\ell}$	45°
$\phi_{1,\ell}$	45°
$\theta_{1,\ell}$	45°
ν	400 Hz
τ_ℓ	1 ns

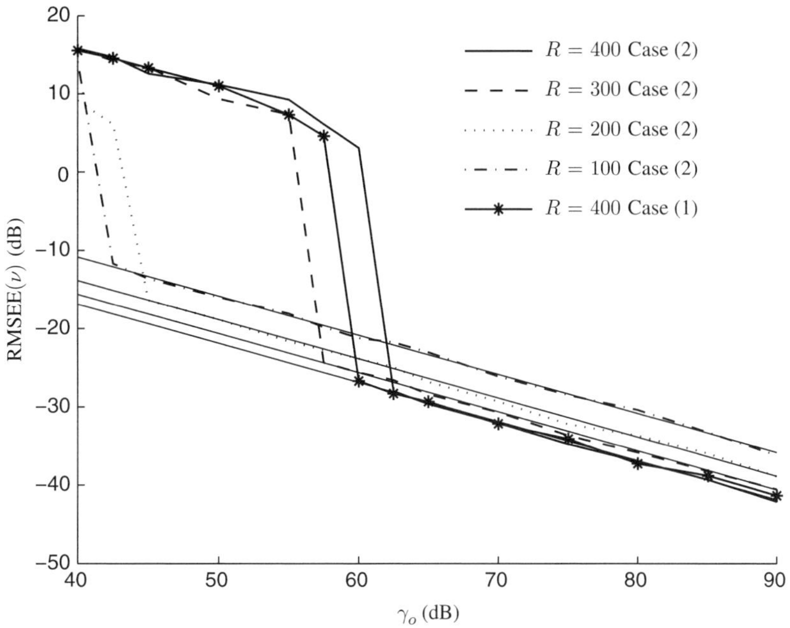

Figure 5.14 RMSEEs of ν versus the SNR γ_o with the repetition rate R as a parameter for Cases (1) and (2). The thin solid straight lines denote the root Cramér–Rao lower bound. The empirical RMSEEs are calculated using 100 Monte-Carlo simulation runs. RMSEE, root mean-square estimation error

This demonstrates that the argument derived with the ideal condition in Section 5.5.2 is applicable in reality. However, a horizontal gap of about 2.5 dB between the thresholds of these two curves illustrates that the estimation scheme performs better in the ideal conditions.

The existence of the threshold in the curves can be explained by looking at the histograms of the estimates $\hat{\nu}$. It appears that when γ_o is below the threshold, the estimates $\hat{\nu}$ are distributed around the true Doppler frequency ν' as well as $\nu' + n/T_{cy}$, $n = \pm 1, \pm 2$,... This shows that the multi-lobe characteristic of $|F(\nu - \nu')|$ leads to the large estimation errors. It is therefore of great importance to select the parameters of the TDM mode in such a way that the threshold is below the prescribed minimum SNR of estimated waves. Notice that the SNR γ_o in the horizontal axis of Figure 5.14 is related to the SNR at the output of the Rx antennas γ_S according to

$$\gamma_S = \gamma_o - 10 \log 10(IM_2M_1KN_s)$$

where $10 \log 10(IM_2M_1KN_s)$ is calculated to be 46.9 dB. So the performance of the Doppler frequency estimator versus γ_S can be obtained by shifting the curves in Figure 5.14 to the left by 46.9 dB.

5.5.6.2 Impact of Neglecting Doppler Shift in Sensing Periods

As reported in Section 5.5.2 Assumption (2), the phase shift due to the Doppler frequency within the sounding interval T_{sc} is neglected in the derivation of the Doppler frequency estimate. To investigate the impact of this approximation, the RMSEE of $\hat{\nu}$ has been assessed in Case (1) for $\nu' = 0$ Hz and $\nu' = 98039.22$ Hz$(= 1/(2T_r))$. The results are depicted in Figure 5.15. It can be observed that as ν' increases, the RMSEE curve is slightly shifted to the right. The observation coincides with a theoretical analysis reported in Appendix A.1, which shows that the impact of the phase change during T_{sc} due to the Doppler frequency can be interpreted as an effective SNR decrease according to

$$\gamma_o(\nu') = |\text{sinc}(\nu' T_s)|^2 \gamma_o$$

where $\gamma_o(\nu')$ denotes the effective SNR versus the true Doppler frequency. Calculation shows that the SNR degradation when $\nu' = 1/(2T_r)$ is 0.912 dB.

From the above, we can conclude that the impact of neglecting the phase shift due to the Doppler frequency in T_{sc} is actually negligible. Therefore the model and estimation scheme presented here are effectively applicable to estimate Doppler frequency in the range $[-\frac{1}{2T_r}, +\frac{1}{2T_r}]$. To remove the degradation, the estimation scheme must be modified to include the rotation of the Doppler phasor within T_{sc}. Notice that in this case, by taking more than one sample within T_{sc}, the estimation range for the Doppler frequency can be further extended to plus/minus half the sample rate.

5.5.6.3 Impact of Switching Modes on Performance

The theoretical investigations of the study case reported in Section 5.5.3 show that the switching mode strongly affects the sidelobes of the objective function of the Doppler frequency and the DoA ML estimates. As a consequence, the switching mode will also affect the robustness of the estimators toward noise because this robustness directly depends on the magnitudes of the sidelobes.

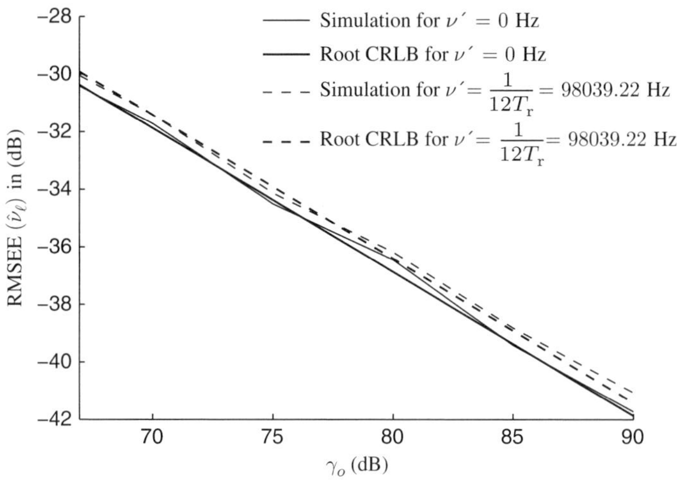

Figure 5.15 RMSEEs of $\hat{\nu}$ versus the SNR γ_o with the true Doppler frequency as a parameter. CRLB, Cramér–Rao lower bound

We define the normalized sidelobe level (NSL) associated with a switching mode to be the magnitude of the highest sidelobe of the corresponding objective function. It is obvious that objective functions with NSL equal to 1 have multiple maxima and therefore lead to an ambiguity in the estimation of Doppler frequency and DoA, whereas objective functions with NSL less than 1 have a unique maximum.

We show by means of Monte-Carlo simulations that the NSL associated with a switching mode can be used as a figure of merit of this switching mode for the optimisation of the performance of the Doppler frequency and the DoA ML estimates. The parameter settings used here are the same as those used in the case study (see Table 5.2). Figure 5.16 depicts the RMSEEs of the ML estimates $\hat{\nu}$ and $\hat{\psi}$ versus the output SNR γ_o [Yin et al. 2003b] for four switching modes, leading to NSLs equal to 0.28, 0.58, 0.80, and 0.85, respectively. The symbol P in the above expression denotes the transmitted signal power. The RMSEEs are compared to the corresponding individual CRLBs calculated by Fleury et al. [1999] assuming parallel SIMO channel sounding.

As shown in Figure 5.16, all curves exhibit the same behavior: when γ_o is larger than a certain threshold, γ_o^{th}, the RMSEEs of $\hat{\nu}$ and $\hat{\psi}$ are close to the corresponding CRLBs. When $\gamma_o < \gamma_o^{th}$, the RMSEEs increase dramatically, as has already been shown [Yin et al. 2003b]. Further simulations show that γ_o^{th} increases along with the NSL. This behavior can be explained as follows: the probability of the event that the maximum of any sidelobe of the objective function is higher than the maximum of its mainlobe is larger when these sidelobes have high magnitudes. Notice that the threshold effect is well-known in non-linear estimation such as frequency estimation [Rife and Boorstyn 1974].

We can use the RMSEE curve of the Doppler frequency ML estimate, under the hypothesis that all other parameters but the complex gain of the path are known, as a benchmark for the Doppler frequency ML estimate performance when all path parameters are unknown. This curve is indeed a lower bound for the RMSEE curve of the latter Doppler frequency estimates. Monte Carlo simulations not reported here show that this benchmark curve exhibits a threshold

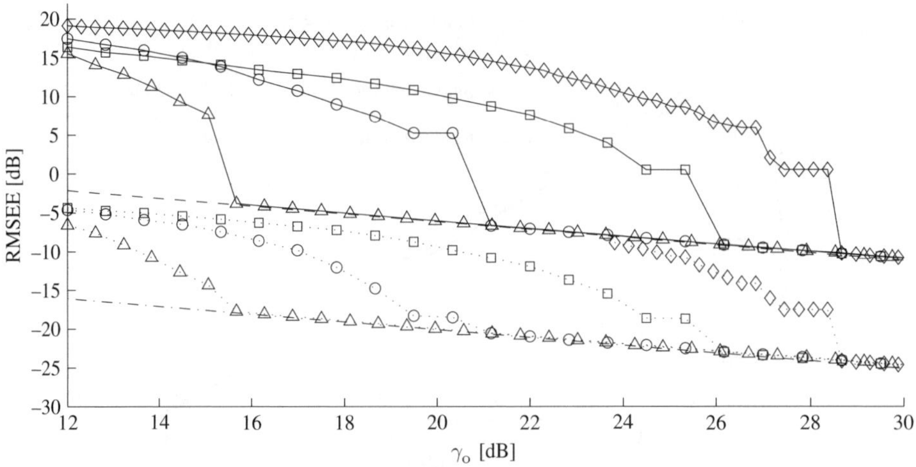

Figure 5.16 RMSEEs of $\hat{\nu}$ (solid curves) and $\hat{\psi}$ (dotted curves) versus γ_o computed using the settings given in Table 5.2 for different switching modes. The dashed and the dash-dotted lines represent the CRLBs of $\hat{\nu}$ and $\hat{\psi}$ respectively. The curves with symbols $\diamond$, $\square$, $\bigcirc$, $\triangle$ have been obtained using three cycle-independent switching modes and one cycle-dependent switching mode, leading to $\text{NSL} = 0.85, 0.80, 0.58$, and 0.28, respectively

$\gamma_o^{\text{th}} = 15$ dB and is close to the CRLB of $\hat{\nu}$ for $\gamma_o > \gamma_o^{\text{th}}$. From Figure 5.16 we observe that the threshold γ_o^{th} of the RMSEE curve of $\hat{\nu}$ obtained for the switching mode leading to $\text{NSL} = 0.28$ is 0.5 dB away from that of the benchmark curve. Hence the former threshold is close to the minimum achievable. This observation confirms that the NSL is a suitable figure of merit for the selection of "good" switching modes: those leading to ML estimates close to the optimum.

5.5.6.4 Experimental Investigations

In this section, we describe experiments on the impact of the switching mode on the objective function used in the SAGE algorithm to estimate the Doppler frequency and DoA of propagation paths based on measurement data. The measurements were performed with the TDM-MIMO channel sounder PROPSound [Stucki 2001]. The Tx array consisted of three conformal subarrays of eight dual-polarized patches uniformly spaced on a cylinder together with a uniform rectangular 2×2 subarray of 4 dual-polarized patches placed on top of the cylinder ($M_1 = 54$). At the Rx a 4×4 planar array with 16 dual-polarized patches was used ($M_2 = 32$). The spacing between the Rx array elements and the elements of the four Tx subarrays was half a wavelength. The selected carrier frequency was 2.45 GHz. The sounding signal was a PN sequence of length $K = 255$ chips, with chip duration $T_c = 10$ ns. The sensing interval coincided with one period of the PN sequence: $T_{sc} = KT_c = 2.55$ µs. The transmitted power was 100 mW.

The Rx array was mounted outside a window on the third floor of the Elektrobit AG building in Bubikon, Switzerland. The Tx array was mounted on the roof of a van moving at approximately 8 m/s away from the building. The measurements were performed twice along the same

Table 5.4 Settings of the channel sounder for measurement Scenarios I and II

Parameters	Scenario I	Scenario II
Switching mode at Array 2	Patch-wise identity switching mode	Patch-wise optimized switching mode
T_r (μs)	3.05	5.10
$T_{\rm cy}$ (ms)	6.2	47.2
Selected DFER (Hz)	$(-\frac{1}{2T_{\rm cy}}, \frac{1}{2T_{\rm cy}}] = (-81.3, 81.3]$	$(-\frac{1}{2T_r}, \frac{1}{2T_r}] = (-98\ 039, 98\ 039]$

route, with different settings of the sounding equipment (see Table 5.4). The van was driving at approximately the same velocity during both measurement recordings to ensure propagation scenarios with almost identical Doppler frequencies. The AoA, EoA and the Doppler frequency of the LOS path can be calculated from the location of the Rx and the position and the velocity of the van to be approximately 5°, 20° and -59 Hz respectively.

The two settings of the sounding equipment were selected in such a way that the maximum Doppler frequency is in the range $(-\frac{1}{2T_{\rm cy}}, \frac{1}{2T_{\rm cy}}]$ in Scenario I and outside this range but in $(-\frac{1}{2T_r}, \frac{1}{2T_r}]$ in Scenario II. These intervals were then selected as the corresponding DFERs for the two scenarios. As explained later, the switching mode at the Tx is irrelevant in the situation at hand. At the Rx, we apply a patch-wise identity switching mode in Scenario I and a patch-wise optimized switching mode in Scenario II. The term "patch-wise" indicates that the two elements of each patch are always switched consecutively. This is done to mitigate the phase-noise effect for accurate polarization estimation.

The SAGE algorithm is applied to the measurement data to estimate the individual parameter vectors of $L = 4$ propagation paths using $I = 4$ measurement cycles. The parameter estimates of the four paths are initialized successively with a non-coherent maximum likelihood (NC-ML) technique [Yin et al. 2003a]. Once the initialization is completed, the E- and M-steps of the SAGE are performed, as described by Fleury et al. [2002a]. It can be shown that the objective function used for the joint initialization of $\hat{\nu}_\ell$ and $\hat{\mathbf{\Omega}}_{2,\ell}$ after the initial delay estimate $\hat{\tau}_\ell(0)$ has been computed is similar to the absolute value of Eq. (5.52) with $\tau_\ell = \hat{\tau}_\ell(0)$ and $\hat{x}_\ell(t) = y(t) - \sum_{\ell'=1}^{\ell-1} s(t; \hat{\boldsymbol{\theta}}'_{\ell'}(0))$. Since at that stage, the DoD of the ℓth path has not yet been estimated, the NC-ML technique is used to initialize $\hat{\nu}_\ell$ and $\hat{\mathbf{\Omega}}_{2,\ell}$ jointly. The switching mode at the Tx is irrelevant when this method is applied. Hence, we can use the initialization procedure of the SAGE algorithm to experimentally investigate scenarios similar to the case study described in Section 5.5.4. The differences between the experimental scenarios and the case study are as follows:

1. the SIMO antenna system considered in the case study is replaced by a MIMO system in the experimental scenario
2. a uniform planar array with dual-polarized elements is used instead of a uniform linear array
3. the array elements are not isotropic
4. in the calculation of $\hat{x}_\ell(t)$, the contribution of the waves other than the ℓth one were either not or were only partially cancelled.

In the sequel we restrict our attention to the LOS path indexed $\ell = 1$. To visualize the behavior of the objective function versus ν_1, we compute

$$F(\nu_1) \doteq \max_{\mathbf{\Omega}_{2,1}} |z(\nu_1, \mathbf{\Omega}_{2,1}; \hat{x}_1 = y)|^2$$

with $z(\nu_1, \mathbf{\Omega}_{2,1}; y)$ given in Eq. (5.52). Notice that $T_i(\check{\mathbf{\Omega}}_{2,1}, \check{\nu}_1)$ (see Eq. (5.51)) depends on the real response of the Rx array; that is, it includes the radiation patterns of the elements in the array. Inserting into Eq. (5.52) the noise term omitted in the definition of $F(\nu_1)$ we obtain

$$\begin{aligned} F(\nu_1) &= \max_{\check{\mathbf{\Omega}}_{2,1}} |\sum_{i=1}^{I} R_i(\check{\nu}_1) T_i(\check{\mathbf{\Omega}}_{2,1}, \check{\nu}_1)|^2 \\ &= \max_{\check{\mathbf{\Omega}}_{2,1}} |G(\check{\nu}_1) T(\check{\mathbf{\Omega}}_{2,1}, \check{\nu}_1)|^2 \\ &= |T'(\check{\nu}_1)|^2 \cdot |G(\check{\nu}_1)|^2 \end{aligned} \tag{5.60}$$

with $T'(\check{\nu}_1) \doteq \max_{\check{\mathbf{\Omega}}_{2,1}} T(\check{\mathbf{\Omega}}_{2,1}, \check{\nu}_1)$. The second line follows in a similar way to Eq. (5.54), since the switching mode is cycle-independent, so the switching mode only affects $F(\nu_1)$ via $|T'(\check{\nu}_1)|^2$.

The right-hand expression in Eq. (5.60) will be useful for understanding the behavior of $F(\nu_1)$ computed from the measurement data. This function is plotted versus ν_1 in the range $(-81.3\ \text{Hz}, 81.3\ \text{Hz}]$ for both scenarios (see Figure 5.17 (top). The pulse-train-like behavior of the curves is due to the factor $|G(\check{\nu}_1)|^2$ in Eq. (5.60), which is periodic with period $1/T_{\text{cy}}$. The maximum of $F(\nu_1)$ in Scenario I, with DFER $(-\frac{1}{2T_{\text{cy}}}, \frac{1}{2T_{\text{cy}}}]$, is located at -52 Hz. In Scenario II, with DFER $(-\frac{1}{2T_r}, \frac{1}{2T_r}]$, the maximum of $F(\nu_1)$ is located at -81 Hz. Notice that these values are the initial Doppler frequency estimates of the LOS path returned by the SAGE algorithm. After four iterations of the algorithm the Doppler frequency estimates of the LOS path have converged to -52.5 Hz, and the AoA and EoA estimates equal $4.6°$ and $27°$ respectively in Scenario I. In Scenario II the Doppler frequency estimate converges to -60 Hz, and the AoA and EoA estimates equal $5.3°$ and $18.7°$ respectively. All these values are in accordance with the theoretically calculated values. The deviation between the two sets of estimates is due to the difference in the velocities and the positions of the van during the measurement recordings.

The pulse-train-behavior of $F(\nu_1)$ due to $|G(\nu_1)|^2$ makes it difficult to visualize the effect of the switching mode (embodied in $|T'(\check{\nu}_1)|^2$) on the former function when $\check{\nu}$ is in the range $(-\frac{1}{2T_r}, \frac{1}{2T_r}]$. To circumvent this problem, we compute an approximation of $|T'(\check{\nu}_1)|$ from $F(\nu_1)$ as follows: $\text{PE}(F(\nu_1))$ is a pseudo-envelope (PE) obtained by dividing the range of ν_1 into multiple bins with equal width $\frac{1}{T_{\text{cy}}}$ and connecting the maxima of $F(\nu_1)$ within each bin using linear interpolation. The bottom panel of Figure 5.17 shows the computed PE curves for both scenarios. For Scenario I, $\text{PE}(F(\nu_1))$ remains close to 1 over the entire range $(-\frac{1}{2T_r}, \frac{1}{2T_r}]$. This behavior is due to the identity switching mode used for the 4×4 planar array. In Scenario II, $\text{PE}(F(\nu_1))$ exhibits a dominant lobe and multiple sidelobes with significantly lower amplitude. The width of the main lobe is in accordance with the analytically derived value of $\frac{2}{M_2}\frac{1}{T_r}$ for the separation between the zero points of the main lobe.

When the DFER is extended to $(-\frac{1}{2T_r}, \frac{1}{2T_r}]$ in Scenario I, the maximum of $F(\nu_1)$ is located at -97.604 kHz in the initialization step (as shown in the bottom panel of Figure 5.17), and

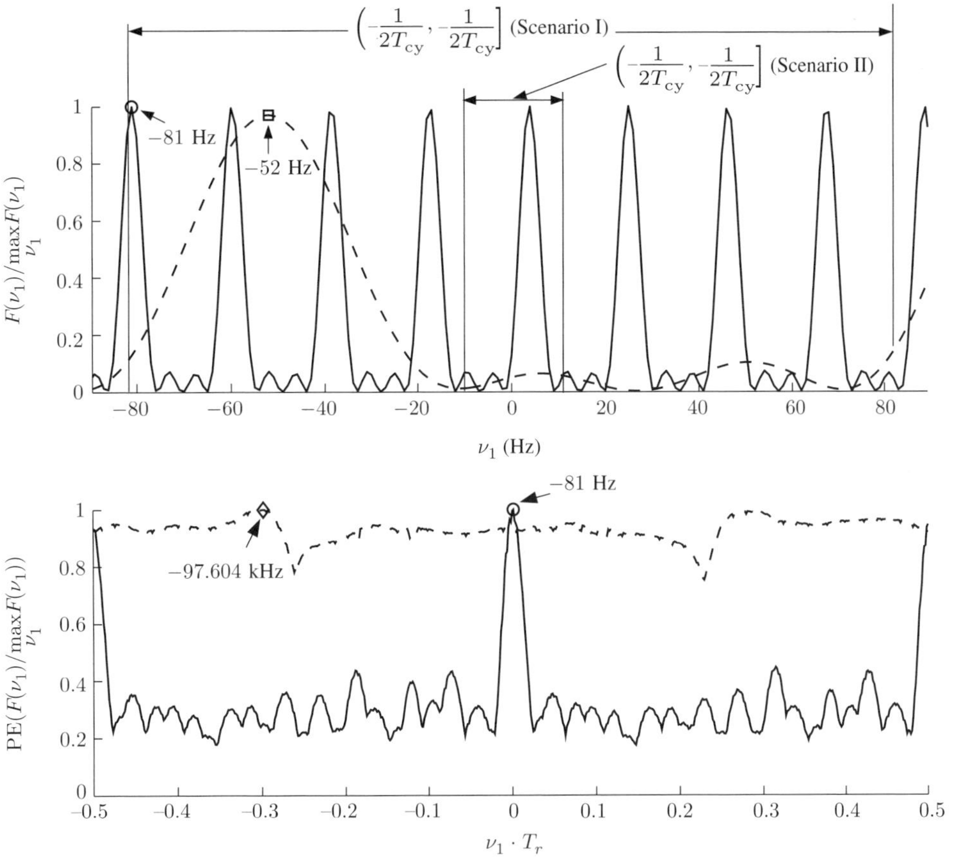

Figure 5.17 Normalized $F(\nu_1)$ (top) and pseudo-envelope $\mathrm{PE}(F(\nu_1))$ (bottom) computed from the measurement data obtained in Scenario I (dashed lines) and Scenario II (solid lines). The marks □ and ◇ denote the maxima of $F(\nu_1)$ in Scenario I when the DFER is respectively $(-\frac{1}{2T_{cy}}, \frac{1}{2T_{cy}}]$ and extended to $(-\frac{1}{2T_r}, \frac{1}{2T_r}]$. The mark ∘ denotes the maximum of $F(\nu_1)$ in Scenario II (DFER $= (-\frac{1}{2T_r}, \frac{1}{2T_r}]$)

stays at this value after four iterations. The AoA and EoA estimates are respectively 70° and 2°. These estimates are obviously artifacts that result from the identity switching mode used at the Rx array.

Notice that the high sidelobes at the boundary of the DFER are due to the patch-wise switching of the arrays. When the Doppler frequency is very low compared to the switching rate, as is the case here, the resulting phase-shift due to the Doppler frequency between consecutive sensing intervals of the elements of a patch is close to zero, which leads to an effective doubling of T_r. As a result, the graph of $\mathrm{PE}(F(\nu_1))$ exhibits two segments of similar shape, as shown in the bottom panel of Figure 5.17.

These investigations show experimentally the ambiguity effect that occurs when the DFER is extended to $(-\frac{1}{2T_r}, \frac{1}{2T_r}]$ and the identity switching mode and a planar array are used. It also demonstrates that this problem is avoided by appropriately selecting the switching mode.

Bibliography

Almers P, Wyne S, Tufvesson F and Molisch A 2005 Effect of random walk phase noise on MIMO measurements *Vehicular Technology Conference, 2005. VTC 2005-Spring. 2005 IEEE 61st*, vol. 1, pp. 141–145.

Baum DS and Bölcskei H 2004 Impact of phase noise on MIMO channel measurement accuracy *Proc. 2004 IEEE Vehicular Technology Conference*, vol. 3, pp. 1614–1618.

Bonek E, Czink N, Holappa VM, Alatossava M, Hentilä L, Nuutinen J and Pal A 2006 Indoor MIMO measurements at 2.55 and 5.25 GHz – a comparison of temporal and angular characteristics *Proceedings of the 15th IST Mobile Summit*, Myconos, Greece.

Czink N 2007 *The random-cluster model – a stochastic MIMO channel model for broadband wireless communication systems of the 3rd Generation and beyond* PhD thesis Technology University of Vienna, Department of Electronics and Information Technologies.

Czink N, Bonek E, Yin X and Fleury BH 2005 Cluster angular spreads in a MIMO indoor propagation environment *Proceedings of the 16th IEEE International Symposium on Personal, Indoor and Mobile Radio Communications (PIMRC'05)*, vol. 1, pp. 664–668, Berlin, Germany.

Czink N, Galdo GD, Yin X and Meklenbrauker C 2006 A novel environment characterisation metric for clustered MIMO channels used to validate a SAGE parameter estimator *Proceedings of the 15th IST Mobile & Wireless Communication Summit, Myconos, Greece.*

Fessler JA and Hero AO 1994 Space-alternating generalized expectation-maximization algorithm. *IEEE Transactions on Signal Processing* **42**(10), 2664–2677.

Fleury BH, Jourdan P and Stucki A 2002a High-resolution channel parameter estimation for MIMO applications using the SAGE algorithm *Proceedings of International Zurich Seminar on Broadband Communications*, vol. 30, pp. 1–9.

Fleury BH, Tschudin M, Heddergott R, Dahlhaus D and Pedersen KL 1999 Channel parameter estimation in mobile radio environments using the SAGE algorithm. *IEEE Journal on Selected Areas in Communications* **17**(3), 434–450.

Fleury BH, Yin X, Jourdan P and Stucki A 2003 High-resolution channel parameter estimation for communication systems equipped with antenna arrays *Proceedings of the 13th IFAC Symposium on System Identification (SYSID)*, vol. ISC-379, Rotterdam, The Netherlands.

Fleury BH, Yin X, Rohbrandt KG, Jourdan P and Stucki A 2002b High-resolution bidirection estimation based on the sage algorithm: Experience gathered from field experiments. *Proc. XXVIIth General Assembly of the Int. Union of Radio Scientists (URSI).*

Fleury BH, Yin X, Rohbrandt KG, Jourdan P and Stucki A 2002c Performance of a high-resolution scheme for joint estimation of delay and bidirection dispersion in the radio channel *Proceedings of the IEEE Vehicular Technology Conference (VTC-Spring)*, vol. 1, pp. 522–526.

Fuhl J, Rossi JP and Bonek E 1997 High-resolution 3-D direction-of-arrival determination for urban mobile radio. *IEEE Transactions on Antennas and Propagation* **45**(4), 672–682.

Heneda K, Takada J and Kobayashi T 2005 Double directional ultra wideband channel characterization in a line-of-sight home environment. *IEICE - Transactions on Fundamentals of Electronics, Communications and Computer Sciences* **E88**(9), 2264–2271.

IRCC 1986 Report 580: Characterization of frequency and phase noise. Technical Report pp. 142–150, International Radio Consultative Committee.

ITU 2009 ITU-R M.2135-1: Guidelines for evaluation of radio interface technologies for IMT-Advanced (12/2009). Technical report, ITU.

Karedal J, Wyne S, Almers P, Tufvesson F and Molisch AF 2004 UWB channel measurements in an industrial environment. *Globecom '04. IEEE Global Telecommunications Conference, 2004.* **6**, 3511–3516.

Käske M, Schneider C, Sommerkorn G and Thomä RS 2009 Part II: Reference campaign quality check for channel sounding measurements. *COST 2100 Temporary Document TD(09)777, Braunschweig, Germany, Feb. 16-18.*

Kattenbach R and Weitzel D 2000 Wideband channel sounder for time-variant indoor radio channels *Proceedings of AP2000 Millennium Conference on Antennas and Propagation*, vol. 55, pp. 190–196, Davos, Switzerland.

Krim H and Viberg M 1996 Two decades of array signal processing research: the parametric approach. *IEEE Transactions on Signal Processing* **13**, 67–94.

Kyösti P, Meinilä J, Hentilä L, Zhao X, Jämsä T, Schneider C, Narandzić M, Milojević M, Hong A, Ylitalo J, Holappa VM, Alatossava M, Bultitude R, de Jong Y and Rautiainen T 2007 WINNER II Channel Models D1.1.2 V1.1.

Liu L, Oestges C, Poutanen J, Haneda K, Vainikainen P, Quitin F, Tufvesson F and Doncker P 2012 The COST 2100 MIMO channel model. *IEEE Transactions on Wireless Communications* **19**(6), 92–99.

MacCartney G and Rappaport T 2014 73 GHz millimeter wave propagation measurements for outdoor urban mobile and backhaul communications in New York City *Proceedinsg of IEEE International Conference on Communications (ICC)*, pp. 4862–4867, Sydney.

MEDAV GmbH 2001 *Manual W701W1.096: RUSK MIMO: Broadband vector channel sounder for MIMO channels*.

Moon T 1997 The expectation-maximization algorithm. *IEEE Signal Processing Magazine* **13**(6), 47–60.

Oestges C 2005 Some open questions on dual-polarized channel modeling. *COST 273*.

Pedersen C and Pedersen T 2004 *On spatio-temporal sampling in channel sounding* Master's thesis Aalborg University.

Pedersen T, Pedersen C, Yin X and Fleury B 2008a Optimization of spatiotemporal apertures in channel sounding. *IEEE Transactions on Signal Processing* **56**(10), 4810–4824.

Pedersen T, Pedersen C, Yin X, Fleury BH, Pedersen RR, Bozinovska B, Hviid A, Jourdan P and Stucki A 2004 Joint estimation of Doppler frequency and directions in channel sounding using switched Tx and Rx arrays *Proceedings of IEEE Global Telecommunications Conference (Globecom)*, vol. 4, pp. 2354–2360.

Pedersen T, Yin X and Fleury BH 2008b Estimation of MIMO channel capacity from phase-noise impaired measurements *GLOBECOM - IEEE Global Telecommunications Conference*, pp. 3308–3313, New Orleans.

Rappaport T, Sun S, Mayzus R, Zhao H, Azar Y, Wang K, Wong G, Schulz J, Samimi M and Gutierrez F 2013 Millimeter wave mobile communications for 5G cellular: It will work!. *IEEE Access* **1**, 335–349.

Richter A 2004 RIMAX – a flexible algorithm for channel parameter estimation from channel sounding measurements. Technical Report TD-04-045, COST 273, Athens, Greece.

Richter A, Landmann M and Thomä RS 2003 Maximum likelihood channel parameter estimation from multidimensional channel sounding measurements *Proceedings of the 57th IEEE Semiannual Vehicular Technology Conference (VTC)*, vol. 2, pp. 1056–1060.

Rife DC and Boorstyn RR 1974 Single tone parameter estimation from discrete-time observations. *IEEE Transactions on Information Theory* **20**(5), 591–598.

Roy R and Kailath T 1989 ESPRIT – estimation of signal parameters via rotational invariance techniques. *IEEE Transactions on Acoustics, Speech, and Signal Processing* **37**(7), 984–995.

Salous S, Fillipides P and Hawkins I 2002 Architecture of multichannel sounder for multiple antenna applications. Technical Report TD-02-002, COST-273.

Sanudin R, Noordin N, El-Rayis A, Haridas N, Erdogan A and Arslan T 2011a Analysis of DOA estimation for directional and isotropic antenna arrays *Antennas and Propagation Conference (LAPC), 2011 Loughborough*, pp. 1–4.

Sanudin R, Noordin N, El-Rayis A, Haridas N, Erdogan A and Arslan T 2011b Capon-like DOA estimation algorithm for directional antenna arrays *Antennas and Propagation Conference (LAPC), 2011 Loughborough*, pp. 1–4.

Schmidt RO 1986 Multiple emitter location and signal parameter estimation. *IEEE Transactions on Antennas and Propagation* **AP-34**(3), 276–280.

Sommerkorn G, Käske M, Schneider C and Thomä R 2012 Transmission loss and shadow fading analysis depending on antenna characteristics. *IC1004 Lyon Meeting, TD(12)04035*.

Steinbauer M, Hampicke D, Sommerkorn G, Schneider A, Molisch A, Thoma R and Bonek E 2000 Array measurement of the double-directional mobile radio channel *Proceedings of IEEE 51st Vehicular Technology Conference (VTC-Spring)*, vol. 3, pp. 1656–1662.

Steinbauer M, Molisch A and Bonek E 2001 The double-directional radio channel. *IEEE Antennas and Propagation Magazine* **43**(4), 51–63.

Stoica P and Moses RL 1997 *Introduction to Spectral Analysis*. Prentice Hall.

Stucki A 2001 PropSound system specifications document: Concept and specifications. Technical report, Elektrobit AG, Switzerland.

Taparugssanagorn A, Alatossava M, Holappa VM and Ylitalo J 2007a Impact of channel sounder phase noise on directional channel estimation by SAGE. *IET Microwaves, Antennas and Propagation* **1**(3), 803–808.

Taparugssanagorn A and Ylitalo J 2005 Reducing the impact of phase noise on the MIMO capacity estimation *Proceedings of Wireless Personal Multimedia Communications (WPMC)*, vol. 5.

Taparugssanagorn A, Yin X, Ylitalo J and Fleury BH 2007b Phase noise mitigation in channel parameter estimation for TDM MIMO channel sounding. *Signals, Systems and Computers, 2007. ACSSC 2007. Conference Record of the Forty-First Asilomar Conference on* pp. 656–660.

Wallace J, Jensen M, Swindlehurst A and Jeffs B 2003 Experimental characterization of the MIMO wireless channel: data acquisition and analysis. *IEEE Transactions on Wireless Communications* **2**(2), 335–343.

Xu H, Rappaport T, Boyle R and Schaffner J 1999 38 GHz wideband point-to-multipoint radio wave propagation study for a campus environment *Vehicular Technology Conference, 1999 IEEE 49th*, vol. 2, pp. 1575–1579.

Xu H, Rappaport T, Boyle R and Schaffner J 2000 Measurements and models for 38-GHz point-to-multipoint radiowave propagation. *IEEE Journal on Selected Areas in Communications* **18**(3), 310–321.

Yin X, Fleury. B, Jourdan P and Stucki. A 2003a Polarization estimation of individual propagation paths using the SAGE algorithm. *Personal, Indoor and Mobile Radio Communications, 2003. PIMRC 2003. 14th IEEE Proceedings on* **2**, 1795–1799.

Yin X, Fleury BH, Jourdan P and Stucki A 2003b Doppler frequency estimation for channel sounding using switched multiple transmit and receive antennae *Proceedings of the IEEE Global Communications Conference (Globecom'03)*, vol. 4, pp. 2177–2181.

Yin X, Hu Y and Zhong Z 2012 Dynamic range selection for antenna-array gains in high-resolution channel parameter estimation *Wireless Communications and Signal Processing (WCSP), 2012 International Conference on*, pp. 1–5.

Yin X, Ling C and Kim MD 2016 Experimental multipath cluster characteristics of 28 GHz propagation channel in office environments. *IEEE Access* **PP**(99), 1–1.

Zetik R, Thomä R and Sachs J 2003 Ultra-wideband real-time channel sounder design and application. Technical Report TD-03-201, COST 273.

Zwick T, Hampicke D, Richter A, Sommerkorn G, Thoma R and Wiesbeck W 2004 A novel antenna concept for double-directional channel measurements. *IEEE Transactions on Vehicular Technology* **53**(2), 527–537.

6

Deterministic Channel-parameter Estimation

As briefly mentioned in Chapter 5, channel-parameter estimation methods can be categorized into two groups: spectral-based approaches and parametric approaches [Krim and Viberg 1996]. Methods in the former group estimate the channel parameters via finding the maxima (or minima) of spectrum-like functions of the dispersion parameters. These methods are computationally attractive as the maxima(or minima)-searching is performed in one dimension for all paths. Parametric approaches use the signals at the array output to estimate the parameters of an underlying parametric model characterizing the effect of the propagation channel on the transmitted signal. They have better estimation accuracy and higher resolution than spectral-based techniques, but the computational complexity is usually high due to the multi-dimensional searches required.

In this chapter, some widely adopted spectral-based and parametric algorithms used for estimating channel parameters from measurement data are introduced. In Section 6.1, the Bartlett beamforming method [Bartlett 1948], which is a common spectral-based method for estimating channel power spectra, is described. In Section 6.2, a subspace-based spectral method, the multiple signal identification classification (MUSIC) algorithm [Schmidt 1986], is introduced. Section 6.3 describes the principle of two subspace-based parametric methods: the estimation of signal parameters via rotational invariance technique (ESPRIT) [Roy and Kailath 1989] and the propagator method [Marcos et al. 1994; 1995]. In Section 6.4, the maximum-likelihood estimation method is elaborated. In Section 6.5, we introduce the EM and the SAGE algorithms used for channel parameter estimation [Fleury et al. 1999]. In Section 6.6, Richter's maximum-likelihood estimation method (RiMAX) is briefly reviewed [Richter et al. 2003]. In Section 6.7, techniques relying on the multi-layer evidence framework using Bayes' theorem are introduced. In Section 6.8, the implementation of the extended Kalman filter (EKF) is described. This approach can be used to track time-variant channel parameters. Finally, a modified particle-filter-based technique is described in Section 6.9. This technique is more readily applicable than the EKF for tracking channel variations with low complexity.

Propagation Channel Characterization, Parameter Estimation and Modelling for Wireless Communications, First Edition.
Xuefeng Yin and Xiang Cheng.

All these algorithms and methods make use of a deterministic model that describes the channel composition in instantaneous snapshots. Thus, we call them deterministic channel parameter estimation methods. In Chapter 7, estimation methods based on statistical channel models describing the spectral behavior of the channel components are introduced. For notational convenience, we call these methods statistical channel parameter estimation methods.

6.1 Bartlett Beamformer

Spectral-based methods have a common feature, namely a smooth spectrum with respect to a specific parameter – delay, angles (i.e. azimuth or elevation) of arrival, and so on – is computed based on observations. By either maximizing or minimizing the spectral height over the parameter(s), it is possible to obtain estimates of the parameters that represent the characteristics of the components in the propagation channel.

The spectral-based methods include the periodogram, the correlogram, the Bartlett beamformer [Bartlett 1948], the Capon beamformer [Capon 1969], and the MUSIC algorithm [Schmidt 1986], as well as many variants of these methods. Generally speaking, spectral-based methods can also be grouped into two classes: non-parametric and parametric. Typical non-parametric methods are the periodogram, the correlogram, the Blackman–Tukey method, the refined Blackman–Tukey method based on windows, and other extended periodogram methods such as the Bartlett method, the Welch method, the Daniell method, and so on.

In this section, we briefly introduce the Bartlett beamforming method, a widely used spectral-based method. Some experimental results are presented to illustrate the performance of method when applied to real measurement data.

In order to understand the advance represented by the Bartlett beamformer, it is necessary first to understand the periodogram and correlogram spectral estimators. The periodogram power-spectral estimation method can be represented by the following equation:

$$\hat{p}_{\mathrm{p}}(\omega) = \frac{1}{N}\left|\sum_{n=1}^{N} y(n)e^{-j\omega n}\right|^2 \tag{6.1}$$

where $y(n)$ is the received signal observation for the nth sampling instance in time or location in space, $e^{-j\omega n}$ represents the system response at the nth sample when the signal component has a frequency of ω. Its obvious that $e^{-j\omega n}$ has a very regular formulation, which does not include the intrinsic system responses for the nth sample.

The correlogram spectral estimation method is

$$\hat{p}_{\mathrm{c}}(\omega) = \sum_{k=(-N-1)}^{N-1} \hat{r}(k)e^{-j\omega k} \tag{6.2}$$

where $\hat{r}(k)$ represents the autocorrelation function of $y(n)$.

The beamforming method is based on the assumption that the array response, which is also called the steering vector $\mathbf{c}(\theta)$, is known. The beamforming method involves the design of a spatial filter that satisfies the following two conditions:

1. For a given direction of arrival (DoA) θ, the filter passes undistorted signals.
2. The filter attenuates signals from all other DoAs.

These two conditions correspond to the constraint that the weights $\boldsymbol{h}$ of the multiple spatial "taps" of the spatial filter should satisfy the condition [see Stoica and Moses 1997, p. 236] that

$$\min_{\boldsymbol{h}} \boldsymbol{h}^{\mathrm{H}}\boldsymbol{h} \text{ subject to } \boldsymbol{h}^{\mathrm{H}}\boldsymbol{c}(\theta) = 1$$

It can be shown that in such a case,

$$\boldsymbol{h} = \frac{\boldsymbol{c}(\theta)}{\boldsymbol{c}(\theta)^{\mathrm{H}}\boldsymbol{c}(\theta)} \tag{6.3}$$

Thus when these optimal weights are applied to the spatial filter, we obtain for the output signal of the spatial filter:

$$\begin{aligned} \mathrm{E}\{|y(t)|^2\} &= \boldsymbol{h}^{\mathrm{H}}\boldsymbol{R}\boldsymbol{h} \\ &= \boldsymbol{h}^{\mathrm{H}}\mathrm{E}\{\boldsymbol{y}(t)\boldsymbol{y}(t)^{\mathrm{H}}\} \\ &= \frac{\boldsymbol{c}(\theta)^{\mathrm{H}}\mathrm{E}\{\boldsymbol{y}(t)\boldsymbol{y}(t)^{\mathrm{H}}\}\boldsymbol{c}(\theta)}{\boldsymbol{c}(\theta)^{\mathrm{H}}\boldsymbol{c}(\theta)} \end{aligned} \tag{6.4}$$

This equation shows that when the DoA of the impinging signal is θ, the output signal has maximum power only when the weights of the filter satisfy Eq. (6.3). Thus when the DoA of the impinging signal is unknown, we can vary the weights of the filter by letting θ be within a certain range. Therefore, a pseudo-power spectrum $p(\theta)$ with respect to θ can be computed. By selecting the specific value of θ that leads to the peak value of $p(\theta)$, an estimate of the DoA is obtained. Based on this rationale, the power spectrum computed can be expressed as

$$p_{\mathrm{B}}(\theta) = \frac{\boldsymbol{c}(\theta)^{\mathrm{H}}\widehat{\boldsymbol{R}}\boldsymbol{c}(\theta)}{\boldsymbol{c}(\theta)^{\mathrm{H}}\boldsymbol{c}(\theta)} \tag{6.5}$$

where $\widehat{\boldsymbol{R}}$ is the covariance matrix of the received signal, which can be calculated according to

$$\begin{aligned} \widehat{\boldsymbol{R}} &= \mathrm{E}\{\boldsymbol{y}(t)\boldsymbol{y}(t)^{\mathrm{H}}\} \\ &= \frac{1}{N}\sum_{n=1}^{N} \boldsymbol{y}_n(t)\boldsymbol{y}_n(t)^{\mathrm{H}} \end{aligned} \tag{6.6}$$

An example of using the beamforming method to estimate the direction of departure is illustrated next. The measurement data used was collected in a 2005 campaign jointly conducted by a commercial business called Elektrobit and the Technology University of Vienna. They using the wideband MIMO sounder PROPSound in a building of Oulu University. Figure 5.8(a) is a photograph of the 50-element antenna array used in the Tx during the measurements. On the receiver side, a 32-element antenna array is used. In total, $50\times32=1600$ spatial subchannels are measured in one measurement cycle.

Figure 6.1 depicts the power direction (i.e. azimuth and elevation) of the departure spectrum estimated using the beamforming method. The signals emitted by 50 Tx antennas and received from the first Rx antenna are used. The responses of the antennas are separated into vertical and horizontal polarizations. In this example, both vertical and horizontal polarization array responses are considered. It can be seen from Figure 6.1 that the power spectra are not exactly the same. The spectrum estimated using vertical polarization has more fluctuations than when

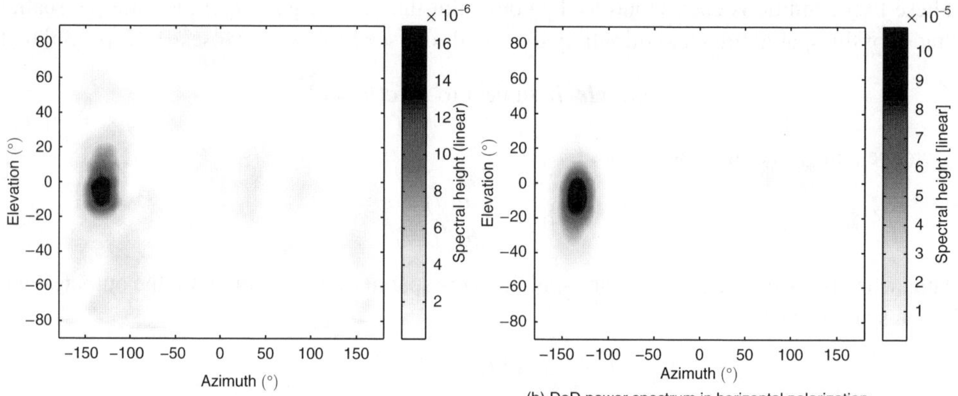

(a) DoD power spectrum in vertical polarization.

(b) DoD power spectrum in horizontal polarization.

Figure 6.1 DoD power spectra calculated using the beamforming method, with 50 Tx antennas and the No. 1 Rx antenna

horizontal polarization is considered. Furthermore, it can be seen that the power spectrum estimated using horizontal polarization has a higher spectral height than when using vertical polarization. It is obvious that the array response may have a significant impact on the estimated power spectrum.

6.2 The MUSIC Algorithm

The MUSIC algorithm was originally proposed by Schmidt [1981, 1986] and Bienvenu and Kopp [1983]. It was introduced in the field of array processing, but has been applied since then in other applications. To explain the principle of the algorithm, we consider the simple model

$$\boldsymbol{Y} = \boldsymbol{C}\boldsymbol{F} + \boldsymbol{W} \tag{6.7}$$

where $\boldsymbol{Y} \in \mathbb{C}^{M\times 1}$ denotes the output signals of the M-element Rx array, $\boldsymbol{C} \doteq [\boldsymbol{c}(\phi_1)\ \boldsymbol{c}(\phi_2)\ \ldots\ \boldsymbol{c}(\phi_D)] \in \mathbb{C}^{M\times D}$, with $\boldsymbol{c}(\phi) \in \mathbb{C}^{M\times 1}$ denoting the array response versus the AoA ϕ and ϕ_d, $d = 1, \ldots, D$ representing the AoAs of the D propagation paths. The vector $\boldsymbol{F} \in \mathbb{C}^{M\times 1}$ consists of the complex path weights and $\boldsymbol{W} \in \mathbb{C}^{M\times 1}$ denotes the temporal-spatial white circularly symmetric Gaussian noise with component variance σ_w^2.

The vector $\boldsymbol{Y}$ can be visualized as a vector in M-dimensional space. The individual columns $\boldsymbol{c}(\phi_d)$, $d \in [1, \ldots, D]$ of $\boldsymbol{C}$ are "mode" vectors [Schmidt 1986]. It is apparent that the vector $\boldsymbol{Y}$ in the case with $\sigma_w^2 = 0$ is a linear combination of the mode vectors. Thus, the signal-only components in $\boldsymbol{Y}$ are confined to the range space of $\boldsymbol{C}$.

Figure 6.2 shows the idea behind the MUSIC algorithm using a simple example, in which $M = 3$, $\boldsymbol{C} \doteq [\boldsymbol{c}(\phi_1), \boldsymbol{c}(\phi_2)]$, $\phi_1 \neq \phi_2$, and $\boldsymbol{e}_1$, $\boldsymbol{e}_2$, $\boldsymbol{e}_3$ are the eigenvectors calculated from the covariance matrix. The range space of $\boldsymbol{C}$ is a two-dimensional subspace of $\mathbb{C}^3$. The vector $\boldsymbol{Y}$ lies in three-dimensional space. The steering vector $\boldsymbol{c}(\phi)$ – the continuum of all possible mode vectors – lies within three-dimensional space. In this example, $\boldsymbol{c}(\phi_1)$ and $\boldsymbol{c}(\phi_2)$ jointly determine a two-dimensional space that coincides with the estimated signal subspace spanned

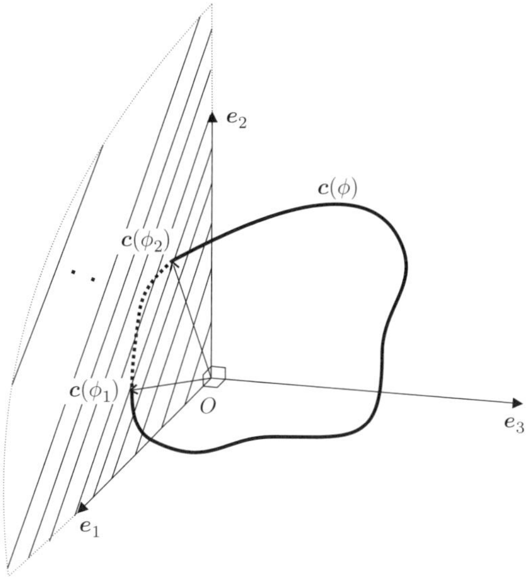

Figure 6.2 The signal subspace and the noise subspace. Vectors $\boldsymbol{c}(\phi_1)$ and $\boldsymbol{c}(\phi_2)$ represent two steering vectors. The eigenvectors $\boldsymbol{e}_1$ and $\boldsymbol{e}_2$ are an orthonormal basis of the range space of the matrix $[\boldsymbol{c}(\phi_1)\ \boldsymbol{c}(\phi_2)]$

by $\boldsymbol{e}_1$ and $\boldsymbol{e}_2$. Since the estimated signal subspace is orthogonal to the estimated noise space, the steering vectors $\boldsymbol{c}(\phi_1)$ and $\boldsymbol{c}(\phi_2)$ are orthogonal to $\boldsymbol{e}_3$. Thus the projection between the steering vector and the estimated noise eigenvector can be formulated as a criterion for parameter estimation. It is obvious that the performance of the MUSIC algorithm depends on the accuracy of the estimated signal and noise subspaces.

Note that the true steering vectors may not exist in the estimated signal subspace in some circumstances. In this case, the projection between the true steering vector and the estimated noise eigenvector is never equal to zero. The pseudo-spectrum may fail to exhibit peaks in the true direction. This occurs, for instance, in the case where the entries in the matrix $\boldsymbol{F}$ are highly correlated [Krim and Proakis 1994] or the propagation paths are characterized by parameters with differences less than the intrinsic resolution of the equipment [Krim and Proakis 1994].

The standard MUSIC algorithm consists of the following steps [Schmidt 1986]:

1. Calculate the sample covariance matrix and its eigenvalue decomposition.
2. Find the orthonormal basis of the estimated noise subspace; the number of specular path components D in the received signal needs to be either known in advance or estimated otherwise.
3. Calculate the pseudo-spectrum: the inverse of the Euclidean distance (squared) between the estimated noise subspace and the steering vector $\boldsymbol{c}(\phi)$ with respect to ϕ.
4. Find the D arguments of the pseudo-spectrum leading to the D highest peaks in the pseudo-spectrum.

The MUSIC algorithm can be easily extended to jointly estimate multi-dimensional parameters, such as delay, angular parameters, and Doppler frequency. A typical example of the extension is the joint angle and delay estimation (JADE) MUSIC algorithm [Vanderveen et al. 1997].

In these extensions of the MUSIC algorithm, the (spatial) steering vector $\boldsymbol{c}(\phi)$ in the standard MUSIC algorithm [Schmidt 1986] is replaced by a space-time, space-frequency, or space-time-frequency response vector. For example, consider joint estimation of direction ϕ, delay τ, and Doppler frequency ν. A space-time-frequency response vector $\boldsymbol{u}(\phi, \tau, \nu)$ can be calculated to be $\boldsymbol{u}(\phi, \tau, \nu) = \boldsymbol{c}(\phi) \otimes \boldsymbol{g}(\tau) \otimes \boldsymbol{h}(\nu)$, where $\otimes$ denotes the Kronecker product and $\boldsymbol{c}(\cdot)$, $\boldsymbol{g}(\cdot)$ and $\boldsymbol{h}(\cdot)$ represent the responses of the spatial array, temporal array, and frequency array respectively. The vector $\boldsymbol{u}(\phi, \tau, \nu)$ denotes a multi-dimensional space-time-frequency manifold. The received signal needs to be manipulated in a way such that the indices of the components in the received signal vector are consistent with the indices of the components in the vector $\boldsymbol{u}(\phi, \tau, \nu)$. In such cases, the dimensions of the sample covariance matrix can be large. Provided the space-time-frequency manifold does not have any ambiguities, the MUSIC algorithm can be applied using a similar procedure as in the standard MUSIC algorithm.

Theoretically, the number of paths that the MUSIC algorithm for joint estimation of multi-dimensional parameters can estimate is up to the product of the numbers of samples in each individual dimension. It is obvious that this number could be much larger than that for the standard MUSIC algorithm, which estimates the parameters in only one dimension.

When applied to processing measurement data, the MUSIC algorithm requires the fast fading condition: the complex weights of the propagation paths must vary quickly from one snapshot to another. This condition is necessary to maintain the non-singularity of the signal covariance matrix. However, in a time-invariant environment, this fast-fading condition may not be satisfied. The performance of the MUSIC algorithms, including the extended MUSIC algorithms for multi-dimensional parameter estimation, are expected to degrade due to their inaccurate estimates of the covariance matrix in such a case.

6.3 The ESPRIT and Propagator Methods

6.3.1 ESPRIT

The ESPRIT algorithm [Roy and Kailath 1989] and the propagator method [Marcos et al. 1994, 1995] are two classical algorithms based on the shift-invariance property. Both algorithms exploit a translational rotational invariance among signal subspaces induced by an array. In these algorithms, parameter estimates can be computed analytically. Thus when applicable, these methods exhibit significant computational advantages over methods that rely on solving optimization problems by exhaustive searching.

As discussed in the previous section, use of the MUSIC algorithm requires a knowledge of the array manifold, while the ESPRIT does not. However, the ESPRIT algorithm needs multiple sensor doublets. The elements in each doublet must have identical radiation patterns and be separated by known constant spacings. Apart from these requirements, the radiation patterns can be arbitrary.

The fundamental idea of the ESPRIT algorithm is as follows. The underlying antenna array is divided into two subarrays. Each subarray consists of the same number of elements. The element spacing in each subarray and the spacing between the subarrays are known. The mth element in the first subarray and the mth element in the second compose a “doublet”. In this case, the array response $\boldsymbol{C}_1(\phi)$ of the first array and the array response $\boldsymbol{C}_2(\phi)$ of the second can be related by $\boldsymbol{C}_1(\phi) = \boldsymbol{C}_1(\phi)\boldsymbol{\Phi}(\phi)$, where

$$\boldsymbol{\Phi}(\phi) = \mathrm{diag}[\exp\{j\psi_1\}, \ldots, \exp\{j\psi_D\}]$$

Here, ψ_d is the phase difference between the signals received at the two elements in each doublet for the dth path. This phase difference is a known function of the AoA ϕ_d. It can be shown that the columns in $\boldsymbol{C}_1(\phi)$ and the columns in $\boldsymbol{C}_2(\phi)$ span the same space. Based on this property, the estimate of the matrix $\boldsymbol{\Phi}(\phi)$ can be obtained from the estimated signal subspace computed using the sample covariance matrix. Since the relation between the elements of $\boldsymbol{\Phi}(\phi)$ and ϕ is known, the estimate of ϕ can be calculated from the estimate of $\boldsymbol{\Phi}(\phi)$ in close-form.

The implementation of the ESPRIT algorithm based on a sample covariance matrix is as follows:

1. Calculate the sample covariance matrix of the signals at the output of the M-element array, and compute the eigenvalue decomposition of the sample covariance matrix. We assume that M is even.
2. Estimate the number of the specular path components, find the orthonormal basis of the estimated signal subspace, and decompose the basis into two parts – say, $\boldsymbol{E}_x$ and $\boldsymbol{E}_y$ – which contain the first $M/2$ rows and the other $M/2$ rows, respectively.
3. Compute the eigenvalue decomposition of the matrix

$$\begin{bmatrix} \boldsymbol{E}_x^{\mathrm{H}} \\ \boldsymbol{E}_y^{\mathrm{H}} \end{bmatrix} [\boldsymbol{E}_x \; \boldsymbol{E}_y] = \boldsymbol{E}\boldsymbol{\Lambda}\boldsymbol{E}^{\mathrm{H}}$$

4. Decompose the resultant unitary orthonormal matrix $\boldsymbol{E}$ into four $D \times D$ matrices:

$$\boldsymbol{E} = \begin{bmatrix} \boldsymbol{E}_{11} & \boldsymbol{E}_{12} \\ \boldsymbol{E}_{21} & \boldsymbol{E}_{22} \end{bmatrix}$$

5. Calculate the eigenvalues of $\boldsymbol{E}_{12}\boldsymbol{E}_{22}^{-1}$ or $\boldsymbol{E}_{21}\boldsymbol{E}_{11}^{-1}$ and compute the azimuth estimates according to either

$$\hat{\phi}_d = \cos^{-1}\left\{\frac{\lambda \arg\{\lambda_d(\boldsymbol{E}_{12}\boldsymbol{E}_{22}^{-1})\}}{2\pi\Delta}\right\} \text{ or } \hat{\phi}_d = \cos^{-1}\left\{\frac{-\lambda \arg\{\lambda_d(\boldsymbol{E}_{21}\boldsymbol{E}_{11}^{-1})\}}{2\pi\Delta}\right\}$$

   where λ represents the wavelength, $\arg(\cdot)$ denotes the complex argument, $\lambda_d(\cdot)$ is the dth eigenvalue of the given matrix, and Δ is the distance between the two antennas in one doublet.

Similar to the MUSIC algorithm, the performance of the ESPRIT algorithm in parameter estimation depends on the accuracy of the estimation of the signal subspace. If the signals contributed by different propagation paths are correlated or if the difference of the path parameters is less than the intrinsic resolution of the measurement equipment, the ESPRIT algorithm fails to resolve the paths accurately. The unitary ESPRIT algorithm [Haardt 1995], which incorporates spatial smoothing techniques, is used if the path components are correlated.

The ESPRIT algorithm can also be extended to estimate multi-dimensional parameters of specular paths. For example, the two-dimensional unitary ESPRIT algorithm [Fuhl et al. 1997, Haardt et al. 1995] is used for joint estimation of the azimuth and elevation at one end of the link. A three-dimensional unitary ESPRIT algorithm has been proposed by Richter et al. [2000] to estimate the delay, DoA, and DoD of individual paths.

6.3.2 The Propagator Algorithm

The propagator method is similar to the ESPRIT algorithm as it also makes use of the identity $\boldsymbol{C}_1(\phi) = \boldsymbol{C}_1(\phi)\boldsymbol{\Phi}(\phi)$. However, the computational complexity is lower than the ESPRIT as the propagator method does not require two eigenvalue decomposition operations as required by the ESPRIT algorithm. The reader is referred to papers by Marcos et al. [1994; 1995] for detail descriptions of the algorithm.

6.4 Maximum-likelihood Method

The maximum-likelihood (ML) approach is a standard method for estimating channel parameters. In this section, this method is briefly introduced by using the channel model where the propagation paths are only characterized by their directions of arrival (DoAs). This simplified model can be replaced by other more complicated channel models when deriving the ML estimators of the parameters of the model.

A narrowband $1 \times N$ single-input, multiple-output (SIMO) scenario is considered, and the propagation paths between the Tx and Rx have different DoA Ω. Following the nomenclature of Yin et al. [2003], the narrowband representation of the impulse responses $\boldsymbol{h} \in \mathcal{C}^N$ of the SIMO channel can be written as

$$\boldsymbol{h} = \sum_{\ell=1}^{L} \alpha_\ell \boldsymbol{c}(\Omega_\ell) + \boldsymbol{w} \tag{6.8}$$

where ℓ is the index of the specular propagation paths, L represents the number of paths, α_ℓ and Ω_ℓ are, respectively, the complex attenuation and the direction of arrival (DoA) of the ℓth path, $\boldsymbol{w}$ represents standard white Gaussian noise with a spectral height of N_o. The DoA Ω_ℓ is a unit vector uniquely determined by the azimuth of arrival (AoA) $\phi \in [-\pi, \pi]$ and the elevation of arrival $\theta \in [0, \pi]$ as

$$\Omega = [\sin(\theta)\cos(\phi), \sin(\theta)\sin(\phi), \cos(\theta)] \tag{6.9}$$

In Eq. (6.8), $\boldsymbol{c}(\Omega)$ represents the array response at a given DoA. The parameters of interest for estimation in the generic channel model Eq. (6.8) are

$$\boldsymbol{\Theta} = (\alpha_1, \Omega_1, \alpha_2, \Omega_2, \ldots, \alpha_N, \Omega_N) \tag{6.10}$$

The ML estimators $\widehat{\boldsymbol{\Theta}}_{\mathrm{ML}}$ of the model parameters can be calculated by maximizing the log-likelihood function of the $\boldsymbol{\Theta}$ given the observations of $\boldsymbol{h}$:

$$\Lambda(\boldsymbol{\Theta}) = \log p[\boldsymbol{\Theta}|\boldsymbol{h}] \tag{6.11}$$

where $p(\boldsymbol{\theta})$ denotes the likelihood function of $\boldsymbol{\Theta}$. Applying the assumption that the noise component $\boldsymbol{w}$ is white Gaussian, it can be shown that

$$\Lambda(\boldsymbol{\Theta}) = -N\log(2\pi\sigma_w) - \frac{1}{2N\sigma_w^2}\left\|\hat{\boldsymbol{x}}_\ell^{[i+1]} - \sum_{\ell=1}^{L}\alpha_\ell \mathbf{c}(\Omega_\ell)\right\|^2$$

The estimators $\widehat{\boldsymbol{\Theta}}_{\mathrm{ML}}$ can be obtained as

$$\widehat{\boldsymbol{\Theta}}_{\mathrm{ML}} = \arg\max_{\boldsymbol{\Theta}} \Lambda(\boldsymbol{\Theta}) \tag{6.12}$$

In practice, the complexity of the optimization problem to be solved in Eq. (6.12) is prohibitive because of the need for exhaustive searching in multiple dimensions.

The performance of the ML estimator can be evaluated by using the Cramér–Rao bound. Fleury et al. [1999] provided the derivation of the Cramér–Rao bound, and readers are referred to this paper for the details.

6.5 The SAGE Algorithm

The ML estimation method provides the optimum unbiased parameter estimates from a statistical perspective. However, it is computationally cumbersome due to the exhaustive multi-dimensional searches required for calculation of the estimates. As an alternative, the SAGE algorithm was proposed by Fessler and Hero [1994] as a low-complexity approximation of the ML estimation. In recent years, this algorithm has been successfully applied for different purposes, such as parameter estimation in channel sounding [Fleury et al. 1999] and joint data-detection and channel-estimation in the receivers of wireless communication systems [Hu et al. 2008, Kocian et al. 2003].

The SAGE algorithm updates the estimates of the unknown parameters sequentially by alternating among subsets of these parameters [Fessler and Hero 1994]. To explain the idea of the algorithm, we introduce the following notation:

y: the observed signal
$\boldsymbol{\theta}$: the parameter vector belonging to a p-dimensional space
$\boldsymbol{\theta}_{\mathrm{S}}$: the entries in $\boldsymbol{\theta}$ with indices specified in a subset of $\{1,\ldots,p\}$
$\boldsymbol{\theta}_{\tilde{\mathrm{S}}}$: the entries in $\boldsymbol{\theta}$ with indices listed in the complement of S
X^{S}: the hidden-data space selected for $\boldsymbol{\theta}_{\mathrm{S}}$
x^{S}: a realization of X^{S}
S^i: the index set selected in the ith iteration of the SAGE algorithm.

The space X^{S} associated with $\boldsymbol{\theta}_{\mathrm{S}}$ is admissible hidden data if the following condition is satisfied [Fessler and Hero 1994]:

$$f(y, x^{\mathrm{S}}; \boldsymbol{\theta}) = f(y|x^{\mathrm{S}}; \boldsymbol{\theta}_{\tilde{\mathrm{S}}}) f(x^{\mathrm{S}}; \boldsymbol{\theta}) \tag{6.13}$$

The above equation implies that the conditional distribution $f(y|x^{\mathrm{S}}; \boldsymbol{\theta})$ coincides with $f(y|x^{\mathrm{S}}; \boldsymbol{\theta}_{\tilde{\mathrm{S}}})$.

In the SAGE algorithm proposed for channel parameter estimation [Fleury et al. 1999], the parameter subsets S^i contain one element. Thus the multiple-dimensional maximization in the ML estimation reduces to a one-dimensional search in each SAGE iteration. The number of elements in S^i can be larger than one, depending on the definition of the hidden-data space X^{S^i}. For example, when two paths are closely-spaced with separation below the resolution of the equipment, the admissible hidden data should be defined to embody the sum of the signal contributions of the two paths. In this case, S^i contains more than one element and the maximization step becomes computationally more "expensive".

Figure 6.3 shows the flow graph of the SAGE algorithm. One iteration of the SAGE algorithm consists of two major steps: the expectation (E-) step and the maximization (M-) step. In the E-step, the expectation of the log-likelihood function of admissible hidden data for the current parameter vector $\boldsymbol{\theta}_S$ is computed based on the observation and the parameter estimates from the previous iteration. This expectation is an objective function that is maximized with respect to the parameter vector $\boldsymbol{\theta}_S$ in the M-step. To further reduce the complexity, in the case where the parameter vector $\boldsymbol{\theta}_S$ contains more than one entry the *coordinate-wise updating*

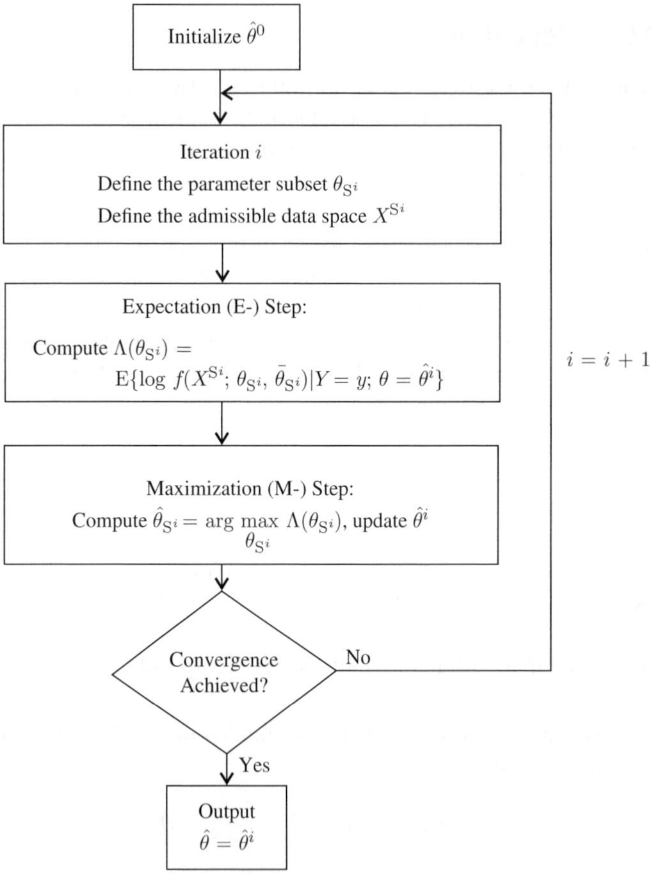

Figure 6.3 Flow graph of the SAGE algorithm

procedure [Fleury et al. 1999] can be used to estimate these parameter entries sequentially. Thus, the multiple-dimensional optimization problem is solved using 1-dimensional searches. This coordinate-wise updating procedure still belongs to the SAGE framework [Fleury et al. 1999]; in other word, updating each parameter entry can be viewed as one SAGE iteration.

The SAGE algorithm was first applied for channel parameter estimation by Fleury et al. [1999] for estimating the delay, Doppler frequency, and direction (azimuth and elevation) of arrival of individual propagation paths. Fleury et al. [2002a,b] extended the SAGE algorithm to include the estimation of the directions of departure of paths and used it to estimate path parameters with the measurement data collected with a multi-antenna array installed in both the transmitter and the receiver. This variant of the SAGE algorithm is called the initialization-and-search-improved SAGE (ISI-SAGE) algorithm. The acronym ISI stresses the fact that the initialization and search procedures of the SAGE algorithm are optimized to speed up its convergence and enhance its capability of detecting weak paths. In order to accurately model the dispersion characteristics of the channels that incorporate the effect of polarizations, the ISI-SAGE algorithm has been extended [Fleury et al. 2003, Yin et al. 2003] to include estimation of the polarization matrices of waves propagating from the Tx site to the Rx site in a MIMO system.

The rest of this section is organized as follows. In Section 6.5.1, we describe the signal model which uses 14 parameters to characterize a single specular propagation path. Then, in Section 6.5.2 the SAGE algorithm derived for estimating channel parameters associated with the novel signal model is presented, with the main steps in the derivation of the algorithm elaborated. In addition, in Section 6.5.3 the incoherent initialization technique is introduced; this can be included into the estimation scheme based on the derived SAGE algorithm for parameter initialization with low complexity.

6.5.1 Signal model

As shown in Figure 6.4 the Rx and Tx antenna arrays have dual antenna configuration. Each of the dual antennas transmits/receives signals in two polarizations at the same time. To differentiate the two polarizations in the underlying model, one of them is referred to as the main polarization, specifying the dominant direction of the signal field pattern. The other is correspondingly referred to the complementary polarization.

In order to describe the wave polarization, a polarization matrix $\boldsymbol{A}_\ell$ is introduced, which is composed of the complex weights for the attenuations along the propagation paths. The signal model describing the contribution of the ℓth wave to the output of the MIMO system reads

$$\boldsymbol{s}(t;\boldsymbol{\theta}_\ell) = \exp(j2\pi\nu_\ell t)\left[\boldsymbol{c}_{2,1}(\boldsymbol{\Omega}_{2,\ell})\ \ \boldsymbol{c}_{2,2}(\boldsymbol{\Omega}_{2,\ell})\right]\begin{bmatrix}\alpha_{\ell,1,1} & \alpha_{\ell,1,2}\\ \alpha_{\ell,2,1} & \alpha_{\ell,2,2}\end{bmatrix} \cdot \left[\boldsymbol{c}_{1,1}(\boldsymbol{\Omega}_{1,\ell})\ \ \boldsymbol{c}_{1,2}(\boldsymbol{\Omega}_{1,\ell})\right]^T \boldsymbol{u}(t-\tau_\ell) \tag{6.14}$$

where $\boldsymbol{c}_{i,p_i}(\boldsymbol{\Omega})$ denotes the steering vector of the transmitter array ($i = 1$) with a total of M_1 entries or receiver array ($i = 2$) with a total of M_2 entries, where M_1 and M_2 are the number of antennas in the Tx and Rx respectively. Here p_i, ($p_i = 1, 2$) denotes the polarization index.

Written in matrix form, Eq. (6.14) is reformulated as

$$\boldsymbol{s}(t;\boldsymbol{\theta}_\ell) = \exp(j2\pi\nu_\ell t)\boldsymbol{C}_2(\boldsymbol{\Omega}_{2,\ell})\boldsymbol{A}_\ell\boldsymbol{C}_1^{\mathrm{T}}(\boldsymbol{\Omega}_{1,\ell})\boldsymbol{u}(t-\tau_\ell) \tag{6.15}$$

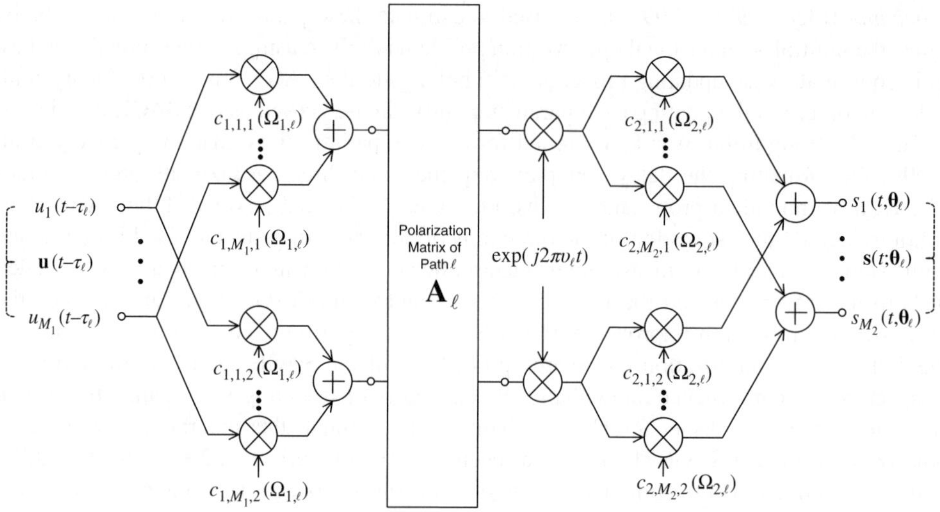

Figure 6.4 Contribution of the ℓth wave to the received signal in a MIMO system incorporating dual-antenna arrays

with

$$\boldsymbol{C}_2(\boldsymbol{\Omega}_{2,\ell}) = \left[\boldsymbol{c}_{2,1}(\boldsymbol{\Omega}_{2,\ell})\ \ \boldsymbol{c}_{2,2}(\boldsymbol{\Omega}_{2,\ell})\right] \tag{6.16}$$

$$\boldsymbol{C}_1(\boldsymbol{\Omega}_{1,\ell}) = \left[\boldsymbol{c}_{1,1}(\boldsymbol{\Omega}_{1,\ell})\ \ \boldsymbol{c}_{1,2}(\boldsymbol{\Omega}_{1,\ell})\right] \tag{6.17}$$

$$\boldsymbol{A}_\ell = \begin{bmatrix} \alpha_{\ell,1,1} & \alpha_{\ell,1,2} \\ \alpha_{\ell,2,1} & \alpha_{\ell,2,2} \end{bmatrix} = [\alpha_{\ell,p_2,p_1}] \tag{6.18}$$

$$\boldsymbol{u}(t) = [u_1(t), \ldots, u_M(t)]^T \tag{6.19}$$

Equation (6.15) can be recast as

$$\begin{aligned} \boldsymbol{s}(t;\boldsymbol{\theta}_\ell) &= \exp(j2\pi\nu_\ell t)\cdot\{[\alpha_{\ell,1,1}\boldsymbol{c}_{2,1}(\boldsymbol{\Omega}_{2,\ell})\boldsymbol{c}_{1,1}^{\mathrm{T}}(\boldsymbol{\Omega}_{1,\ell}) + \alpha_{\ell,1,2}\boldsymbol{c}_{2,1}(\boldsymbol{\Omega}_{2,\ell})\boldsymbol{c}_{1,2}^{\mathrm{T}}(\boldsymbol{\Omega}_{1,\ell}) \\ &\quad + \alpha_{\ell,2,1}\boldsymbol{c}_{2,2}(\boldsymbol{\Omega}_{2,\ell})\boldsymbol{c}_{1,1}^{\mathrm{T}}(\boldsymbol{\Omega}_{1,\ell}) + \alpha_{\ell,2,2}\boldsymbol{c}_{2,2}(\boldsymbol{\Omega}_{2,\ell})\boldsymbol{c}_{1,2}^{\mathrm{T}}(\boldsymbol{\Omega}_{1,\ell})]\boldsymbol{u}(t-\tau_\ell)\} \qquad (6.20) \\ &= \exp(j2\pi\nu_\ell t)\cdot\left(\sum_{p_2=1}^{2}\sum_{p_1=1}^{2}\alpha_{\ell,p_2,p_1}\boldsymbol{c}_{2,p_2}(\boldsymbol{\Omega}_{2,\ell})\boldsymbol{c}_{1,p_1}^{T}(\boldsymbol{\Omega}_{1,\ell})\right)\boldsymbol{u}(t-\tau_\ell) \qquad (6.21) \end{aligned}$$

6.5.1.1 Channel Sounding Technique

Most of the widely-used channel sounders, such as Propsound [Czink 2007], the Medav sounder [Richter and Thomä 2005], and the rBECS sounder [Yin et al. 2012a] are equipped with radio frequency (RF) antenna switches at both Tx and Rx. The timing structures used in these sounder systems are quite similar and are depicted in Figure 6.5.

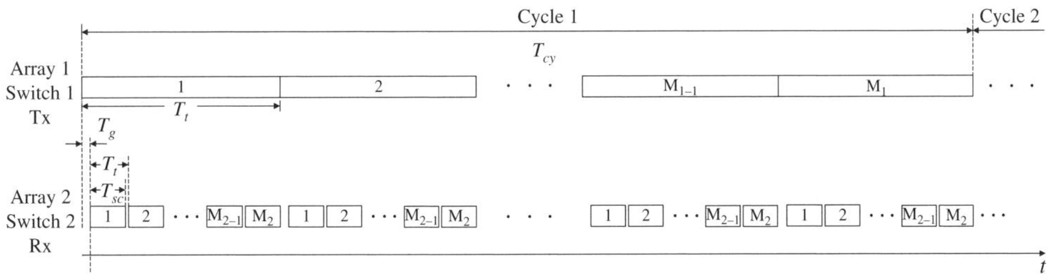

Figure 6.5 Timing structure of sounding and sensing windows

The m_1th antenna element of Array 1 is active during the sounding windows[1]

$$q_{1,m_1}(t) = \sum_{i=1}^{I} q_{T_t}(t - t_{i,m_1} + T_g), m_1 = 1, \ldots, M_1 \tag{6.22}$$

where i denotes the cycle index and

$$t_{i,m_1} = (i-1)T_{cy} + (m_1 - 1)T_t$$

Here $q_{1,m_1}(t)$ is a real function, with value of 1 or 0 corresponding to the active or inactive moments of the m_1th window. Let us define the sounding window vector

$$\boldsymbol{q}_1(t) \doteq [q_{1,1}(t), \ldots, q_{1,M_1}(t)]^T$$

The so-called sensing window

$$q_{T_{sc}}(t - t_{i,m_1,m_2}), m_2 = 1, \ldots, M_2, m_1 = 1, \ldots, M_1$$

corresponds to the case where

- the m_1th Tx antenna is active
- the m_2th Rx antenna is sensing

where

$$t_{i,m_2,m_1} = (i-1)T_{cy} + (m_1 - 1)T_t + (m_2 - 1)T_r$$

The sensing window for the m_2th Rx dual antenna is given by the real function

$$q_{1,m_2}(t) = \sum_{i}^{I} \sum_{m_1=1}^{M_1} q_{T_{sc}}(t - t_{i,m_2,m_1}) \tag{6.23}$$

We can define the sensing window vector

$$\boldsymbol{q}_2(t) \doteq [q_{2,1}(t), \ldots, q_{2,M_2}(t)]^T$$

[1] If another ordering of polarization sounding/sensing is used, the sounding windows $q_1(t)$ and the sensing windows $q_2(t)$ need merely to be appropriately redefined.

as well as

$$q_2(t) = \sum_{i=1}^{I} \sum_{m_2=1}^{M_2} \sum_{m_1=1}^{M_1} q_{T_{\text{sc}}}(t - t_{i,m_2,m_1}) \tag{6.24}$$

6.5.1.2 Transmitted Signal

Making use of the sounding window vector $\boldsymbol{q}_1(t)$, we have the explicit transmitted signal $\boldsymbol{u}(t)$ by concatenating the inputs of the M_1 elements of Array 1

$$\boldsymbol{u}(t) = \boldsymbol{q}_1(t)u(t) \tag{6.25}$$

6.5.1.3 Received Signal

The signal at the output of Switch 2 can be written as

$$Y(t) = \sum_{\ell=1}^{L} \boldsymbol{q}_2^T(t)\boldsymbol{s}(t;\boldsymbol{\theta}_\ell) + \sqrt{\frac{N_o}{2}} q_2(t)W(t) \tag{6.26}$$

with

$$\boldsymbol{s}(t;\boldsymbol{\theta}_\ell) = \exp(j2\pi\nu_\ell t)\boldsymbol{q}_2^{\mathrm{T}}(t)\boldsymbol{C}_2(\boldsymbol{\Omega}_{2,\ell})\boldsymbol{A}_\ell\boldsymbol{C}_1(\boldsymbol{\Omega}_{1,\ell})^{\mathrm{T}}\boldsymbol{q}_1(t-\tau_\ell)u(t-\tau_\ell) \tag{6.27}$$

Implementing Eq. (6.20), we can rewrite

$$\begin{aligned} s(t;\boldsymbol{\theta}_\ell) &= \exp(j2\pi\nu_\ell t) q_2^T(t)\boldsymbol{s}(t;\boldsymbol{\theta}_\ell) \\ &= \exp(j2\pi\nu_\ell t) \cdot \sum_{p_2=1}^{2}\sum_{p_1=1}^{2} \alpha_{\ell,p_2,p_1}\boldsymbol{q}_2^{\mathrm{T}}(t)\boldsymbol{c}_{2,p_2}(\boldsymbol{\Omega}_{2,\ell})\boldsymbol{c}_{1,p_1}^{\mathrm{T}}(\boldsymbol{\Omega}_{1,\ell})\boldsymbol{q}_1(t) \\ &\quad \cdot u(t-\tau_\ell) \end{aligned} \tag{6.28}$$

Equation (6.28) is the extension of the first part of Eq. (7) in the paper by Fleury et al. [2002a] to incorporate polarization. Following the same approach as in this paper, we define the $M_2 \times M_1$ sounding matrices

$$\boldsymbol{U}(t;\tau_\ell) = \boldsymbol{q}_2(t)\boldsymbol{q}_1(t)^{\mathrm{T}}u(t-\tau_\ell) \tag{6.29}$$

With this definition, Eq. (6.28) can be further written as

$$s(t;\boldsymbol{\theta}_\ell) = \exp(j2\pi\nu_\ell t)\sum_{p_2=1}^{2}\sum_{p_1=1}^{2} \alpha_{\ell,p_2,p_1}\boldsymbol{c}_{2,p_2}^{\mathrm{T}}(\boldsymbol{\Omega}_{2,\ell})\boldsymbol{U}(t;\tau_\ell)\boldsymbol{c}_{1,p_1}(\boldsymbol{\Omega}_{1,\ell}) \tag{6.30}$$

We can also express $s(t;\boldsymbol{\theta}_\ell)$ as

$$s(t;\boldsymbol{\theta}_\ell) = \sum_{p_2=1}^{2}\sum_{p_1=1}^{2} s_{p_2,p_1}(t;\boldsymbol{\theta}_\ell) \tag{6.31}$$

where

$$s_{p_2,p_1}(t;\boldsymbol{\theta}_\ell) \doteq \alpha_{\ell,p_2,p_1}\exp(j2\pi\nu_\ell t)\boldsymbol{c}_{2,p_2}^{\mathrm{T}}(\boldsymbol{\Omega}_{2,\ell})\boldsymbol{U}(t;\tau_\ell)\boldsymbol{c}_{1,p_1}(\boldsymbol{\Omega}_{1,\ell}) \tag{6.32}$$

an expression similar to that in Fleury et al. [2002a, Eq (7)].

6.5.2 *The SAGE Algorithm Derived*

In the following, the elements of the SAGE algorithm based on the aforementioned signal models for estimating the unknown channel parameters are introduced.

6.5.2.1 Log-likelihood Function of the Complete/Hidden Data

First, the complete/hidden data is defined as

$$X_\ell(t) = s(t; \boldsymbol{\theta}_\ell) + \sqrt{\beta_\ell}\sqrt{\frac{N_o}{2}} q_2(t) W(t) \tag{6.33}$$

It can be shown that the log-likelihood function of $\boldsymbol{\theta}_\ell$ given the observation $X_\ell(t) = x_\ell(t)$ reads

$$\Lambda(\boldsymbol{\theta}_\ell; x_\ell) \propto 2\mathcal{R}\left\{\underbrace{\int s(t; \boldsymbol{\theta}_\ell)^* x_\ell(t)\mathrm{d}t}_{G_1}\right\} - \underbrace{\int |s(t; \boldsymbol{\theta}_\ell)|^2 \mathrm{d}t}_{G_2} \tag{6.34}$$

After certain manipulations, it can be shown that

$$\Lambda(\boldsymbol{\theta}_\ell; x_\ell) \propto 2\mathcal{R}\{\boldsymbol{\alpha}_\ell^H \boldsymbol{f}(\bar{\boldsymbol{\theta}}_\ell)\} - IPT_{\mathrm{sc}} \cdot \boldsymbol{\alpha}_\ell^H \tilde{\boldsymbol{D}}(\boldsymbol{\Omega}_{2,\ell}, \boldsymbol{\Omega}_{1,\ell})\boldsymbol{\alpha}_\ell \tag{6.35}$$

The calculation of G_1 and G_2 will be elaborated in the following.

6.5.2.2 Computation of G_1

From Eq. (6.31) we can write

$$\begin{aligned} G_1 &= \int \sum_{p_2=1}^{2}\sum_{p_1=1}^{2} s_{p_2,p_1}(t; \boldsymbol{\theta}_\ell)^* x_\ell(t)\mathrm{d}t \\ &= \sum_{p_2=1}^{2}\sum_{p_1=1}^{2} \int s_{p_2,p_1}(t; \boldsymbol{\theta}_\ell)^* x_\ell(t)\mathrm{d}t \end{aligned} \tag{6.36}$$

Inserting Eq. (6.32) into Eq. (6.36) yields

$$\begin{aligned} G_1 &= \sum_{p_2=1}^{2}\sum_{p_1=1}^{2} \int s_{p_2,p_1}(t; \boldsymbol{\theta}_\ell)^* x_\ell(t)\mathrm{d}t \\ &= \sum_{p_2=1}^{2}\sum_{p_1=1}^{2} \int \alpha_{\ell,p_2,p_1}^* \exp(-j2\pi\nu_\ell t)\boldsymbol{c}_{2,p_2}^H(\boldsymbol{\Omega}_{2,\ell})\boldsymbol{U}^*(t; \tau_\ell)\boldsymbol{c}_{1,p_1}^*(\boldsymbol{\Omega}_{1,\ell})x_\ell(t)\mathrm{d}t \\ &= \sum_{p_2=1}^{2}\sum_{p_1=1}^{2} \alpha_{\ell,p_2,p_1}^* \boldsymbol{c}_{2,p_2}^H(\boldsymbol{\Omega}_{2,\ell}) \underbrace{\int \exp(-j2\pi\nu_\ell t)\boldsymbol{U}^*(t; \tau_\ell)x_\ell(t)\mathrm{d}t}_{\doteq \boldsymbol{X}_\ell(\tau_\ell, \nu_\ell)} \boldsymbol{c}_{1,p_1}^*(\boldsymbol{\Omega}_{1,\ell}) \end{aligned} \tag{6.37}$$

with $\boldsymbol{X}_\ell(\tau_\ell, \nu_\ell)$ denoting the $M_2 \times M_1$ dimensional matrix with entries

$$X_{\ell,m_2,m_1}(\tau_\ell, \nu_\ell) = \int \exp(-j2\pi\nu_\ell t) U^*_{m_2,m_1}(t; \tau_\ell) x_\ell(t) \mathrm{d}t$$
$$= \int \exp(-j2\pi\nu_\ell t) q_{2,m_2}(t) q_{1,m_1}(t - \tau_\ell) u^*(t - \tau_\ell) x_\ell(t) \mathrm{d}t$$

We may notice from the timing structure of the sounding and sensing system that, in each cycle; that is, $i = 1, 2, \ldots, I$, only when the m_2th transmitter antenna and the m_1th receiver antennas are active, does the product between $q_{2,m_2}(t)$ and $q_{1,m_1}(t)$ give a non-zero result. Applying this principle yields

$$q_{2,m_2}(t) q_{1,m_1}(t - \tau_\ell) u^*(t - \tau_\ell) = \begin{cases} u^*(t - \tau_\ell) & ; \ t \in [t_{i,m_2,m_1}, t_{i,m_2,m_1} + T_{\mathrm{sc}}] \\ 0 & ; \ \text{otherwise} \end{cases}$$

Therefore

$$X_{\ell,m_2,m_1}(\tau_\ell, \nu_\ell) = \sum_{i=1}^{I} \int_{t_{i,m_2,m_1}}^{t_{i,m_2,m_1}+T_{\mathrm{sc}}} \exp(-j2\pi\nu_\ell t) u^*(t - \tau_\ell) x_\ell(t) \mathrm{d}t \tag{6.38}$$

Using parameter substitution $t' = t - t_{i,m_2,m_1}$ in Eq. (6.38) yields

$$X_{\ell,m_2,m_1}(\tau_\ell, \nu_\ell)$$
$$= \sum_{i=1}^{I} \int_0^{T_{\mathrm{sc}}} \exp(-j2\pi\nu_\ell(t_{i,m_2,m_1} + t')) u^*(t_{i,m_2,m_1} + t' - \tau_\ell) x_\ell(t_{i,m_2,m_1} + t') \mathrm{d}t'$$
$$= \sum_{i=1}^{I} \exp(-j2\pi\nu_\ell t_{i,m_2,m_1}) \int_0^{T_{\mathrm{sc}}} \exp(-j2\pi\nu_\ell t') u^*(t_{i,m_2,m_1} + t' - \tau_\ell) x_\ell(t_{i,m_2,m_1} + t') \mathrm{d}t' \tag{6.39}$$

Here $u(t)$ is a periodic function with period T_{sc}, and t_{i,m_2,m_1} is an integer multiple of T_{sc}, so $u(t_{i,m_2,m_1} + t' - \tau_\ell) = u(t' - \tau_\ell)$. Inserting this expression into Eq. (6.39) and substituting t' with t, we obtain

$$X_{\ell,m_2,m_1}(\tau_\ell, \nu_\ell) = \sum_{i=1}^{I} \exp(-j2\pi\nu_\ell t_{i,m_2,m_1}) \int_0^{T_{\mathrm{sc}}} u^*(t - \tau_\ell) \exp(-j2\pi\nu_\ell t)$$
$$\times x_\ell(t + t_{i,m_2,m_1}) \mathrm{d}t$$

Equation (6.37) can be reformulated as

$$G_1 = \sum_{p_2=1}^{2} \sum_{p_1=1}^{2} \alpha^*_{\ell,p_2,p_1} [\boldsymbol{c}^H_{2,p_2}(\boldsymbol{\Omega}_{2,\ell}) \boldsymbol{X}_\ell(\tau_\ell, \nu_\ell) \boldsymbol{c}_{1,p_1}(\boldsymbol{\Omega}_{1,\ell})^*]$$
$$= \sum_{p_2=1}^{2} \sum_{p_1=1}^{2} \alpha^*_{\ell,p_2,p_1} f_{p_2,p_1}(\bar{\boldsymbol{\theta}}_\ell) \tag{6.40}$$

where $f_{p_2,p_1}(\bar{\boldsymbol{\theta}}_\ell) \doteq \boldsymbol{c}^H_{2,p_2}(\boldsymbol{\Omega}_{2,\ell}) \boldsymbol{X}_\ell(\tau_\ell, \nu_\ell) \boldsymbol{c}_{1,p_1}(\boldsymbol{\Omega}_{1,\ell})^*$ and $\bar{\boldsymbol{\theta}}_\ell = [\boldsymbol{\Omega}_{1,\ell}, \boldsymbol{\Omega}_{2,\ell}, \tau_\ell, \nu_\ell]$.

Defining

$$\boldsymbol{\alpha}_\ell \doteq \mathrm{Vec}(\boldsymbol{A}_\ell^T) = [\alpha_{\ell,1,1}, \alpha_{\ell,1,2}, \alpha_{\ell,2,1}, \alpha_{\ell,2,2}]^T \tag{6.41}$$

$$\boldsymbol{f}(\bar{\boldsymbol{\theta}}_\ell) = \begin{bmatrix} \boldsymbol{c}_{2,1}^{\mathrm{H}}(\boldsymbol{\Omega}_{2,\ell})\boldsymbol{X}_\ell(\tau_\ell,\nu_\ell)\boldsymbol{c}_{1,1}(\boldsymbol{\Omega}_{1,\ell})^* \\ \boldsymbol{c}_{2,1}^{\mathrm{H}}(\boldsymbol{\Omega}_{2,\ell})\boldsymbol{X}_\ell(\tau_\ell,\nu_\ell)\boldsymbol{c}_{1,2}(\boldsymbol{\Omega}_{1,\ell})^* \\ \boldsymbol{c}_{2,2}^{\mathrm{H}}(\boldsymbol{\Omega}_{2,\ell})\boldsymbol{X}_\ell(\tau_\ell,\nu_\ell)\boldsymbol{c}_{1,1}(\boldsymbol{\Omega}_{1,\ell})^* \\ \boldsymbol{c}_{2,2}^{\mathrm{H}}(\boldsymbol{\Omega}_{2,\ell})\boldsymbol{X}_\ell(\tau_\ell,\nu_\ell)\boldsymbol{c}_{1,2}(\boldsymbol{\Omega}_{1,\ell})^* \end{bmatrix} \tag{6.42}$$

we obtain for G_1

$$G_1 = \boldsymbol{\alpha}_\ell^H \boldsymbol{f}(\bar{\boldsymbol{\theta}}_\ell) \tag{6.43}$$

6.5.2.3 Computation of G_2

Inserting Eq. (6.32) into Eq. (6.34) yields for G_2

$$\begin{aligned} G_2 &= \int |s(t;\boldsymbol{\theta}_\ell)|^2 \mathrm{d}t \\ &= \int \left| \sum_{p_2=1}^{2}\sum_{p_1=1}^{2} s_{p_2,p_1}(t;\boldsymbol{\theta}_\ell) \right|^2 \mathrm{d}t \\ &= \int \left| \sum_{p_2=1}^{2}\sum_{p_1=1}^{2} \alpha_{\ell,p_2,p_1} \exp(j2\pi\nu_\ell t) \underbrace{\boldsymbol{c}_{2,p_2}^{\mathrm{T}}(\boldsymbol{\Omega}_{2,\ell})\boldsymbol{U}(t;\tau_\ell)\boldsymbol{c}_{1,p_1}(\boldsymbol{\Omega}_{1,\ell})}_{\doteq V_{\ell,p_2,p_1}(t;\,\boldsymbol{\Omega}_{2,\ell},\boldsymbol{\Omega}_{1,\ell},\tau_\ell)} \right|^2 \mathrm{d}t \\ &= \int \left| \sum_{p_2=1}^{2}\sum_{p_1=1}^{2} \alpha_{\ell,p_2,p_1} V_{\ell,p_2,p_1}(t;\boldsymbol{\Omega}_{2,\ell},\boldsymbol{\Omega}_{1,\ell},\tau_\ell) \right|^2 \cdot \underbrace{|\exp(j2\pi\nu_\ell t)|^2}_{=1} \mathrm{d}t \\ &= \sum_{p_2'=1}^{2}\sum_{p_1'=1}^{2}\sum_{p_2=1}^{2}\sum_{p_1=1}^{2} \alpha_{\ell,p_2',p_1'}^* \alpha_{\ell,p_2,p_1} \underbrace{\int V_{\ell,p_2',p_1'}(t;\boldsymbol{\Omega}_{2,\ell},\boldsymbol{\Omega}_{1,\ell},\tau_\ell)^* V_{\ell,p_2,p_1}(t;\boldsymbol{\Omega}_{2,\ell},\boldsymbol{\Omega}_{1,\ell},\tau_\ell)\mathrm{d}t}_{\doteq D_{p_2',p_1',p_2,p_1}(\boldsymbol{\Omega}_{2,\ell},\boldsymbol{\Omega}_{1,\ell},\tau_\ell)} \end{aligned}$$

$D_{p_2',p_1',p_2,p_1}(\boldsymbol{\Omega}_{2,\ell},\boldsymbol{\Omega}_{1,\ell},\tau_\ell)$ can be further simplified

$$\begin{aligned} & D_{p_2',p_1',p_2,p_1}(\boldsymbol{\Omega}_{2,\ell},\boldsymbol{\Omega}_{1,\ell},\tau_\ell) \\ &= \int V_{\ell,p_2',p_1'}(t;\boldsymbol{\Omega}_{2,\ell},\boldsymbol{\Omega}_{1,\ell},\tau_\ell)^* V_{\ell,p_2,p_1}(t;\boldsymbol{\Omega}_{2,\ell},\boldsymbol{\Omega}_{1,\ell},\tau_\ell)\mathrm{d}t \\ &= \int [\boldsymbol{c}_{2,p_2'}^{\mathrm{T}}(\boldsymbol{\Omega}_{2,\ell})\boldsymbol{U}(t;\tau_\ell)\boldsymbol{c}_{1,p_1'}(\boldsymbol{\Omega}_{1,\ell})]^* \boldsymbol{c}_{2,p_2}^{\mathrm{T}}(\boldsymbol{\Omega}_{2,\ell})\boldsymbol{U}(t;\tau_\ell)\boldsymbol{c}_{1,p_1}(\boldsymbol{\Omega}_{1,\ell})\mathrm{d}t \\ &= \underbrace{\int \boldsymbol{c}_{1,p_1'}^{\mathrm{H}}(\boldsymbol{\Omega}_{1,\ell})\boldsymbol{U}^H(t;\tau_\ell)\boldsymbol{c}_{2,p_2'}^*(\boldsymbol{\Omega}_{2,\ell})\boldsymbol{c}_{2,p_2}^{\mathrm{T}}(\boldsymbol{\Omega}_{2,\ell})\boldsymbol{U}(t;\tau_\ell)\boldsymbol{c}_{1,p_1}(\boldsymbol{\Omega}_{1,\ell})\mathrm{d}t}_{\doteq\eta} \end{aligned} \tag{6.44}$$

where $\boldsymbol{U}(t;\tau_\ell) = \boldsymbol{q}_2(t)\boldsymbol{q}_1(t)^{\mathrm{T}}u(t-\tau_\ell)$. Replacing $\boldsymbol{U}(t;\tau_\ell)$ by this expression in η of Eq. (6.44) yields

$$\eta = \int \underbrace{\boldsymbol{c}^{\mathrm{H}}_{1,p'_1}(\boldsymbol{\Omega}_{1,\ell})\boldsymbol{q}_1(t)}_{\text{a scalar}}\ \underbrace{\boldsymbol{q}_2^{\mathrm{T}}(t)\boldsymbol{c}_{2,p'_2}(\boldsymbol{\Omega}_{2,\ell})}_{\text{a scalar}}\ \underbrace{\boldsymbol{c}^{\mathrm{T}}_{2,p_2}(\boldsymbol{\Omega}_{2,\ell})\boldsymbol{q}_2(t)}_{\text{a scalar}}\ \underbrace{\boldsymbol{q}_1(t)^{\mathrm{T}}\boldsymbol{c}_{1,p_1}(\boldsymbol{\Omega}_{1,\ell})}_{\text{a scalar}}|u(t-\tau_\ell)|^2\mathrm{d}t \tag{6.45}$$

Since the product of each pair of terms in Eq. (6.45) is a scalar, we can rearrange Eq. (6.45) as

$$\eta = \int \boldsymbol{c}^{\mathrm{T}}_{2,p_2}(\boldsymbol{\Omega}_{2,\ell})\boldsymbol{q}_2(t)\boldsymbol{q}_2^{\mathrm{T}}(t)\boldsymbol{c}_{2,p'_2}(\boldsymbol{\Omega}_{2,\ell})\boldsymbol{c}^{\mathrm{H}}_{1,p'_1}(\boldsymbol{\Omega}_{1,\ell})\boldsymbol{q}_1(t)\boldsymbol{q}_1(t)^{\mathrm{T}}\boldsymbol{c}_{1,p_1}(\boldsymbol{\Omega}_{1,\ell})|u(t-\tau_\ell)|^2\mathrm{d}t \tag{6.46}$$

where

$$\begin{aligned}\boldsymbol{q}_2(t)\boldsymbol{q}_2^{\mathrm{T}}(t) &= \left[q_{2,1}(t), q_{2,2}(t), \ldots, q_{2,M_2}(t)\right]^T \left[q_{2,1}(t), q_{2,2}(t), \ldots, q_{2,M_2}(t)\right] \\ &= \mathrm{diag}[|q_{2,1}(t)|^2, \ldots, |q_2, M_2(t)^2] \\ &= \mathrm{diag}[q_{2,1}(t), \ldots, q_2, M_2(t)]\end{aligned} \tag{6.47}$$

with diag[…] representing a diagonal matrix with entries in the argument and $|q_{2,m_2}(t)|^2 = q_{2,m_2}(t)$. Similarly, we have

$$\begin{aligned}\boldsymbol{q}_1(t)\boldsymbol{q}_1(t)^{\mathrm{T}} &= \left[q_{1,1}(t), q_{1,2}(t), \ldots, q_{1,M_1}(t)\right]^T \left[q_{1,1}(t), q_{1,2}(t), \ldots, q_{1,M_1}(t)\right] \\ &= \mathrm{diag}[q_{1,1}(t), \ldots, q_1, M_1(t)]\end{aligned} \tag{6.48}$$

Inserting Eq. (6.47) and Eq. (6.48) into Eq. (6.46) yields

$$\begin{aligned}\eta &= \sum_{m_1=1}^{M_1}\sum_{m_2=1}^{M_2} c_{2,m_2,p_2}(\boldsymbol{\Omega}_{2,\ell})c^*_{2,m_2,p'_2}(\boldsymbol{\Omega}_{2,\ell})c^*_{1,m_1,p'_1}(\boldsymbol{\Omega}_{1,\ell})c_{1,m_1,p_1}(\boldsymbol{\Omega}_{1,\ell}) \\ &\qquad \cdot \underbrace{\int q_{2,m_2}(t)q_{1,m_1}(t-\tau_\ell)|u(t-\tau_\ell)|^2\mathrm{d}t}_{=IPT_{\mathrm{sc}}} \\ &= \left[\sum_{m_1=1}^{M_1} c_{1,m_1,p_1}(\boldsymbol{\Omega}_{1,\ell})c^*_{1,m_1,p'_1}(\boldsymbol{\Omega}_{1,\ell})\right]\left[\sum_{m_2=1}^{M_2} c_{2,m_2,p_2}(\boldsymbol{\Omega}_{2,\ell})c^*_{2,m_2,p'_2}(\boldsymbol{\Omega}_{2,\ell})\right] IPT_{\mathrm{sc}} \\ &= [\boldsymbol{c}^{\mathrm{H}}_{1,p'_1}(\boldsymbol{\Omega}_{1,\ell})\boldsymbol{c}_{1,p_1}(\boldsymbol{\Omega}_{1,\ell})][\boldsymbol{c}^{H}_{2,p'_2}(\boldsymbol{\Omega}_{2,\ell})\boldsymbol{c}_{2,p_2}(\boldsymbol{\Omega}_{2,\ell})]IPT_{\mathrm{sc}}\end{aligned} \tag{6.49}$$

Therefore we obtain for $D_{p'_2,p'_1,p_2,p_1}(\boldsymbol{\Omega}_{2,\ell},\boldsymbol{\Omega}_{1,\ell})$

$$D_{p'_2,p'_1,p_2,p_1}(\boldsymbol{\Omega}_{2,\ell},\boldsymbol{\Omega}_{1,\ell}) = IPT_{\mathrm{sc}} \cdot \tilde{D}_{p'_2,p'_1,p_2,p_1}(\boldsymbol{\Omega}_{2,\ell},\boldsymbol{\Omega}_{1,\ell}) \tag{6.50}$$

where

$$\tilde{D}_{p'_2,p'_1,p_2,p_1}(\boldsymbol{\Omega}_{2,\ell},\boldsymbol{\Omega}_{1,\ell}) \doteq [\boldsymbol{c}^{\mathrm{H}}_{1,p'_1}(\boldsymbol{\Omega}_{1,\ell})\boldsymbol{c}_{1,p_1}(\boldsymbol{\Omega}_{1,\ell})][\boldsymbol{c}^{H}_{2,p'_2}(\boldsymbol{\Omega}_{2,\ell})\boldsymbol{c}_{2,p_2}(\boldsymbol{\Omega}_{2,\ell})]$$

Notice that $D_{p'_2,p'_1,p_2,p_1}(\boldsymbol{\Omega}_{2,\ell},\boldsymbol{\Omega}_{1,\ell})$ and $\tilde{D}_{p'_2,p'_1,p_2,p_1}(\boldsymbol{\Omega}_{2,\ell},\boldsymbol{\Omega}_{1,\ell})$ are functions that depend on $\boldsymbol{\Omega}_{1,\ell}$ and $\boldsymbol{\Omega}_{2,\ell}$ only.

Inserting Eq. (6.50) into Eq. (6.44) yields

$$G_2 = IPT_{\text{sc}} \sum_{p_2'=1}^{2} \sum_{p_1'=1}^{2} \sum_{p_2=1}^{2} \sum_{p_1=1}^{2} \alpha^*_{\ell,p_2',p_1'} \tilde{D}_{p_2',p_1',p_2,p_1}(\boldsymbol{\Omega}_{2,\ell}, \boldsymbol{\Omega}_{1,\ell}) \alpha_{\ell,p_2,p_1}$$
$$= IPT_{\text{sc}} \cdot \boldsymbol{\alpha}_\ell^H \tilde{\boldsymbol{D}}(\boldsymbol{\Omega}_{2,\ell}, \boldsymbol{\Omega}_{1,\ell}) \boldsymbol{\alpha}_\ell \tag{6.51}$$

where

$$\tilde{\boldsymbol{D}}(\boldsymbol{\Omega}_{2,\ell}, \boldsymbol{\Omega}_{1,\ell}) \doteq [\tilde{D}_{p_2',p_1',p_2,p_1}(\boldsymbol{\Omega}_{2,\ell}, \boldsymbol{\Omega}_{1,\ell})]_{(p_2',p_1')=\{1,2\}^2;\ (p_2,p_1)=\{1,2\}^2}$$
$$= \begin{bmatrix} \boldsymbol{c}_{1,1}^H \boldsymbol{c}_{1,1} \boldsymbol{c}_{2,1}^H \boldsymbol{c}_{2,1} & \boldsymbol{c}_{1,1}^H \boldsymbol{c}_{1,2} \boldsymbol{c}_{2,1}^H \boldsymbol{c}_{2,1} & \boldsymbol{c}_{1,1}^H \boldsymbol{c}_{1,1} \boldsymbol{c}_{2,1}^H \boldsymbol{c}_{2,2} & \boldsymbol{c}_{1,1}^H \boldsymbol{c}_{1,2} \boldsymbol{c}_{2,1}^H \boldsymbol{c}_{2,2} \\ \boldsymbol{c}_{1,2}^H \boldsymbol{c}_{1,1} \boldsymbol{c}_{2,1}^H \boldsymbol{c}_{2,1} & \boldsymbol{c}_{1,2}^H \boldsymbol{c}_{1,2} \boldsymbol{c}_{2,1}^H \boldsymbol{c}_{2,1} & \boldsymbol{c}_{1,2}^H \boldsymbol{c}_{1,1} \boldsymbol{c}_{2,1}^H \boldsymbol{c}_{2,2} & \boldsymbol{c}_{1,2}^H \boldsymbol{c}_{1,2} \boldsymbol{c}_{2,1}^H \boldsymbol{c}_{2,2} \\ \boldsymbol{c}_{1,1}^H \boldsymbol{c}_{1,1} \boldsymbol{c}_{2,2}^H \boldsymbol{c}_{2,1} & \boldsymbol{c}_{1,1}^H \boldsymbol{c}_{1,2} \boldsymbol{c}_{2,2}^H \boldsymbol{c}_{2,1} & \boldsymbol{c}_{1,1}^H \boldsymbol{c}_{1,1} \boldsymbol{c}_{2,2}^H \boldsymbol{c}_{2,2} & \boldsymbol{c}_{1,1}^H \boldsymbol{c}_{1,2} \boldsymbol{c}_{2,2}^H \boldsymbol{c}_{2,2} \\ \boldsymbol{c}_{1,2}^H \boldsymbol{c}_{1,1} \boldsymbol{c}_{2,2}^H \boldsymbol{c}_{2,1} & \boldsymbol{c}_{1,2}^H \boldsymbol{c}_{1,2} \boldsymbol{c}_{2,2}^H \boldsymbol{c}_{2,1} & \boldsymbol{c}_{1,2}^H \boldsymbol{c}_{1,1} \boldsymbol{c}_{2,2}^H \boldsymbol{c}_{2,2} & \boldsymbol{c}_{1,2}^H \boldsymbol{c}_{1,2} \boldsymbol{c}_{2,2}^H \boldsymbol{c}_{2,2} \end{bmatrix}$$

In the above expression, $\boldsymbol{\Omega}_{1,\ell}$ and $\boldsymbol{\Omega}_{2,\ell}$ are dropped for notational convenience.

$\tilde{\boldsymbol{D}}(\boldsymbol{\Omega}_{2,\ell}, \boldsymbol{\Omega}_{1,\ell})$ can be expressed as the Kronecker product of two matrices according to

$$\tilde{\boldsymbol{D}}(\boldsymbol{\Omega}_{2,\ell}, \boldsymbol{\Omega}_{1,\ell}) = \begin{bmatrix} \boldsymbol{c}_{2,1}^H \boldsymbol{c}_{2,1} & \boldsymbol{c}_{2,1}^H \boldsymbol{c}_{2,2} \\ \boldsymbol{c}_{2,2}^H \boldsymbol{c}_{2,1} & \boldsymbol{c}_{2,2}^H \boldsymbol{c}_{2,2} \end{bmatrix} \otimes \begin{bmatrix} \boldsymbol{c}_{1,1}^H \boldsymbol{c}_{1,1} & \boldsymbol{c}_{1,1}^H \boldsymbol{c}_{1,2} \\ \boldsymbol{c}_{1,2}^H \boldsymbol{c}_{1,1} & \boldsymbol{c}_{1,2}^H \boldsymbol{c}_{1,2} \end{bmatrix} \tag{6.52}$$

Moreover, each matrix can be also expressed as a product between two matrices:

$$\tilde{\boldsymbol{D}}(\boldsymbol{\Omega}_{2,\ell}, \boldsymbol{\Omega}_{1,\ell}) = \left(\begin{bmatrix} \boldsymbol{c}_{2,1}^H \\ \boldsymbol{c}_{2,2}^H \end{bmatrix} \cdot \begin{bmatrix} \boldsymbol{c}_{2,1} & \boldsymbol{c}_{2,2} \end{bmatrix} \right) \otimes \left(\begin{bmatrix} \boldsymbol{c}_{1,1}^H \\ \boldsymbol{c}_{1,2}^H \end{bmatrix} \cdot \begin{bmatrix} \boldsymbol{c}_{1,1} & \boldsymbol{c}_{1,2} \end{bmatrix} \right)$$
$$= [\boldsymbol{C}_2(\boldsymbol{\Omega}_{2,\ell})^H \boldsymbol{C}_2(\boldsymbol{\Omega}_{2,\ell})] \otimes [\boldsymbol{C}_1(\boldsymbol{\Omega}_{1,\ell})^H \boldsymbol{C}_1(\boldsymbol{\Omega}_{1,\ell})] \tag{6.53}$$

where $\boldsymbol{C}_1(\boldsymbol{\Omega}_{1,\ell})$ and $\boldsymbol{C}_2(\boldsymbol{\Omega}_{2,\ell})$ are defined in Eq. (6.16) and Eq. (6.17) as

$$\boldsymbol{C}_2(\boldsymbol{\Omega}_{2,\ell}) = \begin{bmatrix} \boldsymbol{c}_{2,1}(\boldsymbol{\Omega}_{2,\ell}) & \boldsymbol{c}_{2,2}(\boldsymbol{\Omega}_{2,\ell}) \end{bmatrix}$$
$$\boldsymbol{C}_1(\boldsymbol{\Omega}_{1,\ell}) = \begin{bmatrix} \boldsymbol{c}_{1,1}(\boldsymbol{\Omega}_{1,\ell}) & \boldsymbol{c}_{1,2}(\boldsymbol{\Omega}_{1,\ell}) \end{bmatrix}$$

6.5.2.4 ML Estimation when the Hidden/Complete data is Known

A separable two-step method can be introduced to estimate the channel parameters. At first $\boldsymbol{\alpha}_\ell$ is estimated with respect to a fixed parameter vector $\bar{\boldsymbol{\theta}}_\ell$. The resulting expression of $\boldsymbol{\alpha}_\ell$ is inserted into the log-likelihood function. Then the estimates of $\bar{\boldsymbol{\theta}}_\ell$ are obtained by maximizing the log-likelihood function with respect to $\bar{\boldsymbol{\theta}}_\ell$ only. Finally, the estimate of $\boldsymbol{\alpha}_\ell$ can be calculated with the previously derived formula parameterized by the estimates of $\bar{\boldsymbol{\theta}}_\ell$.

To find out the closed expression of $\boldsymbol{\alpha}_\ell$ with respect to a fixed $\bar{\boldsymbol{\theta}}_\ell$, we take the gradient of $\Lambda(\boldsymbol{\theta}_\ell; x_\ell)$ with respect to $\boldsymbol{\alpha}_\ell$

$$\frac{\partial \Lambda(\boldsymbol{\theta}_\ell; x_\ell)}{\partial \boldsymbol{\alpha}_\ell} = \boldsymbol{f}(\bar{\boldsymbol{\theta}}_\ell) - IPT_{\text{sc}} \cdot \tilde{\boldsymbol{D}}(\boldsymbol{\Omega}_{2,\ell}, \boldsymbol{\Omega}_{1,\ell}) \boldsymbol{\alpha}_\ell \tag{6.54}$$

Setting Eq. (6.54) equal to 0, we obtain an explicit expression of the optimum $\boldsymbol{\alpha}_\ell$ in terms of $\bar{\boldsymbol{\theta}}_\ell$

$$\boldsymbol{\alpha}_\ell = (IPT_{\text{sc}})^{-1}\tilde{\boldsymbol{D}}(\boldsymbol{\Omega}_{2,\ell},\boldsymbol{\Omega}_{1,\ell})^{-1}\boldsymbol{f}(\bar{\boldsymbol{\theta}}_\ell) \tag{6.55}$$

Inserting Eq. (6.55) back into $\Lambda(\boldsymbol{\theta}_\ell; x_\ell)$ gives the log-likelihood with respect to $\bar{\boldsymbol{\theta}}_\ell$ only

$$\begin{aligned}
\Lambda(\bar{\boldsymbol{\theta}}_\ell; x_\ell)
&= \boldsymbol{\alpha}_\ell^H \boldsymbol{f}(\bar{\boldsymbol{\theta}}_\ell) + \boldsymbol{f}(\bar{\boldsymbol{\theta}}_\ell)^H \boldsymbol{\alpha}_\ell - IPT_{\text{sc}} \cdot \boldsymbol{\alpha}_\ell^H \tilde{\boldsymbol{D}}(\boldsymbol{\Omega}_{2,\ell},\boldsymbol{\Omega}_{1,\ell})\boldsymbol{\alpha}_\ell \\
&= (IPT_{\text{sc}})^{-1}\boldsymbol{f}(\bar{\boldsymbol{\theta}}_\ell)^H(\tilde{\boldsymbol{D}}(\boldsymbol{\Omega}_{2,\ell},\boldsymbol{\Omega}_{1,\ell})^{-1})^H\boldsymbol{f}(\bar{\boldsymbol{\theta}}_\ell) + (IPT_{\text{sc}})^{-1}\boldsymbol{f}(\bar{\boldsymbol{\theta}}_\ell)^H\tilde{\boldsymbol{D}}(\boldsymbol{\Omega}_{2,\ell},\boldsymbol{\Omega}_{1,\ell})^{-1} \\
&\quad \times \boldsymbol{f}(\bar{\boldsymbol{\theta}}_\ell) - IPT_{\text{sc}} \cdot (IPT_{\text{sc}})^{-1}\boldsymbol{f}(\bar{\boldsymbol{\theta}}_\ell)^H(\tilde{\boldsymbol{D}}(\boldsymbol{\Omega}_{2,\ell},\boldsymbol{\Omega}_{1,\ell})^{-1})^H \\
&\quad \times \tilde{\boldsymbol{D}}(\boldsymbol{\Omega}_{2,\ell},\boldsymbol{\Omega}_{1,\ell})(IPT_{\text{sc}})^{-1}\tilde{\boldsymbol{D}}(\boldsymbol{\Omega}_{2,\ell},\boldsymbol{\Omega}_{1,\ell})^{-1}\boldsymbol{f}(\bar{\boldsymbol{\theta}}_\ell) \\
&= (IPT_{\text{sc}})^{-1}\boldsymbol{f}(\bar{\boldsymbol{\theta}}_\ell)^H(\tilde{\boldsymbol{D}}(\boldsymbol{\Omega}_{2,\ell},\boldsymbol{\Omega}_{1,\ell})^{-1})^H\boldsymbol{f}(\bar{\boldsymbol{\theta}}_\ell) + (IPT_{\text{sc}})^{-1}\boldsymbol{f}(\bar{\boldsymbol{\theta}}_\ell)^H\tilde{\boldsymbol{D}}(\boldsymbol{\Omega}_{2,\ell},\boldsymbol{\Omega}_{1,\ell})^{-1} \\
&\quad \times \boldsymbol{f}(\bar{\boldsymbol{\theta}}_\ell) - (IPT_{\text{sc}})^{-1}\boldsymbol{f}(\bar{\boldsymbol{\theta}}_\ell)^H(\tilde{\boldsymbol{D}}(\boldsymbol{\Omega}_{2,\ell},\boldsymbol{\Omega}_{1,\ell})^{-1})^H\underbrace{\tilde{\boldsymbol{D}}(\boldsymbol{\Omega}_{2,\ell},\boldsymbol{\Omega}_{1,\ell})\tilde{\boldsymbol{D}}(\boldsymbol{\Omega}_{2,\ell},\boldsymbol{\Omega}_{1,\ell})^{-1}}_{=I}\boldsymbol{f}(\bar{\boldsymbol{\theta}}_\ell) \\
&= (IPT_{\text{sc}})^{-1}\boldsymbol{f}(\bar{\boldsymbol{\theta}}_\ell)^H\tilde{\boldsymbol{D}}(\boldsymbol{\Omega}_{2,\ell},\boldsymbol{\Omega}_{1,\ell})^{-1}\boldsymbol{f}(\bar{\boldsymbol{\theta}}_\ell) \\
&\propto \boldsymbol{f}(\bar{\boldsymbol{\theta}}_\ell)^H\tilde{\boldsymbol{D}}(\boldsymbol{\Omega}_{2,\ell},\boldsymbol{\Omega}_{1,\ell})^{-1}\boldsymbol{f}(\bar{\boldsymbol{\theta}}_\ell)
\end{aligned} \tag{6.56}$$

Hence the ML estimates of $\bar{\boldsymbol{\theta}}_\ell$ and $\boldsymbol{\alpha}_\ell$ are given by

$$(\hat{\bar{\boldsymbol{\theta}}}_\ell)_{\text{ML}} = \arg\max_{\bar{\boldsymbol{\theta}}_\ell} \boldsymbol{f}(\bar{\boldsymbol{\theta}}_\ell)^H\tilde{\boldsymbol{D}}(\boldsymbol{\Omega}_{2,\ell},\boldsymbol{\Omega}_{1,\ell})^{-1}\boldsymbol{f}(\bar{\boldsymbol{\theta}}_\ell) \tag{6.57}$$

$$(\hat{\boldsymbol{\alpha}}_\ell)_{\text{ML}} = (IPT_{\text{sc}})^{-1}\tilde{\boldsymbol{D}}(\boldsymbol{\Omega}_{2,\ell},\boldsymbol{\Omega}_{1,\ell})^{-1}\boldsymbol{f}(\bar{\boldsymbol{\theta}}_\ell) \tag{6.58}$$

6.5.2.5 Conditions for $\tilde{D}(\boldsymbol{\Omega}_{2,\ell},\boldsymbol{\Omega}_{1,\ell})$ to be Invertible

The matrix $\tilde{\boldsymbol{D}}(\boldsymbol{\Omega}_{2,\ell},\boldsymbol{\Omega}_{1,\ell})$ has to be invertible in order for the ML estimates of $\bar{\boldsymbol{\theta}}_\ell$ and $\boldsymbol{\alpha}_\ell$ in Eq. (6.57) and Eq. (6.58) to exist. This section is devoted to analyzing the invertible conditions for $\tilde{\boldsymbol{D}}(\boldsymbol{\Omega}_{2,\ell},\boldsymbol{\Omega}_{1,\ell})$.

According to the properties of the Kronecker product, the determinant of $\tilde{\boldsymbol{D}}(\boldsymbol{\Omega}_{2,\ell},\boldsymbol{\Omega}_{1,\ell})$ can be calculated as

$$\begin{aligned}
\det(\tilde{\boldsymbol{D}}(\boldsymbol{\Omega}_{2,\ell},\boldsymbol{\Omega}_{1,\ell})) &= \det\left(\begin{bmatrix} \boldsymbol{c}_{2,1}^H\boldsymbol{c}_{2,1} & \boldsymbol{c}_{2,1}^H\boldsymbol{c}_{2,2} \\ \boldsymbol{c}_{2,2}^H\boldsymbol{c}_{2,1} & \boldsymbol{c}_{2,2}^H\boldsymbol{c}_{2,2} \end{bmatrix}\right)^2 \cdot \det\left(\begin{bmatrix} \boldsymbol{c}_{1,1}^H\boldsymbol{c}_{1,1} & \boldsymbol{c}_{1,1}^H\boldsymbol{c}_{1,2} \\ \boldsymbol{c}_{1,2}^H\boldsymbol{c}_{1,1} & \boldsymbol{c}_{1,2}^H\boldsymbol{c}_{1,2} \end{bmatrix}\right)^2 \\
&= (\boldsymbol{c}_{2,1}^H\boldsymbol{c}_{2,1}\boldsymbol{c}_{2,2}^H\boldsymbol{c}_{2,2} - \boldsymbol{c}_{2,2}^H\boldsymbol{c}_{2,1}\boldsymbol{c}_{2,1}^H\boldsymbol{c}_{2,2})^2 \\
&\quad \cdot (\boldsymbol{c}_{1,1}^H\boldsymbol{c}_{1,1}\boldsymbol{c}_{1,2}^H\boldsymbol{c}_{1,2} - \boldsymbol{c}_{1,2}^H\boldsymbol{c}_{1,1}\boldsymbol{c}_{1,1}^H\boldsymbol{c}_{1,2})^2 \\
&= (|\boldsymbol{c}_{2,1}|^2|\boldsymbol{c}_{2,2}|^2 - |\boldsymbol{c}_{2,2}^H\boldsymbol{c}_{2,1}|^2)^2 \cdot (|\boldsymbol{c}_{1,1}|^2|\boldsymbol{c}_{1,2}|^2 - |\boldsymbol{c}_{1,2}^H\boldsymbol{c}_{1,1}|^2)^2
\end{aligned} \tag{6.59}$$

where $\boldsymbol{\Omega}_{1,\ell}$ and $\boldsymbol{\Omega}_{2,\ell}$ are dropped for notational convenience.

Notice that the expressions in the parenthesis in Eq. (6.59) satisfy

$$|\boldsymbol{c}_{i,1}|^2|\boldsymbol{c}_{i,2}|^2 - |\boldsymbol{c}_{i,2}^H\boldsymbol{c}_{i,1}|^2 \geq 0, i = 1, 2 \tag{6.60}$$

with equality if and only if

$$\boldsymbol{c}_{i,1} = \gamma \cdot \boldsymbol{c}_{i,2}$$

for some complex number γ. Expression Eq. (6.60) is merely the Schwarz inequality.

Hence, the determinant of $\tilde{\boldsymbol{D}}(\boldsymbol{\Omega}_{2,\ell}, \boldsymbol{\Omega}_{1,\ell})$ vanishes if and only if the steering vectors for the two polarizations of one array are linearly dependent. This will always be the case when the dimension of the steering vectors – the number of elements of the array – is one. This scenario includes the systems with single output or single input: the SISO, SIMO, and MISO cases.

Finally it should be noticed that in case where the steering vectors are orthogonal, the corresponding expression in the parentheses in Eq. (6.59) reduces to $|\boldsymbol{c}_{i,1}|^2|\boldsymbol{c}_{i,2}|^2$. It is further conjectured that the best ML estimates of $\bar{\boldsymbol{\theta}}_\ell$ and $\boldsymbol{\alpha}_\ell$ are obtained in this scenario.

6.5.2.6 Situations leading to a Non-invertible $\tilde{D}(\boldsymbol{\Omega}_{2,\ell}, \boldsymbol{\Omega}_{1,\ell})$

Case 1: The Transmitter Steering Vectors are Linearly Dependent.

It is assumed that the transmitter antenna array has linearly dependent field patterns for both polarizations, which leads to the situation where

$$\boldsymbol{c}_{1,2}(\boldsymbol{\Omega}_{1,\ell}) = \gamma_1 \cdot \boldsymbol{c}_{1,1}(\boldsymbol{\Omega}_{1,\ell})$$

The signal model in this case can be written as

$$\begin{aligned}
\boldsymbol{s}(t;\boldsymbol{\theta}_\ell) &= \exp(j2\pi\nu_\ell t)\begin{bmatrix}\boldsymbol{c}_{2,1}(\boldsymbol{\Omega}_{2,\ell}) & \boldsymbol{c}_{2,2}(\boldsymbol{\Omega}_{2,\ell})\end{bmatrix}\begin{bmatrix}\alpha_{\ell,1,1} & \alpha_{\ell,1,2}\\ \alpha_{\ell,2,1} & \alpha_{\ell,2,2}\end{bmatrix}\begin{bmatrix}\boldsymbol{c}_{1,1}(\boldsymbol{\Omega}_{1,\ell})\\ \gamma\boldsymbol{c}_{1,1}(\boldsymbol{\Omega}_{1,\ell})\end{bmatrix}\boldsymbol{u}(t-\tau_\ell)\\
&= \exp(j2\pi\nu_\ell t)\begin{bmatrix}\boldsymbol{c}_{2,1}(\boldsymbol{\Omega}_{2,\ell}) & \boldsymbol{c}_{2,2}(\boldsymbol{\Omega}_{2,\ell})\end{bmatrix}\begin{bmatrix}\alpha_{\ell,1,1} + \gamma_1 \cdot \alpha_{\ell,1,2}\\ \alpha_{\ell,2,1} + \gamma_1 \cdot \alpha_{\ell,2,2}\end{bmatrix}\boldsymbol{c}_{1,1}(\boldsymbol{\Omega}_{1,\ell})\boldsymbol{u}(t-\tau_\ell)\\
&= \exp(j2\pi\nu_\ell t)\begin{bmatrix}\boldsymbol{c}_{2,1}(\boldsymbol{\Omega}_{2,\ell}) & \boldsymbol{c}_{2,2}(\boldsymbol{\Omega}_{2,\ell})\end{bmatrix}\begin{bmatrix}\alpha'_{\ell,1}\\ \alpha'_{\ell,2}\end{bmatrix}\boldsymbol{c}_{1,1}(\boldsymbol{\Omega}_{1,\ell})\boldsymbol{u}(t-\tau_\ell) \qquad (6.61)
\end{aligned}$$

where

$$\alpha'_{\ell,1} \doteq \alpha_{\ell,1,1} + \gamma_1 \cdot \alpha_{\ell,1,2}$$

$$\alpha'_{\ell,2} \doteq \alpha_{\ell,2,1} + \gamma_1 \cdot \alpha_{\ell,2,2}$$

Since there is no available information – the independent transmit steering vectors – to differentiate the polarization factors within one row, the estimated vector $\boldsymbol{\alpha}'_\ell$, which consists of elements $\alpha'_{\ell,1}$ and $\alpha'_{\ell,2}$, is actually the linear combination of row elements of the polarization matrix with a certain complex factor γ_1.

Case 2: The Receiver Steering Vectors are Linearly Dependent.

We assume in this case that the receiver steering vectors are linearly dependent for both polarizations:

$$\boldsymbol{c}_{2,2}(\boldsymbol{\Omega}_{2,\ell}) = \gamma_2 \cdot \boldsymbol{c}_{2,1}(\boldsymbol{\Omega}_{2,\ell})$$

where γ_2 is some complex number.

The signal model in this case is revised as

$$\begin{aligned}
\boldsymbol{s}(t;\boldsymbol{\theta}_\ell) &= \exp(j2\pi\nu_\ell t)\left[\boldsymbol{c}_{2,1}(\boldsymbol{\Omega}_{2,\ell})\ \ \gamma_2\boldsymbol{c}_{2,1}(\boldsymbol{\Omega}_{2,\ell})\right]\begin{bmatrix}\alpha_{\ell,1,1} & \alpha_{\ell,1,2}\\ \alpha_{\ell,2,1} & \alpha_{\ell,2,2}\end{bmatrix}\begin{bmatrix}\boldsymbol{c}_{1,1}(\boldsymbol{\Omega}_{1,\ell})\\ \boldsymbol{c}_{1,2}(\boldsymbol{\Omega}_{1,\ell})\end{bmatrix}\boldsymbol{u}(t-\tau_\ell)\\
&= \exp(j2\pi\nu_\ell t)\boldsymbol{c}_{2,1}(\boldsymbol{\Omega}_{2,\ell})\begin{bmatrix}\alpha_{\ell,1,1}+\gamma_2\cdot\alpha_{\ell,2,1}\\ \alpha_{\ell,1,2}+\gamma_2\cdot\alpha_{\ell,2,2}\end{bmatrix}^T\begin{bmatrix}\boldsymbol{c}_{1,1}(\boldsymbol{\Omega}_{1,\ell})\\ \boldsymbol{c}_{1,2}(\boldsymbol{\Omega}_{1,\ell})\end{bmatrix}\boldsymbol{u}(t-\tau_\ell)\\
&= \exp(j2\pi\nu_\ell t)\boldsymbol{c}_{2,1}(\boldsymbol{\Omega}_{2,\ell})\left[\alpha''_{\ell,1}\ \ \alpha''_{\ell,2}\right]\begin{bmatrix}\boldsymbol{c}_{1,1}(\boldsymbol{\Omega}_{1,\ell})\\ \boldsymbol{c}_{1,2}(\boldsymbol{\Omega}_{1,\ell})\end{bmatrix}\boldsymbol{u}(t-\tau_\ell) \qquad (6.62)
\end{aligned}$$

where

$$\alpha''_{\ell,1} \doteq \alpha_{\ell,1,1}+\gamma_2\cdot\alpha_{\ell,2,1}$$
$$\alpha''_{\ell,2} \doteq \alpha_{\ell,1,2}+\gamma_2\cdot\alpha_{\ell,2,2}$$

The estimated polarization factors $\alpha''_{\ell,1}$ and $\alpha''_{\ell,2}$, are actually the linear combination of column elements of the polarization matrix with a complex factor γ_2.

Case 3: Both the Transmitter and Receiver Steering Vectors are Linearly Dependent. In this case we have

$$c_{1,2}(\boldsymbol{\Omega}_{1,\ell}) = \gamma_1\cdot c_{1,1}(\boldsymbol{\Omega}_{1,\ell})$$
$$c_{2,2}(\boldsymbol{\Omega}_{2,\ell}) = \gamma_2\cdot c_{2,1}(\boldsymbol{\Omega}_{2,\ell})$$

at the same time. The signal model in this case reads

$$\begin{aligned}
\boldsymbol{s}(t;\boldsymbol{\theta}_\ell) &= \exp(j2\pi\nu_\ell t)\left[\boldsymbol{c}_{2,1}(\boldsymbol{\Omega}_{2,\ell})\ \ \gamma_2\boldsymbol{c}_{2,1}(\boldsymbol{\Omega}_{2,\ell})\right]\begin{bmatrix}\alpha_{\ell,1,1} & \alpha_{\ell,1,2}\\ \alpha_{\ell,2,1} & \alpha_{\ell,2,2}\end{bmatrix}\begin{bmatrix}\boldsymbol{c}_{1,1}(\boldsymbol{\Omega}_{1,\ell})\\ \gamma_1\boldsymbol{c}_{1,1}(\boldsymbol{\Omega}_{1,\ell})\end{bmatrix}\boldsymbol{u}(t-\tau_\ell)\\
&= \exp(j2\pi\nu_\ell t)\boldsymbol{c}_{2,1}(\boldsymbol{\Omega}_{2,\ell})[\alpha_{\ell,1,1}+\gamma_2\cdot\alpha_{\ell,2,1}+\gamma_1\cdot\alpha_{\ell,1,2}+\gamma_2\gamma_1\cdot\alpha_{\ell,2,2}]\\
&\quad\times\boldsymbol{c}_{1,1}(\boldsymbol{\Omega}_{1,\ell})\boldsymbol{u}(t-\tau_\ell)\\
&= \exp(j2\pi\nu_\ell t)\boldsymbol{c}_{2,1}(\boldsymbol{\Omega}_{2,\ell})\alpha'''_\ell\boldsymbol{c}_{1,1}(\boldsymbol{\Omega}_{1,\ell})\boldsymbol{u}(t-\tau_\ell) \qquad (6.63)
\end{aligned}$$

where

$$\alpha'''_\ell \doteq \alpha_{\ell,1,1}+\gamma_2\cdot\alpha_{\ell,2,1}+\gamma_1\cdot\alpha_{\ell,1,2}+\gamma_2\gamma_1\cdot\alpha_{\ell,2,2}$$

The estimated polarization factor α'''_ℓ is the linear combination of the four entries of the polarization matrix, with certain complex coefficients γ_1 and γ_2.

It should be noticed that in the special case with single transmitter antenna and single polarization, the valid polarization factors are actually the product between the transmitter steering vector and the first column in the polarization matrix $\boldsymbol{A}_\ell$.

6.5.2.7 SAGE coordinate updating procedure

Let us define

$$z(\bar{\boldsymbol{\theta}}_\ell;\boldsymbol{x}_\ell) \doteq \boldsymbol{f}(\bar{\boldsymbol{\theta}}_\ell)^H\tilde{\boldsymbol{D}}(\boldsymbol{\Omega}_{2,\ell},\boldsymbol{\Omega}_{1,\ell})^{-1}\boldsymbol{f}(\bar{\boldsymbol{\theta}}_\ell) \qquad (6.64)$$

The coordinate updating procedure is as follows:

- Using the above definition, the updating equations for $\tau_\ell, \theta_{2,\ell}, \phi_{2,\ell}, \theta_{1,\ell}, \phi_{1,\ell}$ and ν_ℓ read

$$\hat{\tau}''_\ell = \arg\max_{\tau_\ell} z(\hat{\phi}'_{1,\ell}, \hat{\theta}'_{1,\ell}, \hat{\phi}'_{2,\ell}, \hat{\theta}'_{2,\ell}, \tau_\ell, \hat{\nu}'_\ell; \hat{x}_\ell)$$

$$\hat{\nu}''_\ell = \arg\max_{\nu_\ell} z(\hat{\phi}'_{1,\ell}, \hat{\theta}'_{1,\ell}, \hat{\phi}'_{2,\ell}, \hat{\theta}'_{2,\ell}, \hat{\tau}''_\ell, \nu_\ell; \hat{x}_\ell)$$

$$\hat{\theta}''_{2,\ell} = \arg\max_{\theta_{2,\ell}} z(\hat{\phi}'_{1,\ell}, \hat{\theta}'_{1,\ell}, \hat{\phi}'_{2,\ell}, \theta_{2,\ell}, \hat{\tau}''_\ell, \hat{\nu}''_\ell; \hat{x}_\ell)$$

$$\hat{\phi}''_{2,\ell} = \arg\max_{\phi_{2,\ell}} z(\hat{\phi}'_{1,\ell}, \hat{\theta}'_{1,\ell}, \phi_{2,\ell}, \hat{\theta}''_{2,\ell}, \hat{\tau}''_\ell, \hat{\nu}''_\ell; \hat{x}_\ell)$$

$$\hat{\theta}''_{1,\ell} = \arg\max_{\theta_{1,\ell}} z(\hat{\phi}'_{1,\ell}, \theta_{1,\ell}, \hat{\phi}''_{2,\ell}, \hat{\theta}''_{2,\ell}, \hat{\tau}''_\ell, \hat{\nu}''_\ell; \hat{x}_\ell)$$

$$\hat{\phi}''_{1,\ell} = \arg\max_{\phi_{1,\ell}} z(\phi_{1,\ell}, \hat{\theta}''_{1,\ell}, \hat{\phi}''_{2,\ell}, \hat{\theta}''_{2,\ell}, \hat{\tau}''_\ell, \hat{\nu}''_\ell; \hat{x}_\ell)$$

- The coefficients α_{ℓ,p_2,p_1} of the polarization matrix $\boldsymbol{A}_\ell$ are updated using Eq. (6.58)

$$(\hat{\boldsymbol{\alpha}}_\ell)''_{\text{ML}} = (IPT_{\text{sc}})^{-1}\tilde{\boldsymbol{D}}(\hat{\boldsymbol{\Omega}}''_{2,\ell}, \hat{\boldsymbol{\Omega}}''_{1,\ell})^{-1}\boldsymbol{f}(\hat{\bar{\boldsymbol{\theta}}}''_\ell)$$

6.5.3 *Initialization of Parameter Estimates for Executing the SAGE Algorithm*

In initialization step, we estimate wave parameters incoherently. The effect of the unknown parameters of the interested wave can be modelled as random variables. The general signal model is written as

$$\boldsymbol{y} = \boldsymbol{s}(t; \beta, \theta) + \boldsymbol{N} \tag{6.65}$$

where β is a random variable $\mathcal{N}(0, \sigma_\beta^2)$, and $\boldsymbol{N}$ denotes a complex circularly symmetric white Gaussian noise vector $\mathcal{N}(\boldsymbol{0}, N_0\boldsymbol{I})$, where $\boldsymbol{0}$ denotes a null vector and $\boldsymbol{I}$ is an identity matrix. The conditional probability density function (PDF) of $\boldsymbol{y}$ given β and θ is calculated to be

$$p(\boldsymbol{y}; \beta, \theta) \propto \exp\left\{\frac{1}{N_0}\left[2\int_{D_0} \mathcal{R}\{\boldsymbol{s}^H(t; \beta, \theta)\boldsymbol{y}(t)\}\text{d}t - \int_{D_0} \| \boldsymbol{s}(t; \beta, \theta)\|^2 \text{d}t\right]\right\}$$

The PDF of $\boldsymbol{y}$ given θ is calculated according to

$$p(\boldsymbol{y}; \theta) = \int p(\boldsymbol{y}; \beta, \theta)p(\beta)\text{d}\beta \tag{6.66}$$

The general representation of $s(t; \boldsymbol{\theta}_\ell)$ for the ℓth wave is given by

$$s_{i,m_2,m_1}(t; \boldsymbol{\theta}_\ell) = \sum_{p_2=1}^{2}\sum_{p_1=1}^{2} \alpha_{\ell,p_2,p_1} c_{2,m_2,p_2}(\boldsymbol{\Omega}_{2,\ell}) c_{1,m_1,p_1}(\boldsymbol{\Omega}_{1,\ell}) \exp\{j2\pi\nu_\ell t_{i,m_2,m_1}\} u(t - \tau_\ell) \tag{6.67}$$

When certain unknown parameters of the wave of interest are not going to be estimated, we assume that they are random variables and their general effect can be formulated as a random variable β in the signal model of Eq. (6.68).

6.5.3.1 Estimate the Delay τ_ℓ of the ℓth Wave

The delay τ_ℓ is estimated first, so the components not relevant to τ_ℓ can be assumed as a random variable β. Since the exact expression of β is as follows:

$$\beta_{i,m_2,m_1} = \sum_{p_2=1}^{2}\sum_{p_1=1}^{2} \alpha_{\ell,p_2,p_1} c_{2,m_2,p_2}(\mathbf{\Omega}_{2,\ell}) c_{1,m_1,p_1}(\mathbf{\Omega}_{1,\ell}) \exp\{j2\pi\nu_\ell t_{i,m_2,m_1}\}$$

which depends on the indexes i, m_2, and m_1, the signal model in Eq. (6.68) can be rewritten as

$$y^{(\ell)}_{i,m_2,m_1}(t) = \beta_{i,m_2,m_1} \cdot u(t + t_{i,m_2,m_1} - \tau_\ell) + N_{i,m_2,m_1} \tag{6.68}$$

where $t \in [0, T_{\text{sc}}]$. Notice that $u(t)$ is a periodic function over T_{sc}. By assuming the interval between two consecutive channels is an integral multiple of T_{sc}, we may write $u(t + t_{i,m_2,m_1} - \tau_\ell)$ as $u(t - \tau_\ell)$, where t is specified in the range $[0, T_{\text{sc}}]$. So the PDF of all the observations for the ℓth wave $y^{(\ell)}_{i,m_2,m_1}$ with $i = 1, \ldots, I, m_2 = 1, \ldots, M_2$ and $m_1 = 1, \ldots, M_1$ can be written as

$$P(\boldsymbol{y}^{(\ell)}; \tau_\ell) = \prod_{i,m_2,m_1} P(y^{(\ell)}_{i,m_2,m_1}; \tau_\ell)$$

where $\boldsymbol{y}^{(\ell)}$ consists of all the observations for the ℓth wave. The log-likelihood function of $\boldsymbol{y}^{(\ell)}$ given τ_ℓ then reads

$$\Lambda(\tau_\ell; \boldsymbol{y}^{(\ell)}) = \sum_{i,m_2,m_1} \Lambda(\tau_\ell; y^{(\ell)}_{i,m_2,m_1})$$

By applying Eq. (6.66) it can be calculate that $P(y^{(\ell)}_{i,m_2,m_1}; \tau_\ell)$ is

$$p(y^{(\ell)}_{i,m_2,m_1}; \tau_\ell) = \left(\frac{\sigma_\beta^2}{N_0}\int_0^{T_{\text{sc}}} |u(t-\tau_\ell)|^2 \mathrm{d}t + 1\right)^{-1/2} \times \exp\left\{\frac{1}{N_0}\frac{\| \int_0^{T_{\text{sc}}} u^*(t-\tau_\ell) y^{(\ell)}_{i,m_2,m_1}(t)\mathrm{d}t\|^2}{\int_0^{T_{\text{sc}}} |u(t-\tau_\ell)|^2 \mathrm{d}t + \frac{N_0}{\sigma_\beta^2}}\right\}$$

So by dropping constant terms the corresponding log-likelihood function yields

$$\Lambda(\tau_\ell; y^{(\ell)}_{i,m_2,m_1}) = -\frac{1}{2}\ln\left(\int_0^{T_{\text{sc}}} |u(t-\tau_\ell)|^2 \mathrm{d}t + \frac{\sigma_\beta^2}{N_0}\right) + \frac{1}{N_0}\frac{\| \int_0^{T_{\text{sc}}} u^*(t + t_{i,m_2,m_1} - \tau_\ell) y^{(\ell)}_{i,m_2,m_1}(t)\mathrm{d}t\|^2}{\int_0^{T_{\text{sc}}} |u(t-\tau_\ell)|^2 \mathrm{d}t + \frac{N_0}{\sigma_\beta^2}}$$

The integral $\int_0^{T_{\text{sc}}} |u(t-\tau_\ell)|^2 \mathrm{d}t$ is a constant value that equals $P_u T_{\text{sc}}$, so the log-likelihood function which is based on all the observations can be written as

$$\Lambda(\tau_\ell; \boldsymbol{y}^{(\ell)}) = -\frac{1}{2} I M_2 M_1 \ln\left(P_u T_{\text{sc}} + \frac{\sigma_\beta^2}{N_0}\right) + \frac{1}{N_0}\frac{1}{P_u T_{\text{sc}} + \frac{N_0}{\sigma_\beta^2}} \sum_{i,m_2,m_1} \left\| \int_0^{T_{\text{sc}}} u^*(t + t_{i,m_2,m_1} - \tau_\ell) y^{(\ell)}_{i,m_2,m_1}(t)\mathrm{d}t \right\|^2$$

By dropping the constant terms, we have

$$\Lambda(\tau_\ell; \boldsymbol{y}^{(\ell)}) = \sum_{i,m_2,m_1} \left\| \int_0^{T_{\rm sc}} u^*(t + t_{i,m_2,m_1} - \tau_\ell) y_{i,m_2,m_1}^{(\ell)}(t) \mathrm{d}t \right\|^2$$

So the estimate of τ_ℓ is obtained by performing the maximization as follows:

$$\hat{\tau}_\ell(0) = \arg\max_{\tau_\ell} \left\{ \sum_{i,m_2,m_1} \left| \int_0^{T_{\rm sc}} u^*(t + t_{i,m_2,m_1} - \tau_\ell) y_{i,m_2,m_1}^{(\ell)}(t) \mathrm{d}t \right|^2 \right\}$$

6.5.3.2 Estimate the Doppler Frequency ν_ℓ of the ℓth Wave

Now we estimate the Doppler frequency of the ℓth wave. Looking at the signal model described in Eq. (6.70) we may notice that the components not relevant to τ_ℓ and ν_ℓ depend on the indices m_2 and m_1. That indicates that the coefficient β in the signal model is constant for specified m_2, m_1 and all the values of i. Therefore we may have the signal model for estimating ν_ℓ as

$$\boldsymbol{y}_{m_2,m_1}^{(\ell)}(t) = \beta_{m_2,m_1} \cdot \boldsymbol{d}_{m_2,m_1}(t; \tau_\ell) + \boldsymbol{N}_{m_2,m_1}$$

where $\boldsymbol{y}_{m_2,m_1}^{(\ell)} \doteq [y_{i,m_2,m_1}^{(\ell)}; i = 1, \ldots, I]^T$, $\boldsymbol{d}_{m_2,m_1}(t; \nu_\ell) \doteq [\exp\{j2\pi\nu_\ell t_{i,m_2,m_1} u(t - \hat{\tau}_\ell(0));$ $i = 1, \ldots, I\}]^T$, and $\boldsymbol{N}_{m_2,m_1} \doteq [N_{i,m_2,m_1}; i = 1, \ldots, I]^T$. Similar to the previous derivation, the probability for $\boldsymbol{y}^{(\ell)}$ given ν_ℓ can be calculated according to

$$P(\boldsymbol{y}^{(\ell)}; \nu_\ell) = \prod_{m_2,m_1} p(\boldsymbol{y}_{m_2,m_1}^{(\ell)}; \nu_\ell)$$

Applying the same procedure as described in the previous section, we obtain the log-likelihood function of ν_ℓ as

$$\begin{aligned}\Lambda(\nu_\ell; \boldsymbol{y}^{(\ell)}) &= \sum_{m_2,m_1} \Lambda(\nu_\ell; \boldsymbol{y}_{m_2,m_1}^{(\ell)}) \\ &= \sum_{m_2,m_1} \left[-\frac{1}{2} \ln \left[\sum_{i=1}^{I} \int_0^{T_{\rm sc}} |\exp\{j2\pi\nu_\ell t_{i,m_2,m_1}\} u(t - \hat{\tau}_\ell(0))|^2 \mathrm{d}t + \frac{N_0}{\sigma_\beta^2} \right] \right. \\ &\quad \left. + \frac{1}{N_0} \frac{\left| \sum_{i=1}^{I} \exp\{-j2\pi\nu_\ell t_{i,m_2,m_1}\} \int_0^{T_{\rm sc}} u^*(t - \hat{\tau}_\ell(0)) y_{i,m_2,m_1}^{(\ell)}(t) \mathrm{d}t \right|^2}{\sum_{i=1}^{I} \int_0^{T_{\rm sc}} |\exp\{j2\pi\nu_\ell t_{i,m_2,m_1}\} u(t - \hat{\tau}_\ell(0))|^2 \mathrm{d}t + \frac{N_0}{\sigma_\beta^2}} \right]\end{aligned} \tag{6.69}$$

The term $\sum_{i=1}^{I} \int_0^{T_{\rm sc}} |\exp\{j2\pi\nu_\ell t_{i,m_2,m_1} u(t - \hat{\tau}_\ell(0))|^2 \mathrm{d}t$ in Eq. (6.69) is constant for any ν_ℓ according to the following calculation:

$$\sum_{i=1}^{I} \int_0^{T_{\rm sc}} |\exp\{j2\pi\nu_\ell t_{i,m_2,m_1}\} u(t - \hat{\tau}_\ell(0))|^2 \mathrm{d}t = I P_u T_{\rm sc}$$

Using the same rationale in estimating τ_ℓ, we drop the constant terms that are not relevant to ν_ℓ and obtain a simplified log-likelihood function as

$$\Lambda(\boldsymbol{y}^{(\ell)};\tau_\ell) = \sum_{m_2,m_1}\left|\sum_{i=1}^{I}\exp\{-j2\pi\nu_\ell t_{i,m_2,m_1}\}\int_0^{T_{\rm sc}} u(t-\hat{\tau}_\ell(0))^* y_{i,m_2,m_1}^{(\ell)}(t)\mathrm{d}t\right|^2$$

The estimate of ν_ℓ is therefore attained by maximizing the log-likelihood function according to

$$\hat{\nu}_\ell(0) = \arg\max_{\nu_\ell}\left\{\sum_{m_2,m_1}\left|\sum_{i=1}^{I}\exp\{-j2\pi\nu_\ell t_{i,m_2,m_1}\}\int_0^{T_{\rm sc}} u(t-\hat{\tau}_\ell(0))^* y_{i,m_2,m_1}^{(\ell)}(t)\mathrm{d}t\right|^2\right\}$$

6.5.3.3 Estimate the Angles of Arrival $\boldsymbol{\Omega}_{2,\ell}$ of the ℓth Wave

The representation of $s(t;\boldsymbol{\theta}_\ell)$ for the ℓth wave can be rearranged as

$$s_{i,m_2,m_1}(t;\boldsymbol{\Omega}_{2,\ell}) = \sum_{p_2=1}^{2}\beta_{m_1,p_2} c_{2,m_2,p_2}(\boldsymbol{\Omega}_{2,\ell})\exp\{j2\pi\hat{\nu}_\ell(0)t_{i,m_2,m_1}\}u(t-\hat{\tau}_\ell(0)) \quad (6.70)$$

where $\beta_{m_1,p_2} \doteq \sum_{p_1=1}^{2}\alpha_{\ell,p_2,p_1}c_{1,m_1,p_1}(\boldsymbol{\Omega}_{1,\ell}), p_2 = 1,2$ are independent random variables with identical property ($\mathcal{N}(0,\sigma_\beta^2)$). From the expression of β_{m_1,p_2} we may find that $\beta_{m_1,p_2}, p_2 = 1,2$ depend only on m_1. Therefore we model the signal in the Eq. (6.68) as

$$\boldsymbol{s}_{m_1}(t;\boldsymbol{\Omega}_{2,\ell}) = [\beta_{m_1,1}\boldsymbol{d}_{2,1,m_1}(\boldsymbol{\Omega}_{2,\ell}) + \beta_{m_1,2}\boldsymbol{d}_{2,2,m_1}(\boldsymbol{\Omega}_{2,\ell})]u(t-\hat{\tau}_\ell(0))$$

where $\boldsymbol{d}_{2,p_2,m_1}(\boldsymbol{\Omega}), p_2 = 1,2$ are defined to be

$$\boldsymbol{d}_{2,p_2,m_1}(\boldsymbol{\Omega}) \doteq \left[c_{2,m_2,p_2}(\boldsymbol{\Omega}_{2,\ell})\exp\{j2\pi\hat{\nu}_\ell(0)t_{i,m_2,m_1}\}; i = 1,\ldots,I; m_2 = 1,\ldots,M_2\right]^T$$

The signal model in Eq. (6.70) can be reformulated to be

$$\boldsymbol{y}_{m_1}(t;\boldsymbol{\Omega}_{2,\ell}) = \boldsymbol{s}_{m_1}(t;\beta_{m_1,1},\beta_{m_1,2},\boldsymbol{\Omega}_{2,\ell}) + \boldsymbol{N}_{m_1}$$

So the PDF of $\boldsymbol{y}_{m_1}$ given $\boldsymbol{\Omega}_{2,\ell}$ can be written as

$$p(\boldsymbol{y}_{m_1},\boldsymbol{\Omega}_{2,\ell}) = \int\!\!\int p(y_{m_1};\beta_{m_1,1},\beta_{m_1,2},\boldsymbol{\Omega}_{2,\ell})p(\beta_{m_1,1})p(\beta_{m_1,2})\mathrm{d}\beta_{m_1,1}\mathrm{d}\beta_{m_1,2}$$

By defining $\boldsymbol{c}'_{2,1}(\boldsymbol{\Omega}) \doteq \frac{1}{\sqrt{\|\boldsymbol{c}_{2,1}(\boldsymbol{\Omega})+\frac{N_0}{\sigma_\beta^2 P u}\|^2}}\boldsymbol{c}_{2,1}(\boldsymbol{\Omega})$ and $\boldsymbol{c}'_{2,2}(\boldsymbol{\Omega}) \doteq \frac{1}{\sqrt{\|\boldsymbol{c}_{2,2}(\boldsymbol{\Omega})+\frac{N_0}{\sigma_\beta^2 P u}\|^2}}\boldsymbol{c}_{2,2}(\boldsymbol{\Omega})$, the log-likelihood function of $\boldsymbol{\Omega}_{2,\ell}$ given $\boldsymbol{y}^{(\ell)}$ can be calculated to be

$$\begin{aligned}
&\Lambda(\boldsymbol{\Omega}_{2,\ell};\boldsymbol{y}^{(\ell)})\\
&= \sum_{m_1=1}^{M_1} \Lambda(\boldsymbol{\Omega}_{2,\ell};\boldsymbol{y}_{m_1}^{(\ell)})\\
&= -\frac{M_1}{2}\ln\left[\left(\|\ \boldsymbol{c}_{2,1}(\boldsymbol{\Omega}_{2,\ell})\|^2+\frac{N_0}{\sigma_\beta^2 P_u}\right)\left(\|\ \boldsymbol{c}_{2,2}(\boldsymbol{\Omega}_{2,\ell})\|^2+\frac{N_0}{\sigma_\beta^2 P_u}\right)\right.\\
&\left.\cdot\left(1-|<\boldsymbol{c}'_{2,2}(\boldsymbol{\Omega}_\ell)|\boldsymbol{c}'_{2,1}(\boldsymbol{\Omega}_\ell)>|^2\right)\right]+\frac{1}{N_0P_u}\sum_{m_1=1}^{M_1}\sum_{k=1}^{2}\\
&\left|\sum_{i=1}^{I}\sum_{m_2=1}^{M_2} v^*_{m_2,k}(\boldsymbol{\Omega}_{2,\ell})\exp\{-j2\pi\hat{\nu}_\ell(0)t_{i,m_2,m_1}\}\int_0^{T_{\rm sc}} y^{(\ell)}_{i,m_2,m_1}(t)u^*(t-\hat{\tau}_\ell(0))\}\right|^2
\end{aligned}$$

where

$$\begin{aligned}
\boldsymbol{v}_1(\boldsymbol{\Omega}) &\doteq \boldsymbol{c}'_{2,1}(\boldsymbol{\Omega}_{2,\ell})\\
\boldsymbol{v}_(2)(\boldsymbol{\Omega}) &\doteq \frac{1}{\sqrt{1-|<\boldsymbol{c}'_{2,2}(\boldsymbol{\Omega}_{2,\ell})|\boldsymbol{c}'_{2,1}(\boldsymbol{\Omega}_{2,\ell})>|^2}}\\
&\quad\times(\boldsymbol{c}'_{2,2}(\boldsymbol{\Omega}_{2,\ell})-<\boldsymbol{c}'_{2,2}(\boldsymbol{\Omega}_{2,\ell})|\boldsymbol{c}'_{2,1}(\boldsymbol{\Omega}_{2,\ell})>\boldsymbol{c}'_{2,1}(\boldsymbol{\Omega}_{2,\ell}))
\end{aligned}$$

Considering that $\|\ \boldsymbol{c}_{2,p_2}(\boldsymbol{\Omega})\|^2, p_2=1,2$ are approximately constant for $\boldsymbol{\Omega}$ and $\frac{N_0}{\sigma_\beta^2 P_u}$ is close to 0 when the signal-to-noise ratio is relatively high, we may neglect the logarithm term and simplify the above expression as

$$\begin{aligned}
\Lambda(\boldsymbol{\Omega}_{2,\ell};\boldsymbol{y}^{(\ell)}) = \sum_{m_1=1}^{M_1}\sum_{k=1}^{2}&\left|\sum_{i=1}^{I}\sum_{m_2=1}^{M_2} v^*_{m_2,k}(\boldsymbol{\Omega}_{2,\ell})\exp\{-j2\pi\hat{\nu}_\ell(0)t_{i,m_2,m_1}\}\right.\\
&\left.\times\int_0^{T_{\rm sc}} y^{(\ell)}_{i,m_2,m_1}(t)u^*(t-\hat{\tau}_\ell(0))\}\right|^2
\end{aligned}$$

where $\boldsymbol{v}_1(\boldsymbol{\Omega})$ and $\boldsymbol{v}_(2)(\boldsymbol{\Omega})$ are redefined by introducing the normalized $\tilde{\boldsymbol{c}}_{2,1}(\boldsymbol{\Omega}) \doteq \frac{1}{\|\boldsymbol{c}_{2,1}(\boldsymbol{\Omega})}\boldsymbol{c}_{2,1}(\boldsymbol{\Omega})$ and $\tilde{\boldsymbol{c}}_{2,2}(\boldsymbol{\Omega}) \doteq \frac{1}{\|\boldsymbol{c}_{2,2}(\boldsymbol{\Omega})}\boldsymbol{c}_{2,1}(\boldsymbol{\Omega})$ as

$$\begin{aligned}
\boldsymbol{v}_1(\boldsymbol{\Omega}) &\doteq \tilde{\boldsymbol{c}}_{2,1}(\boldsymbol{\Omega})\\
\boldsymbol{v}_(2)(\boldsymbol{\Omega}) &\doteq \frac{1}{\sqrt{1-|<\tilde{\boldsymbol{c}}_{2,2}(\boldsymbol{\Omega}_{2,\ell})|\tilde{\boldsymbol{c}}_{2,1}(\boldsymbol{\Omega}_{2,\ell})>|^2}}\\
&\quad\times(\tilde{\boldsymbol{c}}_{2,2}(\boldsymbol{\Omega}_{2,\ell})-<\tilde{\boldsymbol{c}}_{2,2}(\boldsymbol{\Omega}_{2,\ell})|\tilde{\boldsymbol{c}}_{2,1}(\boldsymbol{\Omega}_{2,\ell})>\tilde{\boldsymbol{c}}_{2,1}(\boldsymbol{\Omega}_{2,\ell})).
\end{aligned}$$

6.6 A Brief Introduction to the RiMAX Algorithm

The RiMAX algorithm [Richter 2004, Richter and Thomä 2005, Richter et al. 2003] can be viewed as an extension of the SAGE algorithm based on the specular-scatterer (SS) model [Fleury et al. 1999]. The RiMAX algorithm can be used for joint estimation of the parameters characterizing specular propagation paths and dispersion of distributed diffuse scatterers. The contributions of the distributed diffuse scatterers to the received signal are referred to as dense multipath components (DMCs). The power-delay profile of these components is characterized using a one-sided exponential decay function. This function can be described by three parameters of the DMCs:

- the time of arrival
- the delay spread
- the average power.

In the RiMAX algorithm, unknown parameters are grouped into two sets. One set consists of the parameters of the specular paths and the other contains the parameters characterizing the DMCs. In the M-steps of the algorithm, gradient-based methods, such as the Gauss–Newton or Levenverg–Marquardt [Marquardt 1963] algorithm, are implemented. For each specular path, an approximation of the Hessian is computed as the estimate of the Fisher information matrix of the parameter estimates [Richter and Thomä 2003]. The diagonal elements of the inverse of this matrix provide estimates of the variances of the estimated parameters. In this algorithm, the variance estimates are used to describe the reliability of the corresponding parameter estimates. A specular path is dropped when its parameter estimates are considered to be "unreliable" [Richter et al. 2003]. For detailed information, readers are referred to the literature [Richter 2004, Richter and Thomä 2005, Richter et al. 2003].

6.7 Evidence-framework-based Algorithms

The channel-parameter estimation methods mentioned above, either in spectral-based or parametric, are derived from signal models characterizing the composition of instantaneous channels. In these methods, no assumptions are made about the statistical behavior of the channel parameters to be estimated. Furthermore, issues such as model selection and model-order determination cannot be solved at the same time as channel-parameter estimation is performed.

Recently, the evidence-framework (EF)-based estimation algorithm has started to attract researchers' attention. The algorithm can take into account profile-style characteristics and prior information about the parameter distributions to refine parameter-estimation results. For example, the EF-SAGE algorithm proposed by Shutin et al. [2007] merges the model order estimation into the SAGE iterative scheme, allowing estimation of the number of paths and path parameters simultaneously. Yin et al. [2012b] used a two-layer EF concept to iteratively improve the reasonability of parameter estimation based on prior information on delay spreads. This appears to be a new direction for deriving parameter estimation schemes in the future.

In this section, the maximum-a-posteriori (MAP) channel parameter-estimation approach, which is based on a multi-level EF is systematically introduced. Compared with the high-resolution parameter-estimation algorithms that are widely used at present, the novelty of the algorithm lies in the fact that it provides three levels of inference:

- the parameter estimates for individual channel components
- the parameter estimation for an *a-priori* constraint, which is termed a "regularizer", following statistical terminology
- the selection of the most appropriate method for characterizing the channel components.

6.7.1 Multi-level Evidence Framework

The problem at hand is to estimate the unknown parameters $\boldsymbol{\theta}$ of multiple components in the channel impulse responses (IRs), the parameters $\boldsymbol{\beta}$ specified in a regularizer $\mathcal{R}$, and furthermore to identify the most appropriate method $\mathcal{A}$ for characterization of the channel components. Estimation of $\boldsymbol{\theta}, \boldsymbol{\beta}$, and $\mathcal{A}$ can be performed by solving the following MAP problem:

$$\widehat{\boldsymbol{\theta}}, \hat{\boldsymbol{\beta}}, \hat{\mathcal{A}} = \arg\max_{\boldsymbol{\theta},\boldsymbol{\beta},\mathcal{A}} p(\boldsymbol{\theta}, \boldsymbol{\beta}, \mathcal{A}|\boldsymbol{D}, \mathcal{R}) \tag{6.71}$$

where $\boldsymbol{D}$ represents the received signals in multiple snapshots. Solving Eq. (6.71) directly is prohibitive in practise due to the large computational complexity involved. As an alternative, we may rewrite $p(\boldsymbol{\theta}, \boldsymbol{\beta}, \mathcal{A}|\boldsymbol{D}, \mathcal{R})$ by formulating the unknown parameters into a multi-level EF, and apply the well-known marginal estimation method (MEM) to compute the approximations of the MAP estimates $\widehat{\boldsymbol{\theta}}, \hat{\boldsymbol{\beta}}$, and $\hat{\mathcal{A}}$. The three-level EF is derived as follows. First, the posterior $p(\boldsymbol{\theta}, \boldsymbol{\beta}, \mathcal{A}|\boldsymbol{D}, \mathcal{R})$ is recast as

$$\begin{aligned} p(\boldsymbol{\theta}, \boldsymbol{\beta}, \mathcal{A}|\boldsymbol{D}, \mathcal{R}) &= p(\boldsymbol{\theta}, \boldsymbol{\beta}|\boldsymbol{D}, \mathcal{A}, \mathcal{R})p(\mathcal{A}|\boldsymbol{D}) \\ &= p(\boldsymbol{\theta}|\boldsymbol{D}, \mathcal{A})p(\boldsymbol{\beta}|\mathcal{R}, \boldsymbol{\theta})p(\mathcal{A}|\boldsymbol{D}) \end{aligned} \tag{6.72}$$

where $p(\boldsymbol{\beta}|\mathcal{R}, \boldsymbol{\theta})$ represents the prior probability of $\boldsymbol{\beta}$ under the predefined regularizer $\mathcal{R}$, $p(\boldsymbol{\theta}|\boldsymbol{D}, \mathcal{A})$ denotes the likelihood of $\boldsymbol{\theta}$ given the observation $\boldsymbol{D}$ and a certain characterization method $\mathcal{A}$, and $p(\mathcal{A}|\boldsymbol{D})$ is the posterior probability of $\mathcal{A}$. It is straightforward to show that

$$p(\mathcal{A}|\boldsymbol{D}) \propto p(\boldsymbol{D}|\mathcal{A})p(\mathcal{A}) \tag{6.73}$$

where $p(\boldsymbol{D}|\mathcal{A}) = \int p(\boldsymbol{D}|\mathcal{A}, \boldsymbol{\theta})p(\boldsymbol{\theta}|\mathcal{A})\mathrm{d}\boldsymbol{\theta}$ is called the evidence of $\mathcal{A}$ following the statistical terminology. We further assume that the probability $p(\mathcal{A})$ of $\mathcal{A}$ is uniform. Hence $p(\mathcal{A}|\boldsymbol{D}) \propto p(\boldsymbol{D}|\mathcal{A})$ holds, and $p(\boldsymbol{\theta}, \boldsymbol{\beta}, \mathcal{A}|\boldsymbol{D}, \mathcal{R})$ is rewritten as

$$p(\boldsymbol{\theta}, \boldsymbol{\beta}, \mathcal{A}|\boldsymbol{D}, \mathcal{R}) \propto p(\boldsymbol{\theta}|\boldsymbol{D}, \mathcal{A})p(\boldsymbol{\beta}|\mathcal{R}, \boldsymbol{\theta})p(\boldsymbol{D}|\mathcal{A}) \tag{6.74}$$

The three probabilities on the right-hand-side of Eq. (6.74) are in fact the evidences of the events of interest. The standard MEM algorithm can then be used to maximize the three evidences iteratively and obtain the approximations of the MAP estimates on the left-hand side of Eq. (6.71). It is worth mentioning that the SAGE algorithm can also be embedded in the MEM, for example for maximizing $p(\boldsymbol{\theta}|\boldsymbol{D}, \mathcal{A})$ and $p(\boldsymbol{\beta}|\mathcal{R}, \boldsymbol{\theta})$.

The multi-level EF derived in Eq. (6.74) allows for estimation of channel characteristics in different contexts. We give two examples of applications of the EF. In the first example, the EF together with the MEM is used to estimate the model order and model parameters. The characterization method $\mathcal{A}$ is fixed to a certain generic model, with the model order an unknown parameter for estimation. In the second example, the parameters of stochastic channel models are estimated directly from measurement data. Notice that channel measurements are usually conducted in time-variant environments with transmitter, receiver, or scatterers moving. Conventionally, parameter estimation for channels and establishment of stochastic channel models are conducted separately. In the second example, a proposed EF specifies the regularizer $\mathcal{R}$ as a statistical channel model with $\boldsymbol{\beta}$ being unknown model parameters of interest to be estimated. Estimation of $\hat{\boldsymbol{\beta}}$ is then performed jointly with the estimation of the snapshot-level channel parameters. This method improves the accuracy of the stochastic model parameters.

Notice that the multi-level EF-based method can also be applied for other purposes, such as selection of a parametric model for characterizing individual channel components. In such a case, the possible models may include the specular-path model, the DMC model [Richter and Thomä 2005], and the dispersive-path model [Yin et al. 2006a].

6.7.2 *Example I: Exponential Decay used in Three-level EF*

In this subsection, we show an example of applying the three-level EF for channel parameter estimation. A multipath scenario is considered, in which the transmitter (Tx) and the receiver (Rx) are equipped with M_1- and M_2-element antenna arrays respectively. We assume that in the nth measurement snapshot, the channel IR consists of L_n components. The baseband representation of the received signal $y_{m_1,m_2,n}(t)$ at the output of the m_2th Rx antenna while the m_1th Tx antenna is transmitting reads

$$y_{m_1,m_2,n}(t) = \boldsymbol{w}_n^T \boldsymbol{x}_{m_1,m_2}(t;\boldsymbol{\theta}_n) + z_{m_1,m_2,n}(t),\ t \in [0,T) \tag{6.75}$$

where $\boldsymbol{w}_n = [w_{n,1},\ldots,w_{n,L_n}]^{\mathrm{T}}$ denotes the magnitudes of propagation attenuation experienced by the L_n components, $(\cdot)^{\mathrm{T}}$ represents the transpose operation, $\boldsymbol{x}_{m_1,m_2}(t;\boldsymbol{\theta}_n) \in \mathcal{C}^{L_n}$ describes the complex-valued L_n components with unit magnitude for propagation attenuation, $\boldsymbol{\theta}_n = (\boldsymbol{\theta}_{n,\ell};\ell = 1,\ldots,L_n)$ represents the parameters of the L_n components in the nth snapshot, $\boldsymbol{\theta}_{n,\ell}$ denotes the parameters of the ℓth component, $z_{m_1,m_2,n}(t)$ is the white Gaussian noise component with spectrum height σ_z^2, and T denotes the observation span of the snapshot.

We assume that the characterization of channel components can be chosen from a set consisting of both the specular-path model and a dispersive-path model.

For the specular-path model, $\boldsymbol{\theta}_{n,\ell}$ represents the parameters of a planar wave in multiple dispersion dimensions:

$$\boldsymbol{\theta}_{n,\ell} = (\phi_{n,\ell},\tau_{n,\ell},\boldsymbol{\Omega}_{\mathrm{Tx},n,\ell},\boldsymbol{\Omega}_{\mathrm{Rx},n,\ell},\nu_{n,\ell}) \tag{6.76}$$

where $\phi_{n,\ell}$ is the phase of the propagation attenuation, $\tau_{n,\ell}$ denotes the delay, $\boldsymbol{\Omega}_{\mathrm{Tx},n,\ell}$ and $\boldsymbol{\Omega}_{\mathrm{Rx},n,\ell}$ represent the DoD and DoA respectively, and $\nu_{n,\ell}$ denotes the Doppler frequency. The ℓth element $x_{m_1,m_2,\ell}(t)$ of $\boldsymbol{x}_{m_1,m_2}(t;\boldsymbol{\theta}_n)$ in Eq. (6.75) when the specular-path model is selected, is written as

$$\begin{aligned} x_{m_1,m_2,\ell}(t) &= s_{m_1,m_2}(t;\boldsymbol{\theta}_{n,\ell}) \\ &= \exp\{j\phi_{n,\ell}\}u(t-\tau_{n,\ell})\exp\{j2\pi\nu_{n,\ell}t\}a_{m_1}(\boldsymbol{\Omega}_{\mathrm{Tx},n,\ell})a_{m_2}(\boldsymbol{\Omega}_{\mathrm{Rx},n,\ell}) \end{aligned} \tag{6.77}$$

where $s_{m_1,m_2}(t;\boldsymbol{\theta}_{n,\ell})$ denotes the ℓth specular-path component with unit magnitude for the propagation attenuation, $u(t)$ is the transmitted signal, and $a_{m_1}(\cdot)$ and $a_{m_2}(\cdot)$ represent the responses of the m_1th Tx antenna and of the m_2th Rx antenna respectively.

When a dispersive-path model, such as the generalized-array manifold (GAM) model [Asztely et al. 1998], is used to characterize the components, $x_{m_1,m_2,\ell}(t)$ can be written as

$$x_{m_1,m_2,\ell}(t) = s_{m_1,m_2}(t;\boldsymbol{\theta}_{n,\ell}) + \gamma^{\mathrm{T}} \boldsymbol{s}'_{m_1,m_2}(t;\boldsymbol{\theta})|_{\boldsymbol{\theta}=\boldsymbol{\theta}_{n,\ell}} \tag{6.78}$$

with $\boldsymbol{s}'_{m_1,m_2}(t;\boldsymbol{\theta}) = [\frac{\partial s_{m_1,m_2}(t;\boldsymbol{\theta}_{n,\ell})}{\partial \phi}, \ldots, \frac{\partial s_{m_1,m_2}(t;\boldsymbol{\theta}_{n,\ell})}{\partial \nu}]^{\mathrm{T}}$, and γ being a vector containing the weighting coefficients associated with the components in $\boldsymbol{s}'_{m_1,m_2}(t;\boldsymbol{\theta})$.

Now we try to find a regularizer for characterizing the common behavior of all components. As reported in literature, the attenuation due to propagation increases exponentially with respect to the time of arrival; that is, the delay of an electromagnetic wave [Parsons 2000]. In this example, the regularizer $\mathcal{R}$ can be defined as follows. The magnitude of the complex attenuation $w_{n,\ell}$ should be in the form of

$$w_{n,\ell} = c^{-a+b_{n,\ell}} \tau_{n,\ell}^{-a+b_{n,\ell}} \tag{6.79}$$

where c denotes the speed of light, a represents the exponential attenuation factor, and $b_{n,\ell}$ is a normal-distributed random variable with zero mean and variance σ_b^2. Introducing $b_{n,\ell}$ in Eq. (6.79) is based on the consideration that the power of the signal obtained at specific delays is log-normal distributed [Parsons 2000]. Furthermore, the parameters a and σ_b^2 of the regularizer should be identical for all paths observed in N measurement snapshots obtained in a specific environment.

It is straightforward to show that the a posteriori probability $p(\boldsymbol{\theta}, \boldsymbol{\beta}, \mathcal{A}|\boldsymbol{D}, \mathcal{R})$ in Eq. (6.72) for this example can be calculated as

$$\begin{aligned} p(\boldsymbol{\theta}, a, \mathcal{A}|\boldsymbol{D}, \mathcal{R}) &= (2\pi)^{-(N+L)/2} \sigma_z^{-N} \sigma_b^{-L} \\ &\times \exp\left\{ -\frac{1}{2\sigma_b^2} \sum_{n=1}^{N} \sum_{l=1}^{L_n} (\log w_{n,\ell} - a \log C - a \log \tau_{n,\ell})^2 \right. \\ &\left. - \frac{1}{2\sigma_z^2} \sum_{n=1}^{N} \|\boldsymbol{y}_n - \boldsymbol{w}_n^T \boldsymbol{x}_{n,\mathcal{A}}\|^2 \right\} \end{aligned}$$

where the vector $\boldsymbol{y}_n$ contains the observations in the nth snapshot; that is, $\boldsymbol{y}_n = [y_{m_1,m_2,n}(t); m_1 = 1, \ldots, M_1, m_2 = 1, \ldots, M_2, t \in [0,T)]$. $\boldsymbol{x}_{n,\mathcal{A}}$ is the vector containing $\boldsymbol{x}_{m_1,m_2,n}(t)$ organized in the same manner as those in $\boldsymbol{y}_n$, and the subscript $\mathcal{A}$ of $\boldsymbol{x}_{n,\mathcal{A}}$ indicates that $\boldsymbol{x}_{n,\mathcal{A}}$ is calculated using the characterization method $\mathcal{A}$.

Simulations are conducted to compare the performance of the conventional SAGE algorithm based on the specular-path model [Fleury et al. 1999] and the MEM based on the proposed three-level EF, which we denote with EF-MEM. The characterization method $\mathcal{A}$ is selected from the specular-path model and the GAM-based dispersive-path model. The settings $a = 2, \sigma_b = 0.01$, and $L_n = 20$ are adopted in the simulations. Both the SAGE algorithm and the EF-MEM algorithm are used to estimate $\boldsymbol{\theta}$ and a. The EF-MEM is also applied for determining the appropriate characterization. Notice that for the SAGE algorithm, the estimate of a is computed as $\hat{a} = L^{-1} \sum_{\ell=1}^{L} - \log \hat{w}_\ell / (\log c + \log \hat{\tau}_\ell)$ with $\widehat{(\cdot)}$ denoting the parameter estimates.

Figure 6.6(a) and (b) illustrate the difference between the true magnitudes and delays of paths and their estimates obtained by using the SAGE algorithm and the EF-MEM in a single snapshot. The signal-to-noise ratio (SNR) is 20 dB. The dashed lines in Figure 6.6(a) and

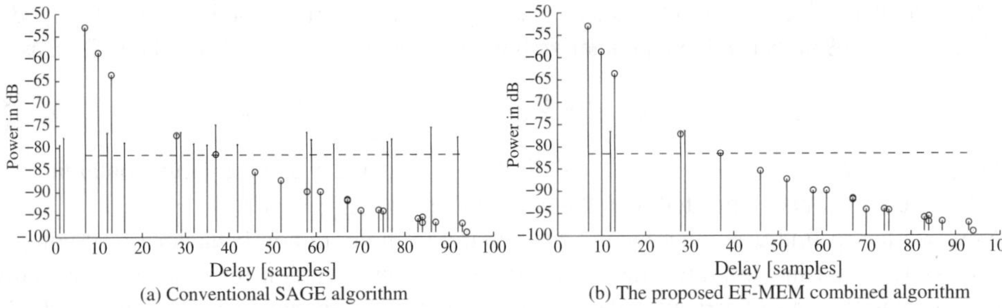

Figure 6.6 Parameter estimates of specular paths obtained using the conventional SAGE algorithm and the proposed EF-MEM. The stems with the markers "∘" and "·" represent the true paths and the estimated paths respectively

(b) indicate the noise spectrum height. It can be observed from Figure 6.6 (a) that the conventional SAGE algorithm returns paths with similar magnitudes for $\tau > 20$ samples. These path estimates are obviously artifacts. The EF-MEM returns fewer paths, all of which exhibit power estimates higher than the noise spectrum height. The improvement obtained by using the EF-MEM is because of the use of the regularizer Eq. (6.79) in the EF-MEM.

Figure 6.7 (a) depicts the root-mean-square estimation errors of a (RMSEE(a)) obtained from 500 Monte-Carlo runs for both the SAGE algorithm and the EF-MEM. It can be seen from Figure 6.7(a) that the EF-MEM outperforms the SAGE algorithm with less RMSEE(a). For the same level of RMSEE(a), more than 10 dB of SNR improvement is achievable when using the EF-MEM. In Figure 6.7(b), the average and the standard deviations of the evidence $P(\boldsymbol{D}|\mathcal{A})$ are depicted for both the specular-path and the dispersive-path models. The vertical bars in Figure 6.7(b) indicate the range $[\bar{P}_{\boldsymbol{D}|\mathcal{A}} + \sigma_1, \bar{P}_{\boldsymbol{D}|\mathcal{A}} - \sigma_2]$ for the SNRs considered, where $\bar{P}_{\boldsymbol{D}|\mathcal{A}}$ denotes the average of $P(\boldsymbol{D}|\mathcal{A})$, and σ_1, σ_2 represent the standard deviations for the evidences larger than and smaller than $\bar{P}_{\boldsymbol{D}|\mathcal{A}}$ respectively. It can be observed from Figure 6.7(b) that $\bar{P}_{\boldsymbol{D}|\mathcal{A}}$ is higher for the specular-path model than for the dispersive-path model. In addition, for the specular-path model, $\bar{P}_{\boldsymbol{D}|\mathcal{A}}$ increases and σ_1, σ_2 decrease when the SNR gets higher. However, for the GAM-based dispersive-path model, when the SNR increases, $\bar{P}_{\mathcal{A}}$ remains constant and the spreads σ_1, σ_2 do not decrease. This result clearly shows that the specular-path model is more suitable for characterization of the channel components than the GAM-based dispersive-path model.

6.7.3 Example II : Delay Spread in a Two-level EF

The composite delay spread is an important large-scale parameter for stochastic channel modeling. For the ith snapshot of propagation channel, the composite delay spread $\sigma_{\tau,i}$ is calculated as the second-central-moment of the power-delay spectral density of the channel [Kyösti et al. 2007]. According to the widely-adopted channel models in the 3GPP standards [3GPP 2007; 2009] and in the WINNER report [Kyösti et al. 2007], delay spread is a random variable following a log-normal distribution $\mathcal{LN}(\bar{\sigma}_\tau, \kappa_{\sigma_\tau})$ with $\bar{\sigma}_\tau$ and κ_{σ_τ} denoting respectively the mean and the standard deviation of $log_{10}(\sigma_{\tau,i})$. Here we are interested at deriving an algorithm to estimate $\tau_{i,\ell}$, $i = 1, \ldots, I$, $\ell = 1, \ldots, L_i$ which yields the delay spreads complying with a log-normal distribution.

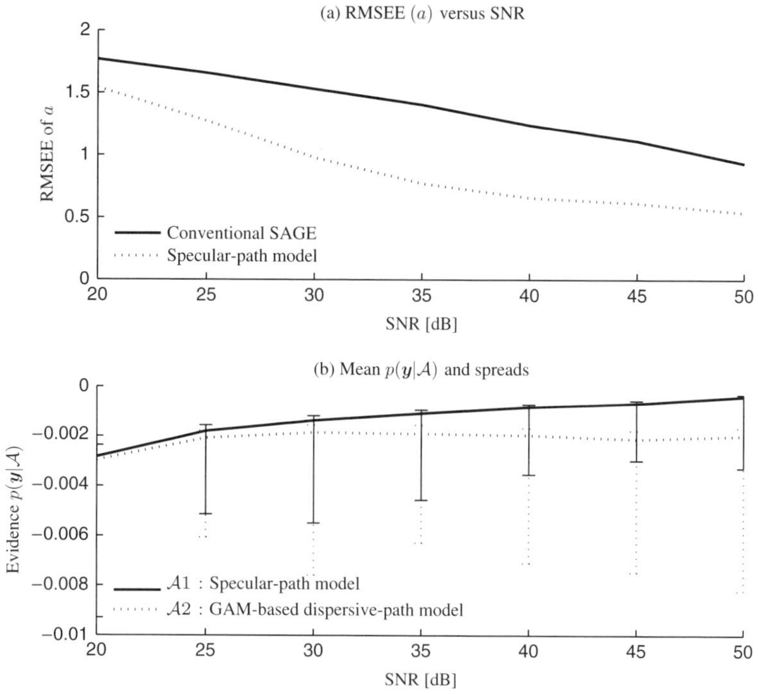

Figure 6.7 (a) RMSEE of a obtained by using the conventional SAGE algorithm and the proposed EF-MEM; (b) mean and spreads of the evidence of the specular-path model and the GAM-based dispersive-path model

The multi-layer EF can be used to collect evidences of events, which in our case are the a-posteriori probabilities of the channel parameters of interest. In this study, we use an EF with two layers, $\mathcal{L}_1$ and $\mathcal{L}_2$, and present iterative max-a-posteriori (MAP) estimators for the channel parameters. For layer $\mathcal{L}_i$, $i = 1, 2$, the notation $\mathcal{H}_i$ is used to represent the predefined generic model, which is considered to be an *a-priori* information, and $\boldsymbol{\theta}_i$ for the parameters in the generic model. The MAP estimators of the model parameters $\boldsymbol{\theta} = [\boldsymbol{\theta}_1, \boldsymbol{\theta}_2]$ can be obtained by maximizing the joint posterior probability $p(\boldsymbol{\theta}|Y, \mathcal{H}_1, \mathcal{H}_2)$:

$$p(\boldsymbol{\theta}|Y, \mathcal{H}_1, \mathcal{H}_2) = p(\boldsymbol{\theta}_1|Y, \mathcal{H}_1)p(\boldsymbol{\theta}_2|\boldsymbol{\theta}_1, \mathcal{H}_2), \tag{6.80}$$

with Y being the realizations of the received signals. By increasing $p(\boldsymbol{\theta}_1|Y, \mathcal{H}_1)$ and $p(\boldsymbol{\theta}_2|\boldsymbol{\theta}_1, \mathcal{H}_2)$ iteratively, the maximum of $p(\boldsymbol{\theta}|Y, \mathcal{H}_1, \mathcal{H}_2)$ can be achieved.

It is worth mentioning that some empirical settings in, for example, the SAGE algorithm, such as the number of paths to estimate and the dynamic range of the path power, can be considered as pre-defined parameters of the generic model and included in the EF. These parameters can be viewed as parts of the specular-path generic model and present as conditions in the layer in which the generic model resides.

As an example where the path number D is considered as a parameter of the generic specular-path model, the first layer $\mathcal{L}_1$ can be defined to be the evidence of the path parameters,

$p(\boldsymbol{\theta}_1|y, D, \mathcal{H}_1)$. Here, $\mathcal{H}_1$ is the specular-path model, $y = [y_1(t), y_2(t), \ldots, y_I(t); t \in [0, T)]$ represents the received signals in a total of I snapshots, each snapshot lasting T seconds, and

$$\boldsymbol{\theta}_1 = [\alpha_{i,\ell}, \tau_{i,\ell}; \ell = 1, \ldots, L_i, i = 1, \ldots, I]$$

contains the delays $\tau_{i,\ell}$ and the complex amplitudes $\alpha_{i,\ell}$ of the ℓth path. Maximizing $p(\boldsymbol{\theta}_1|y, D, \mathcal{H}_1)$ is a problem that can be solved by using the standard maximum likelihood (ML) method or an iterative approximation of the ML method, such as the SAGE algorithm [Fleury et al. 1999]. The evidence in the second layer can be written as

$$p(\bar{\sigma}_\tau, \kappa_{\sigma_\tau}|\sigma_{\tau,i}, i = 1, \ldots, I, \mathcal{H}_2)$$

where $\mathcal{H}_2$ is referred to as the log-normal distribution that the delay spreads usually follow. In our case, the evidence in the layer $\mathcal{L}_2$ is defined to be the inverse of the largest distance between the empirical cumulative probability function (CDF) $F(\sigma_\tau)$ of $\sigma_{\tau,i}, i = 1, \ldots, I$ and the CDF of $\mathcal{LN}(\bar{\sigma}_\tau, \kappa_{\sigma_\tau})$ with $(\bar{\sigma}_\tau, \kappa_{\sigma_\tau})$ determined by fitting $\mathcal{LN}(\bar{\sigma}_\tau, \kappa_{\sigma_\tau})$ to $F(\sigma_\tau)$. When the log-normal distribution has a better fit to the empirical distribution, the evidence of Layer 2 increases. In this example, the value of D may influence the evidences in both layers. Thus, $p(\boldsymbol{\theta}|Y, \mathcal{H}_1, \mathcal{H}_2)$ needs to be maximized with respect to D.

Simulations are conducted to evaluate the performance of the proposed algorithm. The number of paths D is a parameter adjustable when seeking the maximum posterior probability of $\boldsymbol{\theta}$. In the simulation, 100 realizations of impulse response of a wide-sense stationary (WSS) channel are generated. We found that the true delay spreads $\sigma_{\tau,i}, i = 1, \ldots, 100$ do follow the log-normal distribution. In the two-layer EF, we use the log-normal distribution as the a-priori information in $\mathcal{L}_2$. The two-layer EF is implemented for channel parameter estimation using an iterative algorithm with the following steps.

Step 1. Specify initial value for the path number D.

Step 2. Estimate the parameters of individual components in multiple snapshots using the SAGE algorithm.

Step 3. Compute the delay spread and calculate empirical CDF.

Step 4. Conduct hypothesis testing for whether delay spreads follow a log-normal distribution. If false, change the value of D and go back to ***Step 2***; if true, output the final estimation results.

Simulation results are depicted in Figure 6.8, a comparison of the true CDF of the delay spread, the CDF obtained with the proposed algorithm, and the CDFs obtained with $D = 50$ and $D = 100$. It can be seen that the proposed algorithm provides the better estimate of the CDF of delay spread than the conventional methods with predefined D. The appropriate path number identified by using the algorithm equals 295. This result shows that the delay spread statistics can be accurately extracted by using the proposed EF-based parameter estimator.

6.8 Extended Kalman-filter-based Tracking Algorithm

6.8.1 Overview

The temporal behaviors of propagation paths are important for channel characterization in time-variant cases [Czink et al. 2007, Kwakkernaat and Herben 2007]. The time-variant characteristics of path parameters have been considered as an additional degree of freedom for

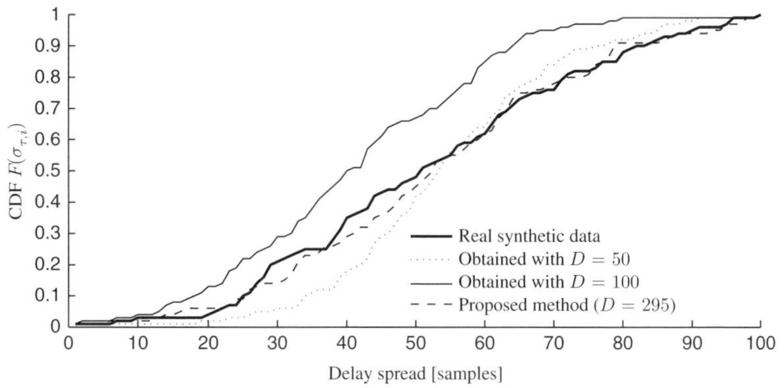

Figure 6.8 The CDFs of the true, estimated delay spreads and the theoretical CDF of a log-normal distribution fitted to the empirical CDF

path clustering [Kwakkernaat and Herben 2007], and the evolution of clusters of paths has been illustrated using measurement data by Czink et al. [2007]. In these studies, path-evolution characteristics are obtained indirectly from the path parameter estimates computed from individual observation snapshots. The estimation methods used – the SAGE algorithm [Czink et al. 2007] and the Unitary ESPRIT [Kwakkernaat and Herben 2007] – are derived under the assumption that the path parameters in different observation snapshots are independent. This (unrealistic) assumption results in a "loss of information" in the estimation of the path evolution in time. Furthermore, due to model-order mismatch and heuristic settings in these algorithms, such as the (usually fixed) dynamic range, a time-variant path may remain undetected in some snapshots. As a result, a time-variant path can be erroneously considered to be several paths. These effects influence the performance of clustering algorithms and the effectiveness of the channel models derived from these results. It is therefore of great importance to use appropriate algorithms to estimate the temporal characteristics of paths directly.

One conventional approach for estimation of the time-evaluation behavior of paths is the extended Kalman filter (EKF) [Salmi et al. 2006]. The EKF was originally introduced as a solution to non-linear estimation and tracking problems. The classical EKF relies on the linearization of non-linear signal models. This linearization is accurate if the changes of the observations are almost linear with respect to time, so the linearization introduces negligible errors. In the papers by Salmi et al. [2006] and Richter et al. [2005], the EKF is used for tracking the delays, DoAs, DoDs, and complex amplitudes of time-variant paths. There are some other methods put forward for tracking of time-variant paths for MIMO channel sounding [Chung and Böhme 2005, Richter et al. 2006, Salmi et al. 2006]. Chung and Böhme [2005] proposed recursive expectation-maximization (EM) and recursive space-alternating generalized EM (SAGE)-inspired algorithms for tracking the azimuths of arrival (AoAs) of paths.

6.8.2 *The Structure of an EKF*

To illustrate the general structure of a tracking algorithm, we outline an EKF based on a state-space model characterizing the time-evolution behavior of the propagation paths and an observation model describing the output signal of the Rx array. Evaluation results are provided for the performance of the EKF under the influences of model mismatch, unknown initial phase, number of observations, and rate of change of path parameters.

6.8.2.1 State (System) Model

For simplicity, we consider the case in which a time-variant propagation path is characterized by its AoA ϕ, elevation of arrival θ, delay τ, the rate of change of these parameters $\Delta(\cdot)$, and the complex attenuation α. The EKF derived can be readily extended to include more parameters. In the case considered here, the state vector $\boldsymbol{\theta}_k$ for the kth observation is defined to be

$$\boldsymbol{\theta}_k = [\phi_k, \Delta\phi_k, \theta_k, \Delta\theta_k, \tau_k, \Delta\tau_k, |\alpha_k|, \angle\alpha_k]^{\mathrm{T}} \in \mathbb{R}^9 \tag{6.81}$$

The transition of the state can be described as

$$\boldsymbol{\theta}_{k+1} = \boldsymbol{F}_k\boldsymbol{\theta}_k + \boldsymbol{n}_k \tag{6.82}$$

where $\boldsymbol{F}_k$ denotes the transition matrix

$$\boldsymbol{F}_k = \begin{bmatrix} 1 & 1 & 0 & 0 & 0 & 0 & 0 & 0 \\ 0 & 1 & 0 & 0 & 0 & 0 & 0 & 0 \\ 0 & 0 & 1 & 1 & 0 & 0 & 0 & 0 \\ 0 & 0 & 0 & 1 & 0 & 0 & 0 & 0 \\ 0 & 0 & 0 & 0 & 1 & 1 & 0 & 0 \\ 0 & 0 & 0 & 0 & 0 & 1 & 0 & 0 \\ 0 & 0 & 0 & 0 & 0 & 0 & 1 & 0 \\ 0 & 0 & 0 & 0 & 0 & 0 & 0 & 1 \end{bmatrix} \tag{6.83}$$

and the noise vector $\boldsymbol{n}_k$ reads

$$\boldsymbol{n}_k = [n_{\phi,k}, n_{\Delta\phi,k}, n_{\theta,k}, n_{\Delta\theta,k}, n_{\tau,k}, n_{\Delta\tau,k}, 0, 0]^{\mathrm{T}}$$

with $n_{x,k}$, $x \in \{\phi, \Delta\phi, \theta, \Delta\theta, \tau, \Delta\tau\}$ representing the noise on the parameter x.

The dynamics of the position parameters is assumed to be caused by the non-zero rate-of-change parameters. The dynamics of the rate-of-change parameters are driven by noise processes due to the movement of scatterers in the environment and the mobility of the transmitter and/or receiver. These noise processes can also be due to diffuse scattering along individual propagation paths. This is reasonable since the radio cross-section of a scatterer can change dramatically in amplitude and phase with respect to incident and departure angles of impinging waves. When these angles are slightly changed from one snapshot to another, a path may become completely non-coherent. When measurement equipment has limited resolution, methods based on the specular-path model will return parameter estimates with heavy-tailed distributions. Based on these considerations, we assume that these noise components are Gaussian distributed and mutually uncorrelated. Furthermore, the covariance matrix $\Sigma_{\boldsymbol{n},k}$ of $\boldsymbol{n}_k$ in Eq. (6.82) is assumed to be known.

6.8.2.2 Observation Model

The signal received at the output of the Rx antenna array can be written:

$$\boldsymbol{y}_{k+1}(t) = \boldsymbol{y}_{\text{tg}}(t; \boldsymbol{\theta}_{k+1}) + \boldsymbol{w}_{k+1}(t) \tag{6.84}$$

where $\boldsymbol{y}_{\text{tg}}(t; \boldsymbol{\theta}_k)$ for a single-path scenario can be written as

$$\boldsymbol{y}_{\text{tg}}(t; \boldsymbol{\theta}_k) = \alpha_k u(t - \tau_k) \exp\{-j2\pi \Delta\tau_k T_s^{-1} f_c t\} \boldsymbol{c}(\boldsymbol{\Omega}_k) \tag{6.85}$$

Here, under the assumption that the array elements have an isotropic radiation pattern, the array response $\boldsymbol{c}(\boldsymbol{\Omega})$ has the form

$$\boldsymbol{c}(\boldsymbol{\Omega}) = \exp\{j2\pi\lambda^{-1}\boldsymbol{R}^{\text{T}}\boldsymbol{\Omega}\} \tag{6.86}$$

where $\boldsymbol{R} = [\boldsymbol{r}_1, \boldsymbol{r}_2, \ldots, \boldsymbol{r}_M]$ with $\boldsymbol{r}_m \in \mathbb{C}^{3\times 1}$ representing the position of the mth antenna in the Cartesian coordinate system. The component $\boldsymbol{w}_{k+1}(t)$ in Eq. (6.84) denotes standard complex white Gaussian noise with the covariance matrix of $\boldsymbol{\Sigma}_{\boldsymbol{w},k+1} = \boldsymbol{I}\sigma_w^2$, where $\boldsymbol{I}$ is a diagonal identity matrix.

6.8.2.3 Extended Kalman Filter

For simplicity, we start with the single-path scenario. Using the state model Eq. (6.82) and the observation model Eq. (6.84), when the $(k+1)$th observation is available, the EKF performs the following operations:

- Prediction of the states:

$$\hat{\boldsymbol{\theta}}_{k+1|k} = \boldsymbol{F}_k \hat{\boldsymbol{\theta}}_k \tag{6.87}$$

  where $(\cdot)_{k+1|k}$ denotes the predicted value of the given argument.
- Prediction of the covariance matrix of states

$$\boldsymbol{P}_{k+1|k} = \boldsymbol{F}_k \boldsymbol{P}_k \boldsymbol{F}_k^{\text{H}} + \boldsymbol{\Sigma}_{\boldsymbol{n},k} \tag{6.88}$$

- Calculation of the Kalman gain matrix

$$\boldsymbol{K}_{k+1} = \boldsymbol{P}_{k+1|k} \boldsymbol{H}(\boldsymbol{\theta}_{k+1|k})^{\text{H}} (\boldsymbol{H}_{k+1} \boldsymbol{P}_{k+1|k} \boldsymbol{H}(\boldsymbol{\theta}_{k+1|k})^{\text{H}} + \Sigma_{\boldsymbol{w},k+1})^{-1} \tag{6.89}$$

  where $\boldsymbol{H}(\boldsymbol{\theta})$ represents the Jacobian matrix of partial derivatives of $\boldsymbol{y}_{\text{tg}}(t; \boldsymbol{\theta})$ with respect to $\boldsymbol{\theta}$. The expressions of the entries of $\boldsymbol{H}(\boldsymbol{\theta})$ are given in the next subsection.
- Update of the state estimation

$$\hat{\boldsymbol{\theta}}_{k+1} = \hat{\boldsymbol{\theta}}_{k+1|k} + \boldsymbol{K}_{k+1}(\boldsymbol{y}_{k+1} - \boldsymbol{y}_{\text{tg}}(\hat{\boldsymbol{\theta}}_{k+1|k})) \tag{6.90}$$

- Update of the covariance matrix of states

$$\boldsymbol{P}_{k+1} = (\boldsymbol{I} - \boldsymbol{K}_{k+1} \boldsymbol{H} \hat{\boldsymbol{\theta}}_{k+1}) \boldsymbol{P}_{k+1|k} \tag{6.91}$$

The received signal $\boldsymbol{y}_{k+1}$ in Eq. (6.90) is written as

$$\boldsymbol{y} = [y_1(t_1), y_2(t_1), \ldots, y_M(t_1), y_1(t_2), y_2(t_2), \ldots, y_M(t_2), y_1(t_3), \ldots, y_M(t_N)]^{\mathrm{T}}$$

The structure for $\boldsymbol{y}_{\mathrm{tg}}(\hat{\boldsymbol{\theta}}_{k+1|k})$ is similar.

An alternative form of the EKF has a different set of operations after Eq. (6.88):

- Calculation of the partial derivative of the log-likelihood function with respect to the parameters $\boldsymbol{\theta}$

$$\boldsymbol{J}(\hat{\boldsymbol{\theta}}_{k+1|k}) = 2\mathcal{R}\{\boldsymbol{H}(\hat{\boldsymbol{\theta}}_{k+1|k})^{\mathrm{H}}\boldsymbol{\Sigma}_{\boldsymbol{w},k+1}^{-1}\boldsymbol{H}(\hat{\boldsymbol{\theta}}_{k+1|k})\} \tag{6.92}$$

- Update the error covariance matrix

$$\boldsymbol{P}_{k+1|k+1} = (\boldsymbol{P}_{k+1|k}^{-1} + \boldsymbol{J}(\hat{\boldsymbol{\theta}}_{k+1|k}))^{-1} \tag{6.93}$$

- Calculate the corrections in the parameter estimates

$$\Delta\hat{\boldsymbol{\theta}} = \boldsymbol{P}_{k+1|k}(\boldsymbol{I} - \boldsymbol{J}(\hat{\boldsymbol{\theta}}_{k+1|k})\boldsymbol{P}_{k+1|k+1})2\mathcal{R}\{\boldsymbol{H}(\hat{\boldsymbol{\theta}}_{k+1|k})^{\mathrm{H}}\boldsymbol{\Sigma}_{\boldsymbol{w},k+1}^{-1}(\boldsymbol{y}_{k+1} - \boldsymbol{y}_{\mathrm{tg}}(\hat{\boldsymbol{\theta}}_{k+1|k}))\} \tag{6.94}$$

- Update the parameter estimates

$$\hat{\boldsymbol{\theta}}_{k+1|k+1} = \hat{\boldsymbol{\theta}}_{k+1|k} + \Delta\hat{\boldsymbol{\theta}} \tag{6.95}$$

The Jacobian matrix $\boldsymbol{H}(\boldsymbol{\theta})$ of the partial derivatives of $\boldsymbol{y}_{\mathrm{tg}}(t;\boldsymbol{\theta})$, $t = t_1, \ldots, t_N$ with respect to $\boldsymbol{\theta}$ is written as

$$\boldsymbol{H}(\boldsymbol{\theta}) = \begin{bmatrix} \frac{\partial \boldsymbol{y}_{\mathrm{tg}}(t_1;\boldsymbol{\theta})}{\partial\phi} & \frac{\partial \boldsymbol{y}_{\mathrm{tg}}(t_1;\boldsymbol{\theta})}{\partial\Delta\phi} & \cdots & \frac{\partial \boldsymbol{y}_{\mathrm{tg}}(t_1;\boldsymbol{\theta})}{\partial\angle\alpha} \\ \frac{\partial \boldsymbol{y}_{\mathrm{tg}}(t_2;\boldsymbol{\theta})}{\partial\phi} & \frac{\partial \boldsymbol{y}_{\mathrm{tg}}(t_2;\boldsymbol{\theta})}{\partial\Delta\phi} & \cdots & \frac{\partial \boldsymbol{y}_{\mathrm{tg}}(t_2;\boldsymbol{\theta})}{\partial\angle\alpha} \\ & & \cdots & \\ \frac{\partial \boldsymbol{y}_{\mathrm{tg}}(t_N;\boldsymbol{\theta})}{\partial\phi} & \frac{\partial \boldsymbol{y}_{\mathrm{tg}}(t_N;\boldsymbol{\theta})}{\partial\Delta\phi} & \cdots & \frac{\partial \boldsymbol{y}_{\mathrm{tg}}(t_N;\boldsymbol{\theta})}{\partial\angle\alpha} \end{bmatrix}$$

Here,

$$\frac{\partial \boldsymbol{y}_{\mathrm{tg}}(t;\boldsymbol{\theta})}{\partial\phi} = \alpha \cdot u(t-\tau)\exp\{-j2\pi\Delta\tau_k T_s^{-1} f_c t\}\frac{\partial \boldsymbol{c}(\boldsymbol{\Omega})}{\partial\phi} \tag{6.96}$$

$$\frac{\partial \boldsymbol{y}_{\mathrm{tg}}(t;\boldsymbol{\theta})}{\partial\theta} = \alpha \cdot u(t-\tau)\exp\{-j2\pi\Delta\tau_k T_s^{-1} f_c t\}\frac{\partial \boldsymbol{c}(\boldsymbol{\Omega})}{\partial\theta} \tag{6.97}$$

$$\frac{\partial \boldsymbol{y}_{\mathrm{tg}}(t;\boldsymbol{\theta})}{\partial\tau} = -j2\pi\alpha \cdot \exp\{-j2\pi\Delta\tau T_s^{-1} f_c t\}\boldsymbol{c}(\boldsymbol{\Omega})\int U(f)\exp\{-j2\pi f\tau\} f \exp\{j2\pi f t\}\mathrm{d}f \tag{6.98}$$

$$\frac{\partial \boldsymbol{y}_{\mathrm{tg}}(t;\boldsymbol{\theta})}{\partial\Delta\tau} = -j2\pi\alpha f_c T_s^{-1} t \cdot u(t-\tau)\exp\{-j2\pi\Delta\tau T_s^{-1} f_c t\}\boldsymbol{c}(\boldsymbol{\Omega}) \tag{6.99}$$

$$\frac{\partial \boldsymbol{y}_{\mathrm{tg}}(t;\boldsymbol{\theta})}{\partial|\alpha|} = \exp\{j\angle\alpha\}\exp\{-j2\pi\Delta\tau_k T_s^{-1} f_c t\}\boldsymbol{c}(\boldsymbol{\Omega}) \cdot u(t-\tau) \tag{6.100}$$

$$\frac{\partial \boldsymbol{y}_{\mathrm{tg}}(t;\boldsymbol{\theta})}{\partial\angle\alpha} = j \cdot \alpha \cdot u(t-\tau)\exp\{-j2\pi\Delta\tau_k T_s^{-1} f_c t\}\boldsymbol{c}(\boldsymbol{\Omega}) \tag{6.101}$$

In Eq. (6.96), if the array elements have an isotropic radiation pattern, the partial derivative $\partial \boldsymbol{c}(\boldsymbol{\Omega})/\partial \phi$ can be calculated to be

$$\frac{\partial \boldsymbol{c}(\boldsymbol{\Omega})}{\partial \phi} = j2\pi\lambda^{-1}\exp\{j2\pi\lambda^{-1}\boldsymbol{R}^{\mathrm{T}}\boldsymbol{\Omega}\}\boldsymbol{R}^{\mathrm{T}}\frac{\partial \boldsymbol{\Omega}}{\partial \phi} \tag{6.102}$$

with $\partial\boldsymbol{\Omega}/\partial\phi = [-\sin(\theta)\sin(\phi), \sin(\theta)\cos(\phi), 0]^{\mathrm{T}}$. Similarly $\partial \boldsymbol{c}(\boldsymbol{\Omega})/\partial \theta$ in Eq. (6.97) is computed to be

$$\frac{\partial \boldsymbol{c}(\boldsymbol{\Omega})}{\partial \theta} = j2\pi\lambda^{-1}\exp\{j2\pi\lambda^{-1}\boldsymbol{R}^{\mathrm{T}}\boldsymbol{\Omega}\}\boldsymbol{R}^{\mathrm{T}}\frac{\partial \boldsymbol{\Omega}}{\partial \theta} \tag{6.103}$$

with $\partial\boldsymbol{\Omega}/\partial\theta = [\cos(\theta)\cos(\phi), \cos(\theta)\sin(\phi), -\sin(\theta)]^{\mathrm{T}}$.

In Eq. (6.98), $U(f)$ denotes the Fourier transform of $u(t)$. The partial derivatives of $\boldsymbol{y}_{\mathrm{tg}}(t;\boldsymbol{\theta})$ with respect to $\Delta\phi$ and $\Delta\theta$ are zero vectors.

6.8.3 *Model Mismatch due to the Linear Approximation*

The model used in the EKF can be different to the (true) effective model. The mismatch can be due to the linear approximation of non-linear models using the Taylor-series expansion operation or the existence of effects in the (true) effective model that are not considered in the model used by the EKF. An example of the latter can be found in the case where the impact of the phase change of the received signal due to the time-variant environment is not considered or is estimated incorrectly in the model used in the EKF.

In the following, we derive the analytical expression of the error between the true signal and the signal calculated with the model used in the EKF. This error is expressed as a function of:

- the deviation of the parameters from the values on which the Taylor-series expansion is performed
- the aperture of the observation, for example, the number of temporal and spatial samples of the received signal
- the estimation error of the initial phase of the signal due to the time-variant property of the environment.

For simplicity, we consider the noise-free case in which a propagation path is characterized by its Doppler frequency only.

The effective received signal is then written as

$$y(t) = \alpha\exp\{j(2\pi\nu t + \phi)\} \tag{6.104}$$

where ν denotes the true Doppler frequency, ϕ represents the initial phase, α denotes the complex amplitude, which is assumed to be constant, and $t \in [0, T]$ is the time instance of observations, with T being the observation interval. For simplicity, we assume that α is known to the receiver. Without loss of generality, $\alpha = 1$ is considered.

The linear model used to approximate $y(t)$ with the first-order Taylor-series expansion of $\exp\{j2\pi\nu t\}$ at $\nu = \nu'$ is written as

$$\tilde{y}(t) = \exp\{j\hat{\phi}\}\cdot[\exp\{j2\pi\nu' t\} + j2\pi(\nu - \nu')t\exp\{j2\pi\nu' t\}] \tag{6.105}$$

where $\hat{\phi}$ denotes an estimate of ϕ. The estimation error of ϕ can be due to a wrong estimate of the Doppler frequency from the previous observations. It can also be because the phase change caused by the non-zero Doppler frequency of the propagation path is not considered in the observation model used in the EKF, as was the case in the papers by Salmi et al. [2006] and Richter et al. [2005].

We compute the normalized square of the difference between $y(t)$ and $\tilde{y}(t)$

$$\Delta \doteq \frac{\int_0^T |y(t) - \tilde{y}(t)|^2 \mathrm{d}t}{\int_0^T |y(t)|^2 \mathrm{d}t} = T^{-1} \int_0^T |y(t) - \tilde{y}(t)|^2 \mathrm{d}t \tag{6.106}$$

It can be shown that

$$\Delta = 2 + \frac{4}{3}\pi^2 T^2 \Delta\nu^2 + 2\cos(\Delta\phi + \Delta\nu 2\pi T) + 4T\pi\Delta\nu \cdot \sin(\Delta\phi)\mathrm{sinc}^2(\Delta\nu T) - 4\cos(\Delta\phi)\mathrm{sinc}(\Delta\nu 2T) \tag{6.107}$$

with $\Delta\nu = \nu - \nu'$ and $\Delta\phi = \phi - \hat{\phi}$. It is evident that Δ is a function of T, $\Delta\nu$, and $\Delta\phi$. In the cases where $\Delta\phi = 0$, Eq. (6.107) becomes

$$\Delta = 2 + \frac{4}{3}\pi^2 T^2 \Delta\nu^2 + 2\cos(\Delta\nu 2\pi T) - 4\mathrm{sinc}(\Delta\nu 2T) \tag{6.108}$$

and for $\Delta\nu = 0$, Eq. (6.107) reduces to

$$\Delta = 2 - 2\cos(\Delta\phi) \tag{6.109}$$

Figure 6.9 illustrates the graphs of Δ with respect to $\Delta\nu$, $\Delta\phi$ and T respectively, with two of these three parameters kept fixed. For Figure 6.9(c), $\Delta\nu = 1\,\mathrm{Hz}$, $\nu' = 10\,\mathrm{Hz}$, $\Delta\phi = 5^\circ$ and the sample rate is 300 Hz. It can be observed from these graphs that the squared error is significant for $\Delta\nu/\nu' > 0.1$ or $\Delta\phi > 12^\circ$. Furthermore, in the case where both $\Delta\nu/\nu'$ and $\Delta\phi$ are small, the squared error increases exponentially with respect to the number of observation samples.

6.8.4 *Tracking Performance and the Initial Phase*

In the following, we use simulations to demonstrate what happens to EKF performance when the initial phase $\Delta\phi \neq 0$. We first conduct the simulation study using the state-space model, first with parameters of Doppler frequency and then again with parameters of AoA. These two simulation studies are used to investigate the effect of failing to consider phase change in the observation model used in the EKF on the estimation and tracking of the Doppler frequency and the AoA.

6.8.4.1 Study Case I: Tracking Doppler Frequency of a Single Path

The space state model is written as

$$\nu_{k+1} = \nu_k + w_k \tag{6.110}$$

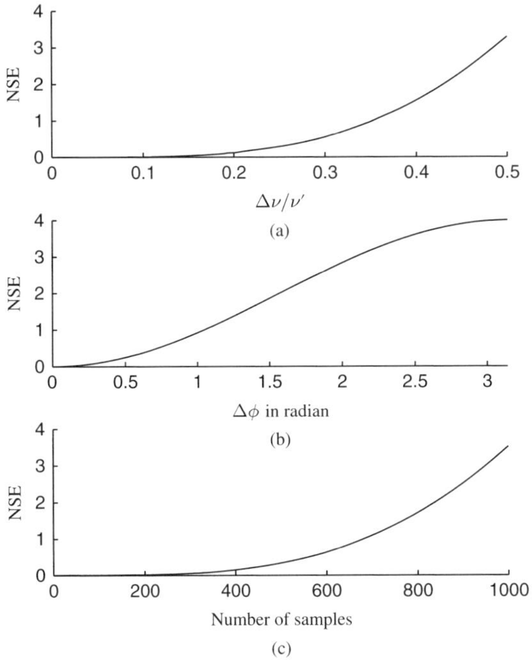

Figure 6.9 Normalized square error (NSE) of the approximation to the effective signal model using the first-order Taylor-series expansion versus (a) the Doppler frequency difference, (b) the phase difference with $T = 1$ s, and (c) the number of observation samples in T

where w_k denotes a real Gaussian noise component with variance σ_w^2. The observation model, which is a narrowband signal in this case, reads

$$y_{k+1}(t) = \alpha \exp\{j2\pi\nu_{k+1}t + \psi_{k+1}\} + n_{k+1}(t) \tag{6.111}$$

where α is the complex amplitude, which is assumed for simplicity to be equal to 1, and ψ_{k+1} denotes the phase change caused by the Doppler frequency. Here

$$\psi_{k+1} = \psi_k + 2\pi\nu_{k+1}T$$

with T denoting the interval between two consecutive snapshots.

The EKF has the following equations:

- Prediction of the state:

$$\hat{\nu}_{k+1|k} = \hat{\nu}_k \tag{6.112}$$

- Prediction of the variance of the state:

$$P_{k+1|k} = P_k + \sigma_\nu^2 \tag{6.113}$$

- Calculation of the derivative of the loglikelihood function with respect to ν

$$J(\hat{\nu}_{k+1|k}) = 2\mathcal{R}\{\boldsymbol{h}(\hat{\nu}_{k+1|k})^{\mathrm{H}}\sigma_w^{-2}\boldsymbol{h}(\hat{\nu}_{k+1|k})\} \tag{6.114}$$

- Update of the error variance

$$P_{k+1|k+1} = [P_{k+1|k}^{-1} + J(\hat{\nu}_{k+1|k})]^{-1} \tag{6.115}$$

- Calculation of the innovation

$$\Delta\hat{\nu} = P_{k+1|k}(1 - J(\hat{\nu}_{k+1|k})P_{k+1|k+1}) \\ 2\mathcal{R}\{\boldsymbol{h}(\hat{\nu}_{k+1|k})^{\mathrm{H}}\sigma_w^{-2}(\boldsymbol{y}_{k+1} - \hat{\boldsymbol{y}}(\hat{\nu}_{k+1|k}))\} \tag{6.116}$$

- Update of the parameter estimates

$$\hat{\nu}_{k+1|k+1} = \hat{\nu}_{k+1|k} + \Delta\hat{\nu} \tag{6.117}$$

Figure 6.10 demonstrates a comparison of the performance of two EKFs in tracking the Doppler frequency of a propagation path. The trajectory of the changes of the Doppler frequency is calculated as

$$\nu_n = \nu_1 + 0.01 \cdot n^2 + v_n \tag{6.118}$$

where ν_n is the Doppler frequency in Hertz in the nth observation snapshot and v_n denotes a Gaussian random process with zero mean and variance equal to 1×10^{-3}. One of the EKFs considers the phase change due to the Doppler frequency and the other does not take into account this phase change. In each observation snapshot, 20 samples are collected. The interval between the starts of two consecutive snapshots is 0.05 s, corresponding to 25 samples. Figure 6.10 shows the original trajectory of the Doppler frequency and the estimates obtained using these two EKFs. The EKF considering the phase change performs much better than the EKF that does not. Figure 6.10(b) shows the absolute estimation errors of the two EKFs.

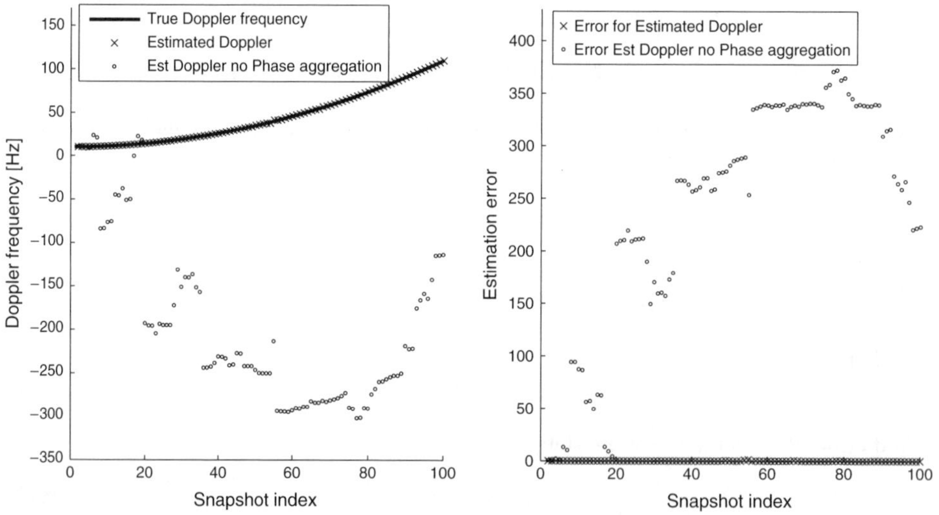

Figure 6.10 A comparison of the performance of the EKFs for (a) tracking the Doppler frequency trajectory in each observation snapshot and (b) the estimation errors obtained by using respectively EKF without and with considering the phase aggregation due to the non-zero Doppler frequency

6.8.4.2 Study Case II: Tracking Azimuth of Arrival of a Single Path

The state-space model in Eq. (6.110) is modified as

$$\phi_{k+1} = \phi_k + w_k \tag{6.119}$$

We consider a uniform linear array (ULA) with M elements. The inter-element spacing is half of the wavelength. The observation model reads

$$\boldsymbol{y}_{k+1} = \alpha \boldsymbol{c}(\phi_{k+1}) \exp\{j\psi_{k+1}\} + \boldsymbol{n}_{k+1} \tag{6.120}$$

where $\psi_{k+1} = \psi_k + 2\pi\nu_{k+1}T$ denotes the phase change caused by the Doppler frequency at the $(k+1)$th observation snapshot, and $\boldsymbol{c}(\phi) = [c_1(\phi), \ldots, c_m(\phi), \ldots, c_M(\phi)]^{\mathrm{T}}$ represents the array response at ϕ. The entry $c_m(\phi)$ of $\boldsymbol{c}(\phi)$ reads $c_m(\phi) = \exp\{j\pi(m-1)\cos(\phi)\}$.

The EKF has the following equations:

- Prediction of the state:

$$\hat{\phi}_{k+1|k} = \hat{\phi}_k \tag{6.121}$$

- Prediction of the variance of the state:

$$P_{k+1|k} = P_k + \sigma_\nu^2 \tag{6.122}$$

- Calculation of the derivative of the log-likelihood function with respect to ϕ

$$J(\hat{\phi}_{k+1|k}) = 2\mathcal{R}\{\boldsymbol{h}(\hat{\phi}_{k+1|k})^{\mathrm{H}}\sigma_w^{-2}\boldsymbol{h}(\hat{\phi}_{k+1|k})\} \tag{6.123}$$

- Update of the error variance

$$P_{k+1|k+1} = [P_{k+1|k}^{-1} + J(\hat{\phi}_{k+1|k})]^{-1} \tag{6.124}$$

- Calculation of the innovation

$$\begin{aligned}\Delta\hat{\phi} =& P_{k+1|k}(1 - J(\hat{\phi}_{k+1|k})P_{k+1|k+1}) \\ & 2\mathcal{R}\{\boldsymbol{h}(\hat{\phi}_{k+1|k})^{\mathrm{H}}\sigma_w^{-2}(\boldsymbol{y}_{k+1} - \hat{\boldsymbol{y}}(\hat{\phi}_{k+1|k}))\}\end{aligned} \tag{6.125}$$

- Update of the parameter estimates

$$\hat{\phi}_{k+1|k+1} = \hat{\phi}_{k+1|k} + \Delta\hat{\phi} \tag{6.126}$$

The simulation study into the impact of not including the initial phase in the EKF is as follows. The receiver antenna array is an 8-element ULA with interelement spacing equal to half the wavelength; the input SNR is 30 dB. The phase changes due to the movement of the target are confined to between 0° and 6°. The trajectory of the AoA and of the Doppler frequency is calculated to be

$$\phi_n = \phi_1 + 0.01 \cdot n^2 + w_n \tag{6.127}$$

where ϕ_n is the AoA in degrees in the nth observation snapshot and w_n denotes a Gaussian random process with zero mean and variance equal to 1×10^{-3}. The Doppler frequency trajectory is calculated using Eq. (6.118). Figure 6.11 shows the simulation results. It can be observed that the EKF when the initial phase is corrected exhibits much less estimation error than the EKF without correction. The latter almost fails to track the variation of the AoA.

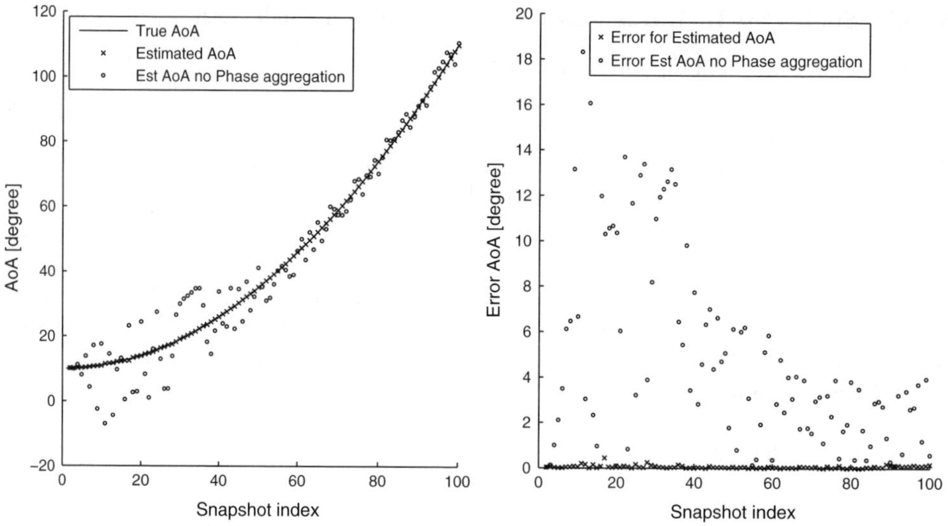

Figure 6.11 A comparison of the performance of the EKFs for (a) tracking the azimuth of arrival trajectory in each observation snapshot and (b) the estimation errors obtained by using respectively EKF without and with considering the phase aggregation due to the non-zero Doppler frequency

6.9 Particle-filter-based Tracking Algorithm

It has been explained in the previous section that when the path parameters fluctuate dramatically, linear approximations based on Taylor-series expansion are significantly inaccurate and that "loss of track" errors may occur. Thus EKF-based tracking algorithms may not work properly. Furthermore, the parameter updating steps in EKF-based algorithms require calculation of the second-order derivatives of the received signal with respect to the path parameters. In channel sounding, these derivatives are computed numerically using the system response gathered from calibration measurements. In the presence of calibration errors, these derivatives may be erroneous and cause significant performance degradation.

In the following we describe a low-complexity particle filtering (PF) algorithm originally proposed by Yin et al. [2008] for tracking the parameters of time-variant propagation paths in MIMO radio channels. In contrast to the EKF and the recursive EM and SAGE-inspired algorithms, the PF can be applied when the observation model is non-linear. Furthermore, it does not require the numerical computation of derivatives. In this section, a state-space model is used to describe the path evolution in delay, AoA, AoD, Doppler frequency, and complex amplitude dimensions. The proposed PF has an additional resampling step, specifically designed for wideband MIMO channel sounding, in which the posterior PDFs of the path states are usually highly concentrated in the multi-dimensional state space. Preliminary investigations using measurement data show that the proposed PF can track paths stably with a small number of particles, say five per path, even when the paths are undetected by the conventional SAGE algorithm.

We start by introducing:

- the state-space model describing the time-evolving path parameters
- the observation model for the received signal in the Rx of the sounding equipment.

For simplicity of presentation, these models are discussed in the context of a single-path scenario. Extension of these models to multiple-path scenarios is straightforward.

6.9.1 State-space Model

We consider a scenario in which the environment consists of time-variant specular paths. The parameters of a path are the delay τ, the AoD ϕ_1, the AoA ϕ_2, the Doppler frequency ν, the rates of change of these parameters, denoted by $\Delta\tau$, $\Delta\phi_1$, $\Delta\phi_2$, and $\Delta\nu$ respectively, as well as the complex amplitude α. The kth observation of the state vector of a path is defined as

$$\boldsymbol{\Omega}_k = [\boldsymbol{P}_k^{\mathrm{T}}, \alpha_k^{\mathrm{T}}, \boldsymbol{\Delta}_k^{\mathrm{T}}]^{\mathrm{T}} \tag{6.128}$$

where $[\cdot]^{\mathrm{T}}$ denotes the transpose operation, $\boldsymbol{P}_k \doteq [\tau_k, \phi_{1,k}, \phi_{2,k}, \nu_k]^{\mathrm{T}}$ represents the "position" parameter vector, $\boldsymbol{\Delta}_k \doteq [\Delta\tau_k, \Delta\phi_{1,k}, \Delta\phi_{2,k}, \Delta\nu_k]^{\mathrm{T}}$ denotes the "rate-of-change" parameter vector, and $\boldsymbol{\alpha}_k \doteq [|\alpha_k|, \arg(\alpha_k)]^{\mathrm{T}}$ is the amplitude vector with $|\alpha_k|$ and $\arg(\alpha_k)$ representing the magnitude and the argument of α_k respectively. The state vector $\boldsymbol{\Omega}_k$ is modelled as a Markov process:

$$p(\boldsymbol{\Omega}_k|\boldsymbol{\Omega}_{1:k-1}) = p(\boldsymbol{\Omega}_k|\boldsymbol{\Omega}_{k-1}), k \in [1, \ldots, K] \tag{6.129}$$

where $\boldsymbol{\Omega}_{1:k-1} \doteq \{\boldsymbol{\Omega}_1, \ldots, \boldsymbol{\Omega}_{k-1}\}$ is a sequence of state values from the 1st to the $(k-1)$th observation, and K denotes the total number of observations. The transition of $\boldsymbol{\Omega}_k$ with respect to k is modelled as

$$\underbrace{\begin{bmatrix} \boldsymbol{P}_k \\ \boldsymbol{\alpha}_k \\ \boldsymbol{\Delta}_k \end{bmatrix}}_{\boldsymbol{\Omega}_k} = \underbrace{\begin{bmatrix} \boldsymbol{I}_4 & \boldsymbol{0}_{4\times 2} & T_k\boldsymbol{I}_4 \\ \boldsymbol{J}_k & \boldsymbol{I}_2 & \boldsymbol{0}_{2\times 4} \\ \boldsymbol{0}_{4\times 4} & \boldsymbol{0}_{4\times 2} & \boldsymbol{I}_4 \end{bmatrix}}_{\boldsymbol{F}_k \doteq} \underbrace{\begin{bmatrix} \boldsymbol{P}_{k-1} \\ \boldsymbol{\alpha}_{k-1} \\ \boldsymbol{\Delta}_{k-1} \end{bmatrix}}_{\boldsymbol{\Omega}_{k-1}} + \underbrace{\begin{bmatrix} \boldsymbol{0}_{4\times 1} \\ \boldsymbol{v}_{\alpha,k} \\ \boldsymbol{v}_{\Delta,k} \end{bmatrix}}_{\boldsymbol{v}_k \doteq} \tag{6.130}$$

where $\boldsymbol{I}_n$ represents the $n \times n$ identity matrix, $\boldsymbol{0}_{b\times c}$ is the all-zero matrix of dimension $b \times c$,

$$\boldsymbol{J}_k = \begin{bmatrix} 0 & 0 & 0 & 0 \\ 0 & 0 & 0 & 2\pi T_k \end{bmatrix}$$

and T_k denotes the interval between the starts of the $(k-1)$th and the kth observation periods. The vector $\boldsymbol{v}_k$ in Eq. (6.130) contains the driving process in the amplitude vector

$$\boldsymbol{v}_{\alpha,k} \doteq [v_{|\alpha|,k}, v_{\arg(\alpha),k}]^{\mathrm{T}} \tag{6.131}$$

and in the rate-of-change parameter vector

$$\boldsymbol{v}_{\Delta,k} \doteq [v_{\Delta\tau,k}, v_{\Delta\phi_1,k}, v_{\Delta\phi_2,k}, v_{\Delta\nu,k}]^{\mathrm{T}} \tag{6.132}$$

The entries $v_{(\cdot),k}$ in Eqs (6.131) and (6.132) are independent Gaussian random variables $v_{(\cdot),k} \sim \mathcal{N}(0, \sigma_{(\cdot)}^2)$.

In this section, we consider the case with $T_k = T, k \in [1, \ldots, K]$. For notational brevity, in what follows we drop the subscript k in $\boldsymbol{F}_k$.

6.9.2 Observation Model

In the kth observation period, the discrete-time signals at the output of the m_2th Rx antenna when the m_1th Tx antenna transmits can be written as

$$\begin{aligned} y_{k,m_1,m_2}(t) &= x_{k,m_1,m_2}(t; \boldsymbol{\Omega}_k) + n_{k,m_1,m_2}(t), \\ & t \in [t_{k,m_1,m_2}, t_{k,m_1,m_2} + T), \\ & m_1 = 1, \ldots, M_1, m_2 = 1, \ldots, M_2 \end{aligned} \tag{6.133}$$

where t_{k,m_1,m_2} denotes the time instant when the m_2th Rx antenna starts to receive signals while the m_1th Tx antenna transmits, T is the sensing duration of each Rx antenna, and M_1 and M_2 represent the total number of Tx antennas and Rx antennas respectively. The signal contribution $x_{k,m_1,m_2}(t; \boldsymbol{\Omega}_k)$ reads

$$\begin{aligned} x_{k,m_1,m_2}(t; \boldsymbol{\Omega}_k) = \alpha_k \exp(j 2\pi \nu_k t) c_{1,m_1}(\phi_{k,1}) c_{2,m_2}(\phi_{k,2}) \\ \cdot u(t - \tau_k) \end{aligned} \tag{6.134}$$

Here, $c_{1,m_1}(\phi)$ and $c_{2,m_2}(\phi)$ represent respectively the response in azimuth of the m_1th Tx antenna, and the response in azimuth of the m_2th Rx antenna, while $u(t - \tau_k)$ denotes the transmitted signal delayed by τ_k. The noise $n_{k,m_1,m_2}(t)$ in Eq. (6.133) is a zero-mean Gaussian process with spectrum height σ_n^2. For notational convenience, we use the vector $\boldsymbol{y}_k$ to represent all the samples received in the kth observation period and $\boldsymbol{y}_{1:k} \doteq \{\boldsymbol{y}_1, \boldsymbol{y}_2, \ldots, \boldsymbol{y}_k\}$ to denote a sequence of observations.

6.9.3 The Proposed Low-complexity Particle Filter

From Eq. (6.129) and Eq. (6.133) we see that the received signal $\boldsymbol{y}_k$ depends only on the current state $\boldsymbol{\Omega}_k$ and is conditionally independent of the other states given $\boldsymbol{\Omega}_k$. Utilizing this property, a PF approach can be used to estimate the posterior PDF $p(\boldsymbol{\Omega}_{1:k} | \boldsymbol{y}_{1:k})$ sequentially [Herman 2002]. The fact that the parameter space is multi-dimensional[2] and the temporal and spatial observation apertures are large in order to achieve high resolution poses a noticeable challenge when using the PF in wideband MIMO sounding. As a result, the posterior PDF $p(\boldsymbol{\Omega}_{1:k} | \boldsymbol{y}_{1:k})$ is highly concentrated in the parameter space. It is then difficult to "steer" the particle sets to the regions where the probability mass is localized. The proposed PF is specifically designed to solve this problem. In this section we first present the algorithm while considering a single-path scenario, and then discuss the extension of the algorithm for tracking multiple paths.

[2] The parameter space has dimension up to 14 in the specular-path scenario [Fleury et al. 2003] and up to 28 in the dispersive-path scenario [Yin et al. 2006b].

6.9.3.1 Initialization of Particle States

We initialize the particle states using the parameter estimates obtained with the conventional SAGE algorithm [Fleury et al. 2003]. This algorithm is based on the assumption that the path parameters at different observation snapshots are independent. Here we use the PF to track $\boldsymbol{\Omega}_k$ from the third observation period. The vector $\boldsymbol{\Omega}_k^i$ of the ith particle has the initial state $\boldsymbol{\Omega}_2^i$. The position parameter vector $\boldsymbol{P}_2^i$ is set to be identical with the parameter estimates obtained with the SAGE algorithm in the second observations. The rate-of-change parameters are calculated by taking the difference between the SAGE estimates obtained at the first and the second observations.

6.9.3.2 Framework of the PF

When a new observation, say $\boldsymbol{y}_k$, is available, the PF performs the following steps.

Step 1: Predict the states of particles and calculate importance weights
The output from the previous observations are the set $\{\boldsymbol{\Omega}_{k-1}^i, w_{k-1}^i\}$, where w_{k-1}^i denotes the importance weight of the ith particle. We first predict the states of all particles for the kth observation period. The rate-of-change parameter vector $\boldsymbol{\Delta}_k^i$ is updated as

$$\boldsymbol{\Delta}_k^i = \boldsymbol{\Delta}_{k-1}^i + \boldsymbol{\Delta w}_k^i, i = 1, \ldots, I \tag{6.135}$$

where I denotes the total number of particles, and the vector $\boldsymbol{\Delta w}_k^i$ is drawn from a $\mathcal{N}(\boldsymbol{0}, \boldsymbol{\Sigma}_{\boldsymbol{w}})$ distribution. The diagonal covariance matrix $\boldsymbol{\Sigma}_{\boldsymbol{w}}$ reads

$$\boldsymbol{\Sigma}_{\boldsymbol{w}} = \mathrm{diag}(\sigma_{\Delta\tau}^2, \sigma_{\Delta\phi_1}^2, \sigma_{\Delta\phi_2}^2, \sigma_{\Delta\nu}^2). \tag{6.136}$$

The values of the diagonal elements $\sigma_{\Delta(\cdot)}^2$ with $(\cdot)$ replaced by τ, ϕ_1, ϕ_2, or ν, are predetermined. The position vector $\boldsymbol{P}_k^i$ is calculated as

$$\boldsymbol{P}_k^i = \boldsymbol{P}_{k-1}^i + \boldsymbol{\Delta}_k^i \tag{6.137}$$

The complex amplitude α_k^i is computed analytically as

$$\alpha_k^i = \frac{(\boldsymbol{s}_k^i)^{\mathrm{H}} \boldsymbol{y}_k}{\| \boldsymbol{s}_k^i \|^2} \tag{6.138}$$

where $(\cdot)^{\mathrm{H}}$ represents the Hermitian transpose, $\| \cdot \|$ denotes the Euclidian norm of the given argument, and the vector $\boldsymbol{s}_k^i$ contains the elements

$$s_{k,m_1,m_2}^i(t; \boldsymbol{P}_k^i) = \exp(j2\pi\nu_k^i t) c_{1,m_1}(\phi_{1,k}^i) c_{2,m_2}(\phi_{2,k}^i) u(t - \tau_k^i), t \in [t_{k,m_1,m_2}, t_{k,m_1,m_2} + T)$$

The importance weights of the particles are calculated as

$$w_k^i = \frac{w_{k-1}^i p(\boldsymbol{y}_k | \boldsymbol{\Omega}_k^i)}{\sum_{i=1}^{I} w_{k-1}^i p(\boldsymbol{y}_k | \boldsymbol{\Omega}_k^i)}, i = 1, \ldots, I \tag{6.139}$$

with

$$p(\boldsymbol{y}_k|\boldsymbol{\Omega}_k^i) \propto \exp\left(-\frac{1}{2\sigma_n^2} \parallel \boldsymbol{y}_k - \alpha_k^i \boldsymbol{s}_k^i \|^2\right) \tag{6.140}$$

Step 2: Additional resampling

In wideband MIMO channel sounding, the number of temporal-spatial samples in one observation period is usually large. As a consequence, significant portions of the posterior PDF $p(\boldsymbol{\Omega}_{1:k}|\boldsymbol{y}_{1:k})$ are concentrated around the modes of the PDF. As the path parameters evolve over time, the particles with predicted states can be too diffuse to "catch" the probability mass. One brute-force solution is to employ a large number of particles. However, the resulting complexity prohibits any practical implementation. This problem can be overcome with low complexity using the methods proposed for vision–based robot localization and tracking, for example assuming a higher noise variance than the true value, distributing particles either uniformly within a subset of the parameter space [Fox et al. 1999] or based on multi-hypothesis [Lenser and Veloso 2000]. However, these methods have the drawback that the weighted particles do not approximate the true posterior density $p(\boldsymbol{\Omega}_{1:k}|\boldsymbol{y}_{1:k})$, and consequently the estimation results can be artifacts.

Here we introduce an additional resampling step in which two techniques are used for allocation of particles without misinterpreting the posterior density. This step is activated when the importance weights of the particles obtained from Eq. (6.139) are all negligible. The first technique uses a part of the observation samples, denoted by $\tilde{\boldsymbol{y}}_k$, to calculate the importance weights. As the number of observation samples in $\tilde{\boldsymbol{y}}_k$ is less than that in $\boldsymbol{y}_k$, the posterior PDF $p(\boldsymbol{\Omega}_k|\tilde{\boldsymbol{y}}_k, \boldsymbol{y}_{1:k-1})$ is less concentrated than the original $p(\boldsymbol{\Omega}_k|\boldsymbol{y}_{1:k})$. Thus the particles can have a higher probability of getting significant importance weights.

The second method computes the importance weights as

$$\tilde{w}_k^i \propto \log p(\boldsymbol{y}_k|\boldsymbol{\Omega}_k^i) + \log w_{k-1}^i \tag{6.141}$$

The obtained set $\{\boldsymbol{\Omega}_k^i, \tilde{w}_k^i\}$ is an estimate of the function $\log p(\boldsymbol{\Omega}_{1:k}|\boldsymbol{y}_{1:k})$. This function exhibits the same modes as $p(\boldsymbol{\Omega}_{1:k}|\boldsymbol{y}_{1:k})$,but also a wider curvature in the vicinities of the modes. So the probability of getting non-negligible importance weights is enhanced.

Based on these two methods, we propose an additional resampling step, which can be implemented according to the following pseudo code.

for $n = 1$ to N **do**

Step 2.1 Select $\tilde{\boldsymbol{y}}_k^n \in \boldsymbol{y}_k$.

Step 2.2 Calculate the importance weights $\tilde{w}_k^i$, $i = 1, \ldots, I$.

if $\{\tilde{w}_k^i\}$ contains non-significant values, e.g. less than $\max\{\tilde{w}_k^i\} - 3$, **then**

Step 2.3 Find the indices $\boldsymbol{A} = \{i^{\mathrm{s}}\}$ of the particles with significant importance weights. Let D denote the number of particles with non-significant weights.

Step 2.4 Generate D new particles with states drawn from $p(\boldsymbol{\Omega}_k|\boldsymbol{\Omega}_{k-1}^{j(\boldsymbol{A}_d)})$, $d = 1, \ldots, D$. Here, $\boldsymbol{A}_d$ denotes the dth element of $\boldsymbol{A}$, and $j(\boldsymbol{A}_d)$ is the index of particle in the $(k-1)$th observation, from which the $\boldsymbol{A}_d$th particle in the kth observation is generated. Replace the particles that have non-significant weights by the new particles.

Step 2.5 Update the importance weights w_{k-1}^i as

$$w_{k-1}^i = J(i)^{-1} w_{k-1}^{j(i)}, \quad i = 1, \ldots, I, \tag{6.142}$$

where $J(i)$ represents the total number of the new particles that are generated using the $j(i)$th particle in the $(k-1)$th observation. Go to Step 2.2.

end if

end for

Step 3: Normal resampling

The operations performed in this step are similar to those shown in the loop in ***Step 2*** except that the importance weights $\tilde{w}_k^i$ are replaced by w_k^i and the observation $\tilde{\boldsymbol{y}}_k^n$ is substituted with $\boldsymbol{y}_k$.

Step 4: Estimate the posterior PDF

The estimate of the posterior pdf can be approximated with the particle states and importance weights

$$\hat{p}(\boldsymbol{\Omega}_k|\boldsymbol{y}_{1:k}) = \sum_{i=1}^{I} w_k^i \delta(\boldsymbol{\Omega}_k - \boldsymbol{\Omega}_k^i). \tag{6.143}$$

This pdf estimates can be used to estimate the expectation of a function of $\boldsymbol{\Omega}_k$. For example, in the single-path scenario, the state vector $\boldsymbol{\Omega}_k$ for the path can be estimated as

$$\hat{\boldsymbol{\Omega}}_k = \sum_{i=1}^{I} \boldsymbol{\Omega}_k^i w_k^i. \tag{6.144}$$

6.9.3.3 Extension to Multi-path Scenarios

We considered the case in which the paths are dispersive in multiple dimensions. In such a case, the states of different paths do not coincide with high probability. The paths can then be tracked individually by using separate PFs. The states of the particles in each PF are initialized with the parameter estimates of a specific path. This method is used to track multiple paths in the experimental investigations introduced in Section 6.9.4.

6.9.4 Experimental Investigation

The measurement data was collected using the Propsound channel sounder. Measurement data was acquired in a burst mode. Each burst has a duration of 16 cycles. Only the data received in the first four cycles in each burst are stored. The sounder setting was described by Yin et al. [2006b, Table 1]. The Tx and the Rx were both equipped with two identical 9-element circular arrays. A diagram of the arrays is shown in the same paper [Yin et al. 2006b, Fig. 2].

We use measurement data acquired in a long corridor. Figure 6.12 depicts the environment and shows the Tx and Rx surroundings. The Rx was fixed at the location marked with a circle

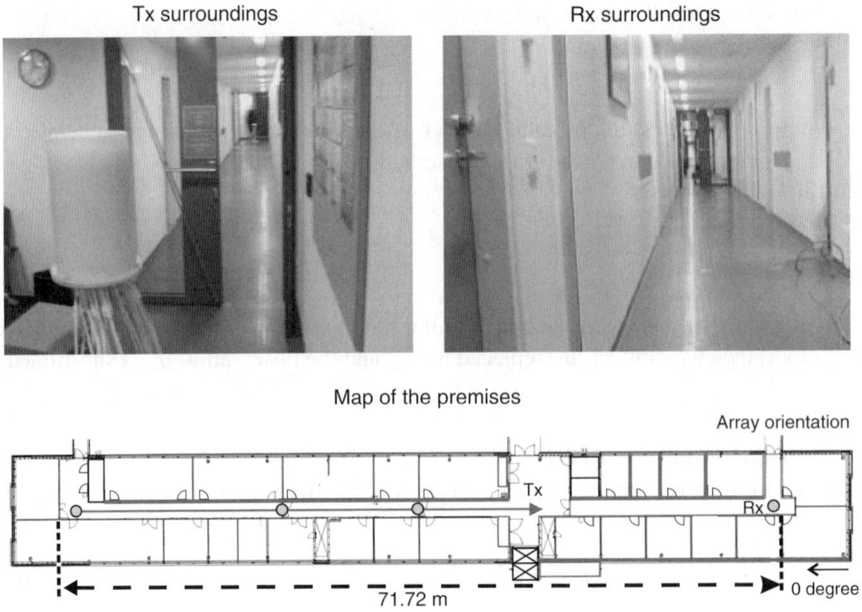

Figure 6.12 Photographs and plan of the environment

on the map. The Tx was moving towards the Rx with a constant speed along the route marked on the map. The Rx was positioned behind a metal door with reinforced glass. There was no line-of-sight path for this scenario. During the measurement, no people were walking. The measurement data was collected in 100 consecutive bursts, in a total time span of 26.93 s. During this period, the Tx moved 7.5 m with a speed of approximately 0.5 m/s. The time-evolution behavior of the propagation paths can be observed from the variation of the power delay profiles (PDPs) of the received signal at different bursts. Figure 6.13 shows the average PDPs calculated from the signals received in the burst, for a total of 50 consecutive bursts. It can be seen that some peaks of the PDPs move with increasing delay, while others exhibit decreasing delay.

We use the proposed PF to track the parameters of three paths. The parameter estimates of three paths obtained with the SAGE algorithm are applied to initialize the states of the particles. The PF uses five particles per path with $\sigma_{\Delta\tau} = 1.5$ ns, $\sigma_{\Delta\phi_1} = \sigma_{\Delta\phi_2} = 4^\circ$ and $\sigma_{\Delta\nu} = 5\,\text{Hz}$. Notice that in experimental scenarios where $\sigma_{\Delta\tau}$, $\sigma_{\Delta\phi_1}$, $\sigma_{\Delta\phi_2}$, and $\sigma_{\Delta\nu}$ are unknown, these parameters can be estimated using ray-tracing techniques based on some assumptions about the motion of the Tx and the Rx. In this preliminary investigation, for simplicity we select these parameters in such a way that they are very likely larger than the true parameters.

Figure 6.14 depicts the trajectories of the parameters of three paths estimated using the PF. In Figure 6.14(a), the trajectories of the path delays are overlapped with the PDPs calculated for the 100 bursts. The estimated delay trajectories are consistent with the time-variations of

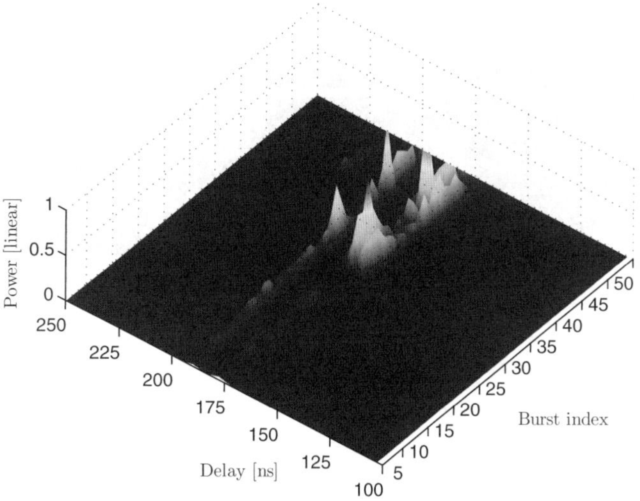

Figure 6.13 Average power-delay profiles of the received signals from 50 bursts

the peaks in the PDPs. In Figure 6.14(b)–(f), the parameter estimates of three paths obtained with the conventional SAGE algorithm are also depicted. The trajectories estimated with the PF match the SAGE estimates for most of the bursts. Furthermore, the filter is able to track a path even in the intverval where the SAGE algorithm failed to detect it (Path 3 is undetected by the SAGE algorithm in the burst interval 82–100).

From Figure 6.14(b) we observe that the delay trajectory of Path 3 fluctuates significantly in the burst interval 80–90. In addition, the Doppler frequency trajectories also exhibit large fluctuations compared to the SAGE estimates. These effects may have the following causes. First, in this preliminary study the PF is used to track paths individually. In the case where a path is close to other paths in the parameter space, the particles may be steered to a wrong position due to interference. The observation of significant fluctuations in the trajectories can also be due to the inappropriate settings of the variance parameters $\sigma^2_{\Delta(\cdot)}$. For instance, the standard deviation $\sigma_{\Delta\nu}$ of Doppler frequency is specified to be 5 Hz. However, this figure is actually much larger than the true value for all tracked paths. Furthermore, the fluctuations of the trajectories may be caused by the fact that the PF uses only five particles for tracking of each path. This is not enough for accurate estimation of the posterior PDF in the dimensions where the intrinsic resolutions of the sounder are low.

To check whether the path evolution behavior estimated by the PF is sensible, in Figure 6.15 we roughly sketch the geometry of the environment and reconstruct three possible propagation paths approximately. Table 6.1 lists some characteristics of these paths drawn from the geometry. We observe that the paths tracked by the PF from the measurement exhibit time-evolution characteristics similar to those of the paths in Figure 6.15. This demonstrates that the proposed PF is applicable for the estimation of the time-variant characteristics of propagation paths.

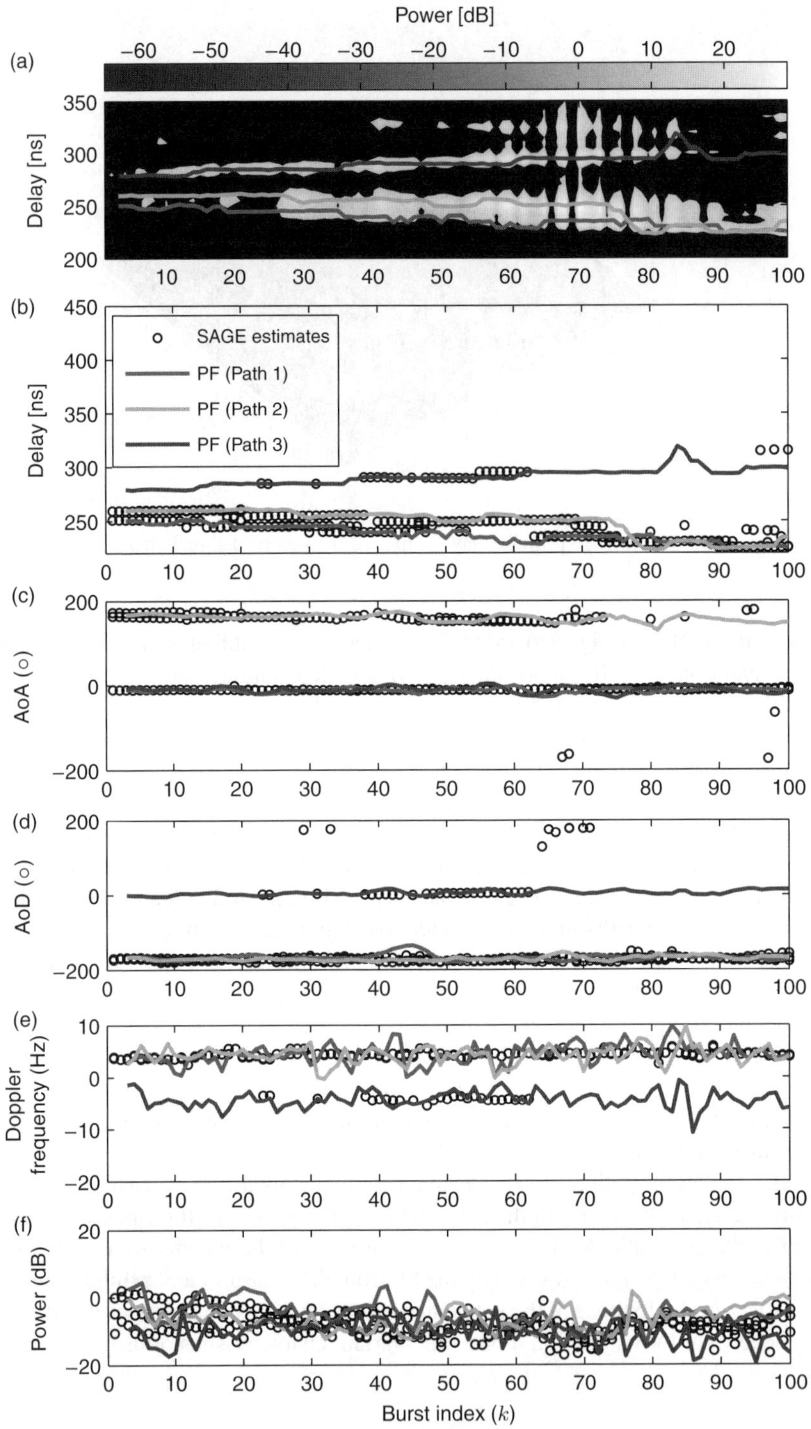

Figure 6.14 Performance of the PF in tracking three time-variant paths. The legend given in (b) applies to (a)–(f). In (a), the PDPs of the signals in 100 bursts are shown in the background

Table 6.1 Parameter values computed from the geometrical figure

Path	Delay rate of change	AoA	AoD	Doppler frequency
No. 1	−0.5 ns/burst	0°	−180°	8 Hz
No. 2	−0.5 ns/burst	170°	−180°	8 Hz
No. 3	+0.5 ns/burst	0°	0°	−8 Hz

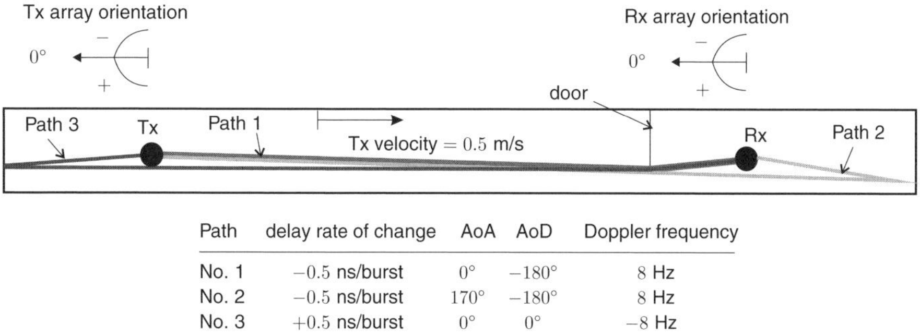

Figure 6.15 Geometries and characteristics of reconstructed propagation paths in the investigated environment

Bibliography

3GPP 2007 TR25.996 v7.0.0: Spatial channel model for multiple input multiple output (MIMO) simulations (release 7). Technical report, 3GPP.

3GPP 2009 TR36.814 v9.0.0: Technical specification group radio access network; further advancements for E-UTRA physical layer aspects (release 9). Technical report, 3GPP.

Asztely D, Ottersten B and Swindlehurst A 1998 Generalised array manifold model for wireless communication channels with local scattering. *IEE Proceedings Radar, Sonar and Navigation* **145**(1), 51–57.

Bartlett M 1948 Smoothing periodograms from time series with continuous spectra. *Nature* **161**, 686–687.

Bienvenu G and Kopp L 1983 Optimality of high resolution array processing using the eigensystem approach. *IEEE Transactions on Acoustics, Speech and Signal Processing* **31**, 1235–48.

Capon J 1969 High-resolution frequency-wavenumber spectrum analysis. *Proceedings of the IEEE* **57**(8), 1408–1418.

Chung P and Böhme JF 2005 Recursive EM and SAGE-inspired algorithms with application to DOA estimation. *IEEE Transactions on Signal Processing* **53**(8), 2664–2677.

Czink N 2007 *The random-cluster model – a stochastic MIMO channel model for broadband wireless communication systems of the 3rd Generation and beyond* PhD thesis Technology University of Vienna, Department of Electronics and Information Technologies.

Czink N, Tian R, Wyne S, Eriksson G, Tufvesson F, Zemen T, Nuutinen J, Ylitalo J, Bonek E and Molisch A 2007 Tracking time-variant cluster parameters in MIMO channel measurements: Algorithm and results *Proceedings of International Conference on Communications and Networking in China (ChinaCOM)*, pp. 1147–1151, Shanghai, China.

Fessler JA and Hero AO 1994 Space-alternating generalized expectation-maximization algorithm. *IEEE Transactions on Signal Processing* **42**(10), 2664–2677.

Fleury BH, Jourdan P and Stucki A 2002a High-resolution channel parameter estimation for MIMO applications using the SAGE algorithm *Proceedings of International Zurich Seminar on Broadband Communications*, vol. 30, pp. 1–9.

Fleury BH, Tschudin M, Heddergott R, Dahlhaus D and Pedersen KL 1999 Channel parameter estimation in mobile radio environments using the SAGE algorithm. *IEEE Journal on Selected Areas in Communications* **17**(3), 434–450.

Fleury BH, Yin X, Jourdan P and Stucki A 2003 High-resolution channel parameter estimation for communication systems equipped with antenna arrays *Proceedings of the 13th IFAC Symposium on System Identification (SYSID)*, vol. ISC-379, Rotterdam, The Netherlands.

Fleury BH, Yin X, Rohbrandt KG, Jourdan P and Stucki A 2002b Performance of a high-resolution scheme for joint estimation of delay and bidirection dispersion in the radio channel *Proceedings of the IEEE Vehicular Technology Conference (VTC-Spring)*, vol. 1, pp. 522–526.

Fox D, Burgard W, Dellart F and Thrun S 1999 Monte Carlo localization: Efficient position estimation for mobile robots *Proceedings of the 16th National Conference on Artificial Intelligence*, Orlando, FL, USA.

Fuhl J, Rossi JP and Bonek E 1997 High-resolution 3-D direction-of-arrival determination for urban mobile radio. *IEEE Transactions on Antennas and Propagation* **45**(4), 672–682.

Haardt, M.; Nossek J 1995 Unitary esprit: how to obtain increased estimation accuracy with a reduced computational burden. *IEEE Transactions on Signal Processing* **43**(5), 1232–1242.

Haardt M, Zoltowski M, Mathews C and Nossek J 1995 2D Unitary ESPRIT for efficient 2D parameter estimation *Proceedings of International Conference on Acoustics, Speech, and Signal Processing (ICASSP)*, vol. 3, pp. 2096–2099.

Herman S 2002 *A particle filtering approach to joint passive radar tracking and target classification* PhD thesis University of Illinois.

Hu B, Land I, Rasmussen L, Piton R and Fleury B 2008 A divergence minimization approach to joint multiuser decoding for coded CDMA. *IEEE Journal on Selected Areas in Communications* **26**(3), 432–445.

Kocian A, B. Hu PS, Rom C, Fleury B and Poulsen E 2003 Iterative joint data detection and channel estimation of DS/CDMA signals in multipath fading using the SAGE algorithm *Proceedings of the 37th Asilomar Conference on Signals, Systems and Computers*, vol. 1, pp. 443–447.

Krim H and Proakis J 1994 Smoothed eigenspace-based parameter estimation. *Automatica* **30**(1), 27–38.

Krim H and Viberg M 1996 Two decades of array signal processing research: the parametric approach. *IEEE Transactions on Signal Processing* **13**, 67–94.

Kwakkernaat M and Herben M 2007 Analysis of clustered multipath estimates in physically nonstationary radio channels *Proceedings of the 17th IEEE International Symposium on Personal, Indoor and Mobile Radio Communications (PIMRC)*, pp. 1–5, Athens, Greece.

Kyösti P, Meinilä J, Hentilä L, Zhao X, Jämsä T, Schneider C, Narandzić M, Milojević M, Hong A, Ylitalo J, Holappa VM, Alatossava M, Bultitude R, de Jong Y and Rautiainen T 2007 WINNER II Channel Models D1.1.2 V1.1.

Lenser S and Veloso M 2000 Sensor resetting localization for poorly modelled mobile robots *Proceedings of IEEE International Conference on Robotics and Automation (ICRA'00)*, vol. 2, pp. 1225–1232.

Marcos S, Marsal A and Benidir M 1994 Performances analysis of the propagator method for source bearing estimation *Proceedings of IEEE International Conference on Acoustics, Speech and Signal Processing (ICASSP)*, vol. IV, pp. 19–22.

Marcos S, Marsal A and Benidir M 1995 The propagator method for source bearing estimation. *Signal Processing* **42**, 121–138.

Marquardt DW 1963 An algorithm for least-squares estimation of nonlinear parameters. *Journal of the Society for Industrial and Applied Mathematics* **11**(2), 431–441.

Parsons JD 2000 *The Mobile Radio Propagation Channel* 2nd edn. John Wiley and Sons.

Richter A 2004 RIMAX – a flexible algorithm for channel parameter estimation from channel sounding measurements. Technical Report TD-04-045, COST 273, Athens, Greece.

Richter A and Thomä RS 2003 Parametric modelling and estimation of distributed diffuse scattering components of radio channels. Technical Report TD-03-198, COST273.

Richter A and Thomä RS 2005 Joint maximum likelihood estimation of specular paths and distributed diffuse scattering *Proceedings of the IEEE 61st Vehicular Technology Conference (VTC-Spring)*, vol. 1, pp. 11–15, Stockholm.

Richter A, Enescu M and Koivunen V 2005 State-space approach to propagation path parameter estimation and tracking *Signal Processing Advances in Wireless Communications, 2005 IEEE 6th Workshop on*, pp. 510–514.

Richter A, Hampicke D, Sommerkorn G and Thoma R 2000 Joint estimation of DoD, time-delay, and DoA for high-resolution channel sounding *Proceedings of IEEE 51st Vehicular Technology Conference (VTC-Spring)*, vol. 2, pp. 1045–1049.

Richter A, Landmann M and Thomä RS 2003 Maximum likelihood channel parameter estimation from multidimensional channel sounding measurements *Proceedings of the 57th IEEE Semiannual Vehicular Technology Conference (VTC)*, vol. 2, pp. 1056–1060.

Richter A, Salmi J and Koivunen V 2006 An algorithm for estimation and tracking of distributed diffuse scattering in mobile radio channels *Proceedings of the 7th IEEE International Workshop on Signal Processing Advances for Wireless Communications (SPAWC'06)*, pp. 1–5.

Roy R and Kailath T 1989 ESPRIT – estimation of signal parameters via rotational invariance techniques. *IEEE Transactions on Acoustics, Speech, and Signal Processing* **37**(7), 984–995.

Salmi J, Richter A and Koivunen V 2006 Enhanced tracking of radio propagation path parameters using state-space modeling *Proceedings of the 2006 European Signal Processing Conference (EUSIPCO)*, pp. 1–5, Florence, Italy.

Schmidt RO 1981 *A signal subspace approach to multiple emitter location and spectral estimation* PhD thesis Stanford University Stanford, CA.

Schmidt RO 1986 Multiple emitter location and signal parameter estimation. *IEEE Transactions on Antennas and Propagation* **AP-34**(3), 276–280.

Shutin D, Kubin G and Fleury BH 2007 Application of the evidence procedure to the estimation of wireless channels. *EURASIP Journal on Advances in Signal Processing*.

Stoica P and Moses RL 1997 *Introduction to Spectral Analysis*. Prentice Hall.

Vanderveen M, Papadias C and Paulraj A 1997 Joint angle and delay estimation (JADE) for multipath signals arriving at an antenna array. *IEEE Communications Letters* **1**, 12–14.

Yin X, Fleury. B, Jourdan P and Stucki. A 2003 Polarization estimation of individual propagation paths using the SAGE algorithm. *Personal, Indoor and Mobile Radio Communications, 2003. PIMRC 2003. 14th IEEE Proceedings on* **2**, 1795–1799.

Yin X, Hu Y and Zhong Z 2012a Dynamic range selection for antenna-array gains in high-resolution channel parameter estimation *Wireless Communications and Signal Processing (WCSP), 2012 International Conference on*, pp. 1–5.

Yin X, Hu Y, Zeng Z, Zhou J, Tong M, Zhong Z and Lu SX 2012b A parameter estimation algorithm for propagation channels based on two-layer evidence framework *2012 IEEE International Symposium on Antennas and Propagation and USNC/URSI National Radio Science Meeting*, Chicago, IL, USA.

Yin X, Pedersen T, Czink N and Fleury B 2006a Parametric characterization and estimation of bi-azimuth dispersion path components *Proceedings of IEEE 7th Workshop on Signal Processing Advances in Wireless Communications. SPAWC'06.*, pp. 1–6, Rome, Italy.

Yin X, Pedersen T, Czink N and Fleury B 2006b Parametric characterization and estimation of bi-azimuth and delay dispersion of individual path components *Antennas and Propagation, 2006. EuCAP 2006. First European Conference on*, pp. 1–8.

Yin X, Steinböck G, Kirkelund GE, Pedersen T, Blattnig P, Jaquier A and Fleury B 2008 Tracking of time-variant radio propagation paths using particle filtering *Proceedings of the International Communications Conference (ICC)*, pp. 920–924, Beijing, China.

7

Statistical Channel-parameter Estimation

In Chapter 6, different estimators of the deterministic parameters of propagation paths were introduced. These methods are widely used to process channel-sounding measurement data for the characterization and modeling of channels in many scenarios. Channel models can be divided into two categories: deterministic and stochastic. The former, for example the tap-delay-line models in the 3GPP channel standards [3GPP 2007], can be used as calibration models for conformity testing of communication technologies and systems. Stochastic channel models are more suitable for generating random channel realizations for performance testing at either the link level or the system level.

Stochastic channel models are usually obtained by first estimating the channel's parameters, such as the composite spreading parameters – the delay spread, Doppler frequency spread, and the parameters of individual propagation paths from individual channel snapshots – and then secondly, extracting the statistics of the channel parameters for modeling. In this chapter, we describe some recently developed parameter estimation algorithms, which can be used to estimate directly the statistical channel parameters from the measurement data.[1]

7.1 A Brief Review of Dispersive Parameter Estimators

In a radio propagation environment, scatterers frequently have a geometrical extent which is small in the view of the receiver (Rx), or local scattering exists around a transmitter (Tx) located far away from the Rx. In both cases, the received signal contributed by each of the scatterers or clusters of local scatterers can be conceived as the sum of contributions originating from multiple subpaths with slightly different directions of arrival. Using the terminology introduced by others [Asztély and Ottersten 1998, Bengtsson and Ottersten 2000, Fleury 2000], we refer to such scatterers or clusters of local scatterers as "slightly distributed scatterers" (SDSs). A quantitative definition of an SDS matching the theoretical framework and purpose

[1] Some of results presented in this chapter have been previously published [Yin et al. 2007a,b, 2008, 2006a, 2011].

Propagation Channel Characterization, Parameter Estimation and Modelling for Wireless Communications, First Edition.
Xuefeng Yin and Xiang Cheng.

of this book is provided in Section 7.2.1. Notice that this definition differs from the one given in the references.

The direction dispersion that an SDS induces in a channel can be characterized as the nominal direction and the spread of the direction power spectrum of the signal contributed by this SDS [Fleury 2000]. For simplicity, we consider horizontal-only propagation. In this case, dispersion is characterized by a nominal azimuth (NA) and an azimuth spread (AS) [Bengtsson and Ottersten 2000, Shahbazpanahi et al. 2001, Trump and Ottersten 1996]. By abuse of terminology, we refer to the azimuth power spectrum, NA and AS as the azimuth power spectrum, the NA and the AS of the SDS.

Conventional high-resolution channel parameter estimation algorithms [Bresler 1986, Fleury et al. 1999, Roy and Kailath 1989, Schmidt 1986, Viberg and Ottersten 1991] are derived from the specular-scatterer (SS) model, which assumes that the propagation environment consists of point scatterers located in the far field of the Tx and Rx arrays [Lee and Xu 1997]. As a result, the contribution of each specular scatterer to the received signal is modeled as a (specular) plane wave. These algorithms are not suitable for the estimation of the NAs and the ASs of SDSs, for two reasons. First, they do not provide with an estimate of the AS. Secondly, the probability of occurrence of large estimation errors is significant [Bengtsson and Völcker 2001].

In recent years, various model-based methods have been proposed for the estimation of the NA and the AS of SDSs. These estimators can be categorized into two groups: estimators based on a distribution model and estimators based on a linear approximation model.

The estimators based on a distribution model rely on a probability distribution characterizing the shape of the azimuth power spectrum of individual SDSs [Besson and Stoica 1999, Han et al. 2006, Meng et al. 1996, Qiang and Zhishun 2007, Ribeiro et al. 2004, Trump and Ottersten 1996, Valaee et al. 1995, Wang and Zoubir 2007, Zoubir et al. 2007]. The uniform distribution [Besson and Stoica 1999, Meng et al. 1996, Sieskul 2006, Valaee et al. 1995, Zoubir et al. 2007], the (truncated) Gaussian distribution [Besson and Stoica 1999, Meng et al. 1996, Qiang and Zhishun 2007, Sieskul 2006, Trump and Ottersten 1996, Wang and Zoubir 2007], the Laplacian distribution [Sieskul 2006], and the von-Mises distribution [Fleury 2000, Ribeiro et al. 2004] have been proposed for horizontal-only propagation. These estimators are not appropriate for applications in channel sounding for two reasons. First, no prior information about the shape of the azimuth power spectrum of the SDS is available. Second, these algorithms exhibit high computational complexity. Their computation demands solution of at least a one-dimensional optimization problem for a certain objective function: the objective function is usually expressed in the form of an integral that needs to be calculated numerically when the antenna array response is non-isotropic.

Some recently proposed algorithms in this category achieve low computational complexity by exploiting the specific structure of the components involved in an SDS or applying a series expansion of the sample covariance matrix of the received signals contributed by an SDS [Han et al. 2006, Qiang and Zhishun 2007, Wang and Zoubir 2007, Zoubir et al. 2007]. However, these properties of the sample covariance matrix do not hold when the array is non-linear and the array elements are non-isotropic. Thus, these estimators are not suitable for the scenarios of interest.

The estimators based on a linear approximation model include those derived using the two-specular-scatterer model [Bengtsson and Ottersten 2000, Souden et al. 2008], the two-SDS model [Shahbazpanahi et al. 2001, Section VI], and the generalized array manifold (GAM) model [Asztély et al. 1997, Jeong et al. 2001, Shahbazpanahi et al. 2001, Shahbazpanahi 2004, Tan et al. 2003]. The two-specular-scatterer model is motivated by the observation that in the incoherently distributed (ID) case, the covariance matrix of the signal contributed by a single SDS is well approximated by the covariance matrix of the signals resulting from two incoherent specular scatterers. The Spread-F algorithm, with F denoting a standard azimuth estimator, such as Root-MUSIC, ESPRIT, MODE, and so on, is derived from this model [Bengtsson and Ottersten 2000]. The algorithm calculates the NA of an individual SDS to be the mean of two azimuth estimates, while the AS estimate is defined to be half the distance between these two azimuth estimates. Pairing of the two azimuth estimates corresponding to the same SDS in a multi-SDS scenario is a major implementation issue, especially in the case where the SDSs are closely spaced in azimuth. Furthermore, a look-up-table solution has to be used to reduce the bias of the AS estimator caused by the discrepancy between the two-specular-scatterer model and the effective model [Bengtsson and Ottersten 2000]. In the case where channel sounding is performed with non-isotropic array responses, this look-up table needs to be generated for multiple entries. For instance, in the case of horizontal-only propagation, the NA, AS, and signal-to-noise ratio (SNR) are entries of the look-up table. The high complexity involved prohibits the implementation of this approach in practice.

The GAM model has been used to derive NA and AS estimators [Asztély et al. 1997, Shahbazpanahi et al. 2001, Shahbazpanahi 2004]. It uses the first-order Taylor series expansion of the array responses to approximate the effective impact of each individual SDS on the received signal. For brevity, we refer to a GAM model using the nth-order Taylor series expansion as an nth-order GAM model. The estimators proposed by Asztély et al. [1997] rely on the MUSIC and the noise subspace fitting algorithms. They cannot be employed in the ID case. The estimators proposed by Shahbazpanahi et al. [2001] make use of the ESPRIT algorithm. They are only applicable when the antenna arrays have two identical antennas in each matched pair [Roy and Kailath 1989], and additionally the distance between the two antennas in each pair is much less than the wavelength [Shahbazpanahi et al. 2001]. This condition is usually not fulfilled in the scenario of interest. The NA and AS estimators derived by Shahbazpanahi [2004] apply the covariance fitting principle. A major implementation issue of these estimators is the knowledge of the second-order derivative of the array response required for the computation of the AS estimate. Calibration errors in the measurement of the array response generate significant discrepancies in the higher-order derivatives of the response. As a consequence, the performance of the estimator is degraded dramatically.

7.2 Dispersive Component Estimation Algorithms

In wireless environments, the individual contributions of distributed scatterers to the received signal can be spread in delay, direction of departure, direction of arrival, Doppler frequency and polarization. In this section, the first-order GAM model proposed by Asztély et al. [1997] is used to approximate the individual contributions of SDSs to the signal received at a multiple-element antenna. A new definition of SDS is proposed based on the two largest eigenvalues of these signal components. With this definition, a distributed scatterer is an SDS when its signal component is closely approximated by the first-order GAM model. A measure is described that quantitatively assesses the degree of the approximation.

As mentioned in Section 7.1, azimuth dispersion caused by an SDS is characterized by means of the NA and the AS. Based on the first-order GAM model, the NA and AS estimators are derived using standard deterministic and stochastic maximum-likelihood (ML) methods, as well as a new extension of the conventional MUSIC algorithm. Due to the discrepancy between the first-order GAM model and the effective signal model, the AS estimators are biased. Instead of reducing the bias by using a multi-dimensional look-up table [Bengtsson and Ottersten 2000], an approach that is usually cumbersome to generate when the antennas used are non-isotropic, an empirical technique is proposed that adaptively selects the size of the array aperture in such a way as to guarantee good agreement between the two models. Simulation results demonstrate the performance improvement achieved with these methods compared to previously published approaches.

In this section, an approximate probability distribution is derived for the error of the ML azimuth estimator used in a scenario with a single SDS. The approach is based on the conventional specular-scatterer model. The result is an extension to the incoherent-SDS case of earlier findings by Asztély and Ottersten [1998] for coherent-SDS. It is shown that in the incoherent-SDS case the probability of occurrence of large estimation errors is significantly reduced by considering more observation samples.

7.2.1 Effective Signal Model

Without loss of generality we restrict our attention to azimuth dispersion generated by SDSs at the Rx site, when the Rx is equipped with an antenna array. We introduce a model describing the effective contribution of SDSs to the signal received at the Rx array. This model is referred to as the "effective signal model" throughput the section. Note that the model and the estimation methods based on it apply equally to the assessment of the direction dispersion by SDSs at the Tx site, provided the Tx is equipped with an antenna array.

In a propagation scenario with a single SDS, the signal at the output of an M-element Rx array is viewed as being composed of the contributions of signals originating from waves propagating along multiple subpaths distributed with respect to the direction of arrival. For simplicity, we assume horizontal-only propagation and neglect all dispersion effects other than dispersion in the azimuth of arrival. We consider a system where the Rx is equipped with M correlators. In the narrowband scenario considered here, the vector $\boldsymbol{y}(t) \in \mathbb{C}^M$ containing the outputs of the correlators at time t is expressed as

$$\boldsymbol{y}(t) = \sum_{\ell=1}^{L} a_\ell(t)\boldsymbol{c}(\bar{\phi} + \tilde{\phi}_\ell) + \boldsymbol{w}(t) \tag{7.1}$$

In Eq. (7.1), the total number L of subpaths originating from the SDS is assumed to be large, $a_\ell(t)$ denotes the complex weight of the ℓth subpath, and $\boldsymbol{c}(\phi) \in \mathbb{C}^M$ represents the array response in azimuth ϕ. The azimuth of the ℓth subpath is decomposed as the sum of a NA $\bar{\phi}$ and a deviation $\tilde{\phi}_\ell$ from $\bar{\phi}$. Throughout the paper, angles are expressed in radians with range $[-\pi, \pi)$ in the theoretical investigations, while they are given in degrees with range $[-180°, +180°)$ in the simulation studies. Addition of angles is defined in such a way that the resultant angle lies in these ranges. The noise vector $\boldsymbol{w}(t)$ in Eq. (7.1) is a circularly symmetric, spatially and temporally white M-dimensional Gaussian process with component spectral height σ_w^2. We assume that a total of N observation samples are considered. These samples are collected at the time instances $t_1, \ldots, t_N$, i.e. $t \in \{t_1, \ldots, t_N\} \subset \mathbb{R}$.

The following assumptions are made:

1. The azimuth deviations $\tilde{\phi}_\ell$, $\ell = 1, \ldots, L$, are independent and identically distributed with zero-mean. The azimuths of the subpaths are concentrated with high probability around the NA $\bar{\phi}$; that is, the azimuth deviations $\tilde{\phi}_\ell$, $\ell = 1, \ldots, L$, are small with high probability.
2. The subpath weight processes $a_1(t), \ldots, a_L(t)$ are uncorrelated zero-mean complex circularly symmetric wide-sense-stationary processes with autocorrelation function

$$R_{a_\ell}(\tau) \doteq \mathrm{E}[a_\ell(t)a_\ell^*(t+\tau)]$$

   Here, $(\cdot)^*$ denotes complex conjugation. Moreover, the subpath weights have equal variance; that is, $R_{a_1}(0) = \cdots = R_{a_L}(0)$.
3. Any two random elements in the set consisting of the azimuth deviations and the subpath weights are uncorrelated.
4. Temporal samples of the subpath weights are uncorrelated: $R_{a_\ell}(t_{n'} - t_n) = 0$, $n \neq n'$.

In a scenario with D SDSs, Eq. (7.1) becomes

$$\boldsymbol{y}(t) = \sum_{d=1}^{D}\sum_{\ell=1}^{L_d} a_{d,\ell}(t)\boldsymbol{c}(\bar{\phi}_d + \tilde{\phi}_{d,\ell}) + \boldsymbol{w}(t) \tag{7.2}$$

where L_d denotes the number of subpaths originating from the dth SDS. We make the additional assumption:

5. The random elements characterizing the signal contributed by each of the D SDSs satisfy Assumptions 1)–4). The probability distributions of the azimuth deviations of distinct SDSs may be different. In addition, any two distinct SDSs are uncorrelated; more specifically, any two random elements related each to distinct SDSs are uncorrelated.

Assumptions 1)–5) correspond to the case with ID sources described by Shahbazpanahi et al. [2001].

7.2.2 Specular–Scatterer Model Estimation

The specular-scatterer (SS) model is widely used in conventional high-resolution parameter estimation algorithms [Bresler 1986, Fleury et al. 1999, Roy and Kailath 1989, Schmidt 1986]. In this section, we briefly introduce the SS model and the ML azimuth estimator derived from it. Then we derive an approximation of the error distribution of this estimator when it is applied in a scenario with one SDS. We consider the ID case with N ($N \geq 1$) independent observation samples. The case with $N = 1$, corresponding to the coherently distributed (CD) case, was discussed previously by Asztély and Ottersten [1998].

7.2.2.1 The SS Model and the ML Azimuth Estimator

The SS model assumes that point scatterers re-radiate impinging waves and that the Rx is in the far field of these scatterers. Thus, specular plane waves are incident at the Rx site. Assuming a scenario with D point scatterers, the output signal vector Eq. (7.1) is approximated by

$$\begin{aligned} \boldsymbol{y}(t) \approx \boldsymbol{y}_{\mathrm{SS}}(t) &\doteq \sum_{d=1}^{D} \alpha_d(t)\boldsymbol{c}(\bar{\phi}_d) + \boldsymbol{w}(t) \\ &= \boldsymbol{C}(\bar{\phi})\alpha(t) + \boldsymbol{w}(t) \end{aligned} \tag{7.3}$$

where $\bar{\phi} \doteq [\bar{\phi}_1, \ldots, \bar{\phi}_D]$, $\boldsymbol{C}(\bar{\phi}) \doteq [\boldsymbol{c}(\bar{\phi}_1), \ldots, \boldsymbol{c}(\bar{\phi}_D)]$, and $\alpha(t) \doteq [\alpha_1(t), \ldots, \alpha_D(t)]^{\mathrm{T}}$. The subscript SS emphasizes that the SS assumption is used to approximate the signals contributed by scatterers. The deterministic ML estimator of $\bar{\phi}$ derived from the SS model (SS-ML) reads [Böhme 1984, Krim and Viberg 1996]:

$$\hat{\bar{\phi}}_{\mathrm{SS-ML}} = \arg \max_{\bar{\phi}} \{\mathrm{tr}[\Pi_{\boldsymbol{C}(\bar{\phi})} \hat{\boldsymbol{\Sigma}}_{\boldsymbol{y}}]\} \tag{7.4}$$

where $\mathrm{tr}[\cdot]$ denotes the trace of the matrix given as an argument, $\Pi_{\boldsymbol{C}(\bar{\phi})} \doteq \boldsymbol{C}(\bar{\phi})\boldsymbol{C}(\bar{\phi})^{\dagger}$ is the projection operator onto the column space of $\boldsymbol{C}(\bar{\phi})$, $\boldsymbol{C}(\bar{\phi})^{\dagger} \doteq [\boldsymbol{C}(\bar{\phi})^{\mathrm{H}}\boldsymbol{C}(\bar{\phi})]^{-1}\boldsymbol{C}(\bar{\phi})^{\mathrm{H}}$ is the pseudo-inverse of $\boldsymbol{C}(\bar{\phi})$, and $\hat{\Sigma}_{\boldsymbol{y}} = \frac{1}{N}\sum_{t=t_1}^{t_N} \boldsymbol{y}(t)\boldsymbol{y}(t)^{\mathrm{H}}$ is the sample covariance matrix. Here, $[\cdot]^{\mathrm{H}}$ denotes Hermitian transposition.

7.2.2.2 Distribution of the Estimation Error

The PDF of the azimuth estimate $\hat{\bar{\phi}}_{\mathrm{SS-ML}}$ is affected by the discrepancy between the effective signal model Eq. (7.2) and the SS model Eq. (7.3). Appendix A.3 shows that in a noiseless single-SDS scenario, the NA estimation error $\breve{\phi} \doteq \hat{\bar{\phi}}_{\mathrm{SS-ML}} - \bar{\phi}$ can be approximated as

$$\breve{\phi} \approx \frac{\sum\limits_{t=t_1}^{t_N} |\alpha(t)|^2 \mathrm{Re}\{\frac{\beta(t)}{\alpha(t)}\}}{\sum\limits_{t=t_1}^{t_N} |\alpha(t)|^2} \tag{7.5}$$

where $\alpha(t) \doteq \sum_{\ell=1}^{L} a_\ell(t)$ and $\beta(t) \doteq \sum_{\ell=1}^{L} \tilde{\phi}_\ell a_\ell(t)$ are computed using the true values of $a_\ell(t)$ and $\tilde{\phi}_\ell$, and $\mathrm{Re}\{\cdot\}$ denotes the real part of the argument. Making use of Assumptions (2) and (3) in Subsection 7.2.1 and invoking the central limit theorem [Shanmugan and Breipohl 1988, Section 2.8.2], $\alpha(t)$ and $\beta(t)$ can be approximated as uncorrelated complex circularly-symmetric Gaussian random processes. Based on this assumption, the PDF of the right-hand side of Eq. (7.5) can be calculated. This PDF provides an approximation of the PDF of $\breve{\phi}$:

$$f_{\breve{\phi}}(\phi) \approx \frac{\Gamma(N+\frac{1}{2})}{\sqrt{\pi}\,\Gamma(N)} \cdot \frac{1}{\sigma_{\breve{\phi}}} \cdot \frac{1}{(1+\phi^2\sigma_{\breve{\phi}}^2)^{(N+\frac{1}{2})}} \tag{7.6}$$

where $\Gamma(\cdot)$ is the gamma function. The PDF in the right-hand side of Eq. (7.6) can be used to compute approximations of the moments of $\breve{\phi}$. For instance,

$$\mathrm{Var}[\breve{\phi}] \approx \frac{\Gamma(N-1)}{2\Gamma(N)} \cdot \sigma_{\tilde{\phi}}^2 \tag{7.7}$$

$$\mathrm{E}[|\breve{\phi}|] \approx \frac{\Gamma(N-\frac{1}{2})}{\sqrt{\pi}\,\Gamma(N)} \cdot \sigma_{\tilde{\phi}} \tag{7.8}$$

The derivations of Eq. (7.6)–(7.8) are given in Appendix A.3. For $N = 1$, the term on the right-hand side of Eq. (7.7) is infinite due to the heavy tails of the PDF on the right-hand side of Eq. (7.6), as reported in [Asztély and Ottersten 1998]. Actually, the variance of $\breve{\phi}$ is always finite, as the NA estimation error $\breve{\phi}$ is confined to the range $[-\pi, \pi)$. However, the fact that the right-hand side of Eq. (7.7) is infinite indicates that the variance of $\breve{\phi}$ is large. As a consequence, large estimation errors occur with significant probability. When $N > 1$, the right-hand side of Eq. (7.7) is finite and decreases as N increases. Thus, in the ID case, the probability of a large estimation error can be reduced by considering more observation samples.

7.2.3 First-order GAM Model Estimation

In this section, we introduce the first-order GAM model and derive estimators of its parameters using standard deterministic and stochastic ML methods, as well as a novel MUSIC algorithm. An AS estimator based on these parameter estimators is also proposed.

7.2.3.1 The First-order GAM Model

The GAM model [Asztély et al. 1997] makes use of the fact that the deviations $\tilde{\phi}_{d,\ell}$ are small with high probability. In the first-order GAM model, the first-order Taylor series expansion of the array response is used to approximate the effective impact of the SDSs on the received signal. We regard a distributed scatterer as an SDS when its contribution to the output signal of the Rx array is closely approximated using the first-order GAM model.

We first consider a single-SDS scenario. The function $\boldsymbol{c}(\bar{\phi} + \tilde{\phi}_\ell)$ in Eq. (7.1) can be approximated by its first-order Taylor series expansion at $\bar{\phi}$. Inserting this approximation for each $\boldsymbol{c}(\bar{\phi} + \tilde{\phi}_\ell)$ in Eq. (7.1) yields the first-order GAM model [Asztély et al. 1997]:

$$\begin{aligned} \boldsymbol{y}(t) \;\approx\; \boldsymbol{y}_{\mathrm{GAM}}(t) &\doteq \sum_{\ell=1}^{L} a_\ell(t)\left[\boldsymbol{c}(\bar{\phi}) + \tilde{\phi}_\ell \boldsymbol{c}'(\bar{\phi})\right] + \boldsymbol{w}(t) \\ &= \alpha(t)\boldsymbol{c}(\bar{\phi}) + \beta(t)\boldsymbol{c}'(\bar{\phi}) + \boldsymbol{w}(t) \end{aligned} \tag{7.9}$$

where $\boldsymbol{c}'(\bar{\phi}) \doteq \frac{\mathrm{d}\boldsymbol{c}(\phi)}{\mathrm{d}\phi}|_{\phi=\bar{\phi}}$. In matrix notation, Eq. (7.9) reads

$$\boldsymbol{y}_{\mathrm{GAM}}(t) = \boldsymbol{F}(\bar{\phi})\xi(t) + \boldsymbol{w}(t) \tag{7.10}$$

with $\boldsymbol{F}(\bar{\phi}) \doteq [\boldsymbol{c}(\bar{\phi})\ \boldsymbol{c}'(\bar{\phi})]$ and $\xi(t) \doteq [\alpha(t), \beta(t)]^{\mathrm{T}}$.

The autocorrelation functions of $\alpha(t)$ and $\beta(t)$ are calculated to be, respectively,

$$R_\alpha(\tau) \doteq \mathrm{E}[\alpha(t)\alpha^*(t+\tau)] = \sum_{\ell=1}^{L} R_{a_\ell}(\tau)$$

and

$$R_\beta(\tau) \doteq \mathrm{E}[\beta(t)\beta^*(t+\tau)] = \sigma_{\tilde{\phi}}^2 \cdot R_\alpha(\tau) \tag{7.11}$$

where $\sigma_{\tilde{\phi}}^2 \doteq \mathrm{E}[\tilde{\phi}_\ell^2]$. Note that $\mathrm{E}[\tilde{\phi}_\ell] = 0$ according to Assumption (1). The parameter $\sigma_{\tilde{\phi}}^2$ is the second-central moment of the azimuth deviation. By denoting the variances of $\alpha(t)$ and $\beta(t)$ with σ_α^2 and σ_β^2 respectively, we conclude from Eq. (7.11) that

$$\sigma_\beta^2 = \sigma_{\tilde{\phi}}^2 \cdot \sigma_\alpha^2 \tag{7.12}$$

This equality can also be obtained using the results given in [Shahbazpanahi et al. 2001, Eqs (49)–(51)]. We refer to the parameter $\sigma_{\tilde{\phi}}$ as the AS of the SDS. Note that as shown by Fleury [2000], the natural figure for characterizing direction dispersion is the direction spread. However, in a scenario with horizontal-only propagation and small azimuth deviations, the direction spread can be approximated by $\sigma_{\tilde{\phi}}$ expressed in radians [Fleury 2000]. For example, in the case where the azimuth power spectrum of an SDS is proportional to the von-Mises PDF [Mardia 1975]

$$f_{\tilde{\phi}_\ell}(\phi) = \frac{1}{2\pi I_0(\kappa)} \exp\{\kappa \cos(\phi - \bar{\phi})\} \tag{7.13}$$

where κ denotes the concentration parameter and $I_0(\cdot)$ represents the modified Bessel function of the first kind and order 0, the approximation is close provided that $\kappa \geq 7$; that is, $\sigma_{\tilde{\phi}} \leq 10°$ [Fleury 2000].

In a scenario with D SDSs, Eq. (7.9) extends to

$$\begin{aligned} \boldsymbol{y}(t) \approx \boldsymbol{y}_{\mathrm{GAM}}(t) &\doteq \left[\sum_{d=1}^{D} \alpha_d(t)\boldsymbol{c}(\bar{\phi}_d) + \beta_d(t)\boldsymbol{c}'(\bar{\phi}_d)\right] + \boldsymbol{w}(t) \\ &= \boldsymbol{B}(\bar{\phi})\gamma(t) + \boldsymbol{w}(t) \end{aligned} \tag{7.14}$$

where $\boldsymbol{B}(\bar{\phi}) \doteq [\boldsymbol{c}(\bar{\phi}_1), \boldsymbol{c}'(\bar{\phi}_1), \ldots, \boldsymbol{c}(\bar{\phi}_D), \boldsymbol{c}'(\bar{\phi}_D)]$ and $\gamma(t) \doteq [\alpha_1(t), \beta_1(t), \ldots, \alpha_D(t), \beta_D(t)]^{\mathrm{T}}$. Under Assumptions (3)–(5) in Subsection 7.2.1, the elements in the vector $\gamma(t)$ are uncorrelated.

7.2.4 Nominal Azimuth Estimators

In this subsection, the standard deterministic and stochastic ML estimation methods as well as a novel MUSIC algorithm are applied using the first-order GAM model to derive estimators of the NAs of SDSs.

7.2.4.1 Deterministic Maximum-likelihood NA Estimator

The deterministic maximum-likelihood (DML) NA estimator based on the first-order GAM model can be derived similarly to the SS–ML azimuth estimator in Eq. (7.4). Assuming that the weight samples $\alpha_d(t)$ and $\beta_d(t), t = t_1, \ldots, t_N, d = 1, \ldots, D$ in Eq. (7.14) are deterministic, the ML estimator of $\bar{\phi}$ is calculated as [Krim and Viberg 1996]:

$$\hat{\bar{\phi}}_{\mathrm{DML}} = \arg \max_{\bar{\phi}} \{\mathrm{tr}[\Pi_{\boldsymbol{B}(\bar{\phi})} \hat{\boldsymbol{\Sigma}}_{\boldsymbol{y}}]\} \tag{7.15}$$

The parameters $\gamma(t), t = t_1, \ldots, t_N$ are estimated as

$$\widehat{(\gamma(t))}_{\mathrm{DML}} = \boldsymbol{B}(\hat{\bar{\phi}})^{\dagger} \boldsymbol{y}(t), t = t_1, \ldots, t_N. \tag{7.16}$$

7.2.4.2 Stochastic Maximum-likelihood NA Estimator

The stochastic maximum-likelihood (SML) azimuth estimator based on the SS model was introduced by Jaffer [1988]. We obtain the SML NA estimator based on the first-order GAM model in a similar manner.

Making use of assumptions (1)–(5) in Section 7.2.1 and invoking the central limit theorem, the weight samples $\alpha_d(t)$ and $\beta_d(t), t = t_1, \ldots, t_N, d = 1, \ldots, D$ are uncorrelated complex circularly-symmetric Gaussian random processes with variances $\sigma^2_{\alpha_d}$ and $\sigma^2_{\beta_d}$, respectively. Let $\boldsymbol{\Omega}$ be the vector containing the parameters to be estimated:

$$\boldsymbol{\Omega} \doteq \left[\sigma^2_w, \bar{\phi}_d, \sigma^2_{\alpha_d}, \sigma^2_{\beta_d}; d = 1, \ldots, D\right] \tag{7.17}$$

The ML estimator of $\boldsymbol{\Omega}$ is a solution to the maximization problem [Krim and Viberg 1996]:

$$\widehat{\boldsymbol{\Omega}}_{\mathrm{SML}} = \arg \max_{\boldsymbol{\Omega}} \{-\ln\ [|\boldsymbol{\Sigma}_{\boldsymbol{y}_{\mathrm{GAM}}}|] - \mathrm{tr}[(\boldsymbol{\Sigma}_{\boldsymbol{y}_{\mathrm{GAM}}})^{-1} \hat{\boldsymbol{\Sigma}}_{\boldsymbol{y}}]\} \tag{7.18}$$

where the covariance matrix $\boldsymbol{\Sigma}_{\boldsymbol{y}_{\mathrm{GAM}}}$ of $\boldsymbol{y}_{\mathrm{GAM}}(t)$ in Eq. (7.14) reads

$$\boldsymbol{\Sigma}_{\boldsymbol{y}_{\mathrm{GAM}}} = \boldsymbol{B}(\bar{\phi}) \boldsymbol{R}_{\gamma} \boldsymbol{B}(\bar{\phi})^{\mathrm{H}} + \sigma^2_w \boldsymbol{I}_M \tag{7.19}$$

Here, $\boldsymbol{I}_M$ denotes the $M \times M$ identity matrix and $\boldsymbol{R}_{\gamma} = \mathrm{diag}(\sigma^2_{\alpha_1}, \sigma^2_{\beta_1}, \ldots, \sigma^2_{\alpha_D}, \sigma^2_{\beta_D})$ is the covariance matrix of $\gamma(t)$. Here, $\mathrm{diag}(\cdot)$ denotes a diagonal matrix with diagonal elements listed as argument.

The maximization operations in Eq. (7.15) and Eq. (7.18) require, respectively, a D-dimensional and a $(3D + 1)$-dimensional search. The high computational complexity of these search procedures prohibits the implementation of $\hat{\bar{\phi}}_{\mathrm{DML}}$ and $\hat{\boldsymbol{\Omega}}_{\mathrm{SML}}$ in real applications. As an alternative, the SAGE algorithm [Fleury et al. 1999, Yin and Fleury 2005] provides a low-complexity approximation of these ML estimators.

7.2.4.3 MUSIC NA Estimator

The standard MUSIC algorithm [Schmidt 1986] derived from the SS model Eq. (7.3) uses the pseudo-spectrum

$$f_{\mathrm{MUSIC}}(\phi) = \frac{\|\ \boldsymbol{c}(\phi)\ \|^2_{\mathrm{F}}}{\|\ \boldsymbol{c}(\phi)^{\mathrm{H}} \boldsymbol{E}_w\ \|^2_{\mathrm{F}}} \tag{7.20}$$

Here, $\|\cdot\|_{\mathrm{F}}$ denotes the Frobenius norm and $\boldsymbol{E}_w$ is an orthonormal basis of the estimated noise subspace calculated from $\hat{\boldsymbol{\Sigma}}_{\boldsymbol{y}}$. The azimuths of the D scatterers are estimated to be the arguments of the pseudo-spectrum corresponding to its D highest peaks.

We propose a natural extension of the standard MUSIC algorithm for the estimation of the NAs of SDSs based on the first-order GAM model. The extension considers the following generalization of the pseudo-spectrum in Eq. (7.20):

$$f_{\mathrm{MUSIC}}(\phi) = \frac{1}{\| \tilde{\boldsymbol{F}}(\phi)^{\mathrm{H}} \boldsymbol{E}_w \|_{\mathrm{F}}^2} \tag{7.21}$$

On the right-hand side of Eq. (7.21), $\tilde{\boldsymbol{F}}(\phi)$ is an orthonormal basis of the space spanned by the columns of $\boldsymbol{F}(\phi)$. The NAs of the D SDSs are estimated to be the arguments of the pseudo-spectrum corresponding to its D highest peaks.

Both the standard MUSIC algorithm and the proposed extension rely on the same principle: that parameter estimates are obtained by minimizing the distance between the subspace spanned by the signal originating from single scatterer and an estimate of this subspace computed from the sample covariance matrix. In the SS case, the signal subspace induced by an SS is spanned by the steering vector $\boldsymbol{c}(\phi)$, while in the SDS scenario the subspace induced by an SDS is spanned by the columns of $\boldsymbol{F}(\phi)$. In that sense, the latter algorithm is a natural extension of the former. Following [Edelman et al. 1998, p. 337], the distance between the subspace spanned by the columns of $\boldsymbol{F}(\phi)$ and the estimated signal subspace coincides with the Frobenius norm of the difference between the projection matrices of the two subspaces. It can be shown that this distance is proportional to the Frobenius norm of the projection of one subspace onto the null space of the other subspace; $\| \tilde{\boldsymbol{F}}(\phi)^{\mathrm{H}} \boldsymbol{E}_w \|_{\mathrm{F}}^2$ in our case. Thus the inverse of the pseudo-spectrum of Eq. (7.21) provides a measure of the distance between the signal subspace spanned by the columns of $\boldsymbol{F}(\phi)$ and the estimated signal subspace. A thorough discussion of the relationships between this extended MUSIC algorithm and other previously published extensions of the standard MUSIC algorithm is given in Subsection 7.2.4.

7.2.4.4 Generalization of the Algorithm

The proposed MUSIC algorithm, which makes use of the pseudo-spectrum of Eq. (7.21), can be generalized to the scenario in which the signals contributed by one component (e.g. a scatterer) span a subspace of any arbitrary dimension. In this case, $\tilde{\boldsymbol{F}}(\phi)$ is an orthonormal basis of the signal subspace. The argument ϕ of $\tilde{\boldsymbol{F}}(\phi)$ may be also multi-dimensional and it is not required that a closed-form expression exists that relates $\tilde{\boldsymbol{F}}(\phi)$ to ϕ. For instance, in the case where azimuth dispersion of an SDS is characterized using a PDF, $\tilde{\boldsymbol{F}}(\phi)$ can be obtained by the eigenvalue decomposition of the covariance matrix calculated using this PDF.

We propose an alternative interpretation of the proposed MUSIC algorithm using the concept of principal angles between subspaces [Golub and Loan 1996], which allows for a comparison with the variant of the MUSIC algorithm published by Christensen et al. [2004]. As shown by [Edelman et al. 1998, p. 337], minimizing the distance between two subspaces is equivalent to minimizing the norm of $\sin(\theta)$, where θ represents the vector containing all principal angles between these two subspaces and $\sin(\cdot)$ is the operator computing the element-wise sin of θ. Thus in our case the NA estimates obtained by maximizing the pseudo-spectrum in Eq. (7.21) in fact minimize $\| \sin(\boldsymbol{\theta}) \|$, where the components of $\boldsymbol{\theta}$ are the principal angles between the subspace spanned by the columns of $\boldsymbol{F}(\phi)$ and the signal subspace estimated from the sample covariance matrix. This is a reasonable approach in the ID case, where the dimension of the signal subspace induced by an SDS is larger than 1.

By contrast, the variant of the MUSIC algorithm proposed by Asztély et al. [1997] computes the NA estimates by maximizing the smallest principal angle between the two-dimensional subspace spanned by the columns of $\boldsymbol{F}(\phi)$ and the estimated signal subspace. This maximization is indeed equivalent to the maximization of the objective function $\lambda_{\min}^{-1}(\boldsymbol{F}(\phi)^{\mathrm{H}}\boldsymbol{E}_w\boldsymbol{E}_w^{\mathrm{H}}\boldsymbol{F}(\phi))$ described by Asztély et al. [1997], with $\lambda_{\min}(\cdot)$ denoting the smallest eigenvalue of the matrix given as argument, to compute the NA estimates [Drmac 2000]. The resulting algorithm is applicable when the dimension of the subspace effectively induced by an SDS is equal to one, for example in the CD case for which the algorithm was initially designed.

The pseudo-spectrum in Eq. (7.21) can be recast as:

$$f_{\mathrm{MUSIC}}(\phi) = \frac{1}{\mathrm{tr}\{\boldsymbol{E}_w^{\mathrm{H}}\boldsymbol{F}(\phi)\boldsymbol{W}(\phi)\boldsymbol{F}(\phi)^{\mathrm{H}}\boldsymbol{E}_w\}} \tag{7.22}$$

where $\boldsymbol{W}(\phi)$ is an azimuth-dependent weighting matrix defined as

$$\boldsymbol{W}(\phi) \doteq \boldsymbol{F}(\phi)^{\dagger}\tilde{\boldsymbol{F}}(\phi)\tilde{\boldsymbol{F}}(\phi)^{\mathrm{H}}(\boldsymbol{F}(\phi)^{\dagger})^{\mathrm{H}} \tag{7.23}$$

At first glance the representation in Eq. (7.22) seems to be similar to the pseudo-spectrum of the weighted MUSIC algorithm [Krim and Viberg 1996, Eq. (37)]. However, the proposed MUSIC algorithm and the standard weighted MUSIC algorithm have fundamental differences. First of all, it is impossible to recast the pseudo-spectrum of Eq. (7.22) in exactly the same form as the pseudo-spectrum of the weighted MUSIC algorithm. More specifically, the weighting matrix in the weighted MUSIC algorithm is inserted between $\boldsymbol{E}_w$ and $\boldsymbol{E}_w^{\mathrm{H}}$, while the weighting matrix is placed between $\boldsymbol{F}(\phi)$ and $\boldsymbol{F}(\phi)^{\mathrm{H}}$ in the proposed MUSIC algorithm. Furthermore, the criteria for the selection of the weighting matrices are fundamentally different. In the standard weighted MUSIC algorithm of Krim and Viberg [1996], the weighting matrix is computed from the eigenvalues and eigenvectors of the sample covariance matrix $\hat{\boldsymbol{\Sigma}}_{\boldsymbol{y}}$ and is constant. By contrast the weighting matrix in Eq. (7.22) is explicitly computed as a function of $\tilde{\boldsymbol{F}}(\phi)$ and, as a consequence, depends on the parameter to be estimated.

The pseudo-spectrum in Eq. (7.21) looks similar to the objective functions maximized in the pseudo-subspace fitting (PSF) method [Bengtsson 1999, Subsection 4.5.1]. However, an essential difference between this method and the proposed MUSIC algorithm is that the latter computes the NA estimates by "scanning" a measure of the distance between a multi-dimensional subspace (induced by a single SDS in our case) and the estimated signal subspace, while in the PSF method, the NA estimates are the values providing the best "fit" between the estimated signal subspace and the subspace spanned by all signals. Thus a one-dimensional search is required in the proposed MUSIC algorithm, while the PSF method requires a multi-dimensional search. Only in a single-SDS scenario is the objective function maximized in the PSF method identical to the pseudo-spectrum of Eq. (7.21) calculated in the proposed MUSIC algorithm.

Christensen et al. [2004] put forward another extension of the standard MUSIC algorithm. Their method considers the projection of the columns of $\boldsymbol{F}(\phi)$ on to $\boldsymbol{E}_w$, and relies on the pseudo-spectrum

$$\frac{\| \boldsymbol{F}(\phi) \|_{\mathrm{F}}^2}{\| \boldsymbol{F}(\phi)^{\mathrm{H}} \boldsymbol{E}_w \|_{\mathrm{F}}^2} \tag{7.24}$$

The inverse of the pseudo-spectrum in Eq. (7.24) corresponds to the distance between the multi-dimensional subspace spanned by the columns of $\boldsymbol{F}(\phi)$ and the signal subspace, if and only if, the columns of $\boldsymbol{F}(\phi)$ are orthonormal for any values of ϕ. This condition is usually not satisfied in real applications. Simulation results also show that the NA estimator derived from Eq. (7.21) outperforms the estimator obtained from Eq. (7.24) in terms of lower root mean-square estimation error.

To the best of our knowledge, the proposed extension of the MUSIC algorithm according to Eq. (7.21) has not yet been reported in the literature. This algorithm is indeed the *natural* extension of the standard MUSIC algorithm to the case where the subspace induced by each individual signal component is multi-dimensional. One application example is the ID case in the SDS scenario considered in this section. Another example is fundamental frequency estimation for signals with a harmonic structure [Christensen et al. 2004].

7.2.5 Azimuth Spread Estimator

Equation (7.12) inspires the following estimator of the AS of an SDS:

$$\widehat{\sigma_{\tilde{\phi}}} = \sqrt{\widehat{\sigma_\beta^2}/\widehat{\sigma_\alpha^2}}. \tag{7.25}$$

The estimates $\widehat{\sigma_\beta^2}$ and $\widehat{\sigma_\alpha^2}$ can be directly obtained for each of the D SDSs from Eq. (7.18) when using the SML estimators, or computed as

$$\widehat{\sigma_\beta^2} = \frac{1}{N}\sum_{t=t_1}^{t_N} |\hat{\beta}(t) - < \hat{\beta}(t) > |^2 \quad \text{and}$$

$$\widehat{\sigma_\alpha^2} = \frac{1}{N}\sum_{t=t_1}^{t_N} |\hat{\alpha}(t) - < \hat{\alpha}(t) > |^2 \tag{7.26}$$

when DML estimation is used. In Eq. (7.26), $\hat{\beta}(t)$ and $\hat{\alpha}(t)$, $t = t_1, \ldots, t_N$ are calculated from Eq. (7.16) and $< \cdot >$ denotes averaging. In the case where the proposed MUSIC algorithm Eq. (7.21) is applied, $\widehat{\sigma_\beta^2}$ and $\widehat{\sigma_\alpha^2}$ can be obtained by applying the least-square covariance matrix fitting method [Johnson and Dudgeon 1993, Section 7.1.2] that we will shortly describe below.

First we rewrite the covariance matrix $\boldsymbol{\Sigma}_{\boldsymbol{y}_{\mathrm{GAM}}}$ in Eq. (7.19) according to

$$\mathrm{vec}(\boldsymbol{\Sigma}_{\boldsymbol{y}_{\mathrm{GAM}}}) = \boldsymbol{D}(\bar{\phi})\boldsymbol{e} \tag{7.27}$$

with $\mathrm{vec}(\cdot)$ denoting the vectorization [Minka 2000]:

$$\boldsymbol{D}(\bar{\phi}) \doteq [\boldsymbol{c}(\bar{\phi}_1) \otimes \boldsymbol{c}(\bar{\phi}_1)^*, \boldsymbol{c}'(\bar{\phi}_1) \otimes \boldsymbol{c}'(\bar{\phi}_1)^*, \ldots, \boldsymbol{c}(\bar{\phi}_D) \otimes \boldsymbol{c}(\bar{\phi}_D)^*,$$
$$\boldsymbol{c}'(\bar{\phi}_D) \otimes \boldsymbol{c}'(\bar{\phi}_D)^*, \mathrm{vec}(\boldsymbol{I}_M)]$$

where $\otimes$ is the Kronecker product, and

$$\boldsymbol{e} \doteq [\sigma_{\alpha_1}^2, \sigma_{\beta_1}^2, \sigma_{\alpha_2}^2, \sigma_{\beta_2}^2, \ldots, \sigma_w^2]^{\mathrm{T}}$$

In the covariance matrix fitting method, the estimate $\hat{\boldsymbol{\Omega}}$ of $\boldsymbol{\Omega}$ in Eq. (7.17) minimizes the Euclidean distance between $\hat{\boldsymbol{\Sigma}}_y$ and $\boldsymbol{\Sigma}_{\boldsymbol{y}_{\text{GAM}}}$. Thus, the identity

$$\left.\frac{\partial \parallel \hat{\boldsymbol{\Sigma}}_y - \boldsymbol{\Sigma}_{\boldsymbol{y}_{\text{GAM}}} \parallel_{\text{F}}^2}{\partial \boldsymbol{e}^{\text{H}}}\right|_{\boldsymbol{e}=\hat{\boldsymbol{e}}} = 0 \tag{7.28}$$

holds for $\boldsymbol{\Omega} = \hat{\boldsymbol{\Omega}}$. Solving Eq. (7.28) yields the close-form expression for the solution $\hat{e}$:

$$\hat{\boldsymbol{e}} = \boldsymbol{D}(\hat{\bar{\phi}})^{\dagger}\text{vec}(\hat{\Sigma}_{\boldsymbol{y}}) \tag{7.29}$$

It is worth mentioning that the AS estimator in Eq. (7.25) does not require knowledge of the PDF of the azimuth deviation. In the case where some assumption is made on the PDF in form of a parametric model, the AS estimate can be used to calculate the model parameters. Shahbazpanahi [2004, (39)–(43)] related the AS $\sigma_{\tilde{\phi}}$ to the parameters controlling the spread of the truncated Gaussian, the Laplacian and the confined uniform distributions. In addition, when the von-Mises PDF is used [Ribeiro et al. 2005], the relation between the AS and the concentration parameter κ of this PDF is approximated according to $\sigma_{\tilde{\phi}} \approx \sqrt{1 - |I_1(\kappa)/I_0(\kappa)|^2}$.

7.2.6 Simulation Studies

In this section we evaluate by means of Monte-Carlo simulations the performance of the proposed estimators. The considered scenarios are simulated as follows. Each individual SDS consists of $L = 50$ subpaths. The subpath weights and azimuth deviations with respect to the NAs are randomly generated according to Assumptions (1)–(5) in Subsection 7.2.1. The NAs of the SDSs lie in the array beam-width. The azimuth deviations are generated according to a von-Mises distribution centered at zero azimuth. The Rx antenna is an eight-element ULA with half-a-wavelength inter-element spacing. Each result (point) shown in Figures 7.1 and 7.2 is calculated from 2000 simulation runs, while the results depicted in Figures 7.3–7.6 are

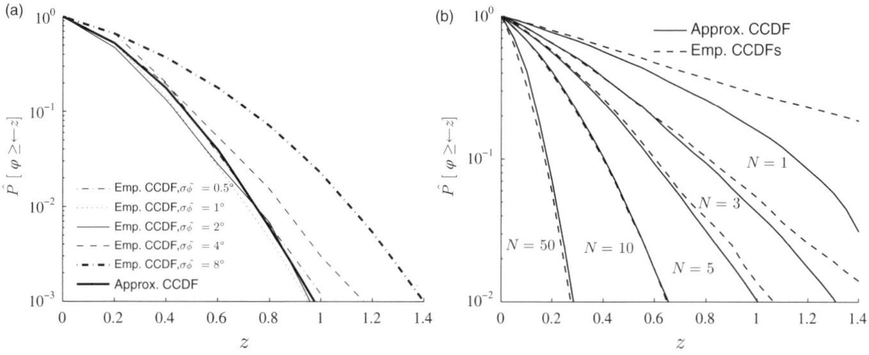

Figure 7.1 Empirical (Emp.) and approximate (Approx.) CCDFs of the absolute normalized estimation error of the SS-ML azimuth estimator applied in an SDS scenario: (a) with $\sigma_{\tilde{\phi}}$ as a parameter, $N = 10$; (b) with N as a parameter, $\sigma_{\tilde{\phi}} = 2°$

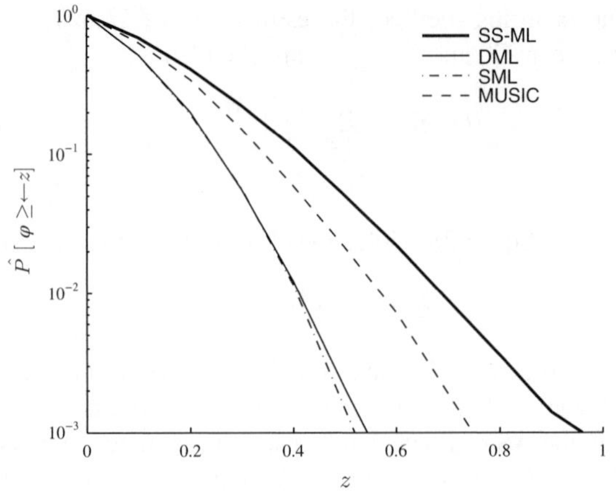

Figure 7.2 Empirical CCDF of the absolute normalization error of the nominal azimuth estimators with $\sigma_{\tilde{\phi}} = 2°$, $\gamma = 25$ dB, $N = 10$

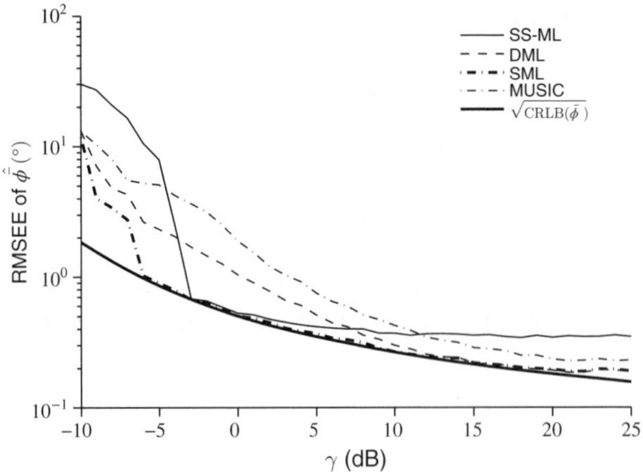

Figure 7.3 RMSEEs of nominal azimuth estimators versus the input SNR γ with $\sigma_{\tilde{\phi}} = 3°$, $N = 50$

calculated from 500 runs each. Note that the MUSIC estimators mentioned in what follows make use of the pseudo-spectrum in Eq. (7.21).

7.2.6.1 Estimation Error of the ML Azimuth Estimator based on the SS Model

We first investigate the accuracy of the approximation in Eq. (7.6) in a noiseless ($\sigma_w^2 = 0$) scenario. Figures 7.1(a) and (b) depict the empirical (estimated) and the approximated complementary cumulative distribution functions (CCDFs) of the absolute normalized estimation error $\varphi \doteq |\check{\phi}|/\sigma_{\tilde{\phi}}$ with respectively the AS $\sigma_{\tilde{\phi}}$ and the number of observation samples N as

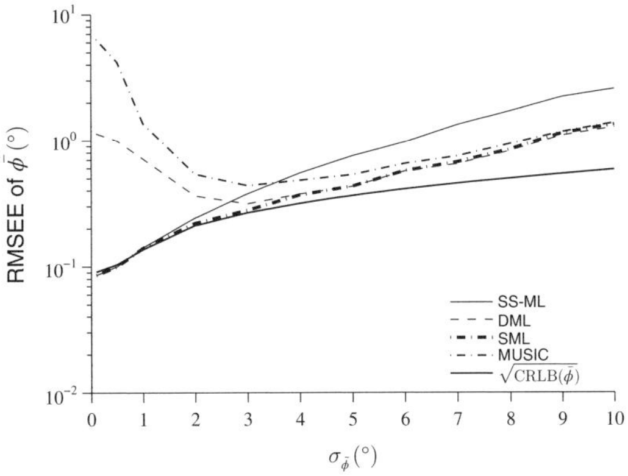

Figure 7.4 RMSEEs of the NA estimators versus the true AS with $\gamma = 10$ dB, $N = 50$

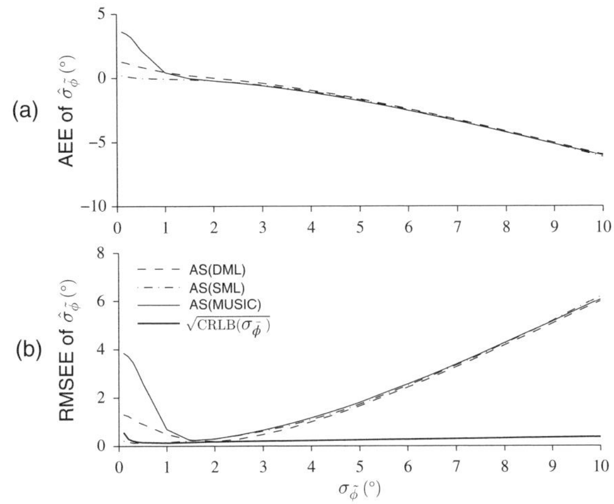

Figure 7.5 AEE (a) and RMSEE (b) of the AS estimators versus the true AS $\sigma_{\tilde{\phi}}$ with the fixed array size $M = 8$, $N = 50$ and $\gamma = 10$ dB

parameters. Using Eq. (7.6), the PDF of φ is approximated according to

$$f_{\varphi}(z) \approx \frac{2\Gamma(N + \frac{1}{2})}{\sqrt{\pi}\,\Gamma(N)} \cdot (1 + z^2)^{-(N+\frac{1}{2})} \tag{7.30}$$

The right-hand PDF in Eq. (7.30) is independent of $\sigma_{\tilde{\phi}}$ and so is the CCDF computed from it. This CCDF is used as an approximation of the CCDF of φ. For short, it is referred to as the "approximate CCDF" below. The results presented in Figures 7.1(a) and (b) are computed using the parameter settings $\bar{\phi} = 0°$, $\sigma_{\tilde{\phi}} \in [0.5°, 8°]$, $N = 10$ and $\bar{\phi} = 0°$, $\sigma_{\tilde{\phi}} = 2°$, $N \in [1, \ldots, 50]$ respectively.

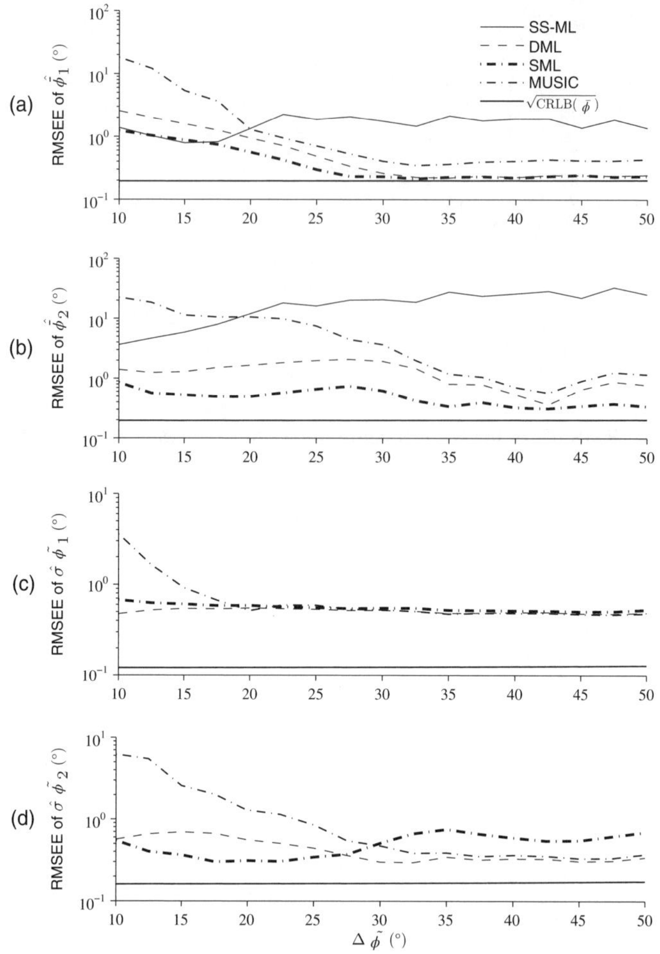

Figure 7.6 RMSEEs of the NA and AS estimators versus the NA separation in a two-SDS scenario, $N = 50$, $\sigma_{\tilde{\phi}_1} = \sigma_{\tilde{\phi}_2} = 3°$, $\gamma_1 = 19$ dB and $\gamma_2 = 10$ dB. The legend given in (a) applies to all subfigures

We observe that the empirical CCDFs are close to the approximate CCDFs when $\sigma_{\tilde{\phi}} < 4°$ (Figure 7.1(a)) and when $\varphi < 0.5$ (Figure 7.1(b)). These observations are in accordance with the fact that the approximation in Eq. (7.6) holds for small $\check{\phi}$. Furthermore, from Figure 7.1(b) the CCDFs decrease when N increases. This indicates that in the ID case, the probability of large estimation errors of the SS-ML azimuth estimator can be reduced by increasing the number of observation samples.

7.2.6.2 Performance of the Nominal Azimuth Estimators

Figure 7.2 depicts the empirical CCDF of the absolute normalized estimation error φ of the NA estimators. The CCDF of φ of the SS ML azimuth estimator is also reported for comparison purposes. The parameter settings are $N = 10$, $\sigma_{\tilde{\phi}} = 2°$, and $\gamma = 25$ dB. It can be observed that the SML and the DML estimators perform similarly, and better than the MUSIC and SS ML estimators. The MUSIC estimator outperforms the SS-ML estimator. The results show that the NA estimators have significantly lower probability of large estimation errors than the conventional SS-ML estimator.

Figures 7.3 and 7.4 depict the root mean-square estimation error (RMSEE) of the NA estimators versus respectively the input SNR γ and the AS. The parameter settings are $\bar{\phi} = 0°$, $N = 50$, $\sigma_{\tilde{\phi}} = 3°$, and $\gamma \in [-10\ \text{dB}, 25\ \text{dB}]$ for Figure 7.3 and $\bar{\phi} = 0°$, $N = 50$, $\sigma_{\tilde{\phi}} \in [0.1°, 10°]$, $\gamma = 10$ dB for Figure 7.4. For comparison purposes, the square root of the Cramér–Rao lower bound (CRLB) for the estimation of the NA ($\sqrt{\text{CRLB}(\bar{\phi})}$) [Ribeiro et al. 2004] is reported. It is observed that the SML estimator outperforms all the other estimators. The RMSEE of this estimator is very close to the root CRLB for small ASs ($\sigma_{\tilde{\phi}} < 3°$). All NA estimators outperform the SS-ML estimator when the SNR and the AS are large.

7.2.6.3 Performance of the Azimuth Spread Estimators

Figures 7.5(a) and (b) depict the performance of the AS estimators. The parameter settings are identical to those used to obtain the results shown in Figure 7.4. In Figure 7.5(b), the root CRLB of $\sigma_{\tilde{\phi}}$ ($\sqrt{\text{CRLB}(\sigma_{\tilde{\phi}})}$) is also reported.

It is observed that the estimators exhibit negative average estimation error (AEE) with increasing magnitude as the AS increases in the range $\sigma_{\tilde{\phi}} > 2°$. This effect is due to the discrepancy between the effective signal model Eq. (7.1) and the first-order GAM model Eq. (7.9). For large ASs, the effective rank of the true signal subspace contributed by a distributed scatterer is larger than two with high probability. In such a case, the first-order GAM model only captures a part of the true signal. Consequently, the AS estimate is less than the true value. When the AS increases, the model discrepancy becomes more significant, so the absolute value of the AEE increases.

We also observe from Figure 7.5a that for $\sigma_{\tilde{\phi}} < 2°$ the AEE of the AS(MUSIC) estimator significantly increases as the AS decreases. This effect is caused by the positive biases of the estimators $\hat{\sigma}^2_\beta$ and $\hat{\sigma}^2_\alpha$ and becomes more pronounced when σ^2_β tends to zero. The same effect causes the similar behaviour in the positive bias of the AS(DML) estimator. The AS(SML) estimator does not exhibit this behaviour since the two estimators $\hat{\sigma}^2_\beta$ and $\hat{\sigma}^2_\alpha$ are unbiased in this case.

7.2.6.4 Performance of the NA and AS Estimators in a Two-SDS Scenario

The investigations reported in this section concern a two-SDS scenario. We assume that the number of SDSs – two in this case – is known. The parameter settings are selected as follows: $\bar{\phi}_1$ and $\bar{\phi}_2$ are symmetric with respect to the array broadside with a certain NA spacing $\Delta\bar{\phi}$, $\sigma_{\tilde{\phi}_1} = \sigma_{\tilde{\phi}_2} = 3°$ and $N = 50$. The input SNRs for SDS1 and SDS2 are 19 dB and 10 dB respectively; that is, we consider a situation with strong SDS power unbalance, say 9 dB. Each element in the pair of computed NA estimates, say $(\hat{\bar{\phi}}', \hat{\bar{\phi}}'')$, is assigned to one of the two SDSs according to

$$(\hat{\bar{\phi}}_1, \hat{\bar{\phi}}_2) = \underset{\substack{(\phi', \phi'') \in \\ \{(\hat{\bar{\phi}}', \hat{\bar{\phi}}''), (\hat{\bar{\phi}}'', \hat{\bar{\phi}}')\}}}{\arg\min} \ \| (\phi', \phi'') - (\bar{\phi}_1, \bar{\phi}_2) \|$$

The performance of the NA and AS estimators is shown in Figure 7.6. The RMSEEs of the NA estimators versus the NA spacing $\Delta\bar{\phi}$ are depicted in Figures 7.6(a) and (b). Figures 7.6(c) and (d) show the RMSEEs of the AS estimators versus $\Delta\bar{\phi}$. It can be seen that the SML NA estimators exhibit the lowest RMSEEs among all estimators. The RMSEEs resulting from using the NA estimators are less than 1° for $\Delta\bar{\phi} > 28°$. Notice that the value $\Delta\bar{\phi} = 28°$ equals approximately twice the intrinsic azimuth resolution of the eight-element ULA used [Fleury et al. 1999]. We observe that the SS-ML azimuth estimator of the second SDS has a constantly large RMSEE even though the two SDSs are well-separated. This indicates that this estimator fails to estimate the NA of the weaker SDSs at low SNR.

7.3 PSD-based Dispersive Component Estimation

Besides the GAM-based channel parameter estimation approaches. which rely on the array-manifold representation of the received signal, the power spectral density (PSD) of the propagation channel can also be estimated from multiple observations of the channel using standard PSD parameter-estimation algorithms.

In recent years, various algorithms have been proposed for the estimation of the dispersive characteristics of individual components in the channel response [Besson and Stoica 1999, Betlehem et al. 2006, Ribeiro et al. 2004, Trump and Ottersten 1996]. These algorithms estimate the PSD of each component. A component PSD can be irregular in real environments, and thus a gross description relying on certain characteristic parameters – for example the center of gravity and spreads – is usually adopted. The algorithms proposed in these contributions estimate these parameters by approximating the shape of the normalized component PSD with a certain PDF, for example of the azimuth of arrival (AoA) [Besson and Stoica 1999, Ribeiro et al. 2004, Trump and Ottersten 1996] or in the AoA and azimuth of departure (AoD) [Betlehem et al. 2006]. The parameter estimates obtained with these algorithms depend on the selected PDFs. However, no rationale behind the selection is given in these papers. Furthermore, the applicability of these PDFs in characterizing the PSD and the performance of the estimation algorithms have not been investigated experimentally with real measurement data.

The maximum-entropy (ME) principle [Jaynes 2003] has been proposed by Yin et al. [2007a, b, 2006a, b] for the selection/derivation of the PSD characterizing the component power distribution. This rationale assumes that each component has a fixed center of gravity and spread, and that no additional information about the PSD exists. The center of gravity and the spreads of a component PSD are described by the first and second moments of the corresponding power distribution. Using the ME principle, we derive a PSD that has fixed first and second moments, while maximizing the entropy of any other constraint. The estimates of the dispersion parameters obtained by modeling the component PSD with this entropy-maximizing PSD provide the "safest" results in the sense that, they are more accurate than the estimates computed using another form of PSD subject to any constraint that is invalid in real situations. Based on this rationale, we derived the ME PSDs to characterize the component power distribution in AoA and AoD [Yin et al. 2006b], in elevation and azimuth [Yin et al. 2007a, 2007b], and in biazimuth (azimuth of arrival and azimuth of departure) and delay [Yin et al. 2006a]. These entropy-maximizing PDFs actually coincide with, respectively, the bivariate von-Mises-Fisher PDF [Mardia et al. 2003], the Fisher–Bingham-5 (FB_5) PDF [Kent 1982], and the extended von-Mises-Fisher PDF [Mardia et al. 2003]. Preliminary investigations using measurement data demonstrate that these characterizations are applicable in real environments.

7.4 Bidirection-delay-Doppler Frequency PSD Estimation

In Section 4.5, the generic PSD model was proposed for characterization of the six-dimensional shape of the bidirection (direction of departure and direction of arrival), delay and Doppler frequency PSD of individual components in the response of a propagation channel. This PSD maximizes entropy under the constraint that the multi-dimensional PSD of each component exhibits a fixed center of gravity, specific spreads, and dependence of the spreads in bidirection, delay, and Doppler frequency. The power spectrum of the channel is modeled as the superposition of the component PSDs multiplied by the component average power.

In this section, the signal model described in Section 4.5 is used to derive the parameter estimator for the PSDs. We adopt the MaxEnt PSD model derived in that section as the generic model for the PSD of individual channel components. More specifically, the component PSD $f_d(\mathbf{\Omega}_1, \mathbf{\Omega}_2, \tau, \nu)$ in Eq. (4.95) is considered to be $f_{\mathrm{MaxEnt}}(\mathbf{\Omega}_1, \mathbf{\Omega}_2, \tau, \nu; \boldsymbol{\theta}_d)$ shown in Eq. (4.67) where $\boldsymbol{\theta}_d$ denotes the component-specific parameters

$$\begin{aligned}\boldsymbol{\theta}_d \triangleq [&\mu_{\phi_{1,d}}, \mu_{\theta_{1,d}}, \mu_{\phi_{2,d}}, \mu_{\theta_{2,d}}, \mu_{\tau_d}, \mu_{\nu_d}, \sigma_{\phi_{1,d}}, \sigma_{\theta_{1,d}}, \sigma_{\phi_{2,d}}, \sigma_{\theta_{2,d}}, \sigma_{\tau_d}, \sigma_{\nu_d}, \rho_{\phi_1\tau,d}, \rho_{\phi_1\theta_1,d},\\ &\rho_{\phi_1\phi_2,d}, \rho_{\phi_1\theta_2,d}, \rho_{\phi_1\tau,d}, \rho_{\phi_1\nu,d}, \rho_{\theta 1\phi_2,d}, \rho_{\theta_1\theta_2,d}, \rho_{\theta_1\tau,d}, \rho_{\theta_1\nu,d}, \rho_{\phi_2\theta_2,d},\\ &\rho_{\phi_2\tau,d}, \rho_{\phi_2\nu,d}, \rho_{\tau\nu,d}]\end{aligned} \tag{7.31}$$

In the scenario where the channel response consists of D components, the power spectrum $P(\mathbf{\Omega}_1, \mathbf{\Omega}_2, \tau, \nu)$ of the channel is parameterized by $\mathbf{\Theta} = (\mathbf{\Theta}_1, \mathbf{\Theta}_2, \ldots, \mathbf{\Theta}_D)$ as shown in Eq. (4.98), where $\mathbf{\Theta}_d = (P_d, \boldsymbol{\theta}_d)$ denotes the parameters of the component d. Using this parametric approach, the estimation of $P(\mathbf{\Omega}_1, \mathbf{\Omega}_2, \tau, \nu)$ is then equivalent to the estimation of the parameters $\mathbf{\Theta}$. In the following, an algorithm is derived to estimate the unknown parameter $\mathbf{\Theta}$ from the output signal of the Rx array in a MIMO system that is described by the signal model presented in Eq. (4.84) in Section 4.5.

7.4.1 Channel Power Spectrum Estimator

The ML estimator of $\mathbf{\Theta}$ [Krim and Viberg 1996] can be derived from the signal model in Eq. (4.84). However, since computation of this estimate necessitates solving a $21D$-dimensional maximization problem, with D being the total number of dispersive components, the resulting complexity prohibits the implementation of the ML estimator in real applications. The SAGE framework [Fessler and Hero 1994] can be used to derive an approximation of the ML estimator of $\mathbf{\Theta}$ with low computational complexity.

As discussed in previous chapters, the SAGE framework has been applied for estimation of the parameters of specular components in a propagation channel [Fleury et al. 1999]. To derive a parameter estimator using the SAGE framework, we need to specify the observable data and the admissible hidden (unobservable) data. As in our case, where the received signal is a superposition of multiple components, the observable data is the received signal and a possible selection of the hidden data is the signal contribution of individual components. The SAGE algorithm updates in an iterative manner the parameter estimates of the admissible hidden data via the expectation (E-) step and the maximization (M-) step. This iteration proceeds until a convergence criterion is achieved. In the following, we present the signal model of the admissible hidden data, the E- and the M-steps, the initialization procedure and the convergence criterion used in the estimator. A pseudo-code representation of this SAGE-based estimator is illustrated in Figure 7.8.

7.4.1.1 Admissible Hidden Data

At the ℓth iteration, the estimates of the parameters in the subset $\boldsymbol{\Theta}_d$ are updated, where d is selected as

$$d = \ell \bmod (D) + 1 \tag{7.32}$$

We consider the hidden data $\boldsymbol{X}_d$

$$\boldsymbol{X}_d \triangleq \boldsymbol{S}_d + \boldsymbol{W}' \tag{7.33}$$

where the noise $\boldsymbol{W}' \in \mathbb{C}^{M_2 M_1 N}$ is a complex, circularly symmetric white Gaussian vector with component variances $\beta \cdot N$, $0 \leq \beta \leq 1$. The parameter β is assumed to be known. The convergence of the parameter estimates increases when a large β is selected. However, our experience from simulation studies show that when the PSDs of two components are close to each other, a small β results in a better estimation result. Effectively, β works as a factor that controls the amount of interference considered in the estimate of the admissible hidden data. This interference is the residual obtained when the other components are subtracted from the received signal. In the case where components are closely spaced, the estimation errors can be significant, especially in the first several SAGE iterations. A small β can limit the interference posed due to the estimation errors.

7.4.1.2 The Expectation Step

In the expectation (E-) step, the objective function

$$Q\left(\boldsymbol{\Theta}_d | \boldsymbol{\Theta}^{[\ell-1]}\right) = \mathrm{E}[\Lambda(\boldsymbol{\Theta}_d; \boldsymbol{X}_d) | \boldsymbol{y}_1, \ldots, \boldsymbol{y}_I, \boldsymbol{\Theta}^{[\ell-1]}] \tag{7.34}$$

is computed. Here, $\boldsymbol{\Theta}^{[\ell]}$ denotes the estimate of $\boldsymbol{\Theta}$ obtained in the ℓth iteration, $\Lambda(\boldsymbol{\Theta}_d; \boldsymbol{X}_d)$ represents the log-likelihood function of $\boldsymbol{\Theta}_d$, and I is the total number of measurement cycles. It can be shown that $Q(\boldsymbol{\Theta}_d | \boldsymbol{\Theta}^{[\ell-1]})$ can be calculated as

$$Q(\boldsymbol{\Theta}_d | \boldsymbol{\Theta}^{[\ell-1]}) = -\ln |\boldsymbol{\Sigma}(\boldsymbol{\Theta}_d)| - \mathrm{tr}[\boldsymbol{\Sigma}(\boldsymbol{\Theta}_d)^{-1} \boldsymbol{\Sigma}_{\boldsymbol{X}_d}^{[\ell-1]}] \tag{7.35}$$

where $|\cdot|$ and $\mathrm{tr}[\cdot]$ represent respectively the determinant and the trace of the matrix given as an argument. The covariance matrix $\boldsymbol{\Sigma}(\boldsymbol{\Theta}_d)$ is calculated as

$$\boldsymbol{\Sigma}(\boldsymbol{\Theta}_d) = \sum_{p_1=1}^{2}\sum_{p_2=1}^{2} P_{d,p_1,p_2} \int_{-\pi}^{+\pi}\int_{-\pi}^{+\pi}\int_{-\infty}^{+\infty}\int_{-\infty}^{+\infty} [\boldsymbol{\zeta}_{p_1}(\tau,\nu,\phi_1,\theta_1)\boldsymbol{\zeta}_{p_1}(\tau,\nu,\phi_1,\theta_1)^{\mathrm{H}}]$$
$$\otimes \boldsymbol{c}_{2,p_2}(\phi_2,\theta_2)\boldsymbol{c}_{2,p_2}(\phi_2,\theta_2)^{\mathrm{H}} f(\phi_1,\theta_1,\phi_2,\theta_2,\tau,\nu;\boldsymbol{\theta}_d)\mathrm{d}\phi_1\mathrm{d}\phi_2\mathrm{d}\tau\mathrm{d}\nu + \beta\cdot N\boldsymbol{I}_{M_2M_1N} \tag{7.36}$$

Here, $(\cdot)^{\mathrm{H}}$ represents the Hermitian transpose, $\otimes$ denotes the Kronecker product, and

$$\zeta_{p_1}(\tau,\nu,\phi_1,\theta_1) \triangleq \boldsymbol{s}(\nu)\otimes\boldsymbol{r}(\tau)\otimes\boldsymbol{c}_{1,p_1}(\phi,\theta) \tag{7.37}$$

with

$$\boldsymbol{s}(\nu) \triangleq [\exp(j2\pi\nu t_1'),\ldots,\exp(j2\pi\nu t_{M1M2}')]^{\mathrm{T}} \tag{7.38}$$

$$\boldsymbol{r}(\tau) \triangleq [r(\tau-t_1),\ldots,r(\tau-t_N)]^{\mathrm{T}} \quad \text{and} \quad r(\tau) = \int_{-\infty}^{+\infty} u(t)u(t-\tau)^*\mathrm{d}t$$

where $t_1',\ldots,t_{M1M2}'$ denote the starting time instance of the subchannels in one cycle. The matrix $\boldsymbol{\Sigma}_{\boldsymbol{X}_d}^{[\ell-1]}$ in Eq. (7.35) is the conditional covariance matrix of $\boldsymbol{X}_d$ given the observations $\boldsymbol{y}_1,\boldsymbol{y}_2,\ldots,\boldsymbol{y}_I$ and under the assumption $\boldsymbol{\Theta}=\boldsymbol{\Theta}^{[\ell-1]}$. It can be shown that

$$\boldsymbol{\Sigma}_{\boldsymbol{X}_d}^{[\ell-1]} = \boldsymbol{\Sigma}(\boldsymbol{\Theta}_d^{[\ell-1]}) - \boldsymbol{\Sigma}(\boldsymbol{\Theta}_d^{[\ell-1]})\boldsymbol{\Sigma}_{\boldsymbol{Y}|\boldsymbol{\Theta}^{[\ell-1]}}^{-1}\boldsymbol{\Sigma}(\boldsymbol{\Theta}_d^{[\ell-1]})$$
$$+ \boldsymbol{\Sigma}(\boldsymbol{\Theta}_d^{[\ell-1]})\boldsymbol{\Sigma}_{\boldsymbol{Y}|\boldsymbol{\Theta}^{[\ell-1]}}^{-1}\hat{\boldsymbol{\Sigma}}_{\boldsymbol{Y}}\boldsymbol{\Sigma}_{\boldsymbol{Y}|\boldsymbol{\Theta}^{[\ell-1]}}^{-1}\boldsymbol{\Sigma}(\boldsymbol{\Theta}_d^{[\ell-1]}) \tag{7.39}$$

where the terms $\hat{\boldsymbol{\Sigma}}_{\boldsymbol{Y}}$ and $\boldsymbol{\Sigma}_{\boldsymbol{Y}|\boldsymbol{\Theta}^{[\ell-1]}}$ are computed as, respectively,

$$\hat{\boldsymbol{\Sigma}}_{\boldsymbol{Y}} = \frac{1}{I}\sum_{i=1}^{I}\boldsymbol{y}_i\boldsymbol{y}_i^{\mathrm{H}} \quad \text{and} \tag{7.40}$$

$$\boldsymbol{\Sigma}_{\boldsymbol{Y}|\boldsymbol{\Theta}^{[\ell-1]}} = \sum_{d=1}^{D}\boldsymbol{\Sigma}(\boldsymbol{\Theta}_d^{[\ell-1]}) + (1-D\beta)N\cdot\boldsymbol{I}_{M_2M_1N} \tag{7.41}$$

7.4.1.3 The Maximization Step

In the maximization (M-) step,

$$\boldsymbol{\Theta}_d^{[\ell]} = \arg\max_{\boldsymbol{\Theta}_d} Q(\boldsymbol{\Theta}_d|\boldsymbol{\Theta}^{[\ell-1]}) \tag{7.42}$$

is calculated. By applying a coordinate-wise updating procedure similar to that used by Fleury et al. [1999], the required multiple-dimensional maximization can be reduced to multiple one-dimensional maximization problems. It can be shown that this coordinate-wise updating still remains within the SAGE framework with the admissible hidden data given in Eq. (7.33).

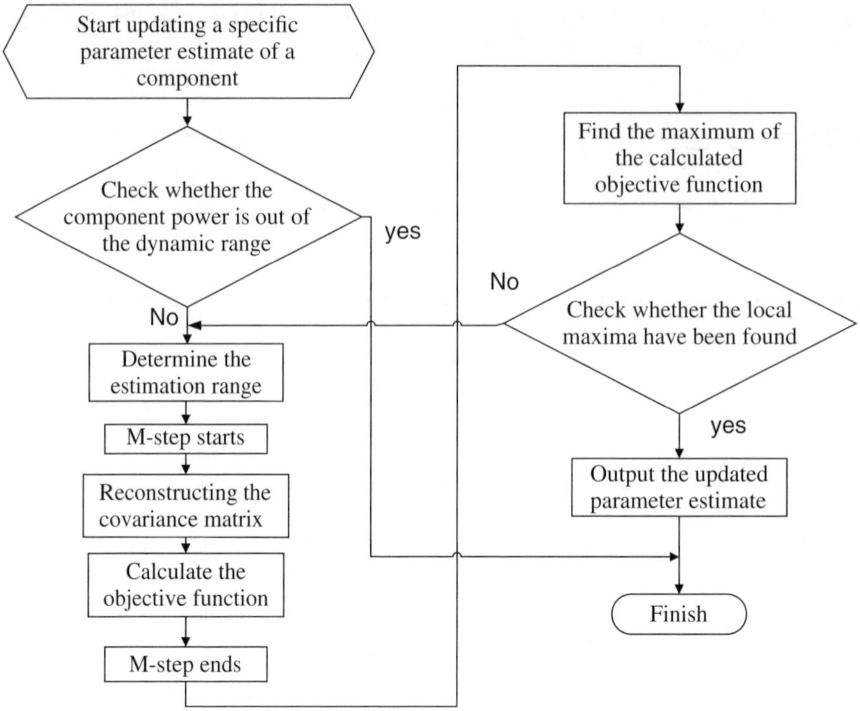

Figure 7.7 Updating the estimate of a specific parameter of the center of gravity of individual components

Figure 7.7 shows the flowchart for updating the estimate of a specific parameter. All updating programs for renewing the parameter estimate have a similar structure. It is worth mentioning that in order to keep the number of computations as small as possible, the estimation range is adapted in so-called "local" iterations. The status of "maximum found" is used to determine whether the calculation should continue or not. This status is updated by checking the location of the maximum and the spacing of the candidate values. When the location of the maximum does not coincide with the border of the estimation range, and the spacing of the candidate values is equal to or less than the minimum spacing specified, the status of "maximum found" is set to 1, which terminates the operation in this program.

The estimation range determination is different with respect to the parameters:

- For parameters with linear supports, such as the delay, the range needs to be set within the overall estimation range.
- For parameters with circular supports, such as the azimuths, the range should be continuous in the circle and be consistent with the values of the steps following which the antenna responses are recorded.
- For the spread parameters, the estimation range should not contain zeros and negative values.
- For the coupling coefficients, we need to keep the candidates within the range $[-1, 1]$. Meanwhile, since the covariance matrix of the spreads in multiple dimensions should be positive

Initialization
- Compute the power spectrum estimate using the Bartlett beamformer;
- Initialize the parameter estimate $\boldsymbol{\Theta}^{[0]}$ according to Subsection 7.4.1

SAGE Iteration
for $\ell = 1, 2, 3, \ldots$ **do**
– Compute the index d of the parameter subset using Eq. (7.32).
– E-step: Compute $Q(\boldsymbol{\Theta}_d|\boldsymbol{\Theta}^{[\ell-1]})$ according to Eq. (7.35)

$$Q(\boldsymbol{\Theta}_d|\boldsymbol{\Theta}^{[\ell-1]}) = -\ln|\boldsymbol{\Sigma}(\boldsymbol{\Theta}_d)| - \mathrm{tr}\left[\boldsymbol{\Sigma}(\boldsymbol{\Theta}_d)^{-1}\boldsymbol{\Sigma}_{\boldsymbol{X}_d}^{[\ell-1]}\right].$$

– M-step: Compute $\boldsymbol{\Theta}_d^{[\ell]}$ using Eq. (7.42)

$$\boldsymbol{\Theta}_d^{[\ell]} = \arg\max_{\boldsymbol{\Theta}_d} Q(\boldsymbol{\Theta}_d|\boldsymbol{\Theta}^{[\ell-1]}).$$

– Update the remaining parameter estimates as

$$\boldsymbol{\Theta}_{d'}^{[\ell]} = \boldsymbol{\Theta}_{d'}^{[\ell-1]}, \quad d' \neq d.$$

– Check if convergence is achieved according to the criterion in Subsection 7.4.1.
end for

Figure 7.8 Pseudo-code representation of the steps of the power spectrum estimator

semi-definite, once the estimates of some of the coupling coefficients are obtained, the other coupling coefficients may not be estimated within the full range of $[-1, 1]$. This consideration is included in the updating programs for the coupling coefficients.

7.4.1.4 Initialization Step

The Bartlett beamformer [Bartlett 1948] and the successive interference cancelation method are used to calculate the initial estimates of the component PSDs. For the dth component PSD, the following steps are performed:

1. The estimate $\widehat{\boldsymbol{\Sigma}}_{\boldsymbol{X}_d}$ of the sample covariance matrix for the dth component is calculated using Eq. (7.39) in the E-step. Note that for $d = 1$, $\widehat{\boldsymbol{\Sigma}}_{\boldsymbol{X}_1} = \widehat{\boldsymbol{\Sigma}}_{\boldsymbol{y}}$ is considered.
2. The bidirectional Bartlett spectra of $\hat{\boldsymbol{\Sigma}}_{\boldsymbol{X}_d}$ with specified delay $\tau \in [\tau_1, \ldots, \tau_N]$ and Doppler frequency are calculated, where N represents the total number of delay samples. These spectra are concatenated to construct an estimate of the six-dimensional bidirection-delay-Doppler frequency spectrum. Then, the maximum of this 6-D Bartlett spectrum is identified. The locations of this maximum in bidirection, delay, and Doppler frequency are used as the initial estimate of the center of gravity of the dth component PSD.
3. To compute the initial estimates of the spreads, the part of the 6-D Bartlett spectrum within the vicinity of the estimated center of gravity is used. We specify a threshold, say 3 dB lower than the maximum, to select the spectrum. The second central moments of the selected part

of the Bartlett spectrum are calculated and used as the initial estimates of the spreads of this component.

4. The average power estimate equals the power of the mode of the spectrum. The coupling coefficients are set to zero.

More specifically, the Bartlett beamformer is used to compute the estimate of the 6-D power spectrum of the channel. These six dimensions include the delay τ, the azimuth of departure ϕ_1, elevation of departure θ_1, azimuth of arrival ϕ_2, elevation of arrival θ_2, and Doppler frequency ν. If we use $\omega = (\tau, \phi_1, \theta_1, \phi_2, \theta_2, \nu)$ to denote the variable vector, the power-spectrum estimate can be obtained as

$$p(\omega) = \frac{\boldsymbol{c}(\boldsymbol{\omega})^{\mathrm{H}}\boldsymbol{\Sigma}_{\boldsymbol{y}}\boldsymbol{c}(\boldsymbol{\omega})}{\boldsymbol{c}(\boldsymbol{\omega})^{\mathrm{H}}\boldsymbol{c}(\boldsymbol{\omega})} \tag{7.43}$$

where $\boldsymbol{\Sigma}_{\boldsymbol{y}}$ denotes the measured sample covariance matrix.

The drawback of using the Bartlett beamformer to do the initialization of the parameter estimates is that the estimated power spectrum can be influenced by the non-isotropic radiation pattern of the antenna array. Furthermore, the sidelobes can be so significant that the estimated paths might be artifacts. The following methods can be used as alternatives to initialize the parameter estimate of individual components:

- Use the specular-path-model-based SAGE algorithm to estimate the parameters of specular paths. These estimates can be used as the center of gravity of individual dispersive components. The drawback of this method is that the estimated paths might belong to the same dispersive component. In such a case, one single dispersive component is estimated as multiple discrete specular paths. When initializing the center of gravity parameters with these specular path parameter estimates, the risk of splitting a dispersive component into multiple smaller dispersive components is high.
- Compute the complete Bartlett power spectrum estimate and use the locations of local maxima as the initial guess of the centers of gravity of the dispersive components. The advantage of this method is that the computational load is tractable as no iterations are necessary. However, due to the low resolution of the Bartlett power spectrum estimate, some of the local maxima may correspond to the sidelobes of the strong components.

7.4.1.5 Convergence Criterion

The convergence of the SAGE algorithm is considered to be achieved when the parameter updates are identical to the previous updates. Practically, when the differences between the estimates obtained from consecutive iterations are less than certain pre-defined values, the changes are considered negligible.

7.4.1.6 Discussion: Rank of the Sample Covariance Matrix

Parameter estimation for component-wise power spectra require that the observed sample covariance matrix have enough rank; it should be larger than the number of the parameters to be estimated. The rank of the observed sample covariance matrix depends on the number of the available independent observations of the channels. For example, in the case where 5 components are to be estimated and each component is characterized by 20 parameters, we will need 100 independent observations of the channel. Sometimes, it is unrealistic to collect a large number of independent snapshots for a WSSUS channel. Thus it might be problematic when using the PSD-SAGE algorithm to estimate the channel power spectrum accurately.

One possible method to relax the requirement for a large number of observations relies on the assumption that *the received signals observed in different delays within one snapshot can be considered as independent observations for different components.* This reduces the minimum number of independent observations required to the number of parameters for one component. Whether this assumption can be applied during estimation depends on the definition of the admissible data.

7.4.1.7 Discussion: Intrinsic Resolution of the Measurement System

The PSD-SAGE algorithm relies on the orthogonal spaces provided by the measurement equipment. When the observation timespan, space-aperture, and bandwidth are limited, the system is not able to resolve the slight differences among multiple components. This property also limits the resolution of the PSD-SAGE algorithm. First of all, we need a sufficient number of independent observations: data samples that provide enough information to enable estimation of multiple parameters of one component. Additionally, we need the system to have the capability to differentiate the details within the component, such that it is possible to estimate the spreads and the shape of the component. It is well-known that for the delay domain, the system is not able to separate two components with delay differences less than one delay sample. Thus, when the dispersive component is smaller than one delay sample, we will not be able to get accurate estimate of its size in delay. Similarly, in the angular domain, the intrinsic resolution of the antenna array is limited. For components larger than the intrinsic resolution, the PSD-SAGE algorithm may return a reasonable result. Simulations by Fleury et al. [1999] show that the SAGE algorithm can estimate closely-spaced components, when they differ in delay by $\frac{1}{5}$ or in angles by $\frac{1}{10}$ of the intrinsic resolution. This might be understood as the effect of interpolation; that is, the array responses measured with refined grids and the frequency responses interpolated at arbitrary frequencies. Verifying by simulation studies the empirical resolution of the PSD-SAGE algorithm in separating dispersive components will be the subject of future research.

7.4.2 Measurement Data Evaluation

Measurement data collected by using ETRI's rBECS the channel sounder is processed using the PSD-SAGE algorithm. Multiple snapshots of observations of the same wide-sense-stationary (WSS) channel are used to calculate the sample covariance matrix. Figure 7.9 depicts the average power-delay profile obtained from the 16 snapshots of data. It can be observed that the channel impulse response (CIR) becomes significant after $\tau = 250$ delay samples. The observation that the dominant part of CIR appears with large delay – beyond 250 delay samples in our case – can be caused by the positive initial delay that results when the receiver is not synchronized with the transmitter of the sounding system. The maximum power of the average PDP equals -11.61 dB. For the study here, we chose the

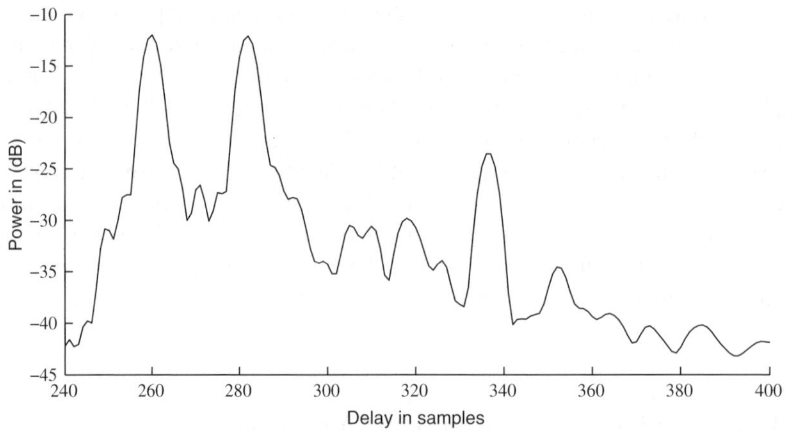

Figure 7.9 The power delay profile of the channel averaged over 16 observation snapshots

effective dynamic range for path power to be 12 dB, which results in the pre-set minimum power equal to −23.61 dB. The delay range of the PDP that yields the power within the range of $[-11.61, -23.61]$ dB is found to be [256, 287] delay samples. We use the CIRs within this range to evaluate the performance of the PSD-SAGE algorithm.

The PSD parameter estimator makes use of the sample covariance matrix, which has the dimension of $(N * M_{Tx} * M_{Rx}) \times (N * M_{Tx} * M_{Rx})$, where N, M_{Tx}, and M_{Rx} represent respectively the number of delay samples, the number of Tx antennas, and the number of Rx antennas. In practice, computers may have difficulties coping with such a large matrix, as it occupies a large amount of memory, so it is necessary to divide the delay range into multiple bins, called "delay bins". For individual bins, the number of dispersive components for estimation can be pre-determined and updated during the execution of the estimation algorithm. The sample covariance matrix used for estimating the PSD of one component can be calculated using the part of the CIRs observed in a specified delay bin. The bin width should be set in such a way that it is not too large with respect to the empirical time-consumption constraint, but not too small either, since observations within a narrow delay range may be just a part of a dispersive component. In our case, we specify that the delay-bin width is close to the span of the main lobe of the autocorrelation function of the transmitted pseudo-noise sequence; that is, the width of two chips. As the oversampling rate is 4 in the receiver, the delay-bin width is set to 8 delay samples.

Parameter estimation using the delay-bin concept is performed in our case as follows.

1. Select a delay range within which the power of the channel is larger than the maximum of the channel-delay profile subtracted from the predefined dynamic range. It may be that the multiple fragments of the PDP are selected, as illustrated in Figure 7.10, which depicts the parts of the PDP with power within the 12-dB dynamic range.
2. Check the maxima and minima of the selected part of the channel PDP. A delay bin can start at the delay giving a local minimum of the PDP and end at the next minimum of the PDP. For the breaks of the PDP, as shown in Figure 7.10, which are caused by truncating in Step 1, we consider these break points as the minima.

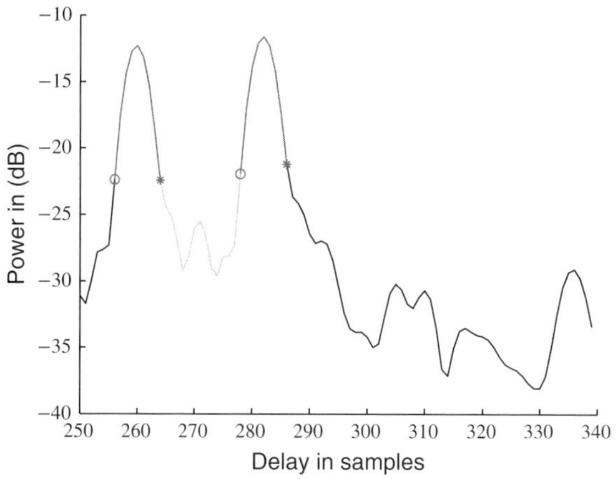

Figure 7.10 The power-delay profile of the channel

3. Resize the width of the bins. It is likely that the bins determined in Step 2 contain only one or two delay samples. In such a case, it is necessary to enlarge the bin width up to, say, 8 delay samples. This is so because only a sufficiently large sample covariance matrix can provide admissible data for estimating the parameters of a PSD.

Notice that the operations in the E-step should avoid any influence from the bin-search method. The covariance matrices for individual components can be reconstructed for small delay-bin width, say up to 8 delay samples. However, the reconstructed covariance matrix of all signals $\hat{\boldsymbol{\Sigma}}_y(\boldsymbol{\Theta})$ needs to be computed by considering the contributions of all components, not just those estimated within the same delay bin.

Table 7.1 Final estimates of the parameters of dispersive components

Comp.	$\bar{\tau}$ (Delay samples)	$\bar{\phi}_1$ (°)	$\bar{\phi}_2$ (°)	$\bar{\theta}_1$ (°)	$\bar{\theta}_2$ (°)
1	259.81	356.00	359.38	−7.50	−6.00
2	260.75	149.25	1.75	−12.00	−6.75
3	281.44	334.75	24.50	−2.00	8.25
4	282.94	334.00	27.00	−28.00	9.00
Comp.	σ_τ (Delay samples)	σ_{ϕ_1} (°)	σ_{ϕ_2} (°)	σ_{θ_1} (°)	σ_{θ_2} (°)
1	0.50	0.72	1.41	0.81	0.71
2	0.50	0.78	6.94	6.00	4.25
3	0.50	3.00	0.99	2.87	2.75
4	0.67	0.93	0.80	6.25	0.50
Comp.	p_{vv}	p_{vh}	p_{hv}	p_{hh}	
1	$3.72e-001$	$1.54e-020$	$7.87e-002$	$3.10e-001$	
2	$1.56e-020$	$7.62e-003$	$1.27e-017$	$2.57e-003$	
3	$2.60e-019$	$2.96e-020$	$2.18e-018$	$1.51e-001$	
4	$1.87e-020$	$3.07e-020$	$7.27e-020$	$1.08e-001$	

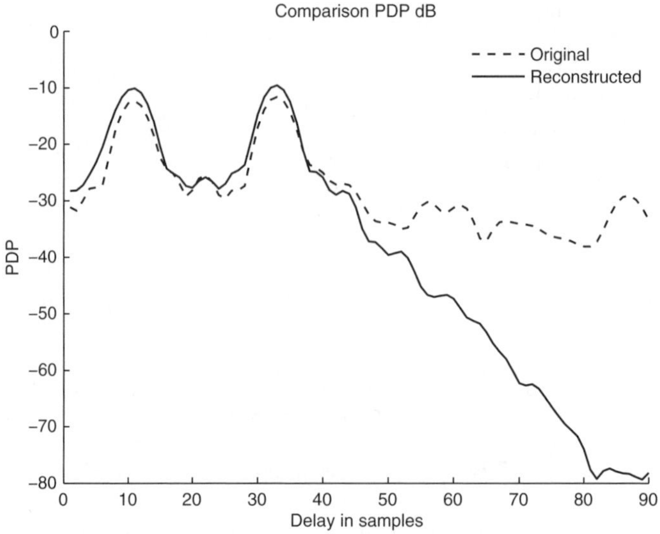

Figure 7.11 Power delay profiles computed from original data and reconstructed using the estimated parameter estimates

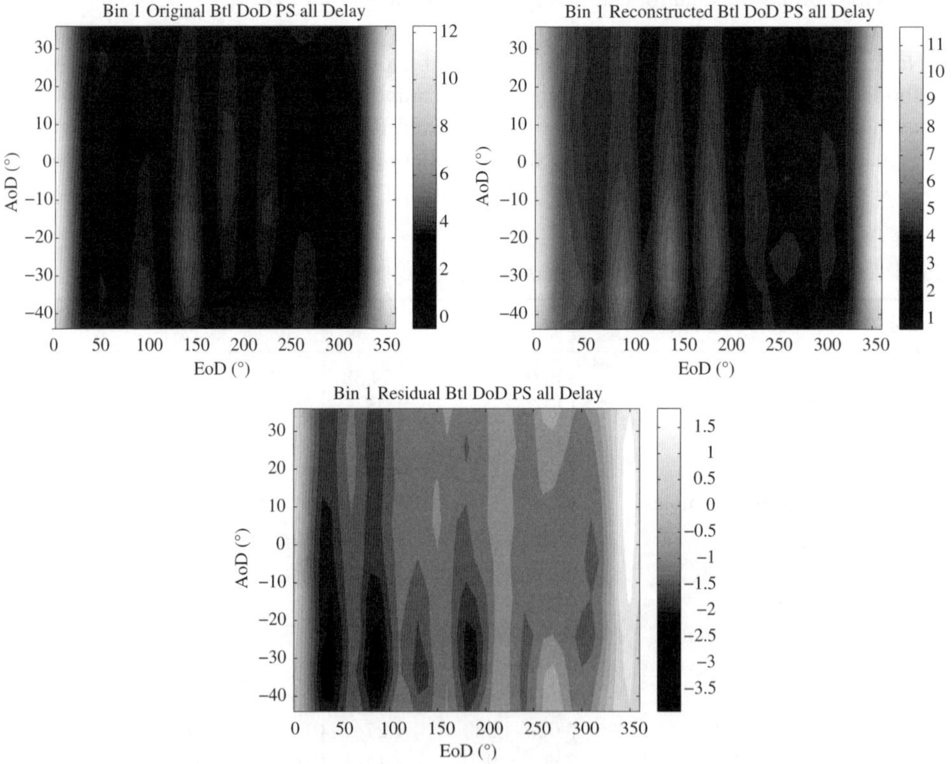

Figure 7.12 Comparison of the DoD Bartlett power spectra for the delay bin no. 1

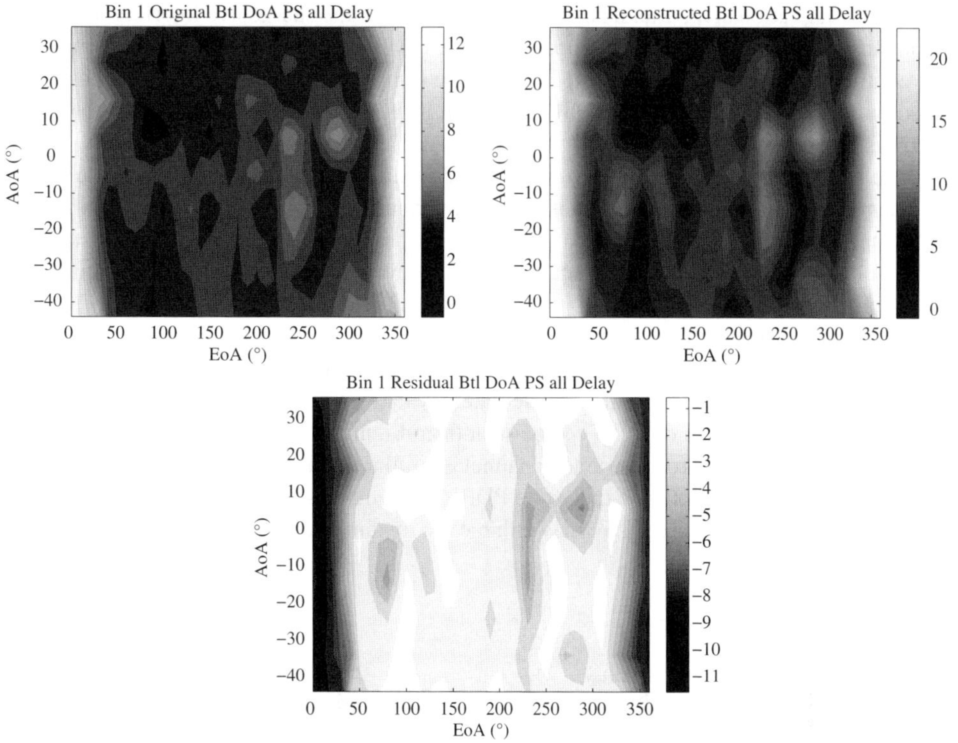

Figure 7.13 Comparison of the DoA Bartlett power spectra for delay bin 1

For the preliminary study here, the PSD-SAGE algorithm is applied to estimate two dispersive components per delay bin. The first bin has relative delay range of $[1, 8]$ delay samples. The Bartlett beamformer is used in the initialization step to estimate the center of gravity of the PSDs of multiple dispersive components. For each delay, the DoD power spectrum is calculated. The maximum of these delay-DoD power spectra is identified. The locations of the maximum in delay and DoD are considered to be the estimates of the center-of-gravity delay and DoD of the PSD for the first dominant dispersive components in the delay bin. With the estimated DoD estimate, the Bartlett beamformer is applied again to compute the DoA spectrum for the delay. Then, the DoA estimate for the same dispersive component is identified by finding the maximum of the DoA power spectrum. Here, since the environment and the measurement equipment were kept fixed, the Doppler frequencies of channel components are assumed to be zero and known in advance. Following the estimation of the center-of-gravity delay, DoD, and DoA for a dispersive component, the parameters characterizing the shape of the PSD are extracted. For simplicity, the correlation coefficients of the PSD model are not estimated in this evaluation. Thus the PSDs of the estimated dispersive components in this case are assumed to be non-tilted in the parameter space.

In the initialization step, after a component is estimated, the successive interference cancellation method is used to obtain the expected sample covariance matrix for the next component. The contribution of an estimated components to the sample covariance matrix is reconstructed based on the generic model of the PSD as shown in Eq. (7.36), and the residual covariance matrix is computed using Eq. (7.39).

In the PSD-SAGE algorithm proposed, the Euclidean distance between the sample covariance matrix and the model-based reconstructed covariance matrix is calculated and minimized with respect to the candidate PSD parameters. The properties of the Euclidean distance as a function of a candidate parameter can be used to evaluate the feasibility or ambiguity of estimating the parameter.

Table 7.1 reports the final estimates of the parameters of the dispersive components after 10 iterations. Figure 7.11 depicts the PDP of the channel computed from the original channel and the reconstructed channel. The latter is calculated by taking the average of the diagonal of the covariance matrix for each delay. It can be seen from Figure 7.11 that in the ranges where the delay bins are located, the reconstructed channel is similar to the original channel. We can also observe the differennces between these two PDPs, the main reason for which is that delay bins are not allocated for these areas. We may enlarge the dynamic range in order to get a complete overview of the channel.

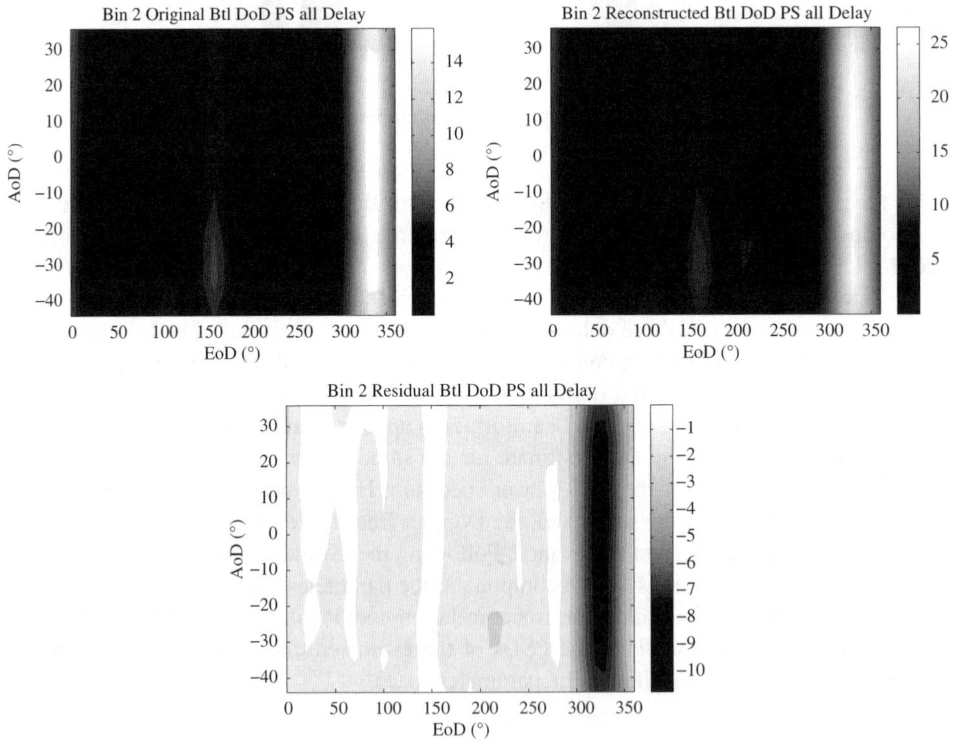

Figure 7.14 Comparison of the DoD Bartlett power spectra for delay bin 2

Figures 7.12 and 7.13 depict, respectively, the DoD and DoA power spectra for the first delay bin. Each figure shows from left to right the spectrum calculated from the original observation data, the reconstructed sample covariance matrix calculated by using the estimated parameters, and the residual signals - the difference between the sample and the reconstructed covariance matrices. It can be observed from Figures 7.12 and 7.13 that the height of the residual spectra are low compared with the originals. This indicates that the PSD estimates for these bins is accurate. Figure 7.14 and 7.15 depicts the DoA and DoD spectra for delay bin 2. The observations are similar to those obtained for the first delay bin.

Figure 7.16 depicts the constellation of the dispersive components estimated by the PSD-SAGE algorithm in the AoA and AoD domains. The contours represent the −3-dB power spectral height compared to the maximum of the component power spectrum. It can be observed from Figure 7.16 that the dispersive components are quite concentrated. The four dispersive components have their AoA spreads equal to 0.72°, 0.78°, 3.00°and 0.93°and AoD spreads equal to 1.41°, 6.94°, 0.99°and 0.80°, respectively. Most of these azimuth spreads are around 1°. These small azimuth spreads may indicate that the components are quite concentrated in the azimuth domain.

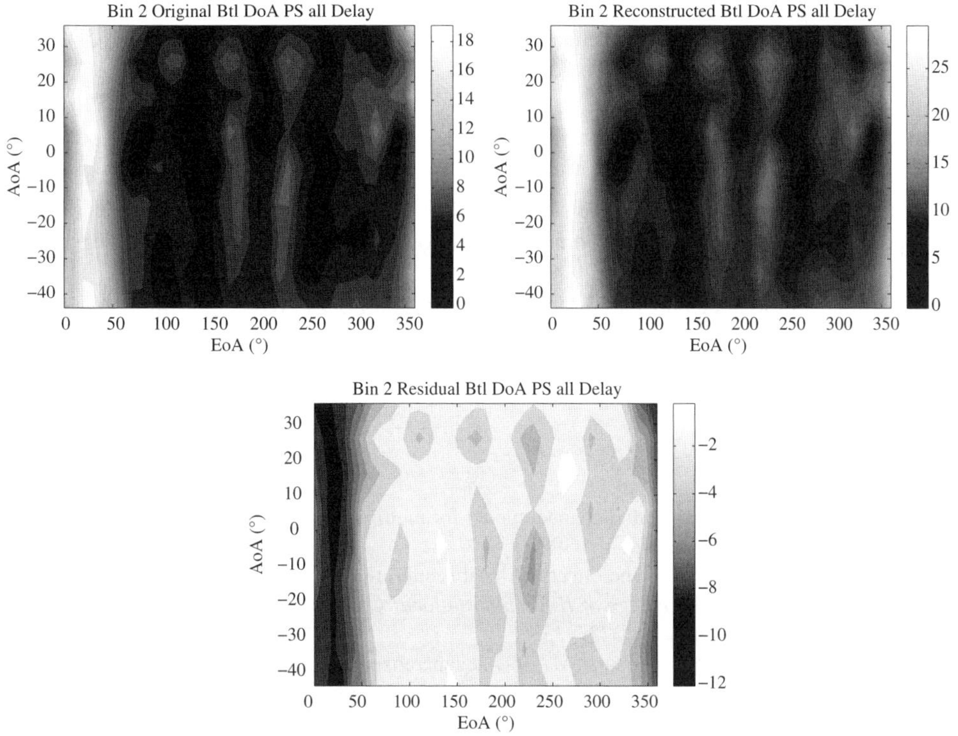

Figure 7.15 Comparison of the DoA Bartlett power spectra for delay bin 2

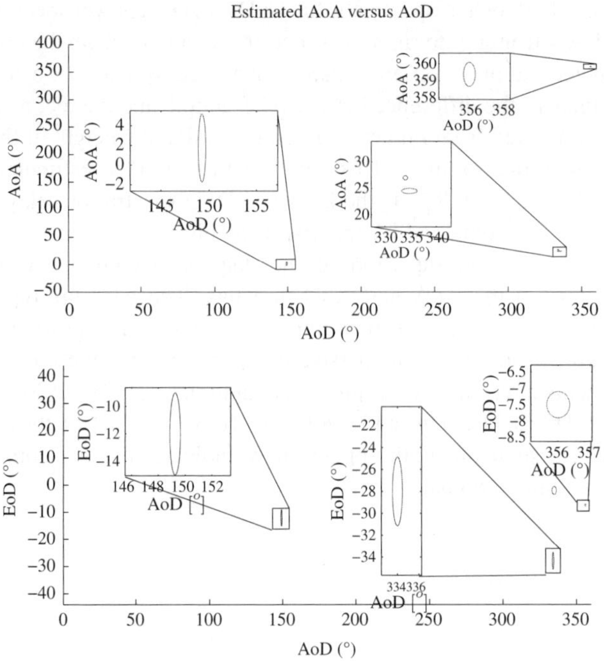

Figure 7.16 Constellation of the dispersive components in AoA–AoD

In summary, these experimental evaluation results demonstrated that:

- The delay-bin-searching approach is feasible for processing measurement data. This approach can be used to reduce computational complexity without significant degradation of the performance.
- In contrast to the specular-path SAGE algorithm, the PSD-SAGE algorithm can estimate a small number of dispersive components per bin. The power spectra of the reconstructed channels based on the estimated parameters are consistent with the measured power spectra.

Bibliography

3GPP 2007 TR25.996 v7.0.0: Spatial channel model for multiple input multiple output (MIMO) simulations (release 7). Technical report, 3GPP.

Asztély D and Ottersten B 1998 The effects of local scattering on direction of arrival estimation with MUSIC and ESPRIT *Proceedings of IEEE International Conference on Acoustics, Speech, Signal Processing (ICASSP)*, vol. 6, pp. 3333–3336.

Asztély D, Ottersten B and Swindlehurst AL 1997 A generalized array manifold model for local scattering in wireless communications *Proceedings of IEEE International Conference on Acoustics, Speech and Signal Processing (ICASSP)*, vol. 5, pp. 4021–4024.

Bartlett M 1948 Smoothing periodograms from time series with continuous spectra. *Nature* **161**, 686–687.

Bengtsson M 1999 Antenna array signal processing for high rank data models. PhD thesis, Royal Institute of Technology, Stockholm, Sweden.

Bengtsson M and Ottersten B 2000 Low-complexity estimators for distributed sources. *IEEE Transactions on Signal Processing* **48**(8), 2185–2194.

Bengtsson M and Völcker B 2001 On the estimation of azimuth distributions and azimuth spectra *Proceedings of the 54th IEEE Vehicular Technology Conference (VTC2001-Fall)*, vol. 3, pp. 1612–1615, Atlantic City, USA.

Besson O and Stoica P 1999 Decoupled estimation of DoA and angular spread for spatially distributed sources. *IEEE Transactions on Signal Processing* **49**, 1872–1882.

Betlehem T, Abhayapala TD and Lamahewa TA 2006 Space-time MIMO channel modelling using angular power distributions *Proceedings of the 7th Australian Communications Theory Workshop*, pp. 165–170, Perth, Australia.

Böhme J 1984 Estimation of source parameters by maximum likelihood and nonlinear regression *Proceedings of IEEE International Conference on Acoustics, Speech, and Signal Processing (ICASSP)*, pp. 7.3.1–7.3.4.

Bresler, Y.; Macovski A 1986 Exact maximum likelihood parameter estimation of superimposed exponential signals in noise. *IEEE Transactions on Acoustics, Speech, and Signal Processing [see also IEEE Transactions on Signal Processing]* **34**, 1081–1089.

Christensen MG, Jensen SH, Andersen S and Jakobsson A 2004 Subspace-based fundamental frequency estimation *Proceedings of the 12th European Signal Processing*.

Drmac Z 2000 On principal angles between subspaces of Euclidean space. *SIAM Journal on Matrix Analysis and Applications* **22**(1), 173–194.

Edelman A, Arias T and Smith S 1998 The geometry of algorithms with orthogonality constraints. *SIAM Journal on Matrix Analysis and Applications* **20**(2), 303–353.

Fessler JA and Hero AO 1994 Space-alternating generalized expectation-maximization algorithm. *IEEE Transactions on Signal Processing* **42**(10), 2664–2677.

Fleury B 2000 First- and second-order characterization of direction dispersion and space selectivity in the radio channel. *IEEE Transactions on Information Theory* **46**(6), 2027–2044.

Fleury BH, Tschudin M, Heddergott R, Dahlhaus D and Pedersen KL 1999 Channel parameter estimation in mobile radio environments using the SAGE algorithm. *IEEE Journal on Selected Areas in Communications* **17**(3), 434–450.

Golub GH and Loan CFV 1996 *Matrix Computations* 3rd edn. The John Hopkins University Press.

Han Y, Wang J and Song X 2006 A low complexity robust parameter estimator for distributed source. *TENCON 2006. 2006 IEEE Region 10 Conference* pp. 1–4.

Jaffer AG 1988 Maximum likelihood direction finding of stochastic sources: a separable solution *Proceedings of IEEE International Conference on Acoustics, Speech and Signal Processing (ICASSP)*, vol. 5, pp. 2893–2896.

Jaynes E 2003 *Probability Theory*. Cambridge University Press.

Jeong J, Sakaguchi K, Takada J and Araki K 2001 Performance analysis of MUSIC and ESPRIT using extended array mode vector in multiple scattering environment *Proceedings of The Fourth International Symposium on Wireless Personal Multimeda Communications (WPMC)*, vol. 1, pp. 277–282.

Johnson D and Dudgeon DE 1993 *Adaptive Signal Processing: Concepts and Techniques*. PTR Prentice Hall.

Kent JT 1982 The Fisher–Bingham distribution on the sphere. *Journal of the Royal Statistical Society, Series B (Methodological)* **44**, 71–80.

Krim H and Viberg M 1996 Two decades of array signal processing research: the parametric approach. *IEEE Transactions on Signal Processing* **13**, 67–94.

Lee D and Xu C 1997 Mechanical antenna downtilt and its impact on system design *Vehicular Technology Conference, 1997, IEEE 47th*, vol. 2, pp. 447–451.

Mardia K, Kent J and Bibby J 2003 *Multivariate Analysis*. Academic Press.

Mardia KV 1975 Statistics of directional data. *Journal of the Royal Statistical Society. Series B (Methodological)* **37**, 349–393.

Meng Y, Stoica P and Wong K 1996 Estimation of the directions of arrival of spatially dispersed signals in array processing. *IEE Proceedings Radar, Sonar and Navigation* **143**(1), 1–9.

Minka TP 2000 Old and new matrix algebra useful for statistics. Technical report, MIT.

Qiang L and Zhishun L 2007 A new independent distributed sources model and doa estimator. *Communications, Circuits and Systems, 2007. ICCCAS 2007. International Conference on* pp. 701–704.

Ribeiro CB, Ollila E and Koivunen V 2004 Stochastic maximum likelihood method for propagation parameter estimation *Proceedings of the 15th IEEE International Symposium on Personal, Indoor and Mobile Radio Communications (PIMRC'06)*, vol. 3, pp. 1839–1843, Helsinki, Finland.

Ribeiro CB, Richter A and Koivunen V 2005 Stochastic maximum likelihood estimation of angle- and delay-domain propagation parameters *Proceedings of the 16th IEEE International Symposium on Personal, Indoor and Mobile Radio Communications (PIMRC'06)*, vol. 1, pp. 624–628, Berlin, Germany.

Roy R and Kailath T 1989 ESPRIT – estimation of signal parameters via rotational invariance techniques. *IEEE Transactions on Acoustics, Speech, and Signal Processing* **37**(7), 984–995.

Schmidt RO 1986 Multiple emitter location and signal parameter estimation. *IEEE Transactions on Antennas and Propagation* **AP-34**(3), 276–280.

Shahbazpanahi S, Valaee S and Bastani M 2001 Distributed source localization using ESPRIT algorithm. *IEEE Transactions on Signal Processing* **49**(10), 2169–2178.

Shahbazpanahi, S.; Valaee SGA 2004 A covariance fitting approach to parametric localization of multiple incoherently distributed sources. *IEEE Transactions on Signal Processing* **52**, 592–600.

Shanmugan KS and Breipohl AM 1988 *Random Signals: Detection, Estimation and Data Analysis*. John Wiley & Sons.

Sieskul, B.T.; Jitapunkul S 2006 An asymptotic maximum likelihood for estimating the nominal angle of a spatially distributed source. *International Journal of Electronics and Communications AEUE* **60**, 279–289.

Souden M, Affes S and Benesty J 2008 A two-stage approach to estimate the angles of arrival and the angular spreads of locally scattered sources. *Signal Processing, IEEE Transactions on [see also Acoustics, Speech, and Signal Processing, IEEE Transactions on]* **56**(5), 1968–1983.

Tan C, Beach M and Nix A 2003 Enhanced-SAGE algorithm for use in distributed-source environments. *Electronics Letters* **39**(8), 697–698.

Trump T and Ottersten B 1996 Estimation of nominal direction of arrival and angular spread using an array of sensors. *Signal Processing* **50**, 57–69.

Valaee S, Champagne B and Kabal P 1995 Parametric localization of distributed sources. *IEEE Transactions on Signal Processing* **43**, 2144–2153.

Viberg M and Ottersten B 1991 Sensor array processing based on subspace fitting. *IEEE Transactions on Signal Processing* **39**(39), 1110–1121.

Wang Y and Zoubir A 2007 Some new techniques of localization of spatially distributed sources. *Signals, Systems and Computers, 2007. ACSSC 2007. Conference Record of the Forty-First Asilomar Conference on* pp. 1807–1811.

Yin X and Fleury BH 2005 Nominal direction estimation for slightly distributed scatterers using the SAGE algorithm *Proc. of the IEEE AP-S International Symposium and USNC/URSI national radio science meeting, Washington DC, US*, pp. 418–421.

Yin X, Liu L, Nielsen D, Czink N and Fleury BH 2007a Characterization of the azimuth-elevation power spectrum of individual path components *Proceedings of the International ITG/IEEE Workshop on Smart Antennas (WSA 2007)*, pp. 1–5, Vienna, Austria.

Yin X, Liu L, Nielsen D, Pedersen T and Fleury B 2007b A SAGE algorithm for estimation of the direction power spectrum of individual path components *Proceedings of Global Telecommunications Conference, 2007. GLOBECOM '07. IEEE*, pp. 3024–3028.

Yin X, Liu L, Pedersen T, Nielsen D and Fleury B 2008 Modeling and estimation of the direction-delay power spectrum of the propagation channel *Proceedings of the 3rd International Symposium on Communications, Control and Signal Processing. ISCCSP 2008.*, pp. 225–230.

Yin X, Pedersen T, Czink N and Fleury B 2006a Parametric characterization and estimation of bi-azimuth and delay dispersion of individual path components *Antennas and Propagation, 2006. EuCAP 2006. First European Conference on*, pp. 1–8.

Yin X, Pedersen T, Czink N and Fleury B 2006b Parametric characterization and estimation of bi-azimuth dispersion path components *Proceedings of IEEE 7th Workshop on Signal Processing Advances in Wireless Communications. SPAWC'06.*, pp. 1–6, Rome, Italy.

Yin X, Zuo Q, Zhong Z and Lu S 2011 Delay-doppler frequency power spectrum estimation for vehicular propagation channels *Antennas and Propagation (EUCAP), Proceedings of the 5th European Conference on*, pp. 3434–3438.

Zoubir A, Wang Y and Charge P 2007 Spatially distributed sources localization with a subspace based estimator without eigendecomposition. *Acoustics, Speech and Signal Processing, 2007. ICASSP 2007. IEEE International Conference on* **2**, 1085–1088.

8

Measurement-based Statistical Channel Modeling

Compared with theoretical and simulation-based channel-modeling methods, measurement-based stochastic channel modeling has the advantage that the channel observations acquired during the measurements are close to the channels experienced by user equipment in realistic application scenarios. However, a statistically sound measurement-based stochastic model is obtained only where certain conditions are met. For example, the measurement sites selected must be representative of the typical propagation scenarios in real applications. In order to build a stochastic model covering all possible random statuses, multiple measurement sites should be considered, with ergodicity satisfied for all kinds of phenomena. Furthermore, compared with the communication equipment, the transmitting and receiving devices used in the measurements are usually configured with more advanced setups, such as wider baseband-signal bandwidth and antenna arrays with large apertures that allow extraction of omnidirectional channel characteristics. The channel components' parameters should be estimated with the influence of the underlying equipment and measurement specifications mitigated. It is necessary to identify and exclude erroneous estimation results, such as ghost paths, from the channel observation set used for stochastic channel modeling. Furthermore, in order to reduce the complexity of the models established, consideration should be given to approaches such as curve-fitting, using existing or standard functions with analytical expressions with as few parameters as possible, or clustering of components.

In this chapter, we first introduce the general channel modeling procedure in Section 8.1. Section 8.2 elaborates the clustering algorithms applied for grouping specular paths. Section 8.3 describes the approaches applicable for separating the observations into multiple segments with different stationarity. Finally, in Section 8.4, some preliminary thoughts and modeling results for the relay channel and cooperative multi-point (CoMP) channel models are presented and discussed.

Propagation Channel Characterization, Parameter Estimation and Modelling for Wireless Communications, First Edition. Xuefeng Yin and Xiang Cheng.

8.1 General Modeling Procedures

Measurement-based stochastic channel modeling is usually conducted following the scheme presented in Figure 8.1, which is similar to that originally introduced in [Kyösti et al. 2007].

8.1.1 Channel Measurement

The first block in the diagram is channel measurement, a step usually called "channel sounding", which is carried out to collect the signals at a receiver site when a transmitter emits signals carried by electromagnetic waves. The objectives of channel measurement is to investigate the mechanism of wave propagation in specified environments or media, and eventually to specify the values of parameters of mathematical models in predefined forms. These data can also be used to assess the applicability of established models to explain certain propagation characteristics.

The channel measurement block in Figure 8.1 has three parts: campaign planning, equipment calibration, and measurement itself. The planning of a campaign determines the necessary configuration and specification of the equipment depending on the purpose of the campaign. Calibration is needed to determine whether the signal transmission power is high enough to obtain a sufficiently large dynamic range of received signals, or to support coverage within the area of interest for measurement. It is also necessary to fully understand the behavior of measurement equipment itself – the inherent system response and its stability for example – since the response of the front-end transmit-and-receive channel and the synchronization between the oscillators in the Tx and Rx are all involved in the measured channel observations. It is necessary to isolate or de-embed the impact of the measurement equipment from the original characteristics of the received signals, which are in fact a mixture of both system and propagation responses.

Calibration can be performed on different aspects of the system depending on the objectives of the measurements. When measurements are conducted with the aim of extracting the

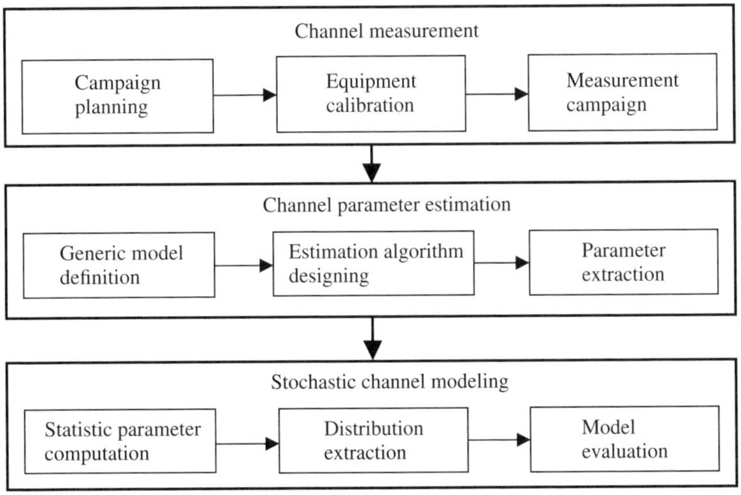

Figure 8.1 The channel modeling approach

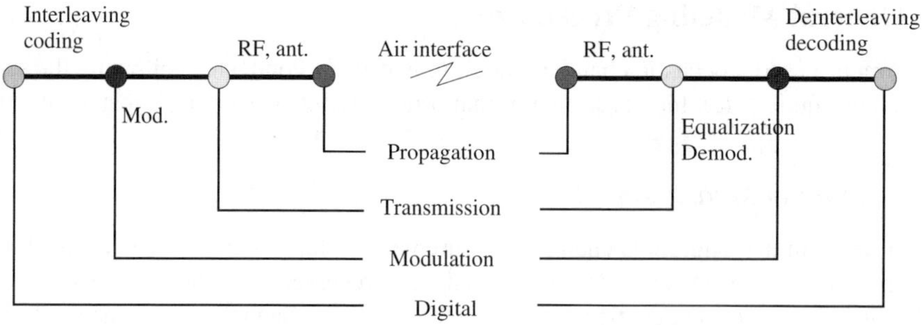

Figure 8.2 The components in an RF channel

wideband characteristics that are represented by the dispersion behavior of the propagation paths, it is necessary to determine several responses:

8.1.1.1 Responses of the Transmitting and Receiving Antennas

It is expected that there will be a complete antenna radiation pattern in the whole sphere centered at the antenna. Only the patterns in the E- and H-plane are insufficient. Furthermore, both the magnitude and the phase of the antenna response are required, since the phase differences among the output of antennas in an array are used to estimate the directional parameters of paths. Moreover, if the sounding signal is transmitted or received through antenna arrays, the responses of the antennas need to include the impact of the coupling effects among the array elements.

8.1.1.2 Responses of the Front-end Radiofrequency Chains in Transmitter and Receiver

When a switch is used to activate the antennas to transmit or receive signals, the radio frequency (RF) chain needs to be calibrated with the switch included. In addition, the cables used to connect the RF chain to the ports of the switch ports should also be considered as parts of the RF chain for calibration.

When parallel sounding is conducted using multiple-antenna arrays, it is necessary to measure the responses of multiple front-end chains. The synchronization status among the multiple chains should also be considered.

Figure 8.2 depicts the components included in the RF chain. It can be seen that the RF channel actually includes many parts, with specific responses. It is necessary to mitigate the impact of the system responses on the estimation of the channel characteristics.

8.1.2 Channel Parameter Estimation

In the second block of Figure 8.1, channel-parameter estimation is carried out. We may split the channel parameters into three classes:

- so-called narrowband parameters, including fading coefficients

- the wideband profiles of the channel, including the channel impulse responses (CIRs) in the time domain and the spatial domain
- the high-resolution channel parameters that are defined for individual components in the channel responses.

The channel parameter estimation block also consists of three steps: generic model selection, algorithm selection, and parameter extraction.

Generic models include the models described in Chapter 4. A popular model is the specular path model, which describes the CIR as the superposition of many specular path components. For time-variant cases, it is necessary to use a path-evolving model. When the number of data samples is small, in order to reduce the errors when extracting the statistical parameters, the power spectral density model, which describes the first and second moments of the channels may be considered.

In the "Estimation algorithm design" step, methods or algorithms for extracting the model parameters are chosen. These algorithms can be selected from those introduced in Chapters 7 and 8, taking into account the total time expenditure on data processing, the dimensions of the channel parameters, and the modeling objectives. At present, the most popular methods are

- the non-parametric approach based on the beamforming technique for computing channel power spectral density
- parametric methods based on, for example, the SAGE or RiMAX principles, for extracting the parameters of generic models, including power spectrum density generic models.

Since the estimation algorithm processes multiple snapshots of measurement data, this step is very time-consuming. Figure 8.3 illustrates an example of the power delay profiles of channels observed from a measurement conducted with carrier frequency of 28 GHz and bandwidth of 500 MHz, and the multipath components estimated by using the SAGE algorithm. An Rx virtual antenna array was used to collect data organized by rotating a directional antenna in azimuth with a total of 36 steps. The half-power beamwidth of the applied directional antenna was 10°. It can be seen that the multipath components estimated can illustrate the dispersion of the channel in more detail than power delay profiles. The paths are also seen to have a clustering effect in the azimuth of arrival and delay domains.

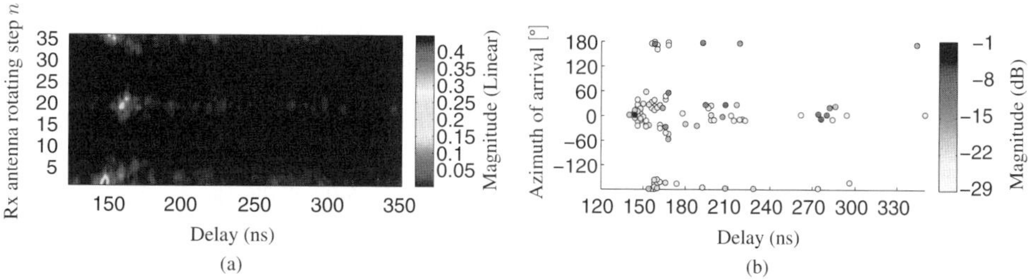

Figure 8.3 (a) Concatenated power delay profiles observed in a LoS scenario from channel measurements at 28 GHz; (b) the scatter plot of multipath components estimated by using the SAGE algorithm. A virtual array formed by rotating an Rx horn antenna in 36 steps along a circle was used to collect the data. The Tx horn antenna was fixed

8.1.3 Stochastic Channel Modeling

In the third block of the flowchart, stochastic model parameters are extracted based on parameters estimated for channel components obtained for instantaneous measurement snapshots. The stochastic model parameters include both large-scale and small-scale parameters. The large-scale parameters include:

- path loss
- K-factor
- correlation matrix of a MIMO channel
- polarization-relevant parameters, such as cross-polarization discrimination and co-polarization power ratio
- composite channel parameters, such as delay spread, angular spreads, Doppler frequency spread
- delay scaling factor describing the concentration of channel components.

The small-scale parameters are those describing the behavior of individual paths, or clusters of paths that have similar values for their geometrical parameters: delay, Doppler frequency and direction. Further discussion of clustering approaches of paths can be found in Section 8.2.

The small-scale channel parameters usually contain intra-cluster and inter-cluster parameters. In different standard channel models, cluster parameters are referred to by slightly different names. An example of the intra-cluster parameters defined in the IEEE 802.11ad standard is illustrated in Figure 8.4 [Erceg et al. 2004], which is an extension of the original Saleh–Valenzuela channel model [Saleh and Valenzuela 1987]. It can be observed from Figure 8.4 that a cluster has a central ray that coincides with the ray exhibiting the largest magnitude in the cluster. The other rays can be called pre- and post-cursor rays, and are characterized by their parameters, including the average power of the rays, the number of rays, and the ray arrival rate.

For clusters consisting of multipath components that have parameters distributed in multi-dimensions, the intra-cluster parameters include the spreads of clusters in multiple dimensions

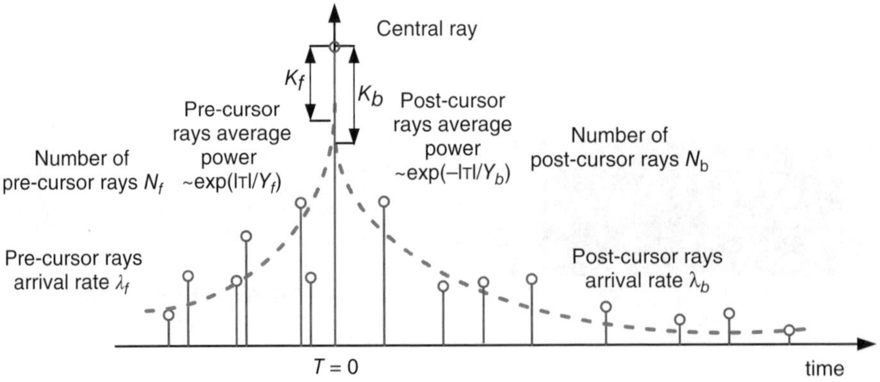

Figure 8.4 A time-domain cluster model proposed in the IEEE 802.11ad standards [Erceg et al. 2004]

and the inter-dependence of those spreads [Czink et al. 2006]. Furthermore, in time-variant cases, clusters can be used to characterize the evolution of channels [Czink et al. 2007]. The birth and death procedure and the dependence among parameters characterizing different clusters are also of interest for modeling [Czink et al. 2007, Yin et al. 2015]. Moreover, as multi-link channel characterization has become more and more popular recently, parameters such as the correlation coefficients of the large-scale and small-scale channel parameters of two or more links have become important in modeling studies [Park et al. 2012, Yin et al. 2012b,c].

In the modeling procedure, three main steps are usually considered and performed sequentially: model parameter calculation, probability distribution extraction, and model evaluation.

In the first step, the parameters of interest are computed for multiple channel snapshots. Consequently, a sufficient number of realizations for the parameters are obtained. It is worth mentioning that an individual channel snapshot may result in multiple instantaneous measurements. It is therefore necessary to apply "data-segmentation" techniques to determine the measurements belonging to the same group of observations for a wide-sense-stationary channel.

In the second step, the empirical occurrence frequency curves of the parameters are extracted, and functions with existing analytical expressions can be used to fit the empirical graphs. It is always important to evaluate whether the fitted PDFs are consistent with their empirical counterparts. Possible evaluation methods include the Akaike Information Criterion [Akaike 1974], the minimum description length [Rissanen 1978], the Kolmogorov–Smirnov test [Peacock 1983], and the Kullback-Leibler distance [Yin et al. 2012d]. Normally we try to avoid using compound PDFs to fit the real data. In order to achieve this, it is necessary to determine whether the snapshots or observations are collected in the same scenario and same type of environment.

In the last step, the applicability of the resulting models are evaluated by checking the accuracy of the prediction of the models with observations taken in the similar scenarios to those covered by the models. After these steps, the output of the modeling procedure can be considered a stochastic channel model. Examples of such models with well-defined application scenarios and parameterizations are:

- the standard channel models, such as 3GPP TR25.996 [3GPP 2007]
- the WINNER II enhanced spatial channel models [Kyösti et al. 2007]
- the IMT-Advanced channel models [ITU 2008].

8.2 Clustering Algorithm based on Specular-path Models

As described in Section 8.1, an important step of modeling wideband channels is to group multipath components into clusters based on their parameter estimates. The statistics of the clusters, such as the center of gravity and the spreads, are used to construct stochastic channel models. We first introduce the widely used clustering approach for channel modeling in Subsection 8.2.1. Then, the details of a clustering algorithm recently proposed for grouping multipath components are provided in Subsection 8.2.2. Subsection 8.2.2.3 illustrates some clustering results based on measurement data.

8.2.1 *Stochastic Cluster-based Channel Modeling*

8.2.1.1 Definition of Cluster and its usage in Channel Modeling

Stochastic cluster-based channel modeling relies on decomposing a channel into multiple clusters of multipath components. The cluster's parameters are considered to be random variables, with PDFs estimated from measurement data. According to Czink [2007], a cluster of paths is a group of paths that have similar parameters. A large number of paths in a channel can be separated into a certain number of clusters by using clustering algorithms.

The concept of a "cluster of multipath components" was coined a long time ago. The cluster delay line model is a typical channel model that makes use of clusters, which represent groups of channel components dispersive in the delay. The 3GPP spatial channel model extends the clusters from purely the delay domain to other dimensions, particularly the spatial domains of direction of departure and of arrival.

The WINNER channel modeling process also makes use of clusters. However, the cluster models are not consistent across different scenarios. For example, clusters that are spread in both the delay and angular domains are used for modeling small-scale propagation characteristics in the following scenarios:

- A1 Indoor office
- A2 Indoor-to-Outdoor
- B1 Urban microcell
- D1 Rural macro-cell.

However, for the "C1 Suburban macro-cell" scenario, the "zero-delay-spread cluster" is used, and this is only dispersive in the angular domain. For scenarios such as "C2 Urban macro-cell", taps instead of clusters are used. Furthermore, clusters were not considered in the channel models constructed for the following scenarios:

- B2, B3, B4 Outdoor to indoor
- B5 Stationary feeder
- C3 Bad urban macro-cell.

8.2.1.2 Dense Multipath Component

A random-cluster modeling (RCM) approach was introduced by Czink [2007]. In contrast to classical cluster-based channel models, the models established using RCM include the description of dense multipath components (DMCs). The existence of DMCs can influence channel diversity, so it is important to include them in the channel modeling process. In the Czink paper, the DMC is considered to exist only in the delay domain. The characteristics of the DMC in the directional domain and in the Doppler frequency domain were not discussed. Obviously, the prerequisite for including spatial-domain DMCs into either the geometry-based stochastic models [Kyösti et al. 2007] or RCMs [Czink 2007] is to find effective approaches to parametric characterization of DMCs in the spatial domains. To our best knowledge, no stable methods are available for extracting spatial DMCs from measurement data.

8.2.1.3 Cluster Lifetime

According to Czink [2007], the RCM uses two time bases to define the lifetime of a cluster: the channel sampling interval and the cluster lifetime interval. The latter can be an integer multiple of the former, since newly born clusters may fade in and dying clusters need to fade out smoothly over time.

8.2.1.4 Clustering Methods

Four methods can be used for clustering paths:

1. hierarchical tree clustering
2. KPowerMeans clustering
3. Gaussian-mixture clustering
4. estimating clusters directly in the impulse response.

Gaussian mixture clustering superimposes a specific structure on the clusters. The cluster parameters are assumed to be Gaussian distributed, showing a higher path density in the centre than on the periphery. For MPC clustering, this approach can be easily combined using path powers. Unfortunately, it turned out that this clustering method has its shortcomings when trying to track clusters, since the clustering results are quite unstable.

Statistical parameter estimation for the power spectra of dispersive paths, as described in Section 8.4, can be considered part of method 4 in the list above. Here, dispersive paths can be viewed as consisting of many non-separable paths, similar to the definition of a cluster.

The RCM scheme proposed by Czink [2007] also involves the evaluation of the channel estimation result through four approaches: mutual information, channel diversity, Demmel condition number of the MIMO channel metrics, and environment characterization metric comparing directly the discrete propagation paths in the channel.

8.2.2 Clustering Algorithms based on Multipath Component Distance

8.2.2.1 Definition of Multipath Component Distance

Multipath component distance (MCD) quantifies the difference between two paths [Czink 2007], scaling differences between the path parameters in different dimensions so that they have a common unit. MCD was first introduced by Steinbauer et al. [2002] to quantify the complete multipath separation of the radio channel.

MCD considers the angular domain and the delay domain. In the angular domain, the MCD between two directions $\mathbf{\Omega}_1$ and $\mathbf{\Omega}_2$ is defined as

$$\mathrm{MCD}_{\mathrm{AoA/AoD},ij} = \frac{1}{2}|\mathbf{\Omega}_1 - \mathbf{\Omega}_2| \tag{8.1}$$

This value has the range of $[0, 1]$.

The MCD of paths in the delay domain can be defined as

$$\mathrm{MCD}_{\tau,ij} = \zeta_\tau \cdot \frac{|\tau_i - \tau_j|}{\Delta\tau_{\max}} \cdot \frac{\tau_{\mathrm{std}}}{\Delta\tau_{\max}} \tag{8.2}$$

where $\Delta\tau_{\max} = \max_{i,j}\{|\tau_i - \tau_j|\}$, τ_{std} is the standard deviation of the delays, and ζ is a scaling factor that can give more "importance" when necessary. Czink [2007] suggested choosing $\zeta = 8$.

Finally the distance between the paths (i, j) is calculated as

$$\text{MCD}_{ij} = \sqrt{\| \text{MCD}_{\mathbf{\Omega}_{\text{Tx}},ij} \|^2 + \| \text{MCD}_{\mathbf{\Omega}_{\text{Rx}},ij} \|^2 + \text{MCD}^2_{\tau,ij}} \tag{8.3}$$

The MCD concept can be extended to include the case where the Doppler frequencies of propagation paths are considered. The MCD in the Doppler frequency domain can be formulated as

$$\text{MCD}_{\nu,ij} = \zeta_\nu \cdot \frac{|\nu_i - \nu_j|}{\Delta\nu_{\max}} \cdot \frac{\nu_{\text{std}}}{\Delta\nu_{\max}} \tag{8.4}$$

where ζ_ν is a scaling factor to give a different "importance" to the MCD difference.

8.2.2.2 A Revised MCD without Weighting Factors

One drawback of the MCD of Czink [2007] is that it relies on these weightings, which are heuristic and are based on experimental experience. To avoid using heuristic factors, we can define a new multipath component distance that relies on the differences between the baseband received signals with the different path parameters. The phasors due to the parameters are compared. The difference of two phasors due to delay can be easily combined with the difference of two phasors due to Doppler frequency or directions. By doing so, it is not necessary to use heuristic settings to specify the importance of parameter distance in different dimensions.

8.2.2.3 Experimental Results for Path Clustering

In the following example of clustering multipath components, the SAGE algorithm was used to process channel measurement data collected using the PROPSound channel sounder at Oulu university. We have determined that the measurement of a wide-sense-stationary channel is composed of three consecutive data bursts. The MCD-based KPower-mean clustering approach as described by Czink [2007] was used to generate clusters.

The MCD importance weighting parameter ζ was heuristically set to 5, as recommended by Czink [2007]. It is necessary to verify how the clustering algorithm performance is influenced by the selection of different values of ζ. In the following, we select the data collected in the scenario "Macrocell, suburban, pedestrian route 1",[1] and choose different values of ζ_τ and ζ_ν, which denote the weighting factor for the MCD in the delay and the Doppler frequency respectively.

Table 8.1 gives the settings used in the SAGE algorithm for processing the data. Figure 8.5 illustrates the constellation of paths and the clustering results with $\zeta_\tau = \zeta_\nu = 1, 5$. The paths with the same color belong to one cluster. From Figure 8.5 it can be seen that for $\zeta_\tau = \zeta_\nu = 5$, those paths that can be well separated in the EoA–AoA domain belong to different clusters, as expected. Paths mixed in the center of the EoA–AoA figure are split into clusters that are well separated in the Doppler frequency and delay domains. Some clusters exhibit large spreads in

[1] Please refer to the appendices in Czink 2007 for a description of the measurement campaign.

Table 8.1 Parameter settings in the SAGE algorithm and the clustering algorithm

Parameter	Setting
No. of paths	20
No. of iterations	5
Dynamic range	10 dB
No. of bursts in one segment	3
No. of clusters	6

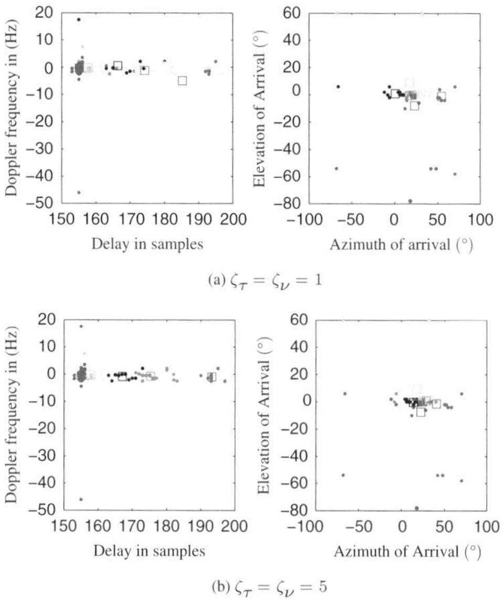

Figure 8.5 Constellation of the paths in the six clusters: Left, in Doppler frequency and delay domains; right, in DoA. $\zeta_\tau = \zeta_\nu = 1, 5$

the EoA–AoA domain, which is due to the limited number of clusters selected in the algorithm. The convergence of the clusters' spreads in multiple dimensions is seen to be achieved after five iterations of the KPower-mean clustering procedure. These results demonstrate that the KPower-mean approach is applicable for clustering multipath components in the measurement data considered.

8.3 Data Segment-length Selection

In order to extract the statistics of channel characteristics, multiple observations of wide-sense-stationary (WSS) channels are needed. To extract information for the small-scale parameters, separate data segments each consisting of several consecutive bursts of received data are processed using parametric estimation techniques, such as the SAGE algorithm. The estimated paths are then grouped into clusters. In order to maintain the stationarity of the channel during each segment, it is necessary to determine the length of the data segment: the number of bursts in our case.

Many measurement campaigns are conducted in time-variant environments. The time-evolution behavior is caused by either the movement of the Tx or of the Rx, or the people walking in the measurement environment. It is necessary to maintain the stationarity of the channel in order to extract the distributions of model parameters that are applicable for describing the channel statistics of the same or similar environments. In this section, we discuss the impact of the number of bursts in one segment on the clusters' behavior. An approach for determining the length of data segment in terms of the number of bursts that was originally introduced by Tian et al. [2010] is revisited.

Determining an appropriate number of data bursts in one segment involves a trade-off between the channel stationarity property and the adequacy of the number of samples collected for the purpose at hand, namely extracting the channel characteristics. Furthermore, the uncertain and fast changing nature of the environments make the selection process more complicated. Due to these uncertainties, it is difficult to come up with a theoretical solution. As an alternative, two heuristic methods can be considered.

Minimum-spread-variation Approach for Data Segmentation

The minimum-spread-variation approach is based on the observation that when the channel is stationary, increasing the number of bursts will lead to a relatively small variation in the spread parameters. Thus by observing when the variation of the spread parameters is a minimum, we can determine the appropriate number of bursts in one segment.

Let us denote with σ_n the spread obtained when n bursts in one segment are considered. The proposed method can be simply written as

$$\hat{n} = \arg\min_{n} |\sigma_{n-1} - \sigma_n| \tag{8.5}$$

The drawback of this method is that the distributions of the paths in the delay or other domains are neglected. Furthermore, this method is highly dependent on the performance of the clustering algorithm.

Kolgomorov–Smirnov Hypothesis-testing Approach for Data Segmentation

The main idea of the Kolgomorov–Smirnov hypothesis-testing Approach is to conduct a hypothesis test to find out whether the samples observed from different settings of the burst number belong to the same distribution. The Kolmogorov–Smirnov test can be used in this case.

The procedure for adopting the Kolmogorov–Smirnov test is as follows:

1. Compute the empirical cumulative distribution function (CDF) of the delays, the Doppler frequencies, the AoAs, and the AoDs of the paths for individual bursts; that is, $F_n(\tau)$, $F_n(\nu), F_n(\phi_{\mathrm{AoA}}), F_n(\phi_{\mathrm{AoD}})$, with $n = 1, \ldots, N$ denoting the indices of the bursts. We

consider the CDFs of the path parameters computed for the first burst as the reference CDFs. We denote the reference CDFs with $F'(\tau)$, $F'(\nu)$, $F'(\phi_{\mathrm{AoA}})$, and $F'(\phi_{\mathrm{AoD}})$.

2. Calculate the distance between the CDFs for the nth burst with $n > 1$ and the reference CDFs. Find the statistic D_{stat}

$$D_{\mathrm{stat}} = \arg \max_{\alpha=\tau,\nu,\phi_{\mathrm{AoA}},\phi_{\mathrm{AoD}}} \{|F_n(\alpha) - F'(\alpha)|\} \tag{8.6}$$

3. Check the hypothesis by comparing D_{stat} and the critical value for D_{stat} at a significance level of 5%. The critical value $D_{\mathrm{stat,critical}}$ can be calculated as

$$D_{\mathrm{stat,critical}} = 1.36 \cdot \sqrt{\frac{M_n M'}{M_n + M'}} \tag{8.7}$$

where M_n denotes the number of paths estimated for the nth burst, and M' represents the number of paths estimated for the reference burst. The null hypothesis that the path estimates obtained from the nth burst have the same distribution as the paths estimated from the reference burst is true when $D_{\mathrm{stat}} > D_{\mathrm{stat,critical}}$ at a significance level of 5%.

If the null hypothesis is true for the nth burst, we consider the channel is WSS for this burst and the reference burst. Then go back to Step 3 with $n = n + 1$.

If the null hypothesis does not hold, the nth burst is considered as a channel statistically different from the reference burst. The bursts with index starting from the index of the reference burst to n are considered as a period within which the channel is WSS. Then Step 4 is performed.

4. We consider the CDFs of the parameters of the paths obtained from the nth burst as the new reference CDFs. Let $n = n + 1$. Go back to perform Step 3.

After performing the above steps for all the bursts, we obtain multiple segments of data, each consisting of a certain number of bursts. For the individual segments, the channel satisfies the WSS condition. The clustering algorithm can then be applied to the estimated paths for generating clusters.

Experimental Evaluation

The proposed methods are evaluated by using measurement data collected in indoor environments. A total of 200 bursts of measurement data are considered. From each burst, the parameters – the delays, the Doppler frequencies, the directions of arrival and directions of departure – of 20 propagation paths are estimated. We implemented the hypothesis-testing-based approach to determine the length of a segment in terms of the number of bursts. Figure 8.6 illustrates the statistic D_{stat} computed as the maximum of the distances between all (marginal) CDFs versus the burst index. When D_{stat} is above the solid line, an old segment is completed and a new data segment is generated. Using the method proposed, for these 200 bursts, we get 53 segments in total.

Figure 8.7 depicts the histogram of the number of bursts per segment. The log-normal PDF which fits the best to the empirical PDF is also shown. The log-normal PDF is with $\mu = 0.50$ and $\sigma = 0.214$. The expected value of the log-normal PDF equals 3.4. It is interesting to see that the number of the bursts in a stationary segment follows a log-normal distribution.

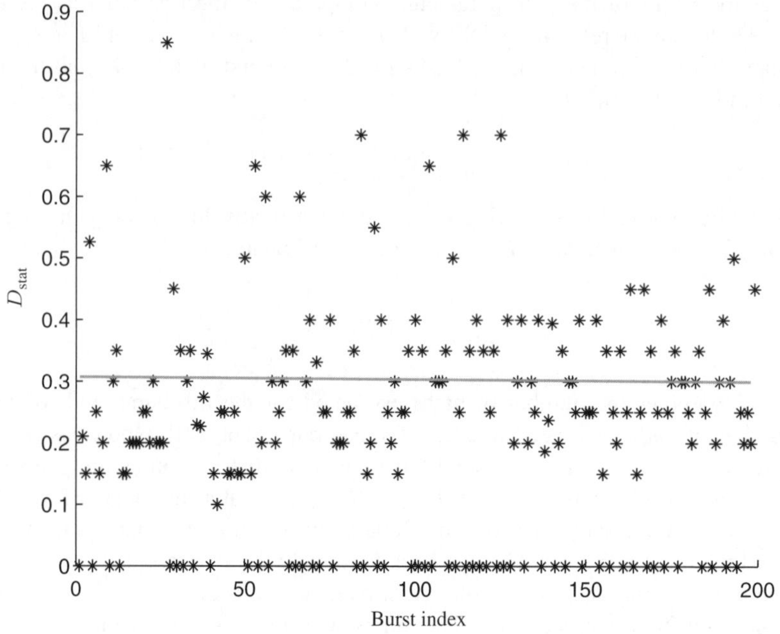

Figure 8.6 D_{stat} versus the burst indices when applying the Kolmogorov–Smirnov test. The solid line represents the critical value for D_{stat} at a significance level of 5%

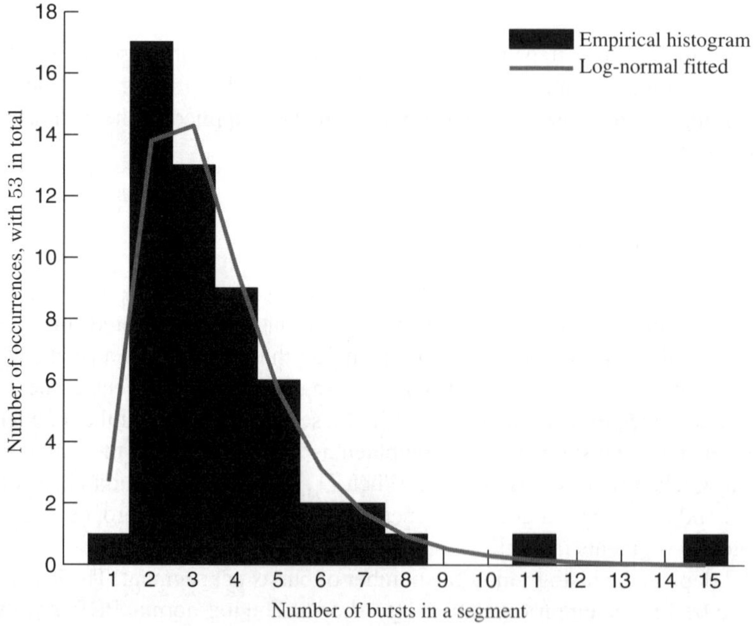

Figure 8.7 The histogram of the number of bursts per segment and a fitted log-normal PDF multipled by the total number of segments

Conclusions

We investigated the methods for determining the length of the segment that is used for clustering paths. The Kolmogorov–Smirnov test based method was used to check whether the distributions of the paths estimated from individual bursts of data are consistent. The bursts exhibiting the same distribution can be considered as observations of the same channel satisfying the WSS assumption. We used measurement data to evaluate the effectiveness of the method. The study shows that the number of bursts in the segment in which the channel is considered stationary follows a log-normal distribution. The expected value is close to the theoretical value calculated from the movement of the Tx and the Rx and the coherence time of the channel.

8.4 Relay and CoMP Channel Modeling

In this subsection stochastic modeling of the cross-correlation of small-scale fading (SSF) is performed for co-existing propagation channels based on measurement data collected using the rBECS system in urban areas in Korea. We define six geometrical parameters, which are considered as variables in the cross-correlation models proposed. These parameters characterize the geographic properties of a three-node cooperative relay system that consists of a base station, a relay station, and a mobile station. The sensitivity of the distribution of SSF cross-correlation coefficients with respect to the proposed variables is assessed. Three of them are identified as sensible variables for modeling.

8.4.1 Introduction

Cooperative wireless communication technologies have been adopted in 4G wireless communication systems, such as the Long-Term-Evolution-Advanced (LTE-A) systems [3GPP 2008; 2009] and World Interoperability for Microwave Access (WiMAX) systems [IEEE 2009]. These technologies, including cooperative relay [Beres and Adve 2008], coordinated beamforming, and joint processing algorithms [Boldi et al. 2009] allow exploitation of macro-spatial diversity gain for capacity enhancement and interference mitigation. The cross-correlation behavior of the channels when these coordinated techniques are activated is essential both for better design of the techniques in advance and optimization of the techniques in operation.

Experimental evidence has shown that propagation environments can generate highly cross-correlated channels in many aspects, such as narrowband fading [Zhou et al. 2011, Zuo et al. 2010] and the composite spread parameters [Yin et al. 2011a,b]. It has been reported that the fading cross-correlation may be attributed to the existence of deterministic components induced by both the line-of-sight (LOS) and dominant non-LOS propagation paths in the two co-existing channels [Zhou et al. 2011, Zuo et al. 2010]. Currently available channel models such as the WINNER II spatial channel models enhanced [Kyösti et al. 2007] and the IMT-Advanced channel models [ITU 2008] mainly focus on characterizing the single-link channels with one transmitter (Tx) and one receiver (Rx), which may be equipped with multiple antennas. As far as we are concerned, the joint properties of multiple co-existing channels have not not been studied thoroughly. Recently, modeling of the correlation of large-scale channel parameters [Jiang et al. 2008, Wang et al. 2010], composite channel parameters [Zhang et al. 2009], and narrowband fading coefficients [Zhou et al. 2011, Zuo et al. 2010] has gained much research attention. However, the correlation of the small-scale characteristics of channels has not been studied. Considering the fact that cooperative techniques are usually applied in the context of wideband communications, it is necessary to investigate methodologies for modeling channel cross-correlation at the small-scale for realistic environments and application scenarios.

We are interested in measurement-based modeling for the cross-correlation of small-scale fading (SSF) in the channels between the relay station (RS) and the mobile station (MS) and between the base station (BS) and the MS. The measurement data utilized was collected by using three wideband channel-sounding systems, which are capable of measuring two co-existing channels simultaneously. The models extracted reveal that the variation of SSF cross-correlation coefficients can be modeled as functions of geometrical variables that characterize the geographic properties of the three-node cooperative relay systems.

The organization of this subsection as follows. In Section 8.4.2, the SSF cross-correlation is defined. The modeling approach adopted is elaborated. Section 8.4.3 describes the measurement equipment, the specifications, and the campaigns, while Section 8.4.4 describes the model extraction procedure and the models obtained.

8.4.2 SSF Cross-correlation and Modeling Methodology

We consider the cross-correlation of the fading of two wideband channels in individual delay bins. The CIR is denoted by $h(\tau)$, with τ representing the time delay of the received signals. The SSF is referred to as the fluctuation of $h(\tau)$ at specific τ. The cross-correlation of the SSF in the BS-MS channel at delay τ_i, denoted by $h_\mathrm{b}(\tau_i)$ and the SSF in the RS-MS channel at τ_j, represented by $h_\mathrm{r}(\tau_j)$ is calculated as

$$C_\mathrm{br}(\tau_i, \tau_j) = \mathcal{R}\{\mathrm{E}[(h_\mathrm{b}(\tau_i) - \bar{h}_\mathrm{b}(\tau_i))(h_\mathrm{r}(\tau_j) - \bar{h}_\mathrm{r}(\tau_j))^*]\} \text{ with } \bar{h}_{\mathrm{b/r}}(\tau_{i/j}) = \mathrm{E}[h_{\mathrm{b/r}}(\tau_{i/j})]$$

where $(\cdot)^*$ represents the complex conjugate, $\mathcal{R}\{\cdot\}$ denotes the real part of a given argument, $\tau_i \in [0, T_\mathrm{b}]$, $\tau_j \in [0, T_\mathrm{r}]$ with T_b and T_r representing respectively the maximum delay of the BS-MS channel and of the RS-MS channel. The cross-correlation coefficient is then computed as

$$c_\mathrm{br}(\tau_i, \tau_j) = C_\mathrm{br}(\tau_i, \tau_j) \cdot (\mathrm{E}[|h_\mathrm{b}(\tau_i) - \bar{h}_\mathrm{b}(\tau_i)|^2]\mathrm{E}[|h_\mathrm{r}(\tau_j) - \bar{h}_\mathrm{r}(\tau_j)|^2])^{-1/2}$$

To mitigate the impact of the white Gaussian noise existing in the channels on the SSF cross-correlation, a power dynamic range can be specified and used to select the portions of $h(\tau)$ considered for modeling.

For notational convenience, we use c_br in the sequel to represent the values of $c_\mathrm{br}(\tau_i, \tau_j)$, $\tau_i \in [0, T_\mathrm{b}]$, $\tau_j \in [0, T_\mathrm{r}]$. The distribution of $c_\mathrm{br}(\tau_i, \tau_j)$, with $\tau_i \in [0, T_\mathrm{b}]$ and $\tau_j \in [0, T_\mathrm{r}]$, and the variation of the distribution with respect to the geometric parameters of a three-node cooperative relay system are of interest for modeling. Figure 8.8 illustrates a diagram of the RS-BS-MS relay system in a wireless environment. Six parameters are proposed as candidates as the geometrical variables of the models:

- Angle of separation θ

$$\theta = \mathrm{acos}\{(2d_\mathrm{rm}d_\mathrm{bm})^{-1}(d_\mathrm{rm}^2 + d_\mathrm{bm}^2 - d_\mathrm{br}^2)\}$$

with d_{rm}, d_{br} and d_{bm} representing the distances between the RS and MS, between the BS and RS, and between the BS and MS, respectively

- Average BS-MS and RS-MS distance $\bar{d}$

$$\bar{d} = (d_{\mathrm{bm}} + d_{\mathrm{rm}})/2$$

- Ratio of the BS-MS distance to the RS-MS distance $\tilde{d}$

$$\tilde{d} = d_{\mathrm{bm}}/d_{\mathrm{rm}}$$

- Ratio between the total length of the BS-MS and RS-MS links and the RS-BS distance $\breve{d}$

$$\breve{d} = (d_{\mathrm{bm}} + d_{\mathrm{rm}})/d_{\mathrm{rb}}$$

- Area S enclosed by the RS-BS-MS triangle

$$S = \sqrt{p(p - d_{\mathrm{bm}})(p - d_{\mathrm{rm}})(p - d_{\mathrm{rb}})}$$

with $p = (d_{\mathrm{bm}} + d_{\mathrm{rb}} + d_{\mathrm{rm}})/2$

- Composite parameter θ_d

$$\theta_d = \theta/\breve{d}$$

For notational convenience, we use $\boldsymbol{\theta} = (\theta, \bar{d}, \tilde{d}, \breve{d}, S, \theta_d)$ to denote the variables.

We are interested at modeling the empirical PDF of c_{br} with respect to individual entries in $\boldsymbol{\theta}$ and evaluating the sensitivity of the PDF to these variables. In order to simplify the modeling

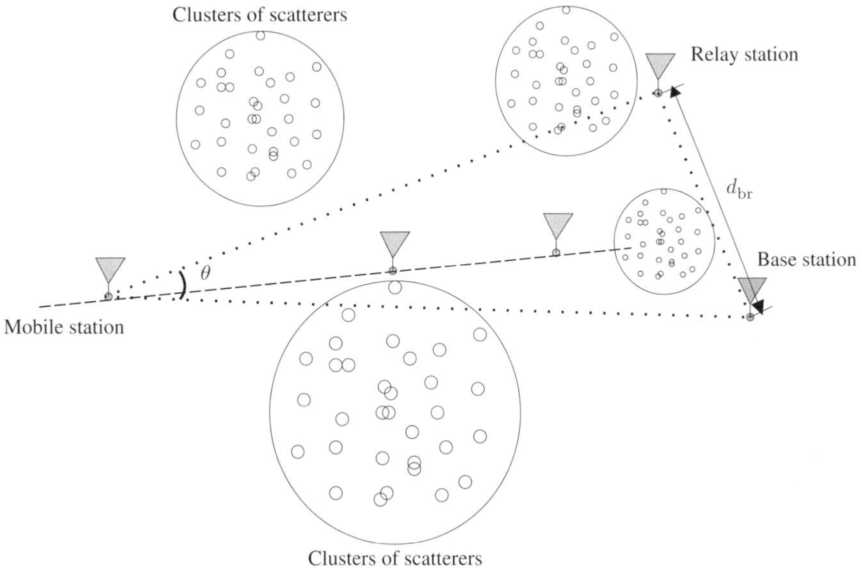

Figure 8.8 Cooperative relay system with a mobile station, a base station, and a relay station

procedure, we first fit the empirical PDF with analytical parametric curves. The parameters obtained will be used as the modeling objects. As will be shown later in the experimental results, we find that

- the mean value of $c_{\rm br}$ is close to 0 regardless of $\boldsymbol{\theta}$
- the empirical PDF $p(c_{\rm br})$ is symmetric with respect to $c_{\rm br} = 0$
- the first-order regression lines fit well to the two symmetric fragments of the PDF.

Figure 8.10 shows some experimental examples of the empirical PDF and the fitted regression lines. The absolute value of the slope s of the regression lines fitted to the PDF can be considered as the parameter characterizing the shape of $p(c_{\rm br})$. In addition, we also select the standard deviation $c_{\rm br,std}$ of $c_{\rm br}$ as a parameter of interest for modeling.

8.4.3 Measurements for Relay-channel Characterization

The measurement data were collected using the rBECS systems, which were designed for measuring multiple co-existing channels, for example in cooperative relay communications. Each rBECS system can be flexibly configured as either a Tx or an Rx. In the measurements considered here, two rBECS systems were used as Txs in the positions of BS and of RS respectively. A third rBECS system was adopted as an Rx in the position of MS. The Rx can acquire signals from the two Txs simultaneously. During the measurements, the carrier frequency was 3.705 GHz and the effective bandwidth was 100 MHz. Both Txs were equipped with a similar 2×4 planar array, and the Rx with a two-ring 16-element circular array [Yin et al. 2012a]. The measurement campaigns were conducted in residential areas in the eastern part of Ilsan City in Korea. Figure 8.9(a) illustrates the environments of three measurement sites. In each site, the measurements of the RS-MS and the BS-MS channels were conducted simultaneously, with the MS moving along predefined routes. Figure 8.9(b) depicts some examples of the Rx routes, denoted R1,. . . ,R6, at one of the measurement sites.

Channel sounding data obtained when the MS moved along 143 measurement routes at all three sites were applied for investigation of the empirical SSF cross-correlation. An individual

Figure 8.9 Ilsan City measurement campaign: (a) environment where the measurements were conducted; (b) routes of the mobile station at one of the measurement sites

measurement route provides observations of around 30 snapshots of the RS-MS and the BS-MS channels. In each snapshot, the CIRs of a total of $8 \times 16 = 128$ spatial subchannels for both RS-MS and BS-MS channels are recorded within 10 consecutive frames in time.

8.4.4 Model Extraction

The SSF cross-correlation coefficients can be directly calculated from the CIRs using correlation methods. However, because the system responses, phase noise attributed to the front-end chains in the Txs and the Rx, and the non-isotropic antenna responses can influence the accuracy of CIR computation, the CIRs obtained may be inconsistent with their true counterparts. In order to remove the impact of the measurement-system responses on the CIRs, the high-resolution channel parameter estimation method, the SAGE algorithm [Fleury et al. 2002], is used to obtain the parameter estimates of multiple specular paths. These parameter estimates were used to reconstruct the CIRs. In this study, the parameters – the delays, Doppler frequencies, directions of arrival (azimuths and elevations), directions of departure, and polarization matrices – are estimated for 30 paths from the data obtained in individual frames. The reconstructed CIRs are obtained by convolving the channel spread function in the delay domain [Fleury et al. 2002] with the autocorrelation function of the transmitted pseudo-noise sequence with bandwidth equal to 100 MHz.

The model extraction procedure consists of the following four steps.

1. The ranges of the entries in $\boldsymbol{\theta}$ are obtained from the location information of the BS, RS, and MS recorded in the measurements. The ranges of these variables are then split equally into N grids.
2. For each grid, $c_{\mathrm{br}}(\tau_i, \tau_j)$ is computed based on $h_{\mathrm{b}}(\tau_i)$ and $h_{\mathrm{r}}(\tau_j)$, $\tau_i \in [0, T_{\mathrm{b}})$, and $\tau_j \in [0, T_{\mathrm{r}})$, which are calculated from the measurements conducted with specific values of $\boldsymbol{\theta}$. Each realization of c_{br} is computed using a certain number of CIRs reconstructed for the channels observed within the channel coherence distance. For each grid, multiple realizations of the function of c_{br} are obtained.
3. The empirical PDF of c_{br} is generated. Regression lines are used to fit the PDF. The slope of the regression line is obtained. The standard deviation of c_{br} is calculated from all the samples of c_{br} collected in individual grids.
4. In order to obtain analytical expressions for the empirical models describing s, and $c_{\mathrm{br,std}}$ with respect to $\boldsymbol{\theta}$, two types of curves are used to fit the empirical samples. These curves are in forms $y = a \times x + b$ and $y = c \times \log_{10} x + d$, with $x \in \boldsymbol{\theta}$, $y \in (s, c_{\mathrm{br,std}})$ and a, b, c, d being the parameters in the models. The most appropriate values for a, b, c, d are obtained by searching over the parameters for the minimum Euclidean distance between the fitted curve and the empirical samples. The most appropriate curve – the one that yields the minimum Euclidean distance among the curves in different forms – is selected as the empirical model.

Figure 8.10(a) depicts the empirical distribution of c_{br} for θ within specific ranges, and Figure 8.10(b) the comparison of the fitted regression lines to the empirical PDF of c_{br}. It can be observed that the regression lines fit well with the empirical PDF of c_{br}. Furthermore, the absolute value of the slopes for the fitted regression lines increase when θ increases, indicating that the components tend to be less correlated when the separation angle increases.

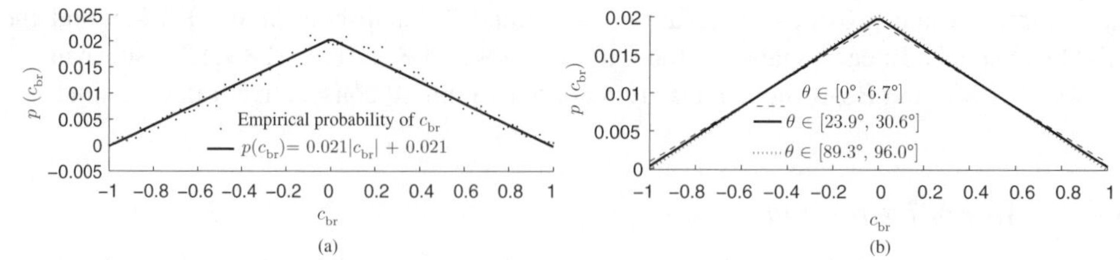

Figure 8.10 (a) Empirical distribution of c_{br} for $\theta \in [101.7°, 108.3°]$; (b) comparison of fitted regression lines and the empirical PDF of c_{br}

Figure 8.11 depicts the contour plots of the empirical PDFs of c_{br} vs. the entries of $\boldsymbol{\theta}$ with $N = 100$. It can be clearly seen from Figure 8.11 that the spread of the PDFs decreases when θ, $|\log_{10}\tilde{d}|$, or S increases. Figures 8.12–8.17 depict for $N = 20$ the scatter plots of the absolute value of the slope s of the first-order regression lines fitted to the empirical PDFs $p(c_{\text{br}})$ and the scatter plots of the standard deviation $c_{\text{br,std}}$ of c_{br} versus the individual entries in $\boldsymbol{\theta}$ respectively. It can be observed from these figures that the absolute value of the slope s and the standard deviation of c_{br} exhibit deterministic trends with respect to four variables – θ, $\log_{10}\tilde{d}$, $\breve{d}$ and S – out of the proposed six variables.

More specifically, it can be seen from Figure 8.12 that $|s|$ increases and $c_{\text{br,std}}$ decreases when the angle of separation θ increases. This is reasonable, as an increasing θ may occur when the MS moves towards the region where the RS and the BS are located. In such cases, the propagation channels from the MS to the RS and to the BS become more different, and consequently, the SSF in the two channels are less correlated.

We observe from Figure 8.14 that $|s|$ and $c_{\text{br,std}}$ exhibit symmetric behavior with respect to $\log_{10}\tilde{d} = 0$. In particular, $|s|$ increases and the standard deviation decreases when $|\log_{10}\tilde{d}|$ increases. Notice that increasing $|\log_{10}\tilde{d}|$ occurs when the MS moves closer to the BS than to the RS, or to the RS than to the BS. The results in Figure 8.14 indicate that in such cases, the SSFs in the two channels are less correlated than in the case where the MS has equal distances to the BS and to the RS.

It can be further observed from Figure 8.15 that $|s|$ decreases and $c_{\text{br,std}}$ increases along with the ratio $\breve{d}$, implying that when the location of the MS is far from the BS and the RS, the SSFs in the two channels are more correlated. This observation is consistent with that obtained from Figure 8.12. It is evident from Figure 8.16 that $|s|$ increases and $c_{\text{br,std}}$ decreases along with the area S. This implies that when the three nodes are separated with a larger distance, the SSFs in the two channels become less correlated. For the parameter $\bar{d}$ and θ_d, we observe from Figures 8.13 and 8.17 that the empirical samples of s and $c_{\text{br,std}}$ are dispersed within ranges that are less than those observed for other variables. Furthermore, the empirical s and $c_{\text{br,std}}$ do not exhibit clear deterministic varying trends with respect to $\bar{d}$ and θ_d compared with the cases for other variables.

In summary, the above results demonstrate that the SSF cross-correlations for two channels follow the PDFs with zero mean and variances monotonically decreasing or increasing, depending on the modeling variables θ, $\log_{10}\tilde{d}$, $\breve{d}$ and S. Furthermore, the shape of $p(c_{\text{br}})$ can be described using two regression lines symmetric with respect to $c_{\text{br}} = 0$. The absolute value of the lines' slopes also increases or decreases monotonically with respect to the aforementioned geometrical variables.

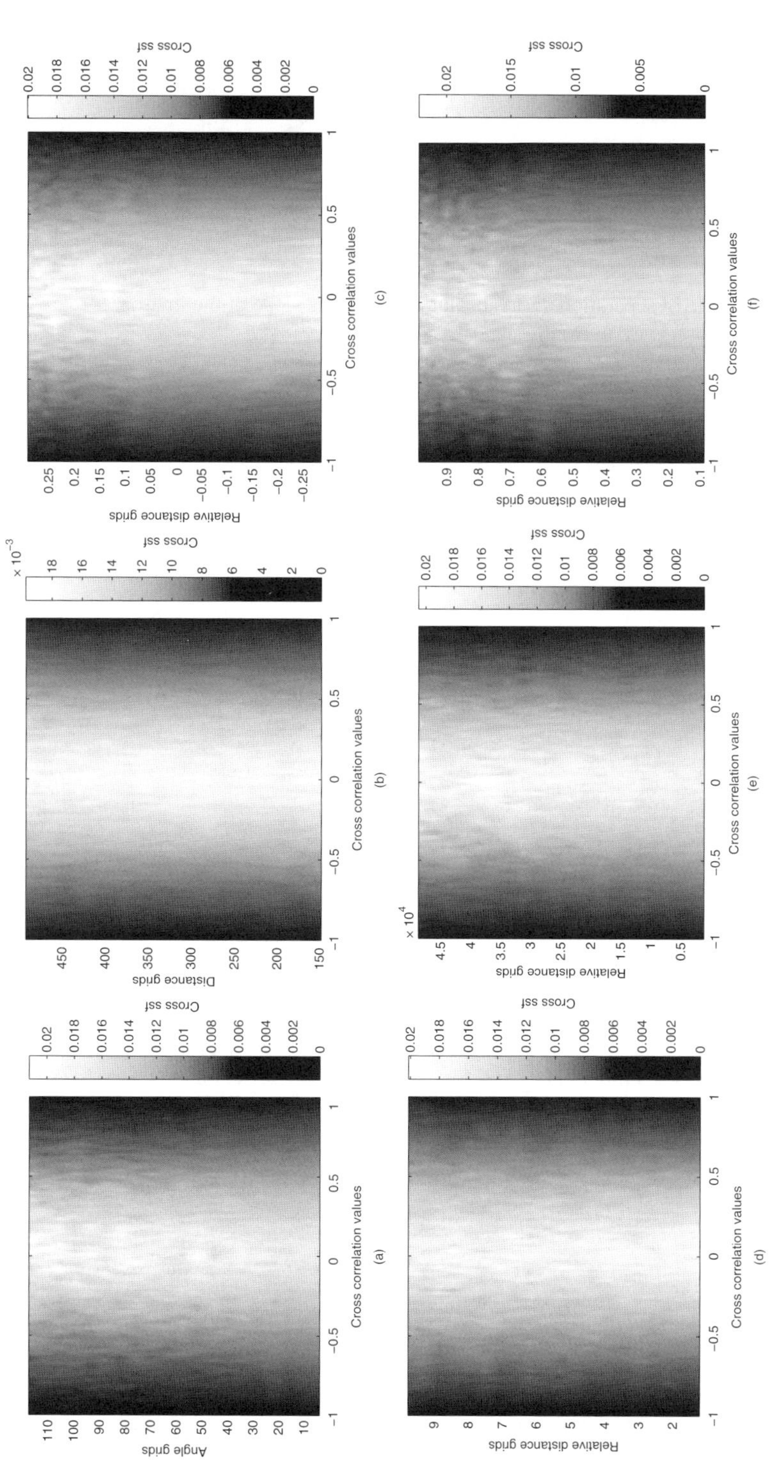

Figure 8.11 Empirical PDFs of the cross-correlation coefficients of the SSF of the BS-MS and RS-MS links vs. the candidates of geometrical variables: (a) the angle of separation θ in degrees; (b) the average distance $\bar{d}$ in meters; (c) the relative distance $\tilde{d}$ represented in logarithm; (d) the ratio $\check{d}$ between $2\bar{d}$ and the RS-BS distance; (e) the BS-RS-MS triangle's area S; (f) the composite parameter defined to be $\theta/\check{d}$

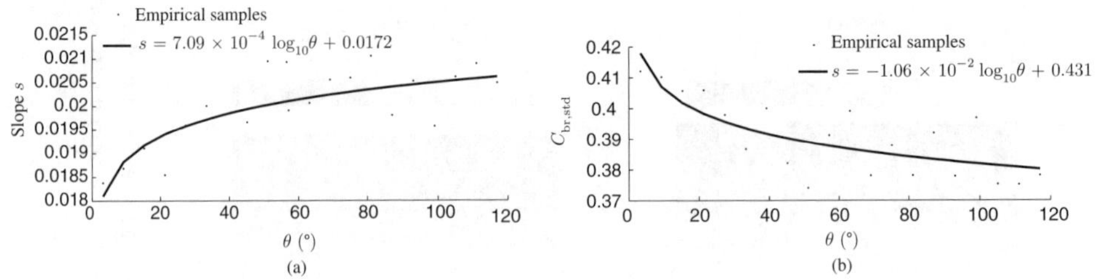

Figure 8.12 The angle of separation θ: (a) absolute value of the slope of the first-order regression line fitted to the empirical $p(c_{\mathrm{br}})$ vs. θ; (b) the standard deviation of c_{br} vs. θ

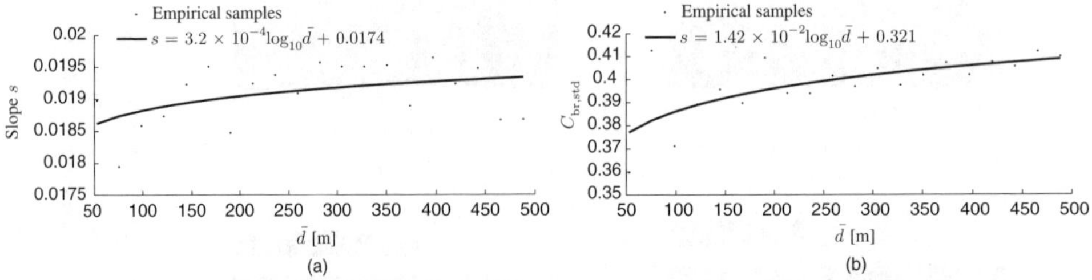

Figure 8.13 The average distance $\bar{d}$: (a) absolute value of the slope of the first-order regression line fitted to the empirical $p(c_{\mathrm{br}})$ vs. $\bar{d}$; (b) standard deviation of (c_{br}) vs. $\bar{d}$

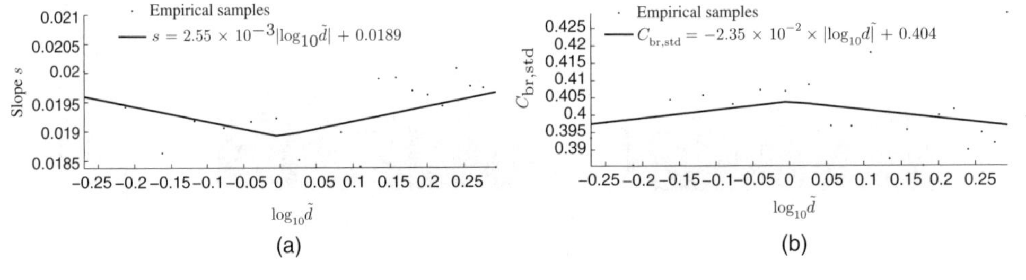

Figure 8.14 Relative distance $\tilde{d}$ in logarithm ($\log_{10}\tilde{d}$): (a) absolute value of the slope of the first-order regression line fitted to the empirical $p(c_{\mathrm{br}})$ vs. $\log_{10}\tilde{d}$; (b) standard deviation of (c_{br}) vs. $\log_{10}\tilde{d}$

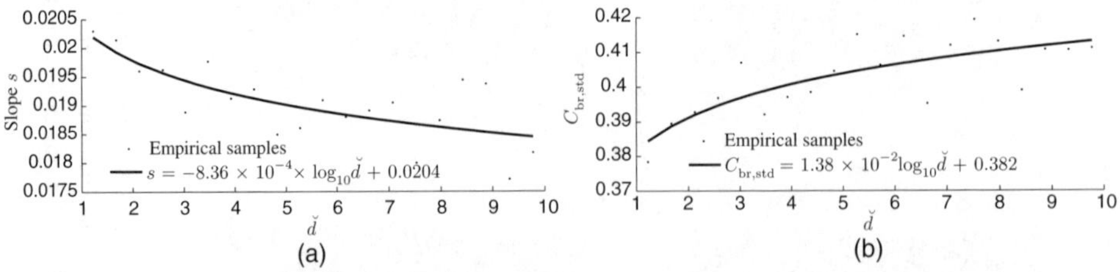

Figure 8.15 The ratio $(d_{\mathrm{rm}} + d_{\mathrm{bm}})/d_{\mathrm{rb}}$: (a) absolute value of the slope of the first-order regression line fitted to the empirical $p(c_{\mathrm{br}})$ vs. the ratio; (b) standard deviation of (c_{br}) vs. the ratio

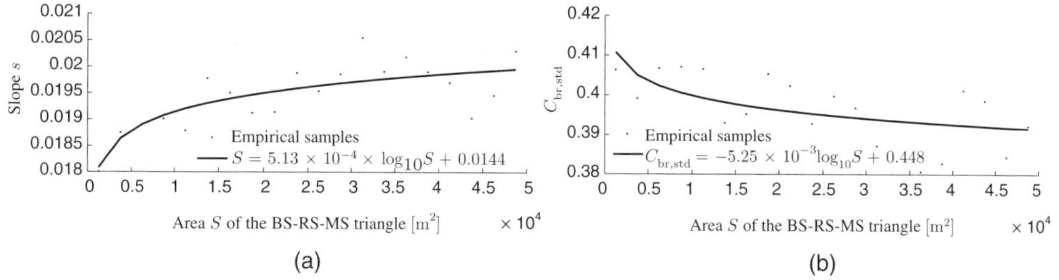

Figure 8.16 Area S of the BS-RS-MS three-node triangle: (a) absolute value of the slope of the first-order regression line fitted to the empirical $p(c_{br})$ vs. area S; (b) standard deviation of (c_{br}) vs. area S

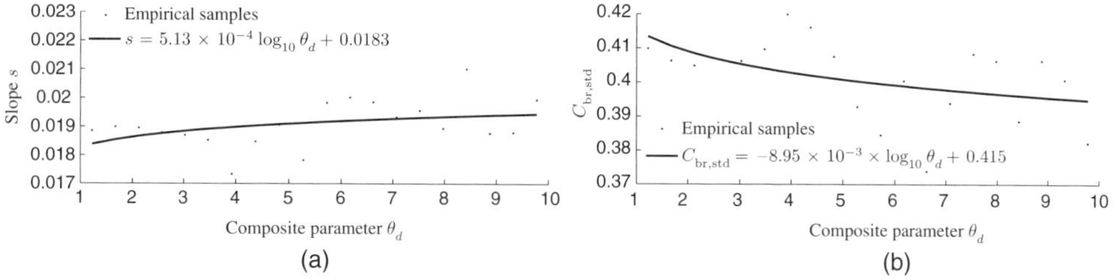

Figure 8.17 The composite parameter θ_d: (a) absolute value of the slope of the first-order regression line fitted to the empirical $p(c_{br})$ vs. θ_d; (b) standard deviation of (c_{br}) vs. θ_d

Bibliography

3GPP 2007 TR25.996 v7.0.0: Spatial channel model for multiple input multiple output (MIMO) simulations (release 7). Technical report, 3GPP.

3GPP 2008 TS 36.300 v8.7.0: Evolved universal terrestrial radio access (EUTRA) and evolved universal terrestrial radio access network (EUTRAN); overall description; stage 2.

3GPP 2009 TR36.814 v9.0.0 (2009) Further advancements for evolved universal terrestrial radio access (EUTRA) physical layer aspects (release 9).

Akaike H 1974 A new look at the statistical model identification. *IEEE Transsaction on Automatic Control* **AC-19**(6), 716–723.

Beres E and Adve R 2008 Selection cooperation in multi-source cooperative networks. *IEEE Transactions on Wireless Communications* **7**(1), 118–127.

Boldi M, Botella C, Boccardi F, Damico V, Hardouin E, Olsson M, Pennanen H, Rost P, Savin V, Svensson T and Tölli A 2009 D1.8 Intermediate report on CoMP. Technical report, Wireless World Initiative New Radio (WINNER+) Document Identifier: protocols, measurements and signalling.

Czink N 2007 The random-cluster model – a stochastic MIMO channel model for broadband wireless communication systems of the 3rd generation and beyond PhD thesis Technische Universitat Wien, Vienna, Austria, FTW Dissertation Series.

Czink N, Cera P, Salo J, Bonek E, Nuutinen JP and Ylitalo J 2006 A framework for automatic clustering of parametric MIMO channel data including path powers *IEEE Vehicular Technology Conference 2006 Fall*, pp. 1–5, Montreal, Canada.

Czink N, Tian R, Wyne S, Tufvesson F, Nuutinen JP, Ylitalo J, Bonek E and Molisch A 2007 Tracking time-variant cluster parameters in MIMO channel measurements *Second International Conference on Communications and Networking in China (CHINACOM)*, pp. 1147–1151.

Erceg V, Soma P, Baum D and Catreux S 2004 Multiple-input multiple-output fixed wireless radio channel measurements and modeling using dual-polarized antennas at 2.5 GHz. *IEEE Transactions on Wireless Communications* **3**(6), 2288–2298.

Fleury BH, Yin X, Rohbrandt KG, Jourdan P and Stucki A 2002 High-resolution bidirection estimation based on the sage algorithm: Experience gathered from field experiments. *Proc. XXVIIth General Assembly of the Int. Union of Radio Scientists (URSI).*

IEEE 2009 802.16 Relay Task Group: IEEE 802.16j proposals.

ITU 2008 Guidelines for evaluation of radio interface technologies for IMT-Advanced.

Jiang L, Thiele L and Jungnickel V 2008 Modeling and measurement of MIMO relay channels *Proceedings of the Vehicular Technology Conference (VTC Spring)*, pp. 419–423.

Kyösti P, Meinilä J, Hentilä L, Zhao X, Jämsä T, Schneider C, Narandzić M, Milojević M, Hong A, Ylitalo J, Holappa VM, Alatossava M, Bultitude R, de Jong Y and Rautiainen T 2007 WINNER II Channel Models D1.1.2 V1.1.

Park JJ, Kim MD, Kwon HK, Chung HK, Yin X and Fu Y 2012 Measurement-based stochastic cross-correlation models of a multilink channel in cooperative communication environments. *ETRI Journal on Wired and Wireless Telecommunication Technologies* **34**(6), 858–868.

Peacock J 1983 Two-dimensional goodness-of-fit testing in astronomy. *Royal Astronomical Society, Monthly Notices* **202**, 615–627.

Rissanen J 1978 Modeling by shortest data description. *Automatica* **14**, 465–471.

Saleh A and Valenzuela R 1987 A statistical model for indoor multipath propagation channel. *IEEE Journal of Selected Areas in Communications* **5**(2), 128–137.

Steinbauer M, Ozcelik H, Hofstetter H, Mecklenbrauker C and Bonek E 2002 How to quantify multipath separation. *IEICE Transactions on Electronics* **E85**(3), 552–557.

Tian L, Yin X and Lu S 2010 Automatic data segmentation based on statistical hypothesis testing for stochastic channel modeling *Personal Indoor and Mobile Radio Communications (PIMRC), 2010 IEEE 21st International Symposium on*, pp. 741–745.

Wang CX, Hong X, Ge X, Cheng X, Zhang G and Thompson J 2010 Cooperative MIMO channel models: A survey. *IEEE Communications Magazine* **48**(2), 80–87.

Yin X, Fu Y, Liang J and Kim MD 2011a Investigation of large- and small-scale fading cross-correlation using propagation graphs *Proceedings of International Symposium on Antennas and Propagation, SB01-1004*, Jeju, Korea.

Yin X, Hu Y and Zhong Z 2012a Dynamic range selection for antenna-array gains in high-resolution channel parameter estimation *Proceedings of International Conference on Wireless Communications and Signal Processing*.

Yin X, Liang J, Fu Y, Kim MD and Chung HK 2011b Preliminary study on angular small-scale cross-correlation of channels in nlos scenarios using propagation graphs *Proceedings of the 5th International Workshop on Broadband MIMO Channel Measurement and Modeling (IWONCMM'11)*, pp. 1–5, Beijing, China.

Yin X, Liang J, Fu Y, Yu J, Zhang Z, Park JJ, Kim MD and Chung HK 2012b Measurement-based stochastic modeling for co-existing propagation channels in cooperative relay scenarios *Future Network Mobile Summit (FutureNetw), 2012*, pp. 1–8.

Yin X, Liang J, Fu Y, Zhang Z, Park J, Kim M and Chung H 2012c Measurement-based stochastic models for the cross-correlation of multi-link small-scale fading in cooperative relay environments *Antennas and Propagation (EUCAP), 2012 6th European Conference on*, pp. 1–5 IEEE.

Yin X, Ling C, Kim MD and Chung HK 2015 Experimental cluster-based channel model for 28 GHz mm-wave propagation in office environments. *Submitted to IEEE Journal on Selected Topics in Signal Processing (Special issue on Signal Processing on Millimeter-wave Communications).*

Yin X, Zeng Z, Cheng X and Zhong Z 2012d Empirical modeling of cross-correlation for spatial-polarimetric channels in indoor scenarios *Personal Indoor and Mobile Radio Communications (PIMRC), 2012 IEEE 23st International Symposium on*, pp. 1677–1681.

Zhang J, Dong D, Nie X, Liu G and Dong W 2009 Propagation characteristics of wideband relay channels in urban environment *Proceedings of The 3rd International Workshop on Broadband MIMO Channel Measurement and Modeling, August 25*, pp. 1–6.

Zhou X, Yin X, Kwak BJ and Chung HK 2011 Experimental investigation of impact of antenna locations on the capacity of wideband distributed antenna systems in indoor environments *Proceedings of the 5th European Conference on Antennas and Propagation (EuCAP)*, pp. 1639–1643, Rome, Italy.

Zuo Q, Yin X and Zeng Z 2010 Spatial correlation characteristics of cooperative multi-point channels in indoor environments *Proceedings of the International Conference on Wireless Communications and Signal Processing (WCSP) 2010*, pp. 1–5, Suzhou, China.

9

In Practice: Channel Modeling for Modern Communication Systems

Based on the knowledge introduced in previous chapters, this chapter will described the development of channel models for various communication scenarios – conventional cellular scenarios, newly emerging vehicle-to-vehicle (V2V) scenarios, and cooperative scenarios – using different modeling approaches. Some new channel models will be described and some interesting observations and conclusions will be given.

9.1 Scenarios for V2V and Cooperative Communications

In general, different communication scenarios result in different channel characteristics. Therefore, to investigate the underlying channel characteristics of any communication system, it is desirable to completely classify the corresponding scenarios for which the communication systems are developed. Unlike newly emerging communication systems, such as V2V communications and cooperative communications systems, the typical scenarios for conventional cellular communication systems have been completely classified and are widely accepted in public: macrocell scenarios, microcell scenarios, and picocell scenarios.

As newly emerging communication systems, the typical scenarios for V2V and cooperative communications are very different to those for conventional cellular systems, resulting in these newly emerging communication systems having unique channel characteristics. This section will introduce and classify typical scenarios for V2V and cooperative communication systems.

9.1.1 V2V Communication Scenarios

Knowledge of the V2V propagation channel for different scenarios is of great importance for the design and analysis of V2V systems. However, due to the large differences between conventional cellular and V2V channels, the understanding of channels gained from conventional cellular systems cannot be directly used for V2V systems. Measurement campaigns have been conducted and others are ongoing to investigate the V2V propagation channels for different application scenarios. Here, we will briefly review and classify some recent V2V scenarios according to carrier frequencies, frequency-selectivity, antennas, environments, Tx/Rx direction of motion, and channel statistics, as shown in Table 9.1.

Propagation Channel Characterization, Parameter Estimation and Modelling for Wireless Communications, First Edition.
Xuefeng Yin and Xiang Cheng.

Table 9.1 Important V2V channel measurements

Measurements	Carrier frequency	Antenna	Frequency-selectivity	Tx/Rx direction of motion	Environments	Channel statistics
Acosta et al. [2004]	2.4 GHz	SISO	wideband	same	SS/EW (pico), LVTD	PDP, DD power profile
Zajic et al. [2009]	2.4 GHz	MIMO	wideband	same	UC/EW (pico), LVTD	STF CF, LCR, SDF PSD
Sen and Matolak [2008]	5 GHz	SISO	wideband	same	UC/SS/EW, (micro/pico), H(L)VTD	amplitude PDF, frequency CF, PDP
Paier et al. [2009]	5.2 GHz	MIMO	wideband	opposite	EW (pico), LVTD	PL, PDP, DD power profile
Cheng et al. [2007]	5.9 GHz	SISO	narrowband	same	SS, (micro/pico) LVTD	PL, CT, amplitude CDF, Doppler PSD
Acosta-Marum and Brunelli [2007]	5.9 GHz	SISO	wideband	same + opposite	UC/SS/EW, (micro/pico) LVTD	amplitude PDF, DD power profile

SS: suburban street; EW: expressway; UC: urban canyon; Micro: microcell; pico: picocell; H(L)VTD: high (low) vehicular traffic density; PDP: power delay profile; DD: Doppler-delay; PSD: power spectrum density; STF: space-time-frequency; CF: correlation function; LCR: level crossing rate; SDF: space-Doppler-frequency; PDF: probability density function; PL: path loss; CDF: cumulative distribution function; CT: coherence time.

9.1.1.1 Carrier Frequencies

Before the IEEE 802.11p standard [IEEE 2007b] was proposed, measurement campaigns were conducted at carrier frequencies outside the 5.9-GHz dedicated short-range communication (DSRC) band. Acosta et al. [2004] and Zajic et al. [2009] carried out V2V measurements at 2.4 GHz: the IEEE 802.11b/g band. Some measurements were also done around the IEEE 802.11a frequency band: at 5 GHz [Sen and Matolak 2008] and at 5.2 GHz [Paier et al. 2009]. Measurements at 5.9 GHz were also presented for narrowband and wideband V2V channels, respectively [Acosta-Marum and Brunelli 2007, Cheng et al. 2007]. From these measurement campaigns, we can observe that the propagation phenomenon in similar environments at different frequencies can vary significantly. Therefore, more measurement campaigns are expected to be conducted at 5.9 GHz for the better design of safety applications for V2V systems following the IEEE 802.11p standard. On the other hand, for the improved design of non-safety applications for V2V systems, measurement campaigns performed at other frequency bands, say 2.4 GHz or 5.2 GHz, are still required.

9.1.1.2 Frequency-selectivity and Antennas

In 1999, the Federal Communications Commission allocated 75 MHz of licensed spectrum, including seven channels, each with approximately 10 MHz instantaneous bandwidth, for DSRC in the USA. Such V2V channels are often frequency-selective (wideband) channels. A narrowband fading channel characterization based on measurement results, such as the one published by Cheng et al. [2007], is not sufficient for such V2V DSRC applications. Wideband measurement campaigns [Acosta et al. 2004, Acosta-Marum and Brunelli 2007, Paier et al. 2009, Sen and Matolak 2008, Zajic et al. 2009] are therefore essential for understanding the frequency-selectivity features of V2V channels and designing high-performance V2V systems.

Most V2V measurement campaigns so far have focused on single-antenna applications, resulting in SISO systems [Acosta et al. 2004, Acosta-Marum and Brunelli 2007, Cheng et al. 2007, Sen and Matolak 2008]. MIMO systems, with multiple antennas at both ends, are promising candidates for future communication systems and are gaining more importance in the IEEE 802.11 standards. MIMO technology is also attractive for V2V systems since multiple antenna elements can easily be placed on large vehicle surfaces. However, until now only a few measurement campaigns have been conducted for MIMO V2V channels [Paier et al. 2009, Zajic et al. 2009] and more will be required to enable future V2V system developments.

9.1.1.3 Environments and Tx/Rx Direction of Motion

Similar to conventional cellular systems, V2V scenarios can be classified as large spatial scale (LSS), moderate spatial scale (MSS), and small spatial scale (SSS) according to the Tx–Rx distance. For LSS scenarios or MSS scenarios, where the Tx–Rx distance is normally larger than 1 km or ranges from 300 m to 1 km, V2V systems are mainly used for broadcasting or "geocasting"; that is, geographic broadcasting [Kosch and Franz 2005]. For SSS scenarios, where the Tx–Rx distance is usually smaller than 300 m, V2V systems can be applied to broadcasting, geocasting, or unicasting. Since most V2V applications fall into MSS or SSS categories, these two scenarios are currently receiving more and more attention, with several measurement campaigns currently taking place [Acosta et al. 2004, Acosta-Marum and Brunelli 2007]. However, there are still a few applications that require communication between vehicles separated by distances larger than 1 km. One example of such LSS V2V applications is V2V decentralized environmental notification, in which vehicles or drivers in a certain area share information about observed events or road features. These applications have not gained much attention and thus no measurement results are available for V2V channels in LSS scenarios.

V2V scenarios can also be categorized in terms of roadside environment and the buildings, bridges, trees, parked cars found there: urban canyon, suburban street, and expressway. Many measurement campaigns have been conducted to study the channel statistics for various types of roadside environments [Acosta et al. 2004, Acosta-Marum and Brunelli 2007, Sen and Matolak 2008]. Due to the unique features of V2V environments, the vehicular traffic density (VTD) also significantly affects the channel statistics, especially for MSS and SSS scenarios. In general, the smaller the Tx–Rx distance, the larger the impact of VTD. Note that V2V channels usually exhibit non-isotropic scattering except in cases of high VTD. To the best of the authors' knowledge, only one measurement campaign has been carried out into the impact of VTD in expressway MSS and SSS scenarios [Sen and Matolak 2008].

The directions of motion of the Tx and Rx also affect channel statistics, through, for example, Doppler effects. Many measurement campaigns have focused on channel characteristics when the Tx and Rx are moving in the same direction [Acosta et al. 2004, Cheng et al. 2007, Sen and Matolak 2008]. A few have investigated channel characteristics when the Tx and Rx are moving in opposite directions [Acosta-Marum and Brunelli 2007, Paier et al. 2009].

In summary, it is desirable to conduct more measurement campaigns for MSS and SSS scenarios with various VTDs when the Tx and Rx are moving in opposite directions. In addition, measurement campaigns for LSS scenarios are indispensable for some V2V applications that need communications between two widely spaced vehicles.

9.1.2 Cooperative Communication Scenarios

Cooperative MIMO groups multiple radio devices to form virtual antenna arrays so that they can cooperate with each other by exploiting the spatial domain of mobile fading channels. Note that the cooperation of these grouped devices does not mean that the base station (BS) and mobile station (MS) have to be in reach of each other. Some devices can be treated as relays, to help communication between the BS and MS. Unlike conventional point-to-point MIMO systems, cooperative MIMO systems consist of multiple radio links, for example BS–BS, BS–relay station (RS), RS–RS, RS–MS, BS–MS, and MS–MS links. Due to the different local scattering environments around BSs, RSs, and MSs, a high degree of link heterogeneity or variation is expected in cooperative MIMO systems. In this chapter, we are interested in the various cooperative MIMO environments, which can be classified based on the physical and application scenarios involved. The physical scenarios include outdoor macrocell, microcell, and picocell scenarios, together with the indoor scenarios. Each physical scenario further includes three application scenarios: BS cooperation, MS cooperation, and relay cooperation. Therefore, 12 cooperative MIMO scenarios are considered in this chapter.

9.2 Channel Characteristics

Knowledge of channel statistics is essential for the analysis and design of a communication system. Unlike the rich and fascinating history of the research of conventional cellular channel characteristics, the investigation of channel characteristics for emerging V2V and cooperative MIMO systems is still in its infancy. This section will introduce the important channel characteristics of V2V and cooperative communication systems.

9.2.1 Channel Characteristics of V2V Communication Systems

As shown in Table 9.1, many different V2V channel statistics have been studied in recent measurement campaigns [Acosta et al. 2004, Acosta-Marum and Brunelli 2007]. Here, we only concentrate on two important statistics: amplitude distribution and Doppler power spectral density (PSD).

Analysis of amplitude distributions has been reported the literature. Acosta-Marum and Brunelli [2007] modeled the amplitude probability density function (PDF) of the received signal as either Rayleigh or Ricean. Cheng et al. [2007] observed that the received amplitude distribution in a dedicated V2V system with a carrier frequency of 5.9 GHz gradually transits from near-Ricean to Rayleighan as the vehicle separation increases. When the line-of-sight (LoS) component is intermittently lost at large distances, the channel fading can become more severe than Rayleighan. A similar conclusion was drawn by Sen and Matolak [2008], who modeled the amplitude PDF as a Weibull distribution. This "worse-than-Rayleigh" fading, termed "severe" fading, is caused by the rapid transitions of multipath components induced by the high speed and low height of the Tx/Rx and fast-moving scatterers [Matolak 2008].

The Doppler PSD has been investigated by several groups [Acosta et al. 2004, Acosta-Marum and Brunelli 2007, Paier et al. 2009, Zajic et al. 2009]. Joint Doppler-delay PSD measurements for wideband V2V channels at 2.4 GHz, 5.2 GHz, and 5.9 GHz were reported by Acosta et al. [2004], Paier et al. [2009], and Acosta-Marum and Brunelli [2007], respectively. Doppler PSDs can vary significantly with different time delays in a wideband V2V channel. Cheng et al. [2007] analyzed the Doppler spread and coherence time of narrowband V2V channels and presented their dependence on both velocity and vehicle separation. Recently, the space-Doppler PSD, which is the Fourier transform of the space-time correlation function in terms of time, was investigated by Zajic et al. [2009] and Paier et al. [2009]. It is worth noting that the Doppler PSD for V2V channels can be significantly different from the traditional U-shaped Doppler PSD for F2M channels.

9.2.2 Channel Characteristics of Cooperative Communication Systems

Several papers have reported measurements of the statistical properties of cooperative MIMO channels for different scenarios. A few indoor cooperative channel measurements have been reported for scenarios where the cooperative nodes are all static [Karedal et al. 2009, Medbo et al. 2004]. Mobile multi-link measurements were presented by Almers et al. [2007] for indoor cooperative MIMO channels. Other groups have addressed outdoor cooperative MIMO channel measurements: Soma et al. [2002] and Ahumada et al. [2005] for static nodes and Kaltenberger et al. [2008] and Zhang et al. [2009] for mobile nodes. All these measurement campaigns concentrated on the investigation of the channel characteristics of individual links for different scenarios, for example path loss, shadow fading, and small-scale fading.

As mentioned before, cooperative MIMO systems include multiple radio links that may exhibit strong correlation. The correlation of multiple links exists due to similarities in their environments arising from common shadowing objects and scatterers; it can significantly affect the performance of cooperative MIMO systems. The investigation of the correlations between different links is rare in the current literature and thus deserves more investigation.

9.3 Scattering Theoretical Channel Models for Conventional Cellular MIMO Systems

The well-known one-ring MIMO F2M regular-shaped geometry-based stochastic model (RS-GBSM) [Abdi and Kaveh 2002] has been widely used for the analysis and design of narrowband MIMO cellular systems for macro-cell scenarios. To meet demand for high-speed communications, wideband MIMO cellular systems have been suggested in many communication standards, leading to an increasing requirement for wideband MIMO F2M channel models.

However, the one-ring structure assumes that effective scatterers are located on a single ring and is therefore overly simplistic and thus unrealistic for modeling wideband channels. Latinovic et al. [2003] extended the narrowband one-ring model to a wideband model by extending the location of effective scatterers on a single-line zero-width ring to a ring with finite width. However, it is not trivial to use this model to match any given or measured PDP since many parameters need adjustments and the process involved is complicated. In addition, the integral expressions of the derived space–time–frequency (STF) correlation functions (CFs) based on this model can only be evaluated numerically : no closed-form expressions were found.

In contrast, Pätzold and Hogstad [2006] extended the one-ring model to a wideband application by dividing the single ring into several segments in terms of different times of arrival (ToAs). This makes the model easier to match to any specific PDPs due to the time-delay line (TDL) structure. However, the one-ring structure was still applied in this model, which makes it unrealistic in that certain ToAs (or propagation delays) are always related to a certain proportion of AoAs. To obtain closed-form expressions of STF CFs, the model of Pätzold and Hogstad also made the unrealistic assumption that the mean AoA and the corresponding angle spread are exactly the same for all the scatterers in different segments.

Neither the Latinovic or the Pätzold model considered the interaction of the AoA, AoD, and ToA, the importance of which was described by Chong et al. [2003]. More importantly, frequency correlation of sub-channels with different carrier frequencies, studied by Wang and Zoubir [2007] for SISO channels, has not yet been investigated for MIMO channels. Chapter 3 showed that the frequency correlation appears when the coherence bandwidth B_c is larger than the frequency separation of different subchannels. This frequency-correlated MIMO channel is commonly encountered in frequency-diversity MIMO communication systems, such as frequency-hopping MIMO systems and MIMO-orthogonal frequency division multiplexing (MIMO-OFDM) systems. For convenience, researchers investigating these systems have typically assumed that different frequency-separated channels are uncorrelated [Jakes 1994], which leads to positively biased performance results. Therefore, accurate theoretical analysis and simulation of frequency-correlated MIMO channels is of great importance for assessing the impact of the frequency correlation on the performance of real frequency-diversity MIMO systems.

The goals of this section are threefold. First, we propose a new wideband MIMO F2M RS-GBSM that represents a reasonable compromise between physical reality and analytical tractability. The proposed model uses a concentric multiple-ring instead of a single ring around the MS, thus avoiding the one-ring structure. Also, to easily match any specified or measured PDP, the model utilizes a virtual confocal multiple-ellipse to construct a TDL structure. The model also has the ability to consider the interaction of the AoA, AoD, and ToA. Secondly, from the proposed model, a closed-form expression of the STF CF between any two subchannels with different carrier frequencies for each time-bin signal is derived. For simplicity, the spatial and frequency correlations were assumed to be independent by previous authors [Latinovic et al. 2003, Pätzold and Hogstad 2006]. In this subsection, however, we derive the STF CF by taking into account the dependency between them, which is a unique characteristic of frequency-correlated MIMO channels [Foersler 2003]. Therefore, the derived STF CF can explicitly relate the frequency correlation to environment parameters, such as the mean AoA and the angle spread of the AoA. Based on the derived CF, we also reveal the inherent frequency correlation within the spatial correlation, which is important for the design of frequency-diversity MIMO systems [Bolcskei et al. 2003, Kalkan and Clarke 1997]. In addition, the wideband model proposed can be reduced to the traditional one-ring model by removing the frequency-selectivity. In this case, the derived STF CF shows a compact closed-form expression that is a generalization of many existing CFs [Abdi and Kaveh 2002, Abdi et al. 2002, Chen et al. 2000, Fulghum et al. 1998, Jakes 1994, Lee 1970]. Since the proposed wideband RS-GBSM is a reference model that assumes an infinite number of effective scatterers, it cannot be implemented directly in practice. Therefore, the third goal of this section is to derive a wideband deterministic sum-of-sinusoids (SoS) simulation model based on the proposed reference model. Closed-form expressions are provided for the STF CF of the simulation model. Similarly, the corresponding narrowband simulation model can be obtained by removing the frequency-selectivity from the wideband simulation model. The statistical properties of our simulation models are verified by comparing with the corresponding statistical properties of the reference models.

9.3.1 A Wideband Multiple-ring-based MIMO Channel Reference Model

A narrowband one-ring MIMO model is suitable for describing a narrowband channel in macrocell scenarios, where the BS is elevated and unobstructed, while the MS is surrounded by a large number of local scatterers. For narrowband systems, since the propagation delays τ'_n of all N ($N \to \infty$) incoming waves are much smaller than the transmitted symbol duration T_s; that is, $\tau'_{\max} = \max\{\tau'_n\}_{n=1}^{N} \ll T_s$, the delay differences caused by different local scatterers positioned randomly around the MS can be neglected in comparison to T_s. Therefore, it is reasonable to use effective scatterers located on a single ring instead of their real positions in order to enable a low-complexity one-ring model to be constructed at the minor expense of accuracy [Abdi et al. 2002]. However, in high data-rate wideband systems, T_s is much smaller than than in narrowband systems. In this case, the propagation delay differences cannot be neglected and thus the channel becomes a wideband channel. Therefore, the one-ring structure violates the basic characteristics of wideband channels as described by Latinovic et al. [2003].

To extend the one-ring model to wideband applications in macrocell scenarios, the primary task is to modify the oversimple one-ring structure. To this end, we replace the single ring of effective scatterers by concentric multiple rings of effective scatterers around the MS, thus capturing the basic characteristics of wideband channels. To make our model better at matching any specified or measured PDP, we utilize confocal multiple virtual ellipses, with the BS and MS located at the foci, to represent the TDL structure, where different delays correspond to different virtual confocal ellipses (in other words, taps). Note that the total number of virtual confocal ellipses L and the values of major axes a of different ellipses are determined according to the specified or measured PDP.

The newly developed structure is shown in Figure 9.1. For clarity, it only presents the effective scatterers from three concentric rings (the total number of concentric rings in the lth tap is Λ_l) belonging to the lth tap ($l = 1, 2, \ldots, L$). Notice that the total number and radii of the concentric multiple rings for different taps (that is, Λ_l and $R_{l,i}$) can be different in terms of different propagation environments for different delays. Therefore, the new wideband model with the appropriate number and major axes of virtual confocal multiple ellipses (that is, L and a), and the appropriate number and radii of concentric multiple rings (Λ_l and $R_{l,i}$) should be suitable for any macrocell scenario. We assume that uniform linear antenna arrays are used with $M_T = M_R = 2$ antennas. The symbols δ_T and δ_R designate the antenna element spacing at the BS and MS, respectively, and D denotes the distance between the BS and MS. The effective scatterers are located on $\sum_{l=1}^{L} \Lambda_l$ rings with radii $R_{l,i}$ ($i = 0, 1, \ldots, \Lambda_l - 1$). It is usually assumed that the assumption $D \gg R_{l,i} \gg \max\{\delta_T, \delta_R\}$ is fulfilled. The multi-element antenna tilt angles are denoted by β_T and β_R. The MS moves with speed v in the direction determined by the angle of motion γ. The angle spread seen at the BS is denoted by $\Theta_{l,i}$, which is related to $R_{l,i}$ and D by $\Theta_{l,i} \approx \arctan(R_{l,i}/D) \approx R_{l,i}/D$.

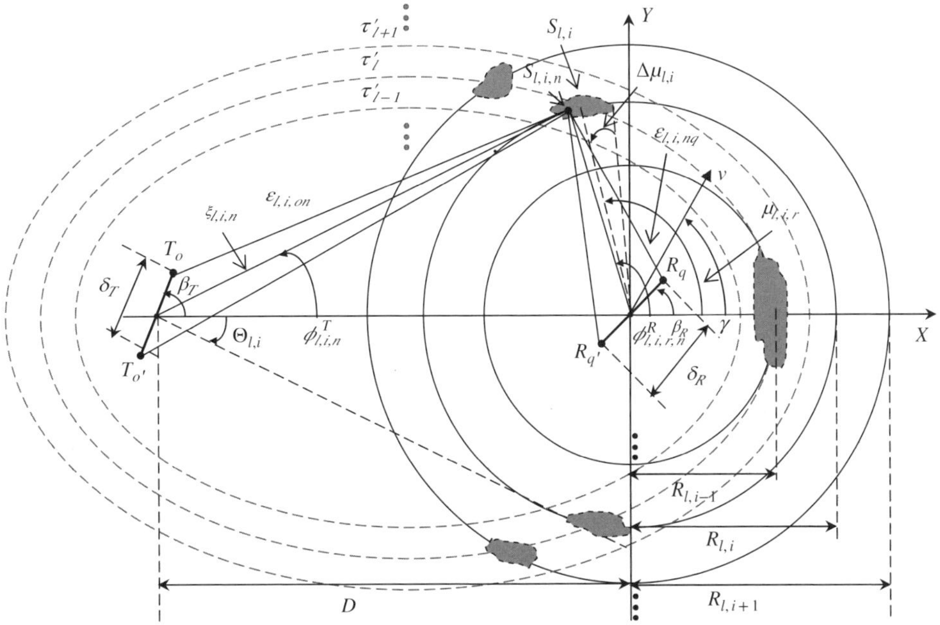

Figure 9.1 A new wideband multiple-ring based MIMO F2M channel model

The received complex impulse response at the carrier frequency f_c for the link $T_o - R_q$ can be expressed as

$$h_{oq}(t,\tau') = \sum_{l=1}^{L} h_{l,oq}(t) \times \delta(\tau' - \tau_l') \tag{9.1}$$

with $o = 1, 2, \ldots, M_T$ and $q = 1, 2, \ldots, M_R$, where $h_{l,oq}(t)$ and τ_l' denote the complex time-variant tap coefficient and the discrete propagation delay of the lth tap, respectively. Similar to the concept of effective scatterers in the narrowband one-ring model, the concept of an effective cluster is introduced in the new wideband multiple-ring model. From Figure 9.1, it is obvious that the position of the effective cluster $S_{l,i}$ is identified by the intersection of the virtual ellipses and multiple rings, and the lth tap includes $2\Lambda_l$ effective clusters. The mean angle of the effective cluster in each tap is $\mu_{l,i,r}$ $(r = 1, 2)$ and the corresponding angle spread is $\Delta\mu_{l,i}$, as illustrated in Figure 9.1. Therefore, the effective cluster can be completely determined by $\mu_{l,i,r}$ and $\Delta\mu_{l,i}$. Note that the angular range of $\mu_{l,i,r}$ in each tap is over $[0, 2\pi)$, which means the effective cluster can be located around the MS over $[0, 2\pi)$ for each tap. The setting of these two parameters follows a fixed rule. To establish this rule, firstly, we need to define the propagation delay subintervals G_l. The propagation delay interval $G = [0, \tau'_{\max}]$ is partitioned into L mutually disjoint subintervals G_l. Here, we utilize the definition of subintervals as [Wang and Zoubir 2007]

$$G_l = \begin{cases} \left[0, \Delta\tau'_{l+1}/2\right), & l = 1 \\ \left[\tau_l' - \Delta\tau_l'/2, \tau_l' + \Delta\tau'_{l+1}/2\right), & l = 2, 3, \ldots, L-1 \\ \left[\tau_l' - \Delta\tau_l'/2, \tau'_{\max}\right], & l = L \end{cases} \tag{9.2}$$

where $\Delta\tau_l' = \tau_l' - \tau'_{l-1}$ and $\tau'_{\max} = 2R_{L-1,\Lambda_{L-1}-1}/c$ (c is the speed of light). The propagation delay τ_l' of the lth tap can be expressed according to the corresponding AoA $\varphi^R_{l,i,r}$ $(r = 1, 2)$ as $\tau_l' \approx \tau'_{\max}(1 + \cos\varphi^R_{l,i,r})/2$ [Pätzold and Hogstad 2006]. Solving this equation for $\varphi^R_{l,i,r}$ gives $\varphi^R_{l,i,r} = \pm\arccos\left(2\tau_l'/\tau'_{\max} - 1\right)$. According to Eq. (9.2), the expression of $\varphi^R_{l,i,r}$ and the geometrical relationship in Figure 9.1, the expression of the mean angle $\mu_{l,i,r}$ and the corresponding angle spread $\Delta\mu_{l,i}$ of the effective cluster in the lth tap are given as

$$\begin{aligned}\mu_{l,i,r} &= \left(\varphi^R_{l,i-1,r} + 2\varphi^R_{l,i,r} + \varphi^R_{l,i+1,r}\right)/4 = \pm\left[\arccos\left(2\tau'_{l-1}/\tau'_{\max} - 1\right)\right.\\ &\quad \left.+2\arccos\left(2\tau_l'/\tau'_{\max} - 1\right) + \arccos\left(2\tau'_{l+1}/\tau'_{\max} - 1\right)\right]/4\end{aligned} \tag{9.3}$$

$$\begin{aligned}\Delta\mu_{l,i} &= \left|\left(\varphi^R_{l,i-1,r} - \varphi^R_{l,i+1,r}\right)/4\right| \\ &= \left[\arccos\left(2\tau'_{l-1}/\tau'_{\max} - 1\right) - \arccos\left(2\tau'_{l+1}/\tau'_{\max} - 1\right)\right]/4\end{aligned} \tag{9.4}$$

Following the definition of the subintervals G_l and the geometrical relationships shown in Figure 9.1, we can determine the effective cluster in each tap according to the propagation delay τ_l'. The time-variant tap coefficient at the carrier frequency f_c of lth tap can be expressed as

$$h_{l,oq}(t) = \lim_{N\to\infty} \frac{1}{\sqrt{N}} \sum_{i=0}^{\Lambda_l-1} \sum_{r=1}^{R_c} \sum_{n=1}^{N} \exp\left\{j\left[\psi_{l,i,n} - 2\pi f_c \tau_{l,i,oq,n} + 2\pi f_D t \cos\left(\varphi^R_{l,i,r,n} - \gamma\right)\right]\right\} \tag{9.5}$$

with $\tau_{l,i,oq,n} = \left(\varepsilon_{l,i,on} + \varepsilon_{l,i,nq}\right)/c$. Here, $\tau_{l,i.oq.n}$ is the travel time of the wave through the link $T_o - S_{l,i,n} - R_q$ scattered by the nth scatterer, $S_{l,i,n}$, R_c is the number of effective clusters for one ring in each tap (here $R_c = 2$), and N is the number of effective scatterers $S_{l,i,n}$ in the effective cluster $S_{l,i}$. The AoA of the wave traveling from the nth scatterer in the effective cluster $S_{l,i}$ towards the MS is denoted by $\varphi^R_{l,i,r,n}$. The phases $\psi_{l,i,n}$ are i.i.d. random variables with uniform distributions over $[0, 2\pi)$ and f_D is the maximum Doppler frequency. As shown by Abdi and Kaveh [2002], the distances $\varepsilon_{l,i,on}$ and $\varepsilon_{l,i,nq}$ can be expressed as functions of $\varphi^R_{l,i,r,n}$ as

$$\varepsilon_{l,i,on} \approx \xi_{l,i,n} - \delta_T[\cos(\beta_T) + \Theta_{l,i}\sin(\beta_T)\sin(\varphi^R_{l,i,r,n})]/2 \tag{9.6a}$$

$$\varepsilon_{l,i,nq} \approx R_{l,i} - \delta_R\cos(\varphi^R_{l,i,r,n} - \beta_R)/2 \tag{9.6b}$$

respectively, where $\xi_{l,i,n} \approx D + R_{l,i}\cos(\varphi^R_{l,i,r,n})$.

Since we assume that the number of effective scatterers in one effective cluster in this reference model tends to infinite (as shown in Eq. (9.5)), the discrete AoA $\varphi^R_{l,i,r,n}$ can be replaced by the continuous expressions $\varphi^R_{l,i,r}$. In the literature, many different scatterer PDFs have been proposed to characterize the AoA $\varphi^R_{l,i,r}$:

- uniform [Salz and Winters 1994]
- Gaussian [Adachi et al. 1986]
- wrapped Gaussian [Schumacher et al. 2002]
- cardioid [Byers and Takawira 2004].

In this chapter, we use the von Mises PDF [Abdi et al. 2002], which can approximate all the PDFs listed. The von Mises PDF is defined as

$$f(\varphi) \stackrel{\Delta}{=} \exp[k\cos(\varphi - \mu)]/2\pi I_0(k) \tag{9.7}$$

where $\varphi \in [-\pi, \pi)$, $I_0(\cdot)$ is the zeroth-order modified Bessel function of the first kind, $\mu \in [-\pi, \pi)$ accounts for the mean value of the angle φ, and k ($k \geq 0$) is a real-valued parameter that controls the angle spread of the angle φ. For $k = 0$ (isotropic scattering), the von Mises PDF reduces to the uniform distribution, while for $k > 0$ (non-isotropic scattering), the von Mises PDF approximates different distributions based on the values of k [Abdi and Kaveh 2002]. To better characterize the AoA in one effective cluster $S_{l,i}$, we further modify the general expression of von Mises PDF as

$$f_c\left(\varphi^R_{l,i,r}\right) = Q_{l,i,r}\exp\left[k_{l,i,r}\cos\left(\varphi^R_{l,i,r} - \mu_{l,i,r}\right)\right]/2\pi I_0\left(k_{l,i,r}\right) \tag{9.8}$$

where $\varphi^R_{l,i,r} \in \left[\mu_{l,i,r} - \Delta\mu_{l,i}, \mu_{l,i,r} + \Delta\mu_{l,i}\right)$ and $Q_{l,i,r}$ is the normalization coefficient. Here we name the PDF in Eq. (9.8) the truncated von Mises PDF. Here, "truncated" means that the range of AoA in this PDF is only defined within a limited interval $[\mu_{l,i,r} - \Delta\mu_{l,i}, \mu_{l,i,r} + \Delta\mu_{l,i})$. Therefore, the expression of the "tapped" PDF of AoA in the lth tap of the proposed wideband multiple-ring channel model is given by

$$f_g\left(\varphi^R_{l,i,r}\right) = \sum_{i=0}^{\Lambda_l - 1}\sum_{r=1}^{R_c}\frac{Q_{l,i,r}\exp\left[k_{l,i,r}\cos\left(\varphi^R_{l,i,r} - \mu_{l,i,r}\right)\right]}{2\pi I_0(k_{l,i,r})} w\left(\varphi^R_{l,i,r}, \mu_{l,i,r} - \Delta\mu_{l,i}, \mu_{l,i,r} + \Delta\mu_{l,i}\right)$$

$$\text{where} \quad w(\varphi, \varphi_l, \varphi_u) = \begin{cases} 1, & \text{if } \varphi_l < \varphi < \varphi_u \\ 0, & \text{otherwise} \end{cases} \tag{9.9}$$

Here, $Q_{l,i,r}$ are computed in such a way that the "tapped" PDF $f_g(\varphi^R_{l,i,r})$ is equal to 1; that is, $\int_{-\pi}^{\pi} f_g(\varphi^R_{l,i,r}) d\varphi^R_{l,i,r} = 1$.

Note that the proposed wideband model allows one to consider the interaction of AoA, AoD, and ToA in a sensible manner. The interaction between the AoA and AoD is obtained in terms of the exact geometrical relationship, while the interaction between the AoA/AoD and ToA is calculated according to the TDL structure, which allows one to investigate the correlation properties in each tap. Therefore, inspired by Chong et al. [2003], the interaction between the AoA/AoD and ToA can be considered via setting the appropriate parameter $k_{l,i,r}$ for the PDF of AoA/AoD in each tap according to the PDF of the ToA.

9.3.2 Generic Space–Time–Frequency CF

In this subsection, from the proposed model we first derive a new generic STF CF for wideband MIMO channels. As shown at the end of this subsection, by removing the frequency-selectivity the derived STF CF can be reduced to a compact closed-form STF CF for narrowband MIMO channels, which includes many existing CFs as special cases.

9.3.2.1 Space–Time–Frequency CF for Wideband MIMO Channels

From the proposed wideband model, we derive the STF CF for each tap. The correlation properties of two arbitrary links $h_{oq}(t, \tau')$ and $h'_{o'q'}(t, \tau')$ at different frequency f_c and f'_c of a MIMO channel are completely determined by the correlation properties of $h_{oq}(t)$ and $h'_{o'q'}(t)$ in each tap since we assume that no correlations exist between the underlying processes in different taps. Therefore, we can restrict our investigations to the following STF CF:

$$\rho_{l,oq;l,o'q'}(\tau, \chi) := \mathbf{E}[h_{l,oq}(t) h'^{*}_{l,o'q'}(t-\tau)] \tag{9.10}$$

where $(\cdot)^*$ denotes the complex conjugate operation and $\mathbf{E}[\cdot]$ designates the statistical expectation operator. Note that this CF is a function of the time separation τ, space separation δ_T and δ_R, and frequency separation $\chi = f'_c - f_c$. As shown in Appendix A.8.1, the closed-form expression of STF CF $\rho_{l,oq;l,o'q'}(\tau, \chi)$ can be presented as

$$\begin{aligned}
\rho_{l,oq;l,o'q'}(\tau, \chi) = & \frac{2}{\pi I_0(k)} \sum_{i=0}^{\Lambda_l - 1} \sum_{r=1}^{R_c} Q_{l,i,r} e^{jC_{l,i}} \Bigg\{ \Delta\mu_{l,i} I_0(A_{l,i,r}) I_0(B_{l,i,r})/2 + I_0(B_{l,i,r}) \sum_{\ell=1}^{\infty} \\
& \times I_\ell(A_{l,i,r}) \sin(\ell \Delta\mu_{l,i}) \cos(\ell \mu_{l,i,r})/\ell + I_0(A_{l,i,r}) \sum_{\ell'=1}^{\infty} (-1)^{\ell'} I_{\ell'}(B_{l,i,r}) \sin(\ell' \Delta\mu_{l,i}) \\
& \times \cos(\ell' \mu_{l,i,r} + \ell' \pi/2)/\ell' + \sum_{\ell=1}^{\infty} (-1)^{\ell} I_\ell(A_{l,i,r}) I_\ell(B_{l,i,r}) [\Delta\mu_{l,i} \cos(\ell\pi/2) \\
& + \sin(2\ell \Delta\mu_{l,i}) \times \cos(2\ell \mu_{l,i,r} + \ell\pi/2)/2\ell] + \sum_{\substack{\ell=1 \\ (\ell \neq q)}}^{\infty} \sum_{q=1}^{\infty} (-1)^{\ell'} I_\ell(A_{l,i,r}) I_{\ell'}(B_{l,i,r})
\end{aligned}$$

$$\left[\sin\left[(\ell+\ell')\,\Delta\mu_{l,i}\right] \times \frac{\cos\left[(\ell+\ell')\,\mu_{l,i,r}+\ell'\pi/2\right]}{\ell+\ell'} + \frac{\sin\left[(\ell-\ell')\,\Delta\mu_{l,i}\right]\cos\left[(\ell-\ell')\,\mu_{l,i,r}-\ell'\pi/2\right]}{\ell-\ell'}\right]\Bigg\} \quad (9.11)$$

where

$$A_{l,i,r} = a_{l,i,r} + j(\mathrm{X}U_{l,i} + y\cos\beta_R + x\cos\gamma) \quad (9.12a)$$

$$B_{l,i,r} = b_{l,i,r} + j(\mathrm{X}V_{l,i} + y\sin\beta_R + z\Theta_{l,i}\sin\beta_T + x\sin\gamma) \quad (9.12b)$$

$$C_{l.i} = z\cos\beta_T + \mathrm{X}T_{l.i} \quad (9.12c)$$

with $x = 2\pi f_D\tau$, $y = 2\pi f_c\delta_R/c$, $z = 2\pi f_c\delta_T/c$, $\mathrm{X} = 2\pi\chi/c$, $a_{l,i,r} = k_{l,i,r}\cos\mu_{l,i,r}$, $b_{l,i,r} = k_{l,i,r}\sin\mu_{l,i,r}$, $T_{l,i} = (\delta_T/2)\cos\beta_T + D + R_{l,i}$, $U_{l,i} = R_{l,i} + (\delta_R/2)\cos\beta_R$, and $V_{l,i} = (\delta_T/2)\Theta_{l,i}\sin\beta_T + (\delta_R/2)\sin\beta_R$. Consequently, the STF CF between $h_{oq}(t,\tau')$ and $h'_{o'q'}(t,\tau')$ can be shown as

$$\rho_{oq,o'q'}(\tau,\chi) = \frac{1}{L}\sum_{l=1}^{\mathrm{L}} \rho_{l,oq;l,o'q'}(\tau,\chi) \quad (9.13)$$

Note that Eqs (9.11) and (9.13) are generic expressions, which apply to the STF CF and the subsequently degenerate CFs (space–time (ST) CF, frequency CF, and so on) differ only in values of $A_{l,i,r}$, $B_{l,i,r}$, and $C_{l,i}$. The corresponding expressions of these three parameters for the degenerate CFs can be easily obtained by setting relevant terms (τ, δ_T and δ_T, and χ) to zero.

9.3.2.2 Space–Time–Frequency CF for Narrowband MIMO Channels

A special case of the proposed model in Eq. (9.1) is given when $L = 1$ and $\Lambda_l = 1$. Consequently, we have $h_{oq}(t,\tau') = h_{oq}(t)\,\delta(\tau')$, which is the complex fading envelope of the narrowband one-ring channel model. To make this evident, we express the complex fading envelope $h_{oq}(t)$ similarly to Eq. (9.5) by removing the subscripts $(\cdot)_l$, $(\cdot)_i$, and $(\cdot)_r$:

$$h_{oq}(t) = \lim_{N\to\infty}\frac{1}{\sqrt{N}}\sum_{n=1}^{N}\exp\left\{j\left[\psi_n - 2\pi f_c\tau_{oq,n} + 2\pi f_D t\cos\left(\varphi_n^R - \gamma\right)\right]\right\} \quad (9.14)$$

with $\tau_{oq,n} = (\varepsilon_{on} + \varepsilon_{nq})/c$, where ε_{on} and ε_{nq} can be expressed as functions of φ_n^R as

$$\varepsilon_{on} \approx \xi_n - \delta_T[\cos(\beta_T) + \Theta\sin(\beta_T)\sin(\varphi_n^R)]/2 \quad (9.15a)$$

$$\varepsilon_{nq} \approx R - \delta_R\cos(\varphi_n^R - \beta_R)/2 \quad (9.15b)$$

where $\xi_n \approx D + R\cos(\varphi_n^R)$. In order to be consistent with the proposed wideband model, the von Mises PDF of Eq. (9.7) is employed to characterize the AoA φ_n^R of the narrowband model. In such a case, the angle spread $\Delta\mu_{l,i} = \pi$, which means the AoA range is over $[0, 2\pi)$. As shown in Appendix A.8.2, the STF CF of the narrowband one-ring model can be obtained from Eq. (9.11) after some manipulation

$$\rho_{oq,o'q'}(\tau,\chi) = e^{jC} I_0\left[\left(A^2+B^2\right)^{1/2}\right]/I_0(k) \tag{9.16}$$

where

$$A = a + j(\mathrm{X}U + y\cos\beta_R + x\cos\gamma) \tag{9.17a}$$

$$B = b + j(\mathrm{X}V + y\sin\beta_R + z\Theta\sin\beta_T + x\sin\gamma) \tag{9.17b}$$

$$C = z\cos\beta_T + \mathrm{X}T \tag{9.17c}$$

with $a = k\cos\mu$, $b = k\sin\mu$, $T = (\delta_T/2)\cos\beta_T + D + R$, $U = R + (\delta_R/2)\cos\beta_R$, and $V = (\delta_T/2)\Theta\sin\beta_T + (\delta_R/2)\sin\beta_R$. The parameters x, y, z, and X are the same as defined in Eq. (9.11). It is worth stressing that Eq. (9.16) is a generic expression, which applies to the STF CF and the subsequently degenerate CFs with differences only in the values of A, B, and C. The corresponding expressions of A, B, and C for the degenerate CFs can be easily obtained from Eq. (9.17) by setting relevant terms to zero.

The derived generic STF CF Eq. (9.16) includes many existing CFs as special cases. For a SISO case, the time CF given by Abdi et al. [2002] is obtained by setting $\delta_T = \delta_R = 0$ and $\chi = 0$ in Eq. (9.16) with $k \neq 0$. If setting $k = 0$ (isotropic scattering) in Eq. (9.16), Clarke's time CF is obtained [Jakes 1994]. For a SIMO case, Lee's ST CF [Lee 1970] is obtained by substituting $\delta_T = 0$, $\chi = 0$, $\beta_R = 0$, and $k = 0$ into Eq. (9.16). For a MISO case, the ST CF of Chen et al. [2000] is obtained by substituting $\delta_R = 0$, $\chi = 0$, and $k = 0$ into Eq. (9.16). If further substituting $f_D = 0$ into Eq. (9.16), the space CF of Fulghum et al. [1998] is obtained. For a MIMO case, the ST CF of Abdi and Kaveh [2002] is obtained by setting $\chi = 0$ in Eq. (9.16) with $k \neq 0$.

9.3.3 MIMO Simulation Models

In this section we propose an efficient deterministic SoS simulation model for wideband MIMO channels, based on the proposed wideband MIMO channel reference model. The proposed wideband simulation model can be further reduced to a narrowband one by removing the frequency-selectivity.

9.3.3.1 A Deterministic Simulation Model for Wideband MIMO Channels

The wideband deterministic simulation model proposed is also based on the TDL structure. The impulse response of the simulation model at the carrier frequency f_c for the $T_o - R_q$ link is again composed of L discrete taps according to

$$\tilde{h}_{oq}(t,\tau') = \sum_{l=1}^{L} \tilde{h}_{l,oq}(t) \times \delta(\tau' - \tau'_l) \tag{9.18}$$

In Eq. (9.18), the complex fading envelope $\tilde{h}_{l,oq}(t)$ is modeled by utilizing only a finite number of scatterers N and keeping all the model parameters fixed as

$$\tilde{h}_{l,oq}(t) = \frac{1}{\sqrt{N}} \sum_{i=0}^{\Lambda_l - 1} \sum_{r=1}^{R_c} \sum_{n=1}^{N} \exp\left\{j\left[\tilde{\psi}_{l,i,n} - 2\pi f_c \tau_{l,i,oq,n} + 2\pi f_D t \cos\left(\tilde{\varphi}^R_{l,i,r,n} - \gamma\right)\right]\right\} \tag{9.19}$$

where the phases $\tilde{\psi}_{l,i,n}$ are simply the outcomes of a random generator uniformly distributed over $[0, 2\pi)$, the discrete AoAs $\tilde{\varphi}^R_{l,i,r,n}$ will be kept constant during simulation, and the other symbol definitions are the same as those in Eq. (9.5). Therefore, we can analyze the properties of the deterministic channel simulator by time averages instead of statistical averages. The STF CF is defined as

$$\tilde{\rho}_{l,oq;l,o'q'}(\tau, \chi) := \left\langle \tilde{h}_{l,oq}(t)\, \tilde{h}'^{*}_{l,o'q'}(t - \tau) \right\rangle \tag{9.20}$$

where $\langle \cdot \rangle$ denotes the time-average operator. Substituting Eq. (9.19) into Eq. (9.20), we can get the closed-form STF CF as

$$\tilde{\rho}_{l,oq;l,o'q'}(\tau, \chi) = \frac{1}{N} \sum_{i=0}^{\Lambda_l - 1} \sum_{r=1}^{R_c} \sum_{n=1}^{N} e^{j\left(C_{l,i} + P_{l,i} \cos \tilde{\varphi}^R_{l,i,r,n} + J_{l,i} \sin \tilde{\varphi}^R_{l,i,r,n}\right)} \tag{9.21}$$

with

$$P_{l,i} = \mathrm{X} U_{l,i} + y \cos \beta_R + x \cos \gamma \tag{9.22a}$$

$$J_{l,i} = \mathrm{X} V_{l,i} + y \sin \beta_R + z\Delta \sin \beta_T + x \sin \gamma \tag{9.22b}$$

where $C_{l,i}$, x, y, z, X, $U_{l,i}$, and $V_{l,i}$ are the same as defined in Eq. (9.11). By analogy to Eq. (9.13), we can further get the STF CF between $\tilde{h}_{oq}(t, \tau')$ and $\tilde{h}'_{o'q'}(t, \tau')$ as

$$\tilde{\rho}_{oq,o'q'}(\tau, \chi) = \frac{1}{L} \sum_{l=1}^{L} \tilde{\rho}_{l,oq;l,o'q'}(\tau, \chi) \tag{9.23}$$

Similar to Eqs (9.11) and (9.13), Eqs (9.21) and (9.23) are generic expressions, which apply to all the CFs of the deterministic simulation model with different $C_{l,i}$, $P_{l,i}$, and $J_{l,i}$. Comparing the expressions of $A_{l,i,r}$ and $B_{l,i,r}$ with $P_{l,i}$ and $J_{l,i}$, respectively, we have $A_{l,i,r} = a_{l,i,r} + jP_{l,i}$ and $B_{l,i,r} = b_{l,i,r} + jJ_{l,i}$. From Eqs (9.21) and (9.23), it is obvious that only $\{\tilde{\varphi}^R_{l,i,r,n}\}_{n=1}^{N}$ needs to be determined for this deterministic simulation model.

As addressed in Chapter 4, the method of equal area (MEA) and the method of exact Doppler spread (MEDS) have been widely used to compute the important parameters of deterministic simulation models for isotropic scattering environments. However, these two methods fail to reproduce the desired statistical properties of the reference model under the condition of non-isotropic scattering [Hogstad et al. 2005, Pätzold and Hogstad 2006]. Therefore, the L_p-norm optimization method [Pätzold 2002] is used to calculate the parameters $\{\tilde{\varphi}^R_{l,i,r,n}\}_{n=1}^{N}$ of the deterministic simulation model based on the corresponding properties of the reference model. The time CF $\rho_{l,oq;l,oq}(\tau)$, frequency CF $\rho_{l,oq;l,oq}(\chi)$, and space CF $\rho_{l,oq;l,o'q'}$ are identified as key properties. Then the optimization method requires the numerical minimization of the following three L_p-norms

$$E_{1,l}^{(p)} := \left\{ \int_0^{\tau_{\max}} \left|\rho_{l,oq;l,oq}(\tau) - \tilde{\rho}_{l,oq;l,oq}(\tau)\right|^p d\tau/\tau_{\max} \right\}^{1/p} \tag{9.24}$$

$$E_{2,l}^{(p)} := \left\{ \int_0^{\chi_{\max}} \left|\rho_{l,oq;l,oq}(\chi) - \tilde{\rho}_{l,oq;l,oq}(\chi)\right|^p d\chi/\chi_{\max} \right\}^{1/p} \tag{9.25}$$

$$E_{3,l}^{(p)} := \left\{ \int_0^{\delta_T^{\max}} \int_0^{\delta_R^{\max}} \left|\rho_{l,oq;l,o'q'} - \tilde{\rho}_{l,oq;l,o'q'}\right|^p d\delta_T d\delta_R/(\delta_T^{\max}\delta_R^{\max}) \right\}^{1/p} \tag{9.26}$$

where $p = 1, 2, \ldots$ Note that τ_{max}, χ_{max}, $\delta_T^{\max}$, and $\delta_R^{\max}$ define the upper limits of the ranges over which the approximations $\tilde{\rho}_{l,oq;l,oq}(\tau) \approx \rho_{l,oq;l,oq}(\tau)$, $\tilde{\rho}_{l,oq;l,oq}(\chi) \approx \rho_{l,oq;l,oq}(\chi)$, and $\tilde{\rho}_{l,oq;l,o'q'} \approx \rho_{l,oq;l,o'q'}$ are of interest. For $\tilde{\rho}_{l,oq;l,oq}(\chi)$ and $\tilde{\rho}_{l,oq;l,o'q'}$, if we replace $\tilde{\varphi}_{l,i,r,n}^{R}$ by $\tilde{\varphi}_{l,i,r,n}^{\prime R}$ and $\tilde{\varphi}_{l,i,r,n}^{\prime\prime R}$, respectively, the three error norms $E_{1,l}^{(p)}$, $E_{2,l}^{(p)}$, and $E_{3,l}^{(p)}$ can be minimized independently.

9.3.3.2 A Deterministic Simulation Model for Narrowband MIMO Channels

Analogous to Section 9.3.2, if we impose $L = 1$ and $\Lambda_l = 1$ on the wideband simulation model in Eq. (9.18), it reduces to a narrowband MIMO channel simulator. It follows that $\tilde{h}_{oq}(t, \tau') = \tilde{h}_{oq}(t)\,\delta(\tau')$ holds. The complex fading envelope $\tilde{h}_{oq}(t)$ of the deterministic simulation model is then given by

$$\tilde{h}_{oq}(t) = \tfrac{1}{\sqrt{N}} \sum_{n=1}^{N} \exp\left\{ j \left[\tilde{\psi}_n - 2\pi f_c \tau_n + 2\pi f_D t \cos\left(\tilde{\varphi}_n^R - \gamma\right) \right] \right\} \tag{9.27}$$

where the phases $\tilde{\psi}_n$ are simply the outcomes of a random generator uniformly distributed over $[0, 2\pi)$, the discrete AoAs $\tilde{\varphi}_n^R$ will be kept constant during simulation, and the other symbol definitions are the same as those in Eq. (9.14). The correlation properties of this narrowband simulation model can be obtained from Eq. (9.21) by simply neglecting the subscripts $(\cdot)_l$, $(\cdot)_i$, and $(\cdot)_r$ in all the affected symbols. Thus

$$\tilde{\rho}_{oq;o'q'}(\tau, \chi) = \frac{1}{N} \sum_{n=1}^{N} e^{j(C + P\cos\tilde{\varphi}_n^R + J\sin\tilde{\varphi}_n^R)} \tag{9.28}$$

with

$$P = XU + y\cos\beta_R + x\cos\gamma \tag{9.29a}$$

$$J = XV + y\sin\beta_R + z\Delta\sin\beta_T + x\sin\gamma \tag{9.29b}$$

where other parameters are the same as defined in Eq. (9.16). Similar to the wideband simulation model, we have $A = a + jP$ and $B = b + jJ$. From Eq. (9.28), it is clear that only $\{\tilde{\varphi}_n^R\}_{n=1}^N$ needs to be determined for this deterministic simulation model. The model parameters $\{\tilde{\varphi}_n^R\}_{n=1}^N$ can be calculated by using the same optimization method as the wideband simulation model. Therefore, by removing the subscript $(\cdot)_l$ in Eqs (9.24)–(9.26), the model parameters can be obtained as follows

$$E_1^{(p)} := \left\{ \int_0^{\tau_{\max}} \left| \rho_{oq;oq}(\tau) - \tilde{\rho}_{oq;oq}(\tau) \right|^p d\tau / \tau_{\max} \right\}^{1/p} \tag{9.30}$$

$$E_2^{(p)} := \left\{ \int_0^{\chi_{\max}} \left| \rho_{oq;oq}(\chi) - \tilde{\rho}_{oq;oq}(\chi) \right|^p d\chi / \chi_{\max} \right\}^{1/p} \tag{9.31}$$

$$E_3^{(p)} := \left\{ \int_0^{\delta_T^{\max}} \int_0^{\delta_R^{\max}} \left| \rho_{oq;o'q'} - \tilde{\rho}_{oq;o'q'} \right|^p d\delta_T d\delta_R / (\delta_T^{\max} \delta_R^{\max}) \right\}^{1/p} \tag{9.32}$$

Similarly, for $\tilde{\rho}_{oq;oq}(\chi)$ and $\tilde{\rho}_{oq;o'q'}$, if we replace $\tilde{\varphi}_n^R$ by $\tilde{\varphi}_n'^R$ and $\tilde{\varphi}_n''^R$, respectively, the three error norms $E_1^{(p)}$, $E_2^{(p)}$, and $E_3^{(p)}$ can be minimized independently.

9.3.4 *Numerical Results and Analysis*

Based on the STF CFs derived for wideband and narrowband MIMO channels in Section 9.3.2, the degenerate CFs are numerically analyzed in detail. In addition, verification of the proposed wideband and narrowband deterministic SoS simulation models is carried out by comparing the correlation properties of the simulation models with those of the corresponding reference models. All the results presented in this section are obtained using the following basic parameters: $f_c = 5\,\text{GHz}$, $f_D = 463\,\text{Hz}$, $D = 2000\,\text{m}$, $\beta_T = \pi/6$, $\beta_R = \pi/3$, and $\gamma = 7\pi/12$.

9.3.4.1 Correlation Properties of Wideband MIMO Channel Models

Without loss of any generality, we constrain our investigation to the correlation properties of the second tap ($l = 2$) with $\Lambda_2 = 4$ and $\left\{R_{2,i}\right\}_{i=0}^{3} = \{50, 100, 400, 750\}$m based on Eq. (9.11). The discrete COST 207 TU channel model with $\{\tau_l'\}_{l=0}^{5} = \{0, 0.2, 0.5, 1.6, 2.3, 5\}$ μs will be applied. For simplicity, we assume that $k_{l,i,r} = k$ for all effective clusters in the tap.

Figures 9.2 and 9.3 show the resulting frequency CF and SF CF with $\delta_T = 0$, respectively. As a good trade off between complexity and performance of the wideband simulation model, $N = 45$ effective scatterers and L_p-norm $p = 2$ are used in both figures. The parameter of L_p-norm $\chi_{\max} = 10\,\text{MHz}$ is used in Figure 9.2, while the parameter of L_p-norm $\delta_R^{\max} = 3\lambda$ is applied in Figure 9.3. From Figure 9.2, it is clear that the trend of frequency CFs decreases with the increase in frequency separation χ. Figure 9.2 also illustrates that the frequency correlations vary according to the environment parameter k that controls the angle spread of the AoA. It can be observed that the frequency correlations increase with the increase of k (that is, with the decrease of angle spread of AoA).

Figure 9.3 shows that the trend of the SF CFs decreases with the increase of the space separation δ_R. Figure 9.3 also depicts the impact of the environment parameter k on the SF correlations. It is obvious that the SF correlations increase with the increase of k. From Figure 9.3, we can also see the impact of frequency separation on spatial correlations. It is clear that the frequency separation decreases the spatial correlation. Therefore, we can conclude that in such a case, the resulting correlation is jointly contributed to by the actual spatial correlation (due

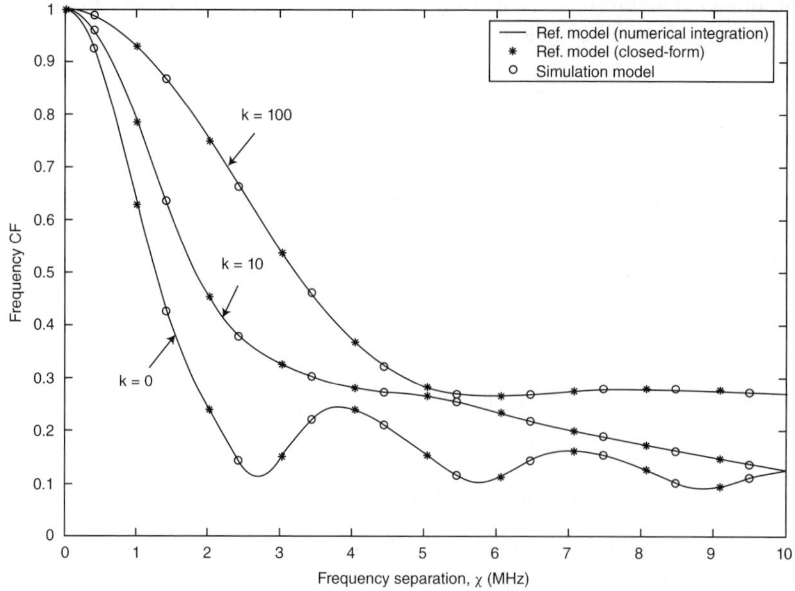

Figure 9.2 The frequency CFs $\left|\rho_{l,oq;l,oq}(\chi)\right|$ (reference model) and $\left|\tilde{\rho}_{l,oq;l,oq}(\chi)\right|$ (simulation model, $N = 45$) for different values of the parameter k

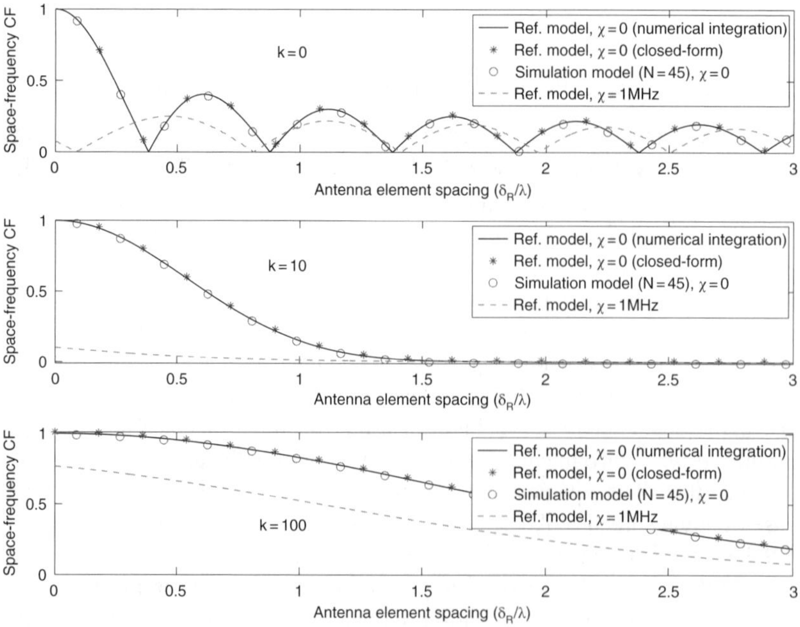

Figure 9.3 The SF CFs $\left|\rho_{l,oq;l,o'q'}\right|$ (reference model) and $\left|\tilde{\rho}_{l,oq;l,o'q'}\right|$ (simulation model, $N = 45$) for different values of the parameter k

to the spatial distance/geometry of arrays only) and inherent frequency correlation. Note that this resulting correlation is important for the appropriate design of STF coded MIMO-OFDM systems [Bolcskei et al. 2003] and is also useful for the sensible utilization of SF diversity [Kalkan and Clarke 1997].

In addition, we depict in both Figures 9.2 and 9.3 the CFs of the reference model – both the closed-form expression and from the numerical integration method – and the simulation model. Clearly, all these results match very well, demonstrating the validity of our derivation and excellent performance of our simulation model.

9.3.4.2 Correlation Properties of Narrowband MIMO Channel Models

In this subsection, we will investigate the correlation properties for narrowband MIMO channels based on Eq. (9.16) and evaluate the performance of the proposed narrowband simulation model. All the results presented in this subsection are obtained using $k = 3$ and $\mu = \pi$.

Figures 9.4(a) and (b) illustrate the SF CFs against the frequency separation and space separation at the BS and MS, respectively. Comparing them, we find that the impact of the normalized antenna spacing at the MS is greater than the one at the BS. This is because that the angular spread Θ at the BS is generally small for macrocell scenarios. Figures 9.4(a) and (b) also show that the trend of the relevant CFs decreases with the increase of the frequency separation χ and space separation δ_T and δ_R.

A plot of the time CF of the reference model is shown in Figure 9.5(a). This figure also depicts the resulting time CF of the simulation model designed with the L_p-norm using $p = 2$ and $\tau_{\max} = 0.08$ s, and $N = 30$ effective scatterers. Figure 9.5(b) illustrates the frequency CF of the reference model and the one of the simulation model, when applying the L_p-norm with $p = 2$ and $\chi_{\max} = 8$ MHz, and $N = 30$ effective scatterers. Figures 9.6(a) and (b) depict the space CF of the reference model and the one of the simulation model with $N = 30$, respectively. The discrete AoAs $\tilde{\varphi}_n''^R$ have been obtained using the L_p-norm, with $p = 2$, $\delta_T^{\max} = 30\lambda$, and $\delta_R^{\max} = 3\lambda$. Figures 9.5 and 9.6 clearly demonstrate that the proposed deterministic simulation model can fit the underlying reference model very well in terms of time, frequency, and space correlation properties.

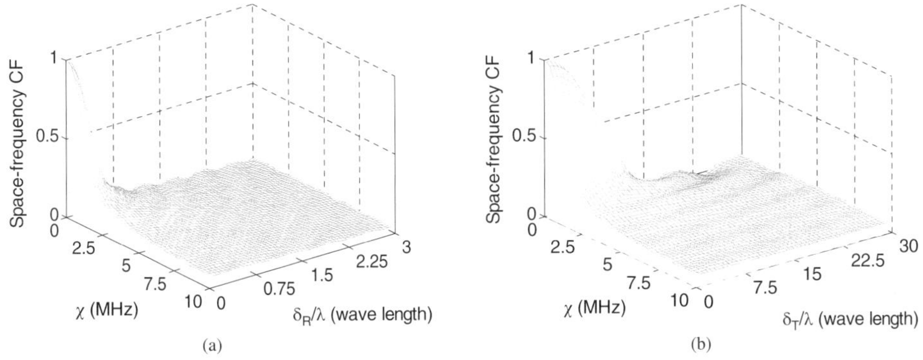

Figure 9.4 (a) The SF CF $|\rho_{oq;o'q'}(\chi)|$ versus the frequency separation χ and the normalized antenna spacing at the MS δ_R with $\delta_T = 0$; (b) the SF CF $|\rho_{oq;o'q'}(\chi)|$ versus the frequency separation χ and the normalized antenna spacing at the BS δ_T with $\delta_R = 0$

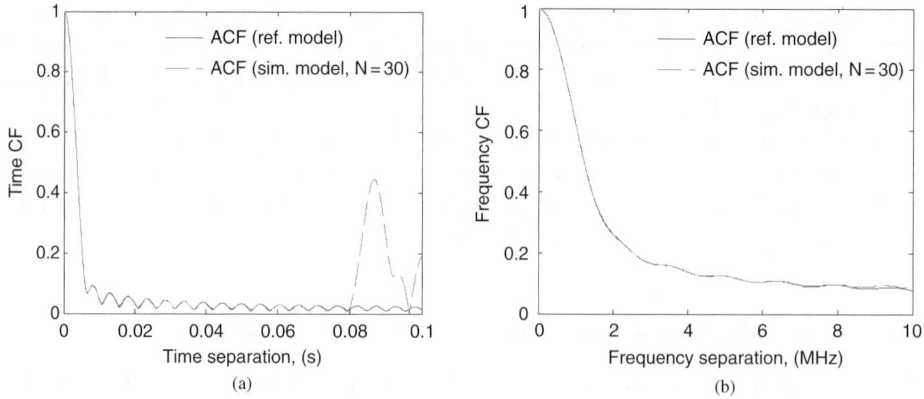

Figure 9.5 (a) The time CF $\left|\rho_{oq;oq}(\tau)\right|$ of the reference model and the time CF $\left|\tilde{\rho}_{oq;oq}(\tau)\right|$ of the corresponding simulation model with $N = 30$; (b) the frequency CF $\left|\rho_{oq;oq}(\chi)\right|$ of the reference model and the time CF $\left|\tilde{\rho}_{oq;oq}(\chi)\right|$ of the corresponding simulation model with $N = 30$

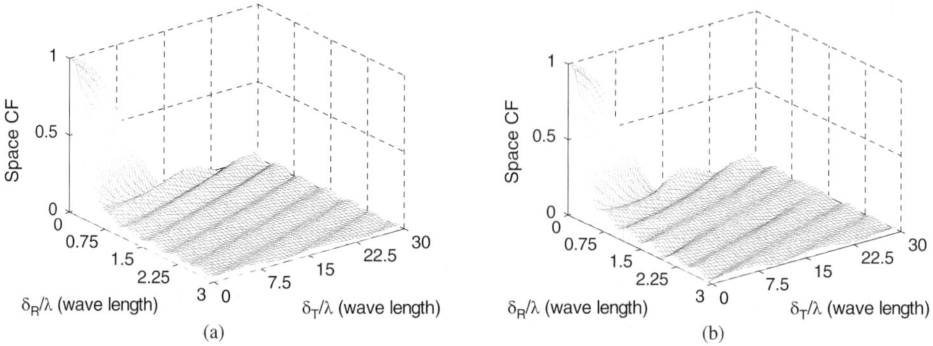

Figure 9.6 (a) The space CF $\left|\rho_{oq;o'q'}\right|$ of the reference model and (b) the space CF $\left|\tilde{\rho}_{oq;o'q'}\right|$ of the corresponding simulation model with $N = 30$

9.3.5 Summary

In Section 9.3, we have extended the narrowband one-ring MIMO model to a new wideband multi-ring MIMO RS-GBSM. According to the TDL structure of our model, the closed-formed expression of the STF CF for each tap has been derived. We have demonstrated that the traditional narrowband one-ring model is actually a special case of the proposed wideband model. Therefore, the derived generic STF CF for wideband MIMO channels can be reduced to the STF CF for narrowband MIMO channels, with a compact closed-form expression, by removing the frequency-selectivity. From the proposed wideband multi-ring MIMO reference model and the traditional narrowband one-ring MIMO reference model, corresponding wideband and narrowband deterministic SoS simulation models have been proposed. Numerical results have revealed the impact of the environment parameter k on frequency correlations and the inherent frequency correlations within spatial correlations. Finally, the excellent agreement of the correlation properties between the reference models and simulation models has validated the utility of the proposed deterministic simulation models. The proposed multi-ring MIMO reference model and deterministic simulation models are very useful for theoretical analysis and practical simulation, respectively, of frequency-correlated MIMO channels such as frequency-hopping MIMO and MIMO-OFDM channels.

9.4 Scattering Theoretical Channel Models for V2V Systems

This section will focus on the development of more real scattering theoretical channel models for MIMO V2V channels in non-isotropic scattering environments.

9.4.1 Modeling and Simulation of MIMO V2V Channels: Narrowband

The two-ring RS-GBSM reference model used in Chapter 4 is oversimple and thus cannot capture some important features of V2V channels. Therefore, as reviewed in Chapter 4, several important RS-GBSMs for narrowband MIMO V2V channels have been provided [Akki 1994, Akki and Haber 1986, Pätzold et al. 2008, Zajic and Stuber 2008b]. However, none of the previously reported RS-GBSMs is sufficiently general to allow it to characterize a wide variety of V2V scenarios, especially picocell scenarios, or to take into account the impact of VTD on channel statistics. Although the Doppler PSD, envelope LCR, and AFD are the most important statistics that distinguish V2V channels from F2M channels, more detailed investigations of these statistics in non-isotropic scattering environments are surprisingly lacking in the open literature. Moreover, frequency correlations of subchannels with different carrier frequencies have not been studied for V2V communications, although OFDM has already been suggested for use in IEEE 802.11p.

Motivated by these gaps in our knowledge, in this chapter we propose a new narrowband RS-GBSM that addresses all these shortcomings of existing RS-GBSMs. Based on the proposed model, some important channel statistics, such as the STF CF, SDF PSD, envelope LCR, and AFD, are derived. The content of the section is as follows.

1. A generic RS-GBSM for narrowband non-isotropic scattering MIMO V2V Ricean fading channels will be introduced. The proposed model can be adapted to a wide variety of scenarios – macro-, micro-, and picocells – by adjusting the model parameters.
2. By distinguishing between the moving cars and the stationary roadside environment in micro- and picocell scenarios, the model becomes the first RS-GBSM to consider the impact of VTD on V2V channel characteristics.
3. This section will present a new general method to derive the exact relationship between the AoA and AoD for any known shape of the scattering region: one-ring, two-ring, or ellipse, in a wide variety of scenarios.
4. This chapter will point out that the widely used CF definition [Abdi and Kaveh 2002, Pätzold et al. 2008, Zajic and Stuber 2008b;b] is incorrect and is actually the complex conjugate of the correct CF definition as given in the book by Papoulis and Pillai [2002].
5. From the proposed model, this chapter will derive the STF CF and the corresponding SDF PSD, which are general and can be reduced to many existing CFs and PSDs (such as those

by Abdi and Kaveh [2002], Akki and Haber [1986], Pätzold et al. [2008], Zajic and Stuber [2008b]). In addition, our analysis shows that the space-Doppler power spectrum density (SD PSD) of a single-bounce two-ring model for non-isotropic scattering MIMO V2V fading channels derived by Zajic and Stuber [2008b] is incorrect.

6. Considering the simplified version (SISO) of our proposed MIMO model, this chapter will derive the envelope LCR and AFD, which include many existing LCRs and AFDs as special cases (for example those by Abdi et al. [2002], Akki [1994], Stüber [2001], Wang et al. [2007]). Our analysis shows several flaws in the derivation and investigation of the LCR and AFD in Zajic et al. 2008 and 2009, revealing some easily neglected but important issues.
7. Based on the derived STF CF, SDF PSD, envelope LCR, and AFD, this chapter will examine in more detail these channel statistics in terms of the important parameters and thus obtain some interesting observations. These observations and conclusions can be considered as guidance for adjusting the important parameters of our model properly and setting up more purposeful V2V channel measurement campaigns in the future. Finally, the theoretical results (Doppler PSDs, LCR, and AFD) and measurement data from Acosta-Marum and Brunelli [2007] are compared. Excellent agreement between the two demonstrates the utility of the proposed model.

9.4.1.1 An Adaptive Model for Non-isotropic Scattering MIMO V2V Ricean Fading Channels

Let us now consider a narrowband single-user MIMO V2V multicarrier communication system with M_T transmit and M_R receive omnidirectional antenna elements. Both the Tx and Rx are equipped with low-elevation antennas. Figure 9.7 illustrates the geometry of the proposed RS-GBSM, which is the combination of a single- and double-bounce two-ring model, a single-bounce ellipse model, and the LoS component. As an example, uniform linear antenna arrays with $M_T = M_R = 2$ were used here. The two-ring model defines two rings of effective scatterers, one around the Tx and the other around the Rx. Suppose there are N_1 effective scatterers around the Tx lying on a ring of radius R_T and the n_1th ($n_1 = 1, \ldots, N_1$) effective scatterer is denoted by $s^{(n_1)}$. Similarly, assume there are N_2 effective scatterers around the Rx lying on a ring of radius R_R and the n_2th ($n_2 = 1, \ldots, N_2$) effective scatterer is denoted by $s^{(n_2)}$. For the ellipse model, N_3 effective scatterers lie on an ellipse with the Tx and Rx located at the foci. The semi-major axis of the ellipse and the n_3th ($n_3 = 1, \ldots, N_3$) effective scatterer are denoted by a and $s^{(n_3)}$, respectively. The distance between the Tx and Rx is $D = 2f$ with f denoting the half length of the distance between the two focal points of the ellipse. The antenna element spacings at the Tx and Rx are designated by δ_T and δ_R, respectively. It is normally assumed that the radii R_T and R_R, and the difference between the semi-major axis a and the parameter f, are all much greater than the antenna element spacings δ_T and δ_R; that is, $\min\{R_T, R_R, a - f\} \gg \max\{\delta_T, \delta_R\}$. The multi-element antenna tilt angles are denoted by β_T and β_R. The Tx and Rx move with speeds υ_T and υ_R in directions determined by the angles of motion γ_T and γ_R, respectively. The AoA of the wave traveling from an effective scatterer $s^{(n_i)}$ ($i \in \{1, 2, 3\}$) toward the Rx is denoted by $\varphi_R^{(n_i)}$. The AoD of the wave that impinges on the effective scatterer $s^{(n_i)}$ is designated by $\varphi_T^{(n_i)}$. Note that $\varphi_{R_q}^{LoS}$ denotes the AoA of a LoS path.

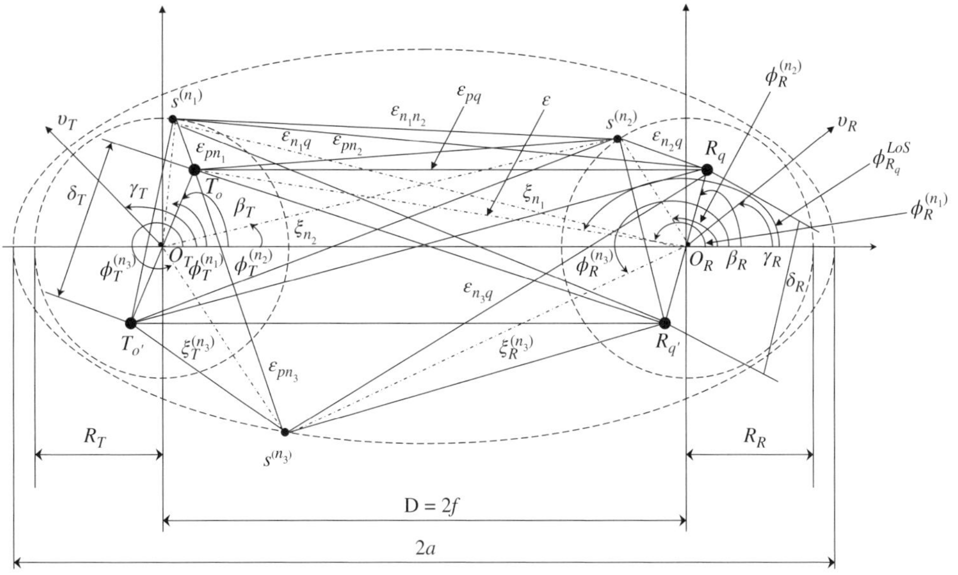

Figure 9.7 A generic channel model combining a two-ring model and an ellipse model with LoS components and single- and double-bounced rays for a MIMO V2V channel ($M_T = M_R = 2$)

The MIMO fading channel can be described by a matrix $\mathbf{H}(t) = [h_{oq}(t)]_{M_R \times M_T}$ of size $M_R \times M_T$. The received complex fading envelope between the oth ($o = 1, \ldots, M_T$) Tx and the qth ($q = 1, \ldots, M_R$) Rx at the carrier frequency f_c is a superposition of the LoS, single-, and double-bounced components, and can be expressed as

$$h_{oq}(t) = h_{oq}^{LoS}(t) + h_{oq}^{SB}(t) + h_{oq}^{DB}(t) \tag{9.33}$$

where

$$h_{oq}^{LoS}(t) = \sqrt{\frac{K_{oq}\Omega_{oq}}{K_{oq}+1}} e^{-j2\pi f_c \tau_{oq}} e^{j\left[2\pi f_{T_{max}} t \cos\left(\pi - \varphi_{R_q}^{LoS} + \gamma_T\right) + 2\pi f_{R_{max}} t \cos\left(\varphi_{R_q}^{LoS} - \gamma_R\right)\right]} \tag{9.34a}$$

$$\begin{aligned} h_{oq}^{SB}(t) &= \sum_{i=1}^{I} h_{oq}^{SB_i}(t) = \sum_{i=1}^{I} \sqrt{\frac{\eta_{SB_i}\Omega_{oq}}{K_{oq}+1}} \lim_{N_i \to \infty} \sum_{n_i=1}^{N_i} \frac{1}{\sqrt{N_i}} e^{j\left(\psi_{n_i} - 2\pi f_c \tau_{oq,n_i}\right)} \\ &\times e^{j\left[2\pi f_{T_{max}} t \cos\left(\varphi_T^{(n_i)} - \gamma_T\right) + 2\pi f_{R_{max}} t \cos\left(\varphi_R^{(n_i)} - \gamma_R\right)\right]} \end{aligned} \tag{9.34b}$$

$$\begin{aligned} h_{oq}^{DB}(t) &= \sqrt{\frac{\eta_{DB}\Omega_{oq}}{K_{oq}+1}} \lim_{N_1, N_2 \to \infty} \sum_{n_1, n_2 = 1}^{N_1, N_2} \frac{1}{\sqrt{N_1 N_2}} e^{j\left(\psi_{n_1,n_2} - 2\pi f_c \tau_{oq,n_1,n_2}\right)} \\ &\times e^{j\left[2\pi f_{T_{max}} t \cos\left(\varphi_T^{(n_1)} - \gamma_T\right) + 2\pi f_{R_{max}} t \cos\left(\varphi_R^{(n_2)} - \gamma_R\right)\right]} \end{aligned} \tag{9.34c}$$

In Eq. (9.34), $\tau_{oq} = \varepsilon_{oq}/c$, $\tau_{oq,n_i} = (\varepsilon_{on_i} + \varepsilon_{n_i q})/c$, and $\tau_{oq,n_1,n_2} = (\varepsilon_{on_1} + \varepsilon_{n_1 n_2} + \varepsilon_{n_2 q})/c$ are the travel times of the waves through the links $T_o - R_q$, $T_o - s^{(n_i)} - R_q$, and $T_o - s^{(n_1)} - s^{(n_2)} - R_q$, respectively. Here, c is the speed of light and $I = 3$. The symbols K_{oq} and Ω_{oq}

designate the Ricean factor and the total power of the $T_o - R_q$ link, respectively. Parameters η_{SB_i} and η_{DB} specify how much the single- and double-bounced rays contribute to the total scattered power $\Omega_{oq}/(K_{oq}+1)$. Note that these energy-related parameters satisfy $\sum_{i=1}^{I}\eta_{SB_i}+\eta_{DB}=1$. The phases ψ_{n_i} and ψ_{n_1,n_2} are i.i.d. random variables with uniform distributions over $[-\pi,\pi)$, and $f_{T_{max}}$ and $f_{R_{max}}$ are the maximum Doppler frequencies with respect to the Tx and Rx, respectively. Note that the AoD $\varphi_T^{(n_i)}$ and AoA $\varphi_R^{(n_i)}$ are independent for double-bounced rays, while they are interdependent for single-bounced rays.

From Figure 9.7, and based on the application of the law of cosines in appropriate triangles, the distances ε_{oq}, ε_{on_i}, ε_{n_iq}, and $\varepsilon_{n_1n_2}$ in Eq. (9.34) for any scenario (macrocell ($D \geq 1000$ m), microcell ($300 < D < 1000$ m), or picocell ($D \leq 300$ m) scenario) can be expressed as

$$\varepsilon_{oq} \approx \varepsilon - k_q\delta_R\cos(\varphi_{R_q}^{LoS}-\beta_R) \tag{9.35}$$

$$\varepsilon_{on_1} \approx R_T - k_o\delta_T\cos(\varphi_T^{(n_1)}-\beta_T) \tag{9.36}$$

$$\varepsilon_{n_1q} \approx \xi_{n_1} - k_q\delta_R\cos(\varphi_R^{(n_1)}-\beta_R) \tag{9.37}$$

$$\varepsilon_{on_2} \approx \xi_{n_2} - k_o\delta_T\cos\left(\varphi_T^{(n_2)}-\beta_T\right) \tag{9.38}$$

$$\varepsilon_{n_2q} \approx R_R - k_q\delta_R\cos\left(\varphi_R^{(n_2)}-\beta_R\right) \tag{9.39}$$

$$\varepsilon_{on_3} \approx \xi_T^{(n_3)} - k_o\delta_T\cos\left(\varphi_T^{(n_3)}-\beta_T\right) \tag{9.40}$$

$$\varepsilon_{n_3q} \approx \xi_R^{(n_3)} - k_q\delta_R\cos\left(\varphi_R^{(n_3)}-\beta_R\right) \tag{9.41}$$

$$\varepsilon_{n_1n_2} \approx D - R_T\cos\varphi_T^{(n_1)} + R_R\cos\varphi_R^{(n_2)} \tag{9.42}$$

where $\varphi_{R_q}^{LoS}\approx\pi$, $\varepsilon\approx D-k_o\delta_T\cos\beta_T$, $\xi_{n_1}=\left(D^2+R_T^2-2D\times R_T\cos\varphi_T^{(n_1)}\right)^{-1/2}$, $\xi_{n_2}=\sqrt{D^2+R_R^2+2DR_R\cos\varphi_R^{(n_2)}}$, $\xi_T^{(n_3)}=\left(a^2+f^2+2af\cos\varphi_R^{(n_3)}\right)/\left(a+f\cos\varphi_R^{(n_3)}\right)$, $\xi_R^{(n_3)}=b^2/\left(a+f\cos\varphi_R^{(n_3)}\right)$, $k_o=(M_T-2o+1)/2$, and $k_q=(M_R-2q+1)/2$. Here b denotes the semi-minor axis of the ellipse and the equality $a^2=b^2+f^2$ holds. As shown in Appendix A.8.3, based on the newly proposed general method to derive the exact relationship between the AoA and AoD for any shape of the scattering region, we have

$$\sin\varphi_R^{(n_1)} = R_T\sin\varphi_T^{(n_1)}/\sqrt{R_T^2+D^2-2R_TD\cos\varphi_T^{(n_1)}} \tag{9.43}$$

$$\cos\varphi_R^{(n_1)} = -\left(D-R_T\cos\varphi_T^{(n_1)}\right)/\sqrt{R_T^2+D^2-2R_TD\cos\varphi_T^{(n_1)}} \tag{9.44}$$

$$\sin\varphi_T^{(n_2)} = R_R\sin\varphi_R^{(n_2)}/\sqrt{R_R^2+D^2+2R_RD\cos\varphi_R^{(n_2)}} \tag{9.45}$$

$$\cos\varphi_T^{(n_2)} = \left(D+R_R\cos\varphi_R^{(n_2)}\right)/\sqrt{R_R^2+D^2+2R_RD\cos\varphi_R^{(n_2)}} \tag{9.46}$$

$$\sin\varphi_T^{(n_3)} = b^2\sin\varphi_R^{(n_3)}/\left(a^2+f^2+2af\cos\varphi_R^{(n_3)}\right) \tag{9.47}$$

$$\cos\varphi_T^{(n_3)} = \left(2af+(a^2+f^2)\cos\varphi_R^{(n_3)}\right)/\left(a^2+f^2+2af\cos\varphi_R^{(n_3)}\right) \tag{9.48}$$

Note that the expressions in Eqs (9.35)–(9.48) are general and suitable for various scenarios. For macro- and microcell scenarios, the assumption $D \gg \max\{R_T, R_R\}$, which is invalid for picocell scenarios, is fulfilled. Then, the general expressions of ξ_{n_1} and ξ_{n_2} can further reduce to the widely used approximate expressions $\xi_{n_1} \approx D - R_T \cos\varphi_T^{(n_1)}$ and $\xi_{n_2} \approx D + R_R \cos\varphi_R^{(n_2)}$. In addition, the general expressions in Eqs (9.43)–(9.46) for the two-ring model can further reduce to the widely used approximate expressions as $\varphi_R^{(n_1)} \approx \pi - \Delta_T \sin\varphi_T^{(n_1)}$ and $\varphi_T^{(n_2)} \approx \Delta_R \sin\varphi_R^{(n_2)}$ with $\Delta_T \approx R_T/D$ and $\Delta_R \approx R_R/D$. Moreover, the relationships of Eqs (9.47) and (9.48) for the ellipse model obtained using our method significantly simplifies the relationships derived from pure ellipse properties, such as in the papers by Patzold and Youssef [2001, Eqs(A1)–(A3)] and Chen and Li [2007, Eqs (27), (28), (32)].

Since the number of effective scatterers is assumed to be infinite; that is, $N_i \to \infty$, the proposed model is actually a mathematical reference model and results in a Ricean PDF. For our reference model, the discrete expressions of the AoA, $\varphi_R^{(n_i)}$, and AoD, $\varphi_T^{(n_i)}$, can be replaced by the continuous expressions $\varphi_R^{(SB_i)}$ and $\varphi_T^{(SB_i)}$, respectively. To characterize AoD $\varphi_T^{(SB_i)}$ and AoA $\varphi_R^{(SB_i)}$ we use the von Mises PDF given in Eq. (9.7).

Note that the proposed model is adaptable to a wide variety of V2V propagation environments by adjusting the model parameters. It turns out that these important model parameters are the energy-related parameters η_{SB_i} and η_{DB}, and the Ricean factor K_{oq}. For a macrocell scenario, the Ricean factor K_{oq} and the energy parameter η_{SB_3} related to the single-bounce ellipse model are very small or even close to zero. The received signal power mainly comes from single- and double-bounced rays of the two-ring model, in which we assume that double-bounced rays carry more energy than single-bounced rays due to the large distance D (larger distance D results in the independence of the AoD and AoA); in other words, $\eta_{DB} > \max\{\eta_{SB_1}, \eta_{SB_2}\} \gg \eta_{SB_3}$. This means that a macrocell scenario can be well characterized using a two-ring model with a negligible LoS component. In contrast to macrocell scenarios, in micro- and picocell scenarios, VTD significantly affects the channel characteristics as described by Sen and Matolak [2008]. To consider the impact of the VTD on channel statistics, we need to distinguish between cars moving around the Tx and Rx and the stationary roadside environment: buildings, trees, parked cars, and so on. Therefore, we use a two-ring model to mimic the moving cars and an ellipse model to depict the stationary roadside environment. Note that ellipse models have been widely used to model F2M channels in micro- and picocell scenarios [Chen and Li 2007, Patzold and Youssef 2001]. However, to the best of the authors' knowledge, this is the first time that an ellipse model has been used to mimic V2V channels. For a low VTD, the value of K_{oq} is large since the LoS component can bear a significant amount of power. Also, the received scattered power is mainly from waves reflected by the stationary roadside environment, which is described by the scatterers located on the ellipse. The moving cars represented by the scatterers located on the two rings are sparse and thus more likely to be single-bounced, rather than double-bounced. This indicates that $\eta_{SB_3} > \max\{\eta_{SB_1}, \eta_{SB_2}\} > \eta_{DB}$ holds. For a high VTD, the value of K_{oq} is smaller than that in the low-VTD scenario. Also, due to the large number of moving cars, the double-bounced rays of the two-ring model bear more energy than the single-bounced rays of two-ring and ellipse models: $\eta_{DB} > \max\{\eta_{SB_1}, \eta_{SB_2}, \eta_{SB_3}\}$. Therefore, microcell and picocell scenarios that take into account VTD can be well characterized by utilizing a combined two-ring and ellipse model with a LoS component.

9.4.1.2 Generic Space–Time–Frequency CF and Space–Doppler-Frequency PSD

In this section, based on the channel model proposed in Eq. (9.33), we will derive the STF CF and the corresponding SDF PSD for a non-isotropic scattering environment.

New Generic Space–Time–Frequency CF

As mentioned in Chapter 2, WSS channels have fading statistics that remain constant over short periods of time or distance (of the order of tens of wavelengths). In the channel model developed here, we have used the WSS assumption. This means that we study the V2V channel over a short distance, when the WSS condition is fulfilled. Under the WSS condition, the normalized STF CF between any two complex fading envelopes $h_{oq}(t)$ and $h'_{o'q'}(t)$ with different carrier frequencies f_c and f'_c, respectively, is defined as [Wang et al. 2007]:

$$\rho_{h_{oq}h'_{o'q'}}(\tau,\chi) = \frac{\mathbf{E}\left[h_{oq}(t)\,h'^{*}_{o'q'}(t-\tau)\right]}{\sqrt{\Omega_{oq}\Omega_{o'q'}}}$$

$$= \rho_{h^{LoS}_{oq}h'^{LoS}_{o'q'}}(\tau,\chi) + \sum_{i=1}^{I}\rho_{h^{SB_i}_{oq}h'^{SB_i}_{o'q'}}(\tau,\chi) + \rho_{h^{DB}_{oq}h'^{DB}_{o'q'}}(\tau,\chi) \quad (9.49)$$

where $(\cdot)^*$ denotes the complex conjugate operation, $\mathbf{E}[\cdot]$ is the statistical expectation operator, $o, o' \in \{1, 2, \ldots, M_T\}$, and $q, q' \in \{1, 2, \ldots, M_R\}$. It should be observed that Eq. (9.49) is a function of time separation τ, space separation δ_T and δ_R, and frequency separation $\chi = f'_c - f_c$. Note that the CF definition in Eq. (9.49) is different from the following definition that is widely used elsewhere [Abdi and Kaveh 2002, Pätzold et al. 2008, Zajic and Stuber 2008a,b]:

$$\tilde{\rho}_{h_{oq}h'_{o'q'}}(\tau,\chi) = \mathbf{E}\left[h_{oq}(t)\,h'^{*}_{o'q'}(t+\tau)\right] / \sqrt{\Omega_{oq}\Omega_{o'q'}} \quad (9.50)$$

The CF definition in Eq. (9.49) is actually the correct one following the CF definition given in Papoulis and Pillai [2002, Eq. (9-51)]. It can easily be shown that the expression Eq. (9.50) equals the complex conjugate of the correct CF in Eq. (9.49); that is, $\tilde{\rho}_{h_{oq}h'_{o'q'}}(\tau,\chi) = \rho^{*}_{h_{oq}h'_{o'q'}}(\tau,\chi)$. It is thus an incorrect definition. Only when $\rho^{*}_{h_{oq}h'_{o'q'}}(\tau,\chi)$ is a real function (no imaginary part), does $\tilde{\rho}_{h_{oq}h'_{o'q'}}(\tau,\chi) = \rho_{h_{oq}h'_{o'q'}}(\tau,\chi)$ hold.

Substituting Eqs (9.34a) and (9.35) into Eq. (9.49), we can obtain the STF CF of the LoS component as

$$\rho_{h^{LoS}_{oq}h'^{LoS}_{o'q'}}(\tau,\chi) = \sqrt{\frac{K_{oq}K_{o'q'}}{(K_{oq}+1)(K_{o'q'}+1)}}\, e^{j\frac{2\pi\chi}{c}\left(D-k_{o'}\delta_T\cos\beta_T+k_{q'}\delta_R\cos\beta_R\right)}$$
$$\times e^{j2\pi(O\cos\beta_T-Q\cos\beta_R)}e^{j2\pi\tau(f_{T_{max}}\cos\gamma_T-f_{R_{max}}\cos\gamma_R)} \quad (9.51)$$

where $O = (o'-o)\,\delta_T/\lambda$, $Q = (q'-q)\,\delta_R/\lambda$, $k'_o = (M_T - 2o' + 1)/2$, and $k'_q = (M_R - 2q' + 1)/2$.

Applying the von Mises PDF to the two-ring model, we obtain $f\left(\varphi_T^{SB_1}\right) = \exp\left[k_T^{TR}\cos\left(\varphi_T^{SB_1}\right)\right] - \mu_T^{TR})] / \left[2\pi I_0\left(k_T^{TR}\right)\right]$ for the AoD $\varphi_T^{SB_1}$ and $f\left(\varphi_R^{SB_2}\right) = \exp\left[k_R^{TR}\cos\left(\varphi_R^{SB_2} - \mu_R^{TR}\right)\right] / \left[2\pi I_0\left(k_R^{TR}\right)\right]$ for the AoA $\varphi_R^{SB_2}$. Substituting Eqs (9.34b) and (9.36)–(9.39) into Eq. (9.49), we can express the STF CF of the single-bounce two-ring model as

$$\rho_{h_{oq}^{SB_{1(2)}} h_{o'q'}^{\prime SB_{1(2)}}}(\tau,\chi) = \frac{\eta_{SB_{1(2)}}}{2\pi I_0\left(k_{T(R)}^{TR}\right)\sqrt{\left(K_{oq}+1\right)\left(K_{o'q'}+1\right)}} \int_{-\pi}^{\pi} e^{k_{T(R)}^{TR}\cos\left(\varphi_{T(R)}^{SB_{1(2)}}-\mu_{T(R)}^{TR}\right)}$$
$$\times e^{j2\pi\tau\left[f_{T_{max}}\cos\left(\varphi_T^{SB_{1(2)}}-\gamma_T\right)+f_{R_{max}}\cos\left(\varphi_R^{SB_{1(2)}}-\gamma_R\right)\right]} e^{j2\pi\left[O\cos\left(\varphi_T^{SB_{1(2)}}-\beta_T\right)+Q\cos\left(\varphi_R^{SB_{1(2)}}-\beta_R\right)\right]}$$
$$\times e^{\frac{j2\pi\chi}{c}\left[R_{T(R)}+\xi_{n_{1(2)}}-k_{o'}\delta_T\cos\left(\varphi_T^{SB_{1(2)}}-\beta_T\right)-k_{q'}\delta_R\cos\left(\varphi_R^{SB_{1(2)}}-\beta_R\right)\right]} d\varphi_{T(R)}^{SB_{1(2)}} \tag{9.52}$$

where the parameters $\sin\varphi_R^{SB_1}$, $\cos\varphi_R^{SB_1}$, $\sin\varphi_T^{SB_2}$, and $\cos\varphi_T^{SB_2}$ follow the expressions in Eqs (9.43)–(9.46), respectively. For the macro- and microcell scenarios, Eq. (9.52) can be further simplified as the following closed-form expression:

$$\rho_{h_{oq}^{SB_{1(2)}} h_{o'q'}^{\prime SB_{1(2)}}}(\tau,\chi) = \eta_{SB_{1(2)}} e^{jC_{T(R)}^{SB_{1(2)}}} \frac{I_0\left\{\sqrt{\left(A_{T(R)}^{SB_{1(2)}}\right)^2+\left(B_{T(R)}^{SB_{1(2)}}\right)^2}\right\}}{\sqrt{\left(K_{oq}+1\right)\left(K_{o'q'}+1\right)}I_0\left(k_{T(R)}^{TR}\right)} \tag{9.53}$$

where

$$A_{T(R)}^{SB_{1(2)}} = k_{T(R)}^{TR}\cos\mu_{T(R)}^{TR} + j2\pi\tau f_{T(R)_{max}}\cos\gamma_{T(R)} + j2\pi O(Q)\cos\beta_{T(R)} - j2\pi\chi X_{A_{T(R)}}/c \tag{9.54a}$$

$$B_{T(R)}^{SB_{1(2)}} = k_{T(R)}^{TR}\sin\mu_{T(R)}^{TR} + j2\pi\tau\left(f_{T(R)_{max}}\sin\gamma_{T(R)} + f_{R(T)_{max}}\Delta_{T(R)}\sin\gamma_{R(T)}\right) + j2\pi\left(O(Q)\sin\beta_{T(R)} + Q(O)\Delta_{T(R)}\sin\beta_{R(T)} - \chi X_{B_{T(R)}}/c\right) \tag{9.54b}$$

$$C_{T(R)}^{SB_1} = \mp 2\pi\tau f_{R(T)_{max}}\cos\gamma_{R(T)} \mp 2\pi Q(O)\cos\beta_{R(T)} + 2\pi\chi X_{C_{T(R)}}/c \tag{9.54c}$$

with $X_{A_T} = R_T - k_{o'}\delta_T\cos\beta_T$, $X_{B_T} = -k_{o'}\delta_T\sin\beta_T - k_{q'}\delta_R\Delta_T\sin\beta_R$, $X_{C_T} = R_T + D - k_{q'}\delta_R\cos\beta_R$, $X_{A_R} = -R_R - k_{q'}\delta_R\cos\beta_R$, $X_{B_R} = -k_{q'}\delta_R\sin\beta_R - k_{o'}\delta_T\Delta_R\sin\beta_T$, and $X_{C_R} = R_R + D + k_{o'}\delta_T\cos\beta_T$.

Applying the von Mises PDF to the ellipse model, we get $f\left(\varphi_R^{SB_3}\right) - \exp\left[k_R^{EL}\cos\left(\varphi_R^{SB_3}-\mu_R^{EL}\right)\right] / \left[2\pi I_0\left(k_R^{EL}\right)\right]$. Performing the substitution of Eq. (9.34b), Eq. (9.40), and Eq. (9.41) into Eq. (9.49), we can obtain the STF CF of the single-bounce ellipse model as

$$\rho_{h_{oq}^{SB_3} h_{o'q'}^{SB_3'}}(\tau,\chi) = \frac{\eta_{SB_3}}{2\pi I_0\left(k_R^{EL}\right)\sqrt{\left(K_{oq}+1\right)\left(K_{o'q'}+1\right)}} \int_{-\pi}^{\pi} e^{k_R^{EL}\cos\left(\varphi_R^{SB_3}-\mu_R^{EL}\right)}$$
$$\times e^{j2\pi\tau\left[f_{T_{max}}\cos\left(\varphi_T^{SB_3}-\gamma_T\right)+f_{R_{max}}\cos\left(\varphi_R^{SB_3}-\gamma_R\right)\right]} e^{j2\pi\left[O\cos\left(\varphi_T^{SB_3}-\beta_T\right)+Q\cos\left(\varphi_R^{SB_3}-\beta_R\right)\right]}$$
$$\times e^{\frac{j2\pi\chi}{c}\left[2a-k_{o'}\delta_T\cos\left(\varphi_T^{SB_3}-\beta_T\right)-k_{q'}\delta_R\cos\left(\varphi_R^{SB_3}-\beta_R\right)\right]} d\varphi_R^{SB_3} \tag{9.55}$$

where the parameters $\sin\varphi_T^{SB_3}$ and $\cos\varphi_T^{SB_3}$ follow the expressions in Eq. (9.47) and Eq. (9.48), respectively.

The substitution of Eq. (9.34c), Eq. (9.36), Eq. (9.39), and Eq. (9.42) into Eq. (9.49) results in the following STF CF for the double-bounce two-ring model

$$\rho_{h_{oq}^{DB} h_{o'q'}^{\prime DB}}(\tau,\chi) = \eta_{DB} e^{jC^{DB}} \frac{I_0\left\{\sqrt{\left(A_T^{DB}\right)^2+\left(B_T^{DB}\right)^2}\right\} I_0\left\{\sqrt{\left(A_R^{DB}\right)^2+\left(B_R^{DB}\right)^2}\right\}}{\sqrt{\left(K_{oq}+1\right)\left(K_{o'q'}+1\right)} I_0\left(k_T^{TR}\right) I_0\left(k_R^{TR}\right)} \tag{9.56}$$

where

$$A_{T(R)}^{DB} = k_{T(R)}^{TR}\cos\mu_{T(R)}^{TR} + j2\pi\tau f_{T(R)_{max}}\cos\gamma_{T(R)} + j2\pi O(Q)\cos\beta_{T(R)} \mp j2\pi\chi\left(R_{T(R)} \mp k_{o'(q')}\cos\beta_{T(R)}\right)/c \tag{9.57a}$$

$$B_{T(R)}^{DB} = k_{T(R)}^{TR}\sin\mu_{T(R)}^{TR} + j2\pi\tau f_{T(R)_{max}}\sin\gamma_{T(R)} + j2\pi O(Q)\sin\beta_{T(R)} + j2\pi\chi k_{o'(q')}\sin\beta_{T(R)}/c \tag{9.57b}$$

$$C^{DB} = 2\pi\chi\left(R_T + R_R + D\right)/c \tag{9.57c}$$

Since the derivations of Eqs (9.51)–(9.53), Eq. (9.55), and Eq. (9.56) are similar, only a brief outline of the derivation of Eq. (9.53) is given in Appendix A.8.4, while the others are omitted for brevity.

The derived STF CF in Eq. (9.49) includes many existing CFs as special cases. If we only consider the two-ring model ($\eta_{SB_3} = 0$) for a V2V channel in a macro- or microcell scenario ($D \gg \max\{R_T, R_R\}$) with the frequency separation $\chi = 0$, then the CF in Eq. (9.49) will be reduced to the CF of Zajic and Stuber [2008b, Eq. (18)], where the time separation τ should be replaced by $-\tau$ since the CF definition Eq. (9.50) is used in that paper. Consequently, the derived STF CF in Eq. (9.49) also includes other CFs listed by Zajic and Stuber [2008b] as special cases, when τ is replaced by $-\tau$. If we consider the one-ring model only around the Rx for a F2M channel in a macrocell scenario ($\eta_{SB_1} = \eta_{SB_3} = \eta_{DB} = f_{T_{max}} = 0$) with non-LoS (NLoS) condition ($K_{oq} = 0$), the derived STF CF in Eq. (9.49) includes the CF Eq. (9.16) in Section 9.3 and, subsequently, other CFs listed in Section 9.3 as special cases, when τ is replaced by $-\tau$. Furthermore, the CF in Wang et al. [2007, Eq. (7)] can be obtained from Eq. (9.49) with $K_{oq} = f_{T_{max}} = \chi = \eta_{SB_3} = \eta_{DB} = 0$. Consequently, the other CFs listed by Wang et al. [2007] can also be obtained from Eq. (9.49).

New Generic Space-Doppler-Frequency PSD

Applying the Fourier transform to the STF CF in Eq. (9.49) in terms of τ, we can obtain the corresponding SDF PSD as

$$\begin{aligned} S_{h_{oq}h'_{o'q'}}(f_D,\chi) &= \mathcal{F}\left\{\rho_{h_{oq}h'_{o'q'}}(\tau,\chi)\right\} = \int_{-\infty}^{\infty}\rho_{h_{oq}h'_{o'q'}}(\tau,\chi)\, e^{-j2\pi f_D\tau}d\tau \\ &= S_{h_{oq}^{LoS}h_{o'q'}^{\prime LoS}}(f_D,\chi) + \sum_{i=1}^{I} S_{h_{oq}^{SB_i}h_{o'q'}^{SB_{i'}}}(f_D,\chi) + S_{h_{oq}^{DB}h_{o'q'}^{\prime DB}}(f_D,\chi) \end{aligned} \tag{9.58}$$

where f_D is the Doppler frequency. The integral in Eq. (9.58) must be evaluated numerically in the case of the single-bounce two-ring and ellipse models. For other cases, we can obtain the following closed-form solutions.

1. In the case of the LoS component, substituting Eq. (9.51) into Eq. (9.58) we have

$$S_{h_{oq}^{LoS} h'^{LoS}_{o'q'}}(f_D, \chi) = \mathcal{F}\left\{\rho_{h_{oq}^{LoS} h'^{LoS}_{o'q'}}(\tau, \chi)\right\} = \sqrt{\frac{K_{oq} K_{o'q'}}{\left(K_{oq}+1\right)\left(K_{o'q'}+1\right)}} e^{j2\pi O \cos\beta_T}$$
$$\times e^{j2\pi Q \cos\beta_R} e^{j\frac{2\pi\chi}{c}\left(D - k_{o'}\delta_T \cos\beta_T + k_{q'}\delta_R \cos\beta_R\right)}$$
$$\times \delta\left(f_D - f_{T_{max}} \cos\gamma_T + f_{R_{max}} \cos\gamma_R\right). \quad (9.59)$$

   where $\delta(\cdot)$ denotes the Dirac delta function.

2. In terms of the single-bounce two-ring model for macro- and microcell scenarios, substituting Eq. (9.53) into Eq. (9.58) we have

$$S_{h_{oq}^{SB_{1(2)}} h'^{SB_{1(2)}}_{o'q'}}(f_D, \chi) = \mathcal{F}\left\{\rho_{h_{oq}^{SB_{1(2)}} h'^{SB_{1(2)}}_{o'q'}}(\tau, \chi)\right\} = \frac{\eta_{SB_{1(2)}} 2 e^{jU_{T(R)}^{SB_{1(2)}}}}{I_0\left(k_{T(R)}^{TR}\right)}$$
$$\times \frac{e^{jO_{T(R)}^{SB_{1(2)}} \frac{D_{T(R)}^{SB_{1(2)}}}{W_{T(R)}^{SB_{1(2)}}}} \cos\left(\frac{E_{T(R)}^{SB_{1(2)}}}{W_{T(R)}^{SB_{1(2)}}} \sqrt{W_{T(R)}^{SB_{1(2)}} - \left(O_{T(R)}^{SB_{1(2)}}\right)^2}\right)}{\sqrt{\left(K_{oq}+1\right)\left(K_{o'q'}+1\right)} \sqrt{W_{T(R)}^{SB_{1(2)}} - \left(O_{T(R)}^{SB_{1(2)}}\right)^2}} \quad (9.60)$$

   where $O_{T(R)}^{SB_{1(2)}} = 2\pi\left(f_D \pm f_{R(T)_{max}} \cos\gamma_{R(T)}\right)$

$$U_{T(R)}^{SB_{1(2)}} = \mp 2\pi Q(O) \cos\beta_{R(T)} + 2\pi\chi\left(R_{T(R)} + D \pm k_{q'(o')}\delta_{R(T)} \cos\beta_{R(T)}\right)/c \quad (9.61a)$$

$$W_{T(R)}^{SB_{1(2)}} = 4\pi^2 f_{T(R)_{max}}^2 + 4\pi^2 f_{R(T)_{max}}^2 \Delta_{T(R)}^2 \sin^2\gamma_{R(T)} + 8\pi^2 f_{T_{max}} f_{R_{max}}$$
$$\times \Delta_{T(R)} \sin\gamma_T \sin\gamma_R \quad (9.61b)$$

$$D_{T(R)}^{SB_{1(2)}} = -j2\pi k_{T(R)}^{TR} J_{T(R)} + 4\pi^2 O(Q)(f_{T(R)_{max}} \cos\left(\beta_{T(R)} - \gamma_{T(R)}\right) + \Delta_{T(R)} f_{R(T)_{max}}$$
$$\times \sin\beta_{T(R)} \sin\gamma_{R(T)}) + 4\pi^2 Q(O)(\Delta_{T(R)} f_{T(R)_{max}} \sin\beta_{R(T)} \sin\gamma_{T(R)}$$
$$+ \Delta_{T(R)}^2 f_{R(T)_{max}} \sin\beta_{R(T)} \sin\gamma_{R(T)}) - 4\pi^2\chi\left(f_{T(R)_{max}} Y_{TD_{T(R)}}\right.$$
$$\left. + f_{R(T)_{max}} \Delta_{T(R)} \sin\gamma_{R(T)} Y_{RD_{T(R)}}\right)/c \quad (9.61c)$$

$$E_{T(R)}^{SB_{1(2)}} = -j2\pi k_{T(R)}^{TR}(f_{T(R)_{max}} \sin\left(\gamma_{T(R)} - \mu_{T(R)}^{TR}\right) + f_{R(T)_{max}} \Delta_{T(R)} \sin\gamma_{R(T)}$$
$$\times \cos\mu_{T(R)}^{TR}) - 4\pi^2 O(Q)(f_{T(R)_{max}} \sin\left(\beta_{T(R)} - \gamma_{T(R)}\right) - \Delta_{T(R)} f_{R(T)_{max}}$$

$$\times \cos\beta_{T(R)} \sin\gamma_{R(T)}) - 4\pi^2 Q(O)\Delta_{T(R)} f_{T(R)_{max}} \sin\beta_{R(T)} \cos\gamma_{T(R)}$$
$$- 4\pi^2 \chi \left(f_{T(R)_{max}} Y_{TE_{T(R)}} + f_{R(T)_{max}} \Delta_{T(R)} \sin\gamma_{R(T)} Y_{RE_{T(R)}} \right) / c \qquad (9.61d)$$

with

$$J_{T(R)} = f_{T(R)_{max}} \cos\left(\gamma_{T(R)} - \mu^{TR}_{T(R)}\right) - f_{R(T)_{max}} \Delta_{T(R)} \sin\gamma_{R(T)} \sin\mu^{TR}_{T(R)} \qquad (9.62a)$$

$$Y_{TD_{T(R)}} = \pm R_{T(R)} \cos\gamma_{T(R)} - k_{o'(q')}\delta_{T(R)} \cos\left(\beta_{T(R)} - \gamma_{T(R)}\right)$$
$$- k_{q'(o')}\delta_{R(T)}\Delta_{T(R)} \sin\beta_{R(T)} \sin\gamma_{T(R)} \qquad (9.62b)$$

$$Y_{TE_{T(R)}} = \pm R_{T(R)} \sin\gamma_{T(R)} + k_{o'(q')}\delta_{T(R)} \sin\left(\beta_{T(R)} \mp \gamma_{T(R)}\right)$$
$$+ k_{q'(o')}\delta_{R(T)}\Delta_{T(R)} \sin\beta_{R(T)} \cos\gamma_{T(R)} \qquad (9.62c)$$

$$Y_{RD_{T(R)}} = -k_{o'(q')}\delta_{T(R)} \sin\beta_{T(R)} - k_{q'(o')}\delta_{R(T)}\Delta_{T(R)} \sin\beta_{R(T)} \qquad (9.62d)$$

$$Y_{RE_{T(R)}} = \pm R_{T(R)} - k_{o'(q')}\delta_{T(R)} \cos\beta_{T(R)} \qquad (9.62e)$$

For the Doppler PSD in Eq. (9.60), the range of Doppler frequencies is limited by $|f_D + f_{R_{max}} \times \cos\gamma_R| \leq \sqrt{W^{SB_1}_T}/(2\pi)$ and $|f_D - f_{T_{max}} \cos\gamma_T| \leq \sqrt{W^{SB_2}_R}/(2\pi)$. Note that the expression of Eq. (9.60) corrects those in the paper by Zajic and Stuber [2008b, Eqs (40),(41)].

3. In the case of the double-bounce two-ring model, substituting Eq. (9.56) into Eq. (9.58) we have

$$S_{h^{DB}_{oq} h'^{DB}_{o'q'}}(f_D, \chi) = \mathcal{F}\left\{\rho_{h^{DB}_{oq} h'^{DB}_{o'q'}}(\tau, \chi)\right\} = \frac{\eta_{DB} e^{jC^{DB}}}{\sqrt{(K_{oq}+1)(K_{o'q'}+1)} I_0\left(k^{TR}_T\right) I_0\left(k^{TR}_R\right)}$$
$$\times 2e^{jO^{DB}\frac{D^{DB}_T}{W^{DB}_T}} \frac{\cos\left(\frac{E^{DB}_T}{W^{DB}_T}\sqrt{W^{DB}_T - (O^{DB})^2}\right)}{\sqrt{W^{DB}_T - (O^{DB})^2}} \odot 2e^{jO^{DB}\frac{D^{DB}_R}{W^{DB}_R}} \frac{\cos\left(\frac{E^{DB}_R}{W^{DB}_R}\sqrt{W^{DB}_R - (O^{DB})^2}\right)}{\sqrt{W^{DB}_R - (O^{DB})^2}}$$
$$(9.63)$$

where $\odot$ denotes convolution, $O^{DB} = 2\pi f_D$, $W^{DB}_{T(R)} = 4\pi^2 f^2_{T(R)_{max}}$

$$D^{DB}_{T(R)} = 4\pi^2 O(Q) f_{T(R)_{max}} \cos\left(\beta_{T(R)} - \gamma_{T(R)}\right) - j2\pi k^{TR}_{T(R)} f_{T(R)_{max}}$$
$$\times \cos\left(\gamma_{T(R)} - \mu^{TR}_{T(R)}\right) - \frac{4\pi^2 \chi f_{T(R)_{max}} Y_{D_{T(R)}}}{c} \qquad (9.64a)$$

$$E^{DB}_{T(R)} = \pm 4\pi^2 O(Q) f_{T(R)_{max}} \sin\left(\beta_{T(R)} - \gamma_{T(R)}\right) \pm j2\pi k^{TR}_{T(R)} f_{T(R)_{max}}$$
$$\times \sin\left(\gamma_{T(R)} - \mu^{TR}_{T(R)}\right) + \frac{4\pi^2 \chi f_{T(R)_{max}} Y_{E_{T(R)}}}{c} \qquad (9.64b)$$

with $Y_{D_T} = R_T \cos\gamma_T - k_{o'}\delta_T \cos(\beta_T - \gamma_T)$, $Y_{E_T} = R_T \sin\gamma_T + k_{o'}\delta_T \sin(\beta_T - \gamma_T)$, $Y_{D_R} = -R_R \cos\gamma_R - k_{q'}\delta_R \cos(\beta_R - \gamma_R)$, and $Y_{E_R} = -R_R \sin\gamma_R + k_{q'}\delta_R$

$\sin(\beta_R - \gamma_R)$. For the Doppler PSD in Eq. (9.63), the range of Doppler frequencies is limited by $|f_D| \le f_{T_{max}} + f_{R_{max}}$. Due to the similar derivations of Eq. (9.59), Eq. (9.60), and Eq. (9.63), Appendix A.8.5 only gives a brief outline of the derivation of Eq. (9.60), while others are omitted here.

Many existing Doppler PSDs are special cases of the SDF PSD of Eq. (9.58). The simplest case is Clarke's Doppler PSD $1/\left(2\pi f_D \sqrt{1-(f_D/f_{R_{max}})^2}\right)$ ($|f_D| \le f_{R_{max}}$) [Stüber 2001], which can be obtained from Eq. (9.58) by setting $K_{oq} = 0$ (NLoS condition), $k_R^{TR} = 0$ (isotropic scattering around the Rx), $\chi = 0$ (no frequency separation), $f_{T_{max}} = \eta_{SB_1} = \eta_{SB_3} = \eta_{DB} = 0$ (fixed Tx, no scattering around the Tx), and applying $D \gg \max\{R_T, R_R\}$ (macro- and microcell scenarios). The Doppler PSD for isotropic V2V fading channels as presented by Akki and Haber [1986, Eq. (41)] can be obtained from Eq. (9.58) by setting $K_{oq} = k_T^{TR} = k_R^{TR} = \delta_T = \delta_R = \chi = \eta_{SB_1} = \eta_{SB_2} = \eta_{SB_3} = 0$ and using $D \gg \max\{R_T, R_R\}$. Similarly, the space-Doppler PSD for a non-isotropic double-bounce two-ring model shown as (42)[1] in [Zajic and Stuber 2008b] can be obtained from Eq. (9.58) by setting $K_{oq} = \chi = \eta_{SB_1} = \eta_{SB_2} = \eta_{SB_3} = 0$ and utilizing $D \gg \max\{R_T, R_R\}$, when D_T^{TR} and D_R^{TR} are replaced by $-D_T^{TR}$ and $-D_R^{TR}$, respectively, due to applying the different CF definitions.

To further demonstrate that the CF definition in Eq. (9.49) is correct, in Appendix A.8.6 we compare the Doppler PSDs with the CFs of Eq. (9.49) and Eq. (9.50), where it is shown that Eq. (9.49) enables the Doppler PSD to capture the underlying physical phenomena of real channels for any scenario, while the widely used expression of Eq. (9.50) is only applicable to scenarios where the Doppler PSD is a real function and symmetrical around the origin, for example Clarke's scenario. In many papers [Abdi and Kaveh 2002, Pätzold et al. 2008, Zajic and Stuber 2008a, b], Eq. (9.50) has been misapplied to non-isotropic F2M or V2V scenarios.

9.4.1.3 Envelope Level Crossing Rate and Average Fade Duration

This section will derive the envelope LCR and AFD for non-isotropic V2V fading channels by using the narrowband SISO V2V channel model, which is a simplified version of the MIMO V2V model above.

Narrowband SISO V2V Reference Model

Figure 9.8 shows the geometry of the corresponding SISO V2V RS-GBSM, which is a simplified version of the above proposed MIMO V2V RS-GBSM. This SISO V2V RS-GBSM is also the combination of a LoS component, a single- and double-bounce two-ring model, and a single-bounce ellipse model. In order to keep consistent, all the parameters shown in Figure 9.8 are the same as those presented in Figure 9.7. Therefore, from the SISO V2V model, the received complex fading envelope is a superposition of the LoS component

[1] Note that the expression of (42) in Zajic and Stuber [2008b] is inaccurate. The corrected expression should replace the terms $-jk_T\cos(\gamma_T - \mu_T)$ and $-jk_R\cos(\gamma_R - \mu_R)$ with $+jk_T\cos(\gamma_T - \mu_T)$ and $+jk_R\cos(\gamma_R - \mu_R)$, respectively.

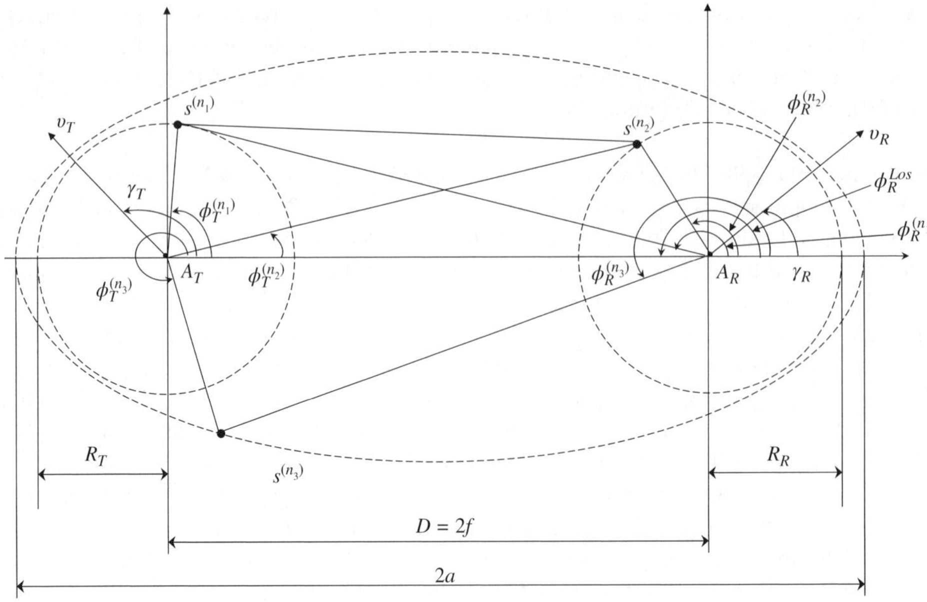

Figure 9.8 A generic channel model combining a two-ring model and an ellipse model with LoS components, single- and double-bounced rays for a SISO V2V channel

and diffuse component, which consists of single-, and double-bounced rays, and can be expressed as

$$h(t) = h^{LoS}(t) + h^{DIF}(t) = h^{LoS}(t) + h^{SB}(t) + h^{DB}(t) \quad (9.65)$$

where

$$h^{LoS}(t) = \sqrt{\frac{K\Omega}{K+1}} e^{j\left[2\pi f_{T_{max}} t \cos\left(\pi - \varphi_R^{LoS} + \gamma_T\right) + 2\pi f_{R_{max}} t \cos\left(\varphi_R^{LoS} - \gamma_R\right)\right]} \quad (9.66a)$$

$$h^{SB}(t) = \sum_{i=1}^{I} h^{SB_i}(t) = \sum_{i=1}^{I} \sqrt{\frac{\eta_{SB_i}\Omega}{K+1}} \lim_{N_i \to \infty} \sum_{n_i=1}^{N_i} \frac{1}{\sqrt{N_i}} \times e^{j\psi_{n_i}} e^{j\left[2\pi f_{T_{max}} t \cos\left(\varphi_T^{(n_i)} - \gamma_T\right) + 2\pi f_{R_{max}} t \cos\left(\varphi_R^{(n_i)} - \gamma_R\right)\right]} \quad (9.66b)$$

$$h^{DB}(t) = \sqrt{\frac{\eta_{DB}\Omega}{K+1}} \lim_{N_1, N_2 \to \infty} \sum_{n_1, n_2 = 1}^{N_1, N_2} \frac{1}{\sqrt{N_1 N_2}} \times e^{j\psi_{n_1, n_2}} e^{j\left[2\pi f_{T_{max}} t \cos\left(\varphi_T^{(n_1)} - \gamma_T\right) + 2\pi f_{R_{max}} t \cos\left(\varphi_R^{(n_2)} - \gamma_R\right)\right]}. \quad (9.66c)$$

Note that Eqs (9.66a)–(9.66c) can be obtained by removing the term related to τ_{oq} in Eq. (9.34a), the term related to τ_{oq,n_i} in Eq. (9.34b), and the term related to τ_{oq,n_1,n_2} in Eq. (9.34c), respectively. Therefore, all the parameters in Eq. (9.66) are the same as those in Eq. (9.34). Since the numbers of effective scatterers N_i tend to be infinite as shown in Eq. (9.66b) and Eq.

(9.66c), the discrete expressions of the AoA, $\varphi_R^{(n_i)}$, and AoD, $\varphi_T^{(n_i)}$, can be replaced by the continuous expressions $\varphi_R^{SB_i}$ and $\varphi_T^{SB_i}$, respectively. To characterize the AoD $\varphi_T^{SB_i}$ and AoA $\varphi_R^{SB_i}$, the von Mises PDF given in Eq. (9.7) is used. Analogous to our MIMO V2V model, for the angle of interest in this SISO V2V model, i.e., the AoD $\varphi_T^{SB_1}$ for the one ring around the Tx, the AoA $\varphi_R^{SB_2}$ for the one ring around the Rx, and the AoA $\varphi_R^{SB_3}$ for the ellipse, we use appropriate parameters of the von Mises PDF as μ_T^{TR} and k_T^{TR}, μ_R^{TR} and k_R^{TR}, and μ_R^{EL} and k_R^{EL}, respectively.

Derivation of Envelope LCR and AFD

In this section, based on the model developed in Eqs (9.65) and (9.66), we will derive the LCR and AFD for a non-isotropic scattering environment. The LCR, $L_\xi(r_l)$, is by definition the average number of times per second that the signal envelope, $\xi(t) = |h(t)|$, crosses a specified level r_l with positive/negative slope. Using the traditional PDF-based method of Youssef et al. [2005], we derive a general expression for the LCR for V2V Ricean fading channels as

$$L_\xi(r_l) = \frac{2r_l}{\pi^{3/2}}\sqrt{\frac{B(K+1)}{b_0}}e^{-K-(K+1)r_l^2}\int_0^{\pi/2}\cosh\left(2\sqrt{K(K+1)}r_l\cos\theta\right) \times\left[e^{-\left(\frac{\varsigma\zeta\sin\theta}{\sqrt{2B}}\right)^2}+\sqrt{\frac{\pi}{2B}}\varsigma\zeta\sin\theta\cdot\operatorname{erf}\left(\frac{\varsigma\zeta\sin\theta}{\sqrt{2B}}\right)\right]d\theta \tag{9.67}$$

where $\cosh(\cdot)$ is the hyperbolic cosine, erf$(\cdot)$ denotes the error function, and parameters B, ζ, and ς are $B = b_2 - b_1^2/b_0$, $\zeta = \sqrt{K\Omega/(K+1)}$, and $\varsigma = E + b_1/b_0$, respectively, with $E = 2\pi f_{T_{max}}\cos(\pi - \varphi_R^{LoS} + \gamma_T) + 2\pi f_{R_{max}}\cos(\varphi_R^{LoS} - \gamma_R)$. Finally, the key parameters b_m $(m = 0, 1, 2)$ are defined as

$$b_m = \frac{d^m\rho_{h^{DIF}}(\tau)}{2j^m d\tau^m}|_{\tau=0} \tag{9.68}$$

with $\rho_{h^{DIF}}(\tau) = \mathbf{E}\left[h^{DIF}(t)\,h^{DIF*}(t-\tau)\right]/\Omega$, where $j^2 = -1$, $\rho_{h^{DIF}}(\tau)$ is the time ACF of the diffuse component $h^{DIF}(t)$ of the complex fading envelope, $(\cdot)^*$ denotes the complex conjugate operation, and $\mathbf{E}[\cdot]$ designates the statistical expectation operator. Note that Eq. (9.67) is actually the same as presented by Pätzold et al. [1998, Eq. (29)]. It is obvious that Eq. (9.67) is the general expression and can reduce to those presented by Zajic et al. [2008, Eq.(6)] and [2009, Eq. (43)] by setting $E = 0$. Since the parameter E refers to the Doppler shift of the LoS component, we can conclude that Eq. (9.67) can be used for a non-isotropic environment with a time-variant LoS component ($E \neq 0$), while (6) in [Zajic et al. 2008] or (43) in Zajic et al. 2009 is suitable only for a non-isotropic environment with a time-invariant LoS component ($E = 0$). Note that (6) in [Zajic et al. 2008] or (43) in Zajic et al. 2009 was misapplied to the LCR in a non-isotropic environment with a time-variant LoS component.

9.4.1.4 Derivation of b_m for a Wide Variety of Scenarios

Based on the above presented SISO V2V model, we now derive the key parameters b_m for a wide variety of scenarios. Substituting Eq. (9.66b) into Eq. (9.68) and setting $m = 0$, we can obtain the parameter b_0 as

$$b_0 = b_0^{SB_1} + b_0^{SB_2} + b_0^{SB_3} + b_0^{DB} = \frac{1}{2(K+1)} \tag{9.69}$$

where $b_0^{SB_1} = \frac{\eta_{SB_1}}{2(K+1)}$, $b_0^{SB_2} = \frac{\eta_{SB_2}}{2(K+1)}$, $b_0^{SB_3} = \frac{\eta_{SB_3}}{2(K+1)}$, and $b_0^{DB} = \frac{\eta_{DB}}{2(K+1)}$. Similarly, we can express the parameters b_1 and b_2 as

$$b_m = b_m^{SB_1} + b_m^{SB_2} + b_m^{SB_3} + b_m^{DB}, \qquad m = 1, 2 \tag{9.70}$$

Considering the von Mises PDFs for the two-ring model, we can express the parameters $b_m^{SB_{1(2)}}$ $(m = 1, 2)$ as

$$b_1^{SB_{1(2)}} = b_0^{SB_{1(2)}} \int_{-\pi}^{\pi} \frac{f_{T_{max}} \cos(\varphi_T^{SB_{1(2)}} - \gamma_T) + f_{R_{max}} \cos(\varphi_R^{SB_{1(2)}} - \gamma_R)}{I_0(k_{T(R)}^{TR})}$$
$$\times e^{k_{T(R)}^{TR} \cos(\varphi_{T(R)}^{SB_{1(2)}} - \mu_{T(R)}^{TR})} d\varphi_{T(R)}^{SB_{1(2)}} \tag{9.71a}$$

$$b_2^{SB_{1(2)}} = b_0^{SB_{1(2)}} \int_{-\pi}^{\pi} \frac{2\pi \left[f_{T_{max}} \cos(\varphi_T^{SB_{1(2)}} - \gamma_T) + f_{R_{max}} \cos(\varphi_R^{SB_{1(2)}} - \gamma_R)\right]^2}{I_0(k_{T(R)}^{TR})}$$
$$\times e^{k_{T(R)}^{TR} \cos(\varphi_{T(R)}^{SB_{1(2)}} - \mu_{T(R)}^{TR})} d\varphi_{T(R)}^{SB_{1(2)}} \tag{9.71b}$$

Based on the general relationships in Eqs (9.43)–(9.46), the parameters $\varphi_R^{SB_1}$ and $\varphi_T^{SB_2}$ in Eqs (9.71a) and (9.71b) can be expressed by $\varphi_T^{SB_1}$ and $\varphi_R^{SB_2}$, respectively. Similarly, considering the von Mises PDF for the ellipse model, we can express the parameters $b_m^{SB_3}$ $(m = 1, 2)$ as

$$b_1^{SB_3} = b_0^{SB_3} \int_{-\pi}^{\pi} \frac{f_{T_{max}} \cos(\varphi_T^{SB_3} - \gamma_T) + f_{R_{max}} \cos(\varphi_R^{SB_3} - \gamma_R)}{I_0(k_R^{EL})}$$
$$\times e^{k_R^{EL} \cos(\varphi_R^{SB_3} - \mu_R^{EL})} d\varphi_R^{SB_3} \tag{9.72a}$$

$$b_2^{SB_3} = b_0^{SB_3} \int_{-\pi}^{\pi} \frac{2\pi \left[f_{T_{max}} \cos(\varphi_T^{SB_3} - \gamma_T) + f_{R_{max}} \cos(\varphi_R^{SB_3} - \gamma_R)\right]^2}{I_0(k_R^{EL})}$$
$$\times e^{k_R^{EL} \cos(\varphi_R^{SB_3} - \mu_R^{EL})} d\varphi_R^{SB_3} \tag{9.72b}$$

According to the general relationship in Eqs (9.47) and (9.48), the parameter $\varphi_T^{SB_3}$ in Eqs (9.72a) and (9.72b) can be expressed by $\varphi_R^{SB_3}$. Considering the von Mises PDFs for the two-ring model and applying the equalities $\int_{-\pi}^{\pi} e^{a \sin c + b \cos c} dc = 2\pi I_0\left(\sqrt{a^2 + b^2}\right)$, $dI_V(z)/dz = [I_{V-1}(z) + I_{V+1}(z)]/2$, and $zI_{V-1}(z) - zI_{V+1}(z) = (V+1)I_V(z)$ [I.S.Gradshteyn and Ryzhik 2000], where $I_V(\cdot)$ is the Vth-order modified Bessel function of the first kind, we can get the following closed-form expressions of the parameters b_m^{DB} $(m = 1, 2)$

$$b_1^{DB} = b_0^{DB} \left[\frac{2\pi f_{T_{max}} \cos(\gamma_T - \mu_T^{TR}) I_1(k_T^{TR})}{I_0(k_T^{TR})} + \frac{2\pi f_{R_{max}} \cos(\gamma_R - \mu_R^{TR}) I_1(k_R^{TR})}{I_0(k_R^{TR})}\right] \tag{9.73a}$$

$$b_2^{DB} = b_0^{DB}\left[4\pi^2 f_{T_{max}}^2 \frac{1+\cos\left(2\left(\gamma_T - \mu_T^{TR}\right)\right) I_2\left(k_T^{TR}\right)}{2I_0\left(k_T^{TR}\right)}\right.$$
$$+ 4\pi^2 f_{R_{max}}^2 \frac{1+\cos\left(2\left(\gamma_R - \mu_R^{TR}\right)\right) I_2\left(k_R^{TR}\right)}{2I_0\left(k_R^{TR}\right)}$$
$$\left. + 8\pi^2 f_{T_{max}} f_{R_{max}} \cos(\gamma_T - \mu_T^{TR})\cos(\gamma_R - \mu_R^{TR})\frac{I_1(k_T^{TR})I_1(k_R^{TR})}{I_0(k_T^{TR})I_0(k_R^{TR})}\right] \tag{9.73b}$$

Numerical integration methods are needed to evaluate the integrals in Eqs (9.71) and (9.72).

Derivation of b_m for Macro- and Microcell Scenarios

For macro- and microcell scenarios, the widely used assumption $D \gg \max\{R_T, R_R\}$ can be used in the derivation of b_m. This results in the closed-form solution of the integrals in Eq. (9.71), while the expressions of other parameters b_m remain unchanged. Note that the application of $D \gg \max\{R_T, R_R\}$ leads to the approximate relationships $\varphi_R^{SB_1} \approx \pi - \Delta_T \sin\varphi_T^{SB_1}$ and $\varphi_T^{SB_2} \approx \Delta_R \sin\varphi_R^{SB_2}$. Based on these two approximate relations and the approximate relations $\sin\chi \approx \chi$ and $\cos\chi \approx 1$ when χ is small, the parameters $b_m^{SB_{1(2)}}$ $(m = 1, 2)$ in Eq. (9.71) can be further simplified as

$$b_m^{SB_1} = b_0^{SB_1}\int_{-\pi}^{\pi} \frac{(2\pi)^{m-1}\left[f_{T_{max}}\cos(\varphi_T^{SB_1} - \gamma_T) + f_{R_{max}}(\Delta_T \sin\varphi_T^{SB_1}\sin\gamma_R - \cos\gamma_R)\right]^m}{I_0(k_T^{TR})}$$
$$\times e^{k_T^{TR}\cos(\varphi_T^{SB_1} - \mu_T^{TR})} d\varphi_T^{SB_1} \tag{9.74a}$$

$$b_m^{SB_2} = b_0^{SB_2}\int_{-\pi}^{\pi} \frac{(2\pi)^{m-1}\left[f_{T_{max}}(\Delta_R \sin\varphi_R^{SB_2}\sin\gamma_T + \cos\gamma_T) + f_{R_{max}}\cos(\varphi_R^{SB_2} - \gamma_R)\right]^m}{I_0(k_R^{TR})}$$
$$\times e^{k_R^{TR}\cos(\varphi_R^{SB_2} - \mu_R^{TR})} d\varphi_R^{SB_2} \tag{9.74b}$$

The integrals in Eqs (9.74a) and (9.74b) can be further simplified and thus the closed-form expressions of parameters $b_m^{SB_{1(2)}}$ can be obtained as

$$b_1^{SB_{1(2)}} = b_0^{SB_{1(2)}}\left\{2\pi\left[f_{T(R)_{max}}\cos(\gamma_{T(R)} - \mu_{T(R)}^{TR}) + f_{R(T)_{max}}\Delta_{T(R)}\sin\mu_{T(R)}^{TR}\sin\gamma_{R(T)}\right]\right.$$
$$\left.\times\frac{I_1(k_{T(R)}^{TR})}{I_0(k_{T(R)}^{TR})} \mp 2\pi f_{R(T)_{max}}\cos\gamma_{R(T)}\right\} \tag{9.75a}$$

$$b_2^{SB_{1(2)}} = b_0^{SB_{1(2)}}\left\{2\pi^2\left(2f_{R(T)_{max}}^2\cos^2\gamma_{R(T)} + f_{T(R)_{max}}^2 + f_{R(T)_{max}}^2\Delta_{T(R)}^2\sin^2\gamma_{R(T)}\right.\right.$$
$$\left. + 2f_{T_{max}}f_{R_{max}}\Delta_{T(R)}\sin\gamma_T\sin\gamma_R\right) \mp 4\pi^2\left[2f_{T_{max}}f_{R_{max}}\cos\gamma_{R(T)}\right.$$
$$\left.\times\cos\left(\gamma_{T(R)} - \mu_{T(R)}^{TR}\right) + f_{R(T)_{max}}^2\Delta_{T(R)}\sin\left(2\gamma_{R(T)}\right)\sin\mu_{T(R)}^{TR}\right]\frac{I_1\left(k_{T(R)}^{TR}\right)}{I_0\left(k_{T(R)}^{TR}\right)}$$
$$+ 2\pi^2\left[f_{T(R)_{max}}^2\cos\left(2\left(\gamma_{T(R)} - \mu_{T(R)}^{TR}\right)\right) - f_{R(T)_{max}}^2\Delta_{T(R)}^2\sin^2\gamma_{R(T)}\cos\left(2\mu_{T(R)}^{TR}\right)\right.$$

$$+f_{T_{max}} f_{R_{max}} \Delta_{T(R)} \sin\gamma_{R(T)} \sin\left(\gamma_{T(R)} - 2\mu_{T(R)}^{TR}\right)\Bigg] \frac{I_2(k_{T(R)}^{TR})}{I_0(k_{T(R)}^{TR})}\Bigg\} \tag{9.75b}$$

In Appendix A.8.7, we only provide a brief outline for the derivation of Eq. (9.75b) since the derivations of Eqs (9.75a) and (9.75b) are similar. The parameters $b_m^{SB_{1(2)}}$ for 3D narrowband and wideband V2V channels were derived in Zajic et al. 2008 and Zajic et al. 2009, respectively. The expressions of $b_m^{SB_{1(2)}}$ given in Zajic et al. 2009 can reduce to the ones in Zajic et al. 2008 by setting $R_{t(r)1} = R_{t(r)2}$. If we further assume elevation-related parameters $\beta_{T(R)} = \Delta_H = 0$ in the parameters $b_m^{SB_{1(2)}}$ in Zajic et al. 2008, the parameters $b_m^{SB_{1(2)}}$ for 2D macro- and microcell scenarios can be obtained. To distinguish this from the parameters $b_m^{SB_{1(2)}}$ in Eq. (9.75), we denote the parameters $b_m^{SB_{1(2)}}$ obtained from the Zajic et al. papers [2008, 2009] as $\tilde{b}_m^{SB_{1(2)}}$.

The difference between the $b_m^{SB_{1(2)}}$ and $\tilde{b}_m^{SB_{1(2)}}$ motivates us to find which derivations are correct. To this end, in Figure 9.9 we compare the LCRs having the same expression Eq. (9.67) but with different expressions for parameters $b_m^{SB_{1(2)}}$:

- $L_{\xi-C}^{TR}(r_l)$ having $b_m^{SB_{1(2)}}$ calculated by Eq. (9.75)
- $L_{\xi-W}^{TR}(r_l)$ having $\tilde{b}_m^{SB_{1(2)}}$
- $L_{\xi-N}^{TR}(r_l)$ having the numerically computed $b_m^{SB_{1(2)}}$ in Eq. (9.74).

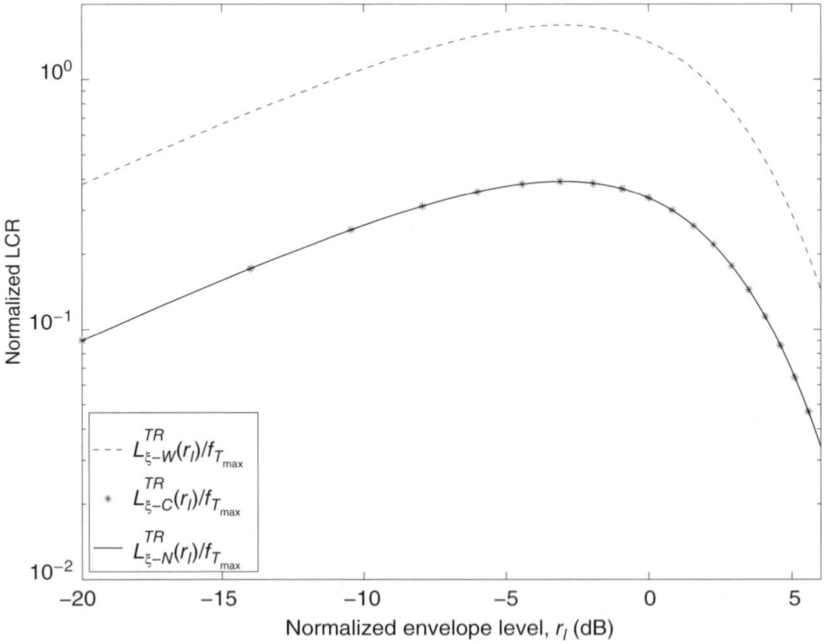

Figure 9.9 Comparison of the LCRs $L_{\xi-W}^{TR}(r_l)$, $L_{\xi-C}^{TR}(r_l)$, and $L_{\xi-N}^{TR}(r_l)$

The parameters used in Figure 9.9 are $f_{T_{max}} = f_{R_{max}} = 500\,\text{Hz}$, $K = 0$, $\Delta_T = \Delta_R = 0.01$, $\mu_T^{TR} = 31.2°$, $k_T^{TR} = 18.2$, $\mu_R^{TR} = 216.3°$, and $k_R^{TR} = 10.6$. Figure 9.9 shows excellent agreement between $L_{\xi-C}^{TR}(r_l)$ and $L_{\xi-N}^{TR}(r_l)$, demonstrating the correctness of our derivation of Eq. (9.75).

It is worth emphasizing that the LCR, $L_{\xi-C}^{TR}(r_l)$, is obtained based on the assumption $D \gg \max\{R_T, R_R\}$. To fulfill this assumption, the values of parameters Δ_T and Δ_R should be chosen carefully due to the relationships $\Delta_T \approx R_T/D$ and $\Delta_R \approx R_R/D$ (in general, the smaller the better). However, the values of parameters Δ_T and Δ_R are chosen to be comparatively large ($\Delta_T = \Delta_R = 0.6$) in [Zajic et al. 2008]. This raises several questions, such as whether $\Delta_T = \Delta_R = 0.6$ violates the assumption $D \gg \max\{R_T, R_R\}$, and if so, how inaccurate the LCR $L_{\xi-C}^{TR}(r_l)$ is and whether any such inaccuracy can be ignored.

To address these questions, we define an error function to measure the error between the LCR, $L_{\xi-G}^{TR}(r_l)$, having the expression of Eq. (9.67) with the general expressions of parameters $b_m^{SB_{1(2)}}$ in Eq. (9.71), and $L_{\xi-C}^{TR}(r_l)$ as $\varepsilon = L^{-1}\sum_{l=1}^{L}\left|L_{\xi-G}^{TR}(r_l) - L_{\xi-C}^{TR}(r_l)\right| / \left|L_{\xi-G}^{TR}(r_l)\right|$, where L is the total number of the investigated specified level r_l. Figure 9.10 illustrates the error ε versus the parameter Δ ($\Delta_T = \Delta_R = \Delta$) using the same parameters as in Figure 9.9. The investigated levels r_l were obtained by taking $L = 200$ equal-distance samples between -20 dB and 5 dB. As expected, it is shown that the inaccuracy of $L_{\xi-C}^{TR}(r_l)$ increases with the increase of the parameter Δ. From Figure 4.4, it is obvious that at $\Delta = 0.6$, we have very high error $\varepsilon = 0.894547$. This demonstrates that the parameters $\Delta_T = \Delta_R = 0.6$ in [Zajic et al. 2008] may result in an extremely inaccurate $L_{\xi-C}^{TR}(r_l)$. Therefore, it is desirable to propose a

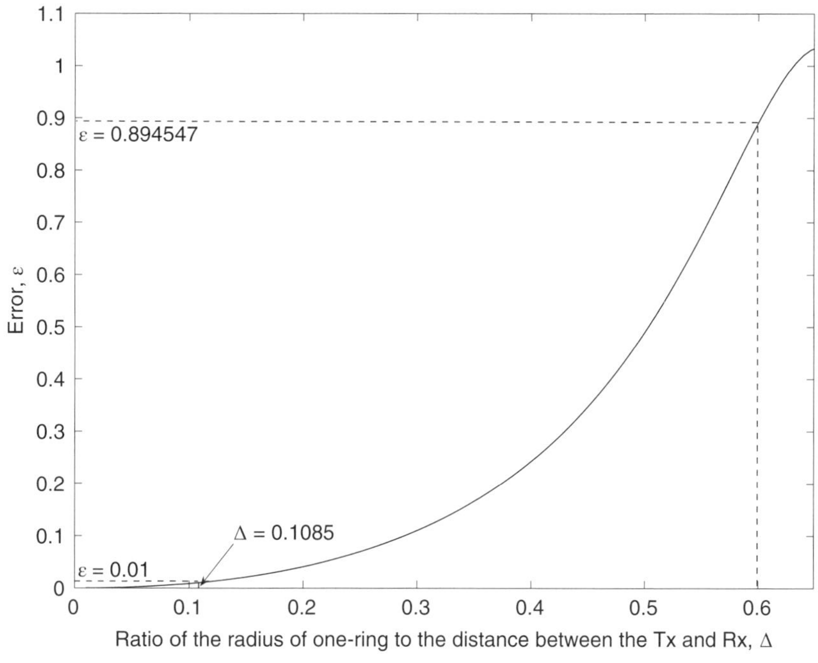

Figure 9.10 The error ε between the LCRs $L_{\xi-G}^{TR}(r_l)$ and $L_{\xi-C}^{TR}(r_l)$

criterion for Δ_T and Δ_R to guarantee the accuracy of $L_{\xi-C}^{TR}(r_l)$ is acceptable. In this chapter, we assume the accuracy of $L_{\xi-C}^{TR}(r_l)$ is acceptable if the error $\varepsilon \le 0.01$. In such a case, the LCR $L_{\xi-C}^{TR}(r_l)$ can be applied under the condition that $\Delta_T = \Delta_R \le 0.1085$. In other words, when $\Delta_T = \Delta_R > 0.1085$, the LCR $L_{\xi-G}^{TR}(r_l)$ can be used instead of $L_{\xi-C}^{TR}(r_l)$.

The AFD, $T_{\xi_-}(r_l)$, is the average time over which the signal envelope, $\xi(t)$, remains below a certain level r_l. In general, the AFD $T_{\xi_-}(r_l)$ for Ricean fading channels is defined by [Pätzold and Laue 1998]:

$$T_{\xi_-}(r_l) = \frac{P_{\xi_-}(r_l)}{L_\xi(r_l)} = \frac{1 - Q\left(\sqrt{2K}, \sqrt{2(K+1)}r_l\right)}{L_\xi(r_l)} \tag{9.76}$$

where $P_{\xi_-}(r_l)$ indicates a cumulative distribution function of $\xi(t)$ with $Q(\cdot,\cdot)$ denoting the generalized Marcum Q function.

Many existing LCRs and AFDs are special cases of the non-isotropic V2V LCR and AFD in Eqs (9.67) and (9.76). The simplest case is Clarke's LCR $\sqrt{2\pi} f_{R_{max}} r_l \exp(-r_l^2)$ and AFD $(\exp(r_l^2) - 1)/(\sqrt{2\pi} f_{R_{max}} r_l)$ [Stüber 2001], which can be obtained from Eqs (9.67) and (9.76), respectively, by setting $K = 0$ (NLoS condition), $k_R^{TR} = 0$ (isotropic scattering around Rx), and $f_{T_{max}} = k_T^{TR} = \eta_{SB_1} = \eta_{SB_3} = \eta_{DB} = 0$ (fixed Tx, no scattering around Tx). Expressions for other LCRs and AFDs based on the isotropic one-ring model only around the Rx, for example those in the papers by Pätzold and Laue [1998] and Rice [1958], can be obtained in a similar way. The LCR and AFD based on the non-isotropic two-ring model with single-bounced rays [Wang et al. 2007] are obtained from Eqs (9.67) and (9.76), respectively, by setting $K = f_{T_{max}} = \eta_{SB_3} = \eta_{DB} = 0$. Finally, the LCR and AFD for isotropic V2V Rayleigh channels [Akki 1994] are obtained from Eqs (9.67) and (9.76), respectively, by setting $K = k_T^{TR} = k_R^{TR} = \eta_{SB_1} = \eta_{SB_2} = \eta_{SB_3} = 0$.

9.4.1.5 Numerical Results and Analysis

Unless otherwise specified, all the results presented in this section are obtained using $f_c = 5.9\,\text{GHz}$, $f_{T_{max}} = f_{R_{max}} = 570\,\text{Hz}$, $D = 300\,\text{m}$, $a = 200\,\text{m}$, and $R_T = R_R = 40\,\text{m}$.

Figures 9.11 and 9.12 illustrate the space and frequency CFs of the single- and double-bounce two-ring model and single-bounce ellipse model for different scenarios. It is obvious that both the space and frequency CFs vary significantly for different scenarios (Scenario a and Scenario b). We also notice that directions of motion (related to the values of γ_T and γ_R) have no impact on the space and frequency CFs.

Figure 9.13 shows normalized Doppler PSDs for different scenarios (Scenario a and Scenario b). For Scenario a, it is clear that no matter what the direction of motion (same or opposite) and the shape of the scattering region (one-ring, two-ring, or ellipse) are, the Doppler PSD of single-bounced rays is similar to the U-shaped PSD of F2M cellular channels,[2] whereas the Doppler PSD of double-bounced rays has a "rounded" shape having a peak in the middle. This indicates that the U-shaped Doppler PSD will appear when high dependency exists between the AoD and AoA, while the "rounded" Doppler PSD will appear when the AoD and

[2] Note that when the Tx and Rx move in the same direction, the Doppler PSD of the single-bounce ellipse model is not an exact U shape. It is reasonable to consider it an approximate U, however, since peaks exist on both the left and right sides of the Doppler PSD instead of in the middle.

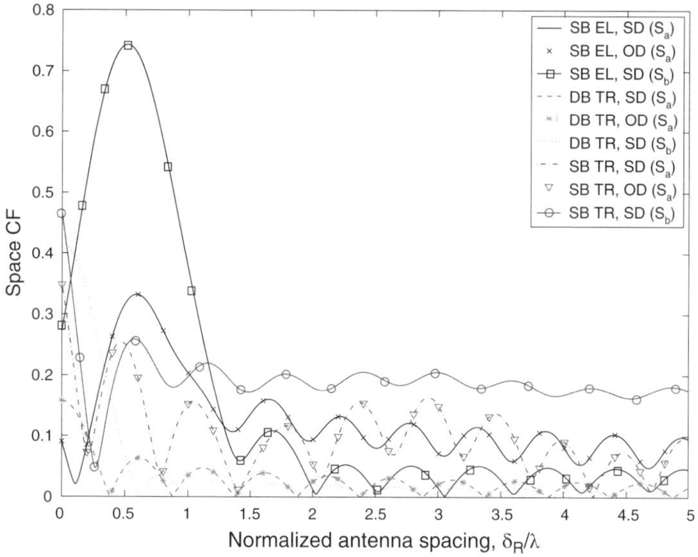

Figure 9.11 Space CFs of the single-bounce (SB) ellipse (EL) model, double-bounce (DB) two-ring (TR) model, and SB TR model for different scenarios ($\tau = 0$, $\chi = 0$, and $\delta_T = 2$). SD: same direction ($\gamma_T = \gamma_R = 0$); OD: opposite direction ($\gamma_T = 0$ and $\gamma_R = \pi$); Scenario a (S_a): $k_T^{TR} = k_R^{TR} = k_R^{EL} = 0$ (isotropic environments); Scenario b (S_b): $k_T^{TR} = k_R^{TR} = k_R^{EL} = 3$ (non-isotropic environments), $\mu_T^{TR} = \pi/4$, and $\mu_R^{TR} = \mu_R^{EL} = 3\pi/4$

AoA are relatively independent. We can also observe that for different directions of motion, the Doppler PSDs of double-bounced rays remain unchanged, while the Doppler PSDs of single-bounced rays change with different ranges of Doppler frequencies. More importantly, we found that the impact of single-bounced rays from different rings (ring around the Tx or Rx) on the Doppler PSD are the same for V2V channels when the Tx and Rx are moving in opposite directions, leading to the U-shaped Doppler PSD for the single-bounce two-ring model. When the Tx and Rx are moving in the same direction, the impact of single-bounced rays from different rings on the Doppler PSD are different in terms of the range of Doppler frequencies, which results in a double-U-shaped Doppler PSD for the single-bounce two-ring model. Therefore, we can conclude that a more realistic V2V channel model should take into account the different contributions from different rings. However, this has not been done in all the existing V2V GBSMs, for example in Zajic and Stuber [2008b].

It is worth mentioning that by setting one terminal fixed (say, $f_{T_{max}} = 0$), our V2V model can reduce to a F2M model. In this case, we studied the Doppler PSD for the corresponding single- and double-bounce two-ring F2M models and single-bounce ellipse F2M model, and found that they have the same U-shaped PSD. For brevity, the results regarding F2M channels are omitted here. These observations indicate that the impact of single- and double-bounced rays on the Doppler PSD are completely different for V2V channels (U-shaped and "rounded", respectively), while they are the same for F2M channels (U-shaped). At the end, the comparison of Scenario a and Scenario b illustrates the significant impact of angle spreads (related

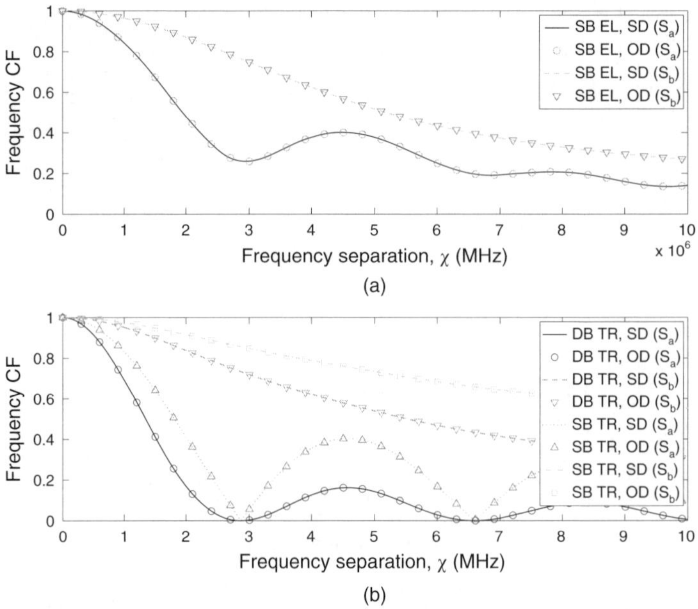

Figure 9.12 Frequency CFs of the single-bounce (SB) ellipse (EL) model, double-bounce (DB) two-ring (TR) model, and SB TR model for different scenarios ($\tau = 0$, $\delta_T = \delta_R = 0$): SD, same direction ($\gamma_T = \gamma_R = 0$); OD, opposite direction. Scenario a (S_a), $k_T^{TR} = k_R^{TR} = k_R^{EL} = 0$ (isotropic environments); Scenario b (S_b), $k_T^{TR} = k_R^{TR} = k_R^{EL} = 3$ (non-isotropic environments), $\mu_T^{TR} = \pi/4$, and $\mu_R^{TR} = \mu_R^{EL} = 3\pi/4$

to the values of k_T^{TR}, k_R^{TR}, and k_R^{EL}) and mean angles (related to the values of μ_T^{TR}, μ_R^{TR}, and μ_R^{EL}) on the Doppler PSD.

Figures 9.14 and 9.15 depict the impact of the antenna element spacing and frequency separation on the Doppler PSD, respectively. Figure 9.14 shows that the space separation introduces fluctuations in the Doppler PSD no matter what the shape of the scattering region is. Figure 9.15 illustrates that the frequency separation only generates fluctuations in the Doppler PSD for the double-bounce two-ring model, while for other cases, the impact of the frequency separation vanishes.

Figures 9.16(a) and (b) show the theoretical Doppler PSDs obtained from the proposed V2V model for different VTDs (low and high) when the Tx and Rx are moving in opposite directions and same direction, respectively. For further comparison, the measured data taken from the paper by Acosta-Marum and Brunelli [2007, Figs 4(a), 4(c)] are also plotted in Figures 9.16(a) and (b), respectively. Acosta-Marum and Brunelli [2007] performed their measurement campaigns at a carrier frequency of 5.9 GHz on an expressway with low VTD in metropolitan Atlanta, Georgia. The maximum Doppler frequencies were $f_{T_{max}} = f_{R_{max}} = 570$ Hz, the distance between the Tx and Rx was approximately $D = 300$ m, and the directions of movement were $\gamma_T = 0$, $\gamma_R = \pi$ (opposite direction; Figure 4(a) in the original paper) and $\gamma_T = \gamma_R = 0$ (same direction; Figure 4(c) in the original paper). Both the Tx and Rx were equipped with one omnidirectional antenna; in other words, in the SISO case.

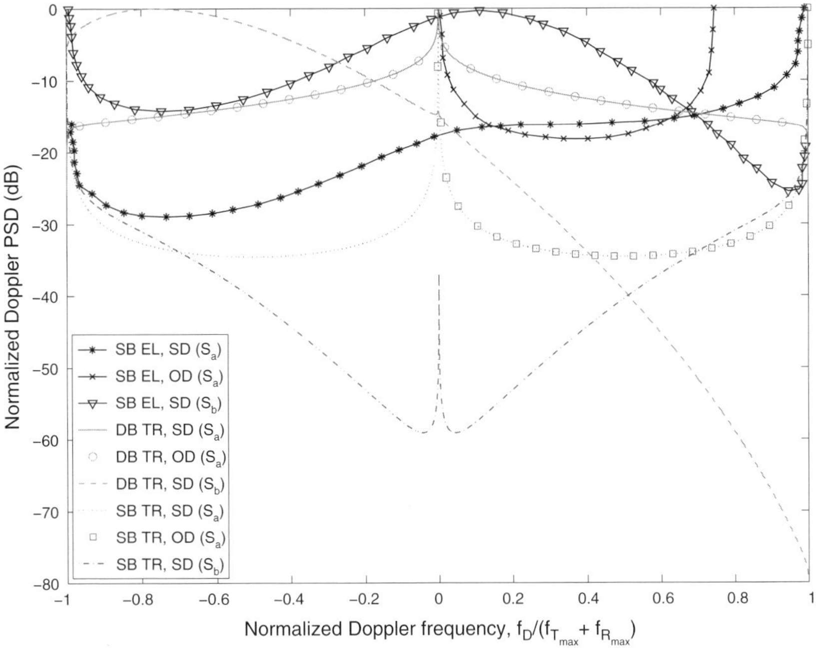

Figure 9.13 Normalized Doppler PSDs of the single-bounce (SB) ellipse (EL) model, double-bounce (DB) two-ring (TR) model, and SB TR model for different scenarios ($\delta_T = \delta_R = 0$, $\chi = 0$): SD, same direction ($\gamma_T = \gamma_R = 0$); OD, opposite direction ($\gamma_T = 0$ and $\gamma_R = \pi$). Scenario a, $(S_a) k_T^{TR} = k_R^{TR} = k_R^{EL} = 0$ (isotropic environments); Scenario b (S_b), $k_T^{TR} = k_R^{TR} = k_R^{EL} = 3$ (non-isotropic environments), $\mu_T^{TR} = \pi/4$, and $\mu_R^{TR} = \mu_R^{EL} = 3\pi/4$

Based on the measured scenarios of Acosta-Marum and Brunelli [2007], we chose the following environment-related parameters: $k_T^{TR} = 6.6$, $k_R^{TR} = 8.3$, $k_R^{EL} = 5.5$, $\mu_T^{TR} = 12.8°$, $\mu_R^{TR} = 178.7°$, and $\mu_R^{EL} = 131.6°$ for Figure 9.16(a), and $k_T^{TR} = 9.6$, $k_R^{TR} = 3.6$, $k_R^{EL} = 11.5$, $\mu_T^{TR} = 21.7°$, $\mu_R^{TR} = 147.8°$, and $\mu_R^{EL} = 171.6°$ for Figure 9.16(b). Considering the constraints of the Ricean factor and energy-related parameters for different propagation scenarios, we choose the following parameters in order to fit the measured Doppler PSDs reported by Acosta-Marum and Brunelli [2007] for the two scenarios with low VTD:

- $K = 2.186$, $\eta_{DB} = 0.005$, $\eta_{SB_1} = 0.252$, $\eta_{SB_2} = 0.262$, and $\eta_{SB_3} = 0.481$ for Figure 9.16(a)
- $K = 3.786$, $\eta_{DB} = 0.051$, $\eta_{SB_1} = 0.335$, $\eta_{SB_2} = 0.203$, and $\eta_{SB_3} = 0.411$ for Figure 9.16(b).

The excellent agreement between the theoretical results and measured data confirms the utility of the proposed model. The environment-related parameters for high VTD in Figures 9.16(a) and (b) are the same as those for low VTD except $k_T^{TR} = k_R^{TR} = 0.6$, which are related to the distribution of moving cars (normally, the smaller the values, the more distributed the moving cars; in other words, the higher the VTD). The Doppler PSDs for high VTD shown

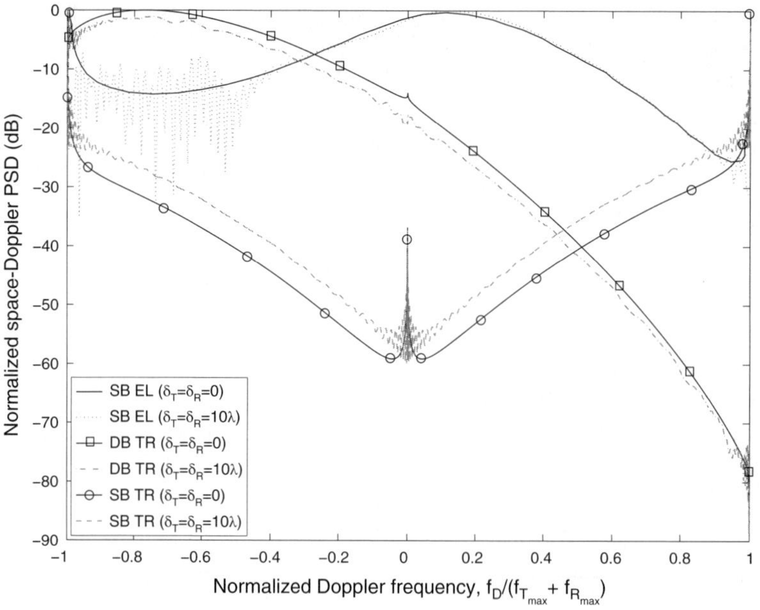

Figure 9.14 Normalized space-Doppler PSDs of the single-bounce (SB) ellipse (EL) model, double-bounce (DB) two-ring (TR) model, and SB TR model for different antenna element spacings in a V2V non-isotropic scattering environment ($k_T^{TR} = k_R^{TR} = k_R^{EL} = 3$, $\mu_T^{TR} = \pi/4$, $\mu_R^{TR} = \mu_R^{EL} = 3\pi/4$) with the Tx and Rx moving in the same direction ($\gamma_T = \gamma_R = 0$)

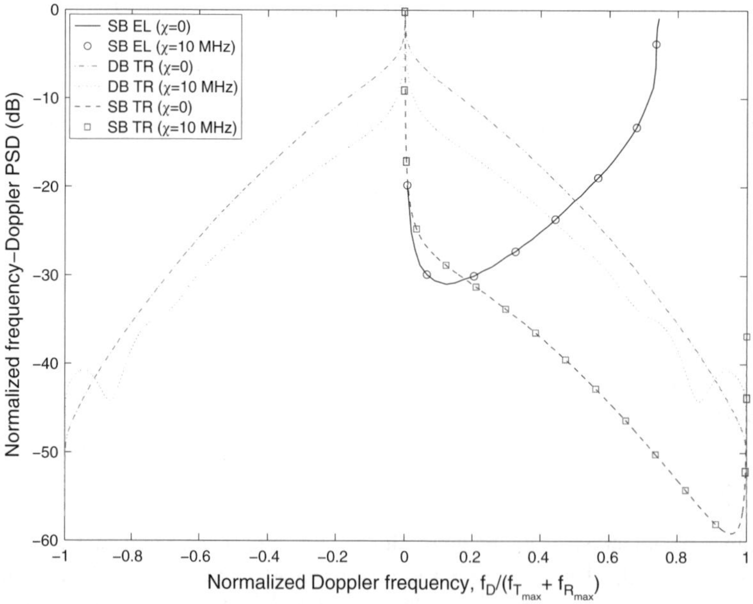

Figure 9.15 Normalized frequency-Doppler PSDs of the single-bounce (SB) ellipse (EL) model, double-bounce (DB) two-ring (TR) model, and SB TR model for different frequency separations in a V2V non-isotropic scattering environment ($k_T^{TR} = k_R^{TR} = k_R^{EL} = 3$, $\mu_T^{TR} = \pi/4$, $\mu_R^{TR} = \mu_R^{EL} = 3\pi/4$) with the Tx and Rx moving in the opposite direction ($\gamma_T = 0$ and $\gamma_R = \pi$)

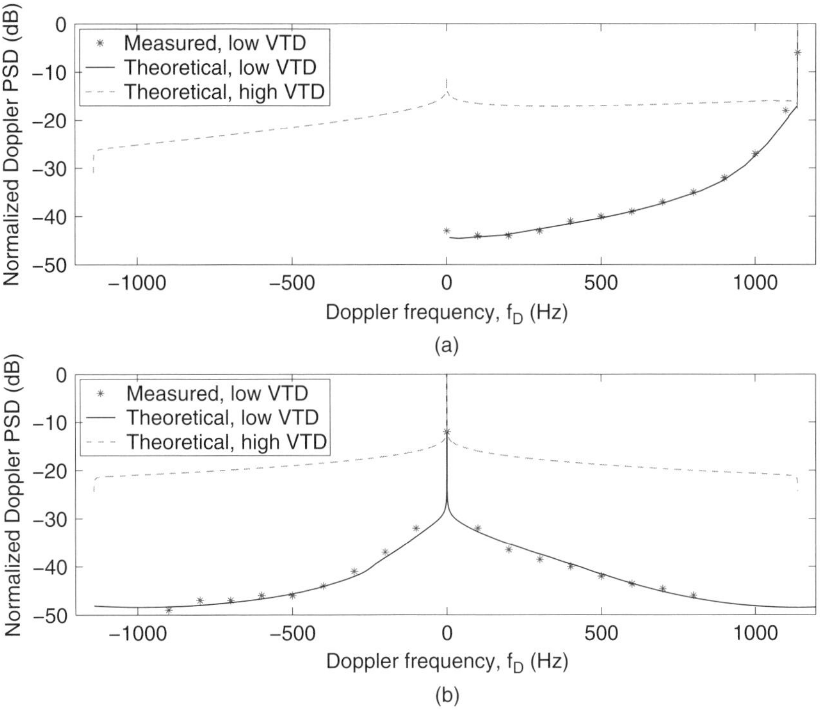

Figure 9.16 Normalized Doppler PSDs of the proposed adaptive model for different SISO picocell scenarios ($\delta_T = \delta_R = 0$, $\chi = 0$): (a) Tx and Rx are moving in opposite directions; (b) Tx and Rx are moving in the same direction. VTD: vehicular traffic density

in Figures 9.16(a) and (b) were obtained with the parameters $K = 0.2$, $\eta_{DB} = 0.715$, $\eta_{SB_1} = \eta_{SB_2} = 0.115$, and $\eta_{SB_3} = 0.055$. Unfortunately, to the best of the authors' knowledge, no measurement results are available regarding the impact of high VTD (say, a traffic jam) on the Doppler PSD.[3]

Comparing the theoretical Doppler PSDs in Figures 9.16(a) and (b), we observe that the VTD significantly affects both the shape and value of the Doppler PSD for V2V channels. The Doppler PSD tends to be more evenly distributed across all Doppler frequencies with a higher VTD. This is because with a high VTD, the received power mainly comes from cars moving around the Tx and Rx and therefore from all directions, while the power of the LoS component is not that significant. This means that the received power for different Doppler frequencies (directions) is more evenly distributed. With low VTD, the received power from the LoS component may be significant, while the power from the moving cars may be small. Therefore, the power tends to be concentrated at some Doppler frequencies.

Figures 9.17(a) and (b) depict the theoretical LCRs and AFDs for different VTDs (low or high) in V2V Ricean fading channels when the Tx and Rx are moving in the same direction, respectively. The parameters $f_c = 5.2$ GHz and $f_{T_{max}} = f_{R_{max}} = 500$ Hz are used. For a further

[3] No such values are reported in the papers by, for example, Acosta et al. [2004], Zajic et al. [2009], Sen and Matolak [2008], Paier et al. [2009], Cheng et al. [2007], Acosta-Marum and Brunelli [2007], Maurer et al. [2002].

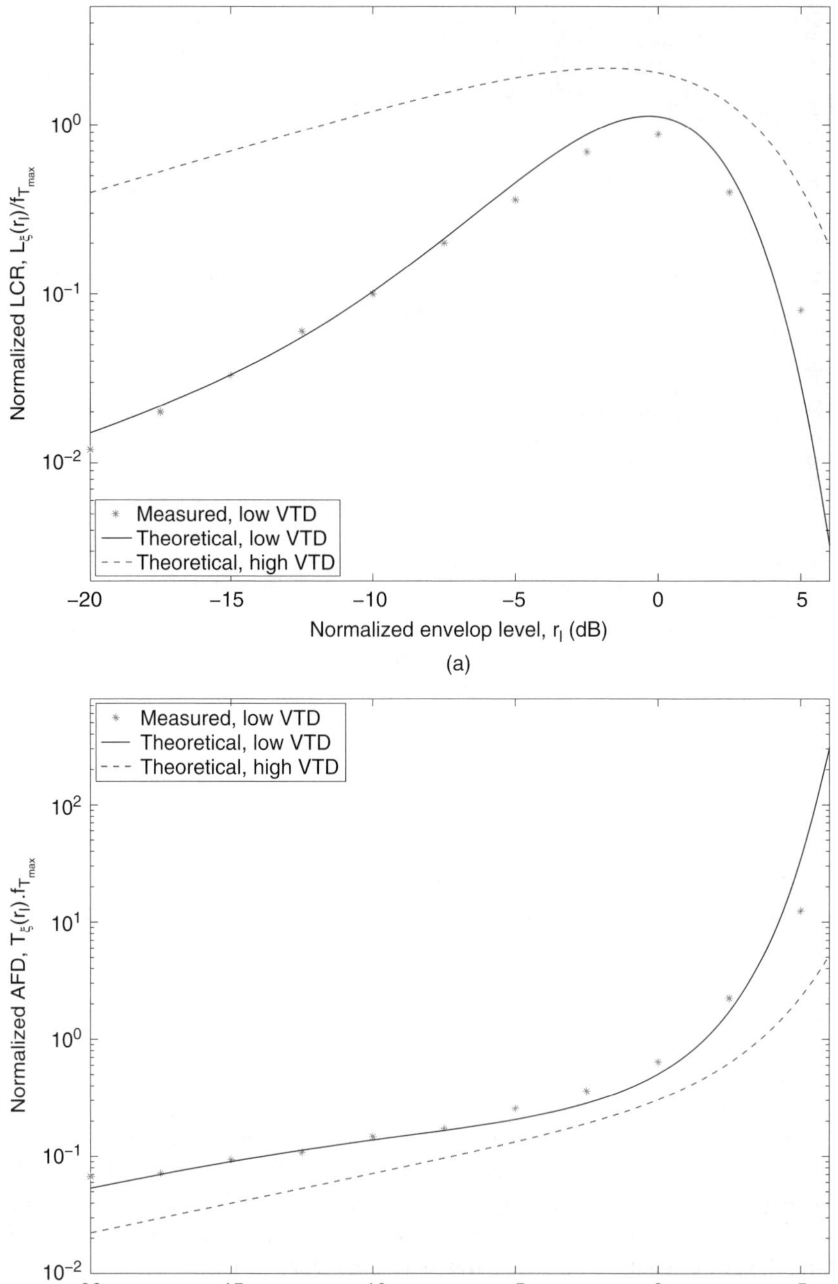

Figure 9.17 (a) LCRs and (b) AFDs of the developed V2V channel model with a low or high vehicular traffic density (VTD) when the Tx and Rx are moving in the same direction

comparison, results taken from the paper by Maurer et al. [2002, Figure 8(b)] are also plotted. Maurer et al. [2002] performed their measurement campaigns on a highway in Germany with low VTD; the directions of movement of the Tx and Rx were $\gamma_T = \gamma_R = 0$ (same direction). Based on the scenarios in their paper, we chose the following environment-related parameters for Figures 9.17(a) and (b): $\mu_T^{TR} = 33.2°$, $k_T^{TR} = 18.2$, $\mu_R^{TR} = 148.6°$, $k_R^{TR} = 13.3$, $\mu_R^{EL} = 148.6°$, and $k_R^{EL} = 8.6$. Considering the constraints of the Ricean factor and energy-related parameters for a propagation scenario with low VTD, we choose the following parameters in order to fit the measured LCR/AFD reported by Maurer et al.: $K = 4.26$, $\eta_{DB} = 0.08$, $\eta_{SB_1} = 0.12$, $\eta_{SB_2} = 0.18$, and $\eta_{SB_3} = 0.62$. The excellent agreement between the theoretical results and measured data confirms the utility of the proposed model. The environment-related parameters for a high VTD in Figures 9.17(a) and (b) are the same as those for a low VTD except $k_T^{TR} = k_R^{TR} = 0.6$. The theoretical LCR and AFD for a high VTD shown in Figures 9.17(a) and (b) were obtained with the parameters $K = 0.56$, $\eta_{DB} = 0.58$, $\eta_{SB_1} = 0.1$, $\eta_{SB_2} = 0.18$, $\eta_{SB_3} = 0.14$. Unfortunately, to the best of the authors' knowledge, no measurement results are available regarding the impact of high VTD (say, a traffic jam) on the LCR and AFD.[4]

Comparing the theoretical LCRs and AFDs in Figures 9.17(a) and (b), respectively, we observe that the VTD significantly affects the LCR and AFD for V2V channels. Moreover, Figure 9.17(a) shows that the fades are shallower when the VTD is lower. We see from Figure 9.17(b) that the AFD tends to be larger with lower VTD. This is because with a high VTD, the received power mainly comes from the cars moving around the TX and Rx and therefore from all directions, while for a low VTD the received power is concentrated in several directions, for example, the directions of the LoS components and/or large stationary scatterers on the roadside.

9.4.1.6 Summary

In Section 9.4.1, we have proposed a generic and adaptive RS-GBSM for non-isotropic scattering MIMO V2V Ricean fading channels. By adjusting some model parameters and with the help of the newly derived general relationship between the AoA and AoD, the proposed model is adaptable to a wide variety of V2V propagation environments. In addition, for the first time, VTD is taken into account in an RS-GBSM for modeling V2V channels. From this model, we have derived the STF CF and the corresponding SDF PSD for non-isotropic scattering environments, where the closed-form expressions are available in the case of the single-bounce two-ring model for macrocell and microcell scenarios, and the double-bounce two-ring model for any scenarios. The envelope LCR and AFD have also been derived in terms of the simplified version (SISO) of the proposed MIMO V2V model. Based on the derived STF CFs and SDF PSDs, we have further investigated the degenerate CFs and PSDs in detail and found that some parameters – the angle spread, direction of motion, antenna element spacing, and so on – have a great impact on the resulting CFs and PSDs. It has also been demonstrated that for V2V isotropic scattering scenarios, no matter what the direction of motion and shape of the scattering region are, single-bounced rays will result in a U-shaped Doppler PSD, while double-bounced rays will result in a "rounded' Doppler PSD. More importantly, we have investigated the impact of the VTD on the Doppler PSD, envelope LCR, and AFD and found that

[4] There are no such results in the paper by Acosta et al. [2004], Zajic et al. [2009], Sen and Matolak [2008], Paier et al. [2009], Cheng et al. [2007], Acosta-Marum and Brunelli [2007], Maurer et al. [2002].

with a lower VTD the Doppler PSD tends to be concentrated on some Doppler frequencies, the envelope LCR is smaller (that is, the fades are shallower) and the AFD tends to be larger. Finally, it has been shown that theoretical Doppler PSDs, envelope LCRs, and AFDs match the data measured by Acosta-Marum and Brunelli [2007] and Maurer et al. [2002], validating the utility of our model.

9.4.2 Modeling and Simulation of MIMO V2V Channels: Wideband

In Section 9.4.1, we proposed a narrowband MIMO V2V RS-GBSM for non-isotropic scattering Ricean fading channels. However, most potential transmission schemes for V2V communications use relatively wide bandwidths; approximately 10 MHz for the IEEE 802.11p standard. The underlying V2V channels exhibit frequency-selectivity since the signal bandwidth is larger than the coherence bandwidth of such channels (normally around 4- 6 MHz [Sen and Matolak 2008]). Therefore, wideband V2V channel models are indispensable.

As mentioned in Chapter 4, only one 3D wideband GBSM has been proposed for MIMO V2V Ricean fading channels [Zajic and Stuber 2009]. However, this model cannot describe the channel statistics for different time delays, which are important for V2V channels [Acosta et al. 2004, Acosta-Marum and Brunelli 2007]. Although the measurement campaign by Sen and Matolak [2008] has demonstrated that the VTD significantly affects the channel statistics for V2V channels, the impact of the VTD is not considered in the wideband model of Zajic and Stuber [2009]. In addition, it is non-trivial to use this model to match any given or measured PDP since many parameters need adjustment via a complicated procedure.

To fill this gap, in the first part of Section 9.4.2, through application of the TDL concept, we propose a new 2D wideband V2V RS-GBSM, which is an extension of the narrowband model proposed in Section 9.4.1 with respect to frequency-selectivity. This wideband model comprises a two-ring model and a multiple confocal ellipse model incorporating LoS, single-, and double-bounced rays. By using the multiple confocal ellipses to construct a TDL structure, the proposed wideband model can be used to investigate the channel statistics for different time delays – per-tap channel statistics – and can also easily match any specified or measured PDP.

In order to take into account the impact of the VTD on channel statistics for every tap in the proposed wideband model while maintaining complexity levels similar to those of the narrowband model described in Section 9.4.1, we first distinguish between the moving cars and the stationary roadside environment, which are described by a two-ring model and a multiple confocal ellipses model, respectively. Then, from an analysis of real V2V communication environments, we generate a novel and simple approach to incorporate the impact of the VTD into every tap of our wideband model. From the proposed model, we derive the ST CF, the corresponding SD PSD, the frequency CF (FCF), and the corresponding PDP. Finally, the resuling theoretical per-tap Doppler PSDs and the measurement data obtained by Acosta-Marum and Brunelli [2007] are compared. Excellent agreement between them demonstrates the utility of the proposed model.

Since the proposed wideband RS-GBSM is actually a reference model that assumes an infinite number of effective scatterers, it cannot be directly implemented in practice. Therefore, accurate V2V channel simulation models play a major role in the practical simulation and performance evaluation of V2V systems. As addressed in Chapter 4, up to now, many V2V channel simulators are limited to isotropic scattering narrowband SISO channels [Patel et al.

2005, Wang et al. 2009, Zajic and Stuber 2006], while only Pätzold et al. [2008] and Zajic and Stuber [2008b] have proposed SoS simulation models for non-isotropic scattering narrowband MIMO V2V channels. So far, only one deterministic SoS simulation model has been proposed for non-isotropic scattering wideband MIMO V2V channels [Zajic and Stuber 2009]. No stochastic SoS simulation models for non-isotropic scattering wideband MIMO V2V channels are available in the current literature.

The second part of Section 9.4.2 derives new deterministic and stochastic wideband MIMO V2V channel simulation models based on the proposed wideband RS-GBSM. First, we propose a new wideband deterministic simulation model. Based on the TDL concept, the proposed deterministic model employs a combination of two-ring and multiple confocal ellipse models that comprise LoS, single-, and double-bounced rays. The statistical properties of the model are verified by simulations. Furthermore, a new parameter computation method, named improved modified method of equal areas (IMMEA), is proposed for deterministic MIMO V2V channel simulators under non-isotropic scattering conditions. Compared with the existing parameter computation methods – the MMEA in [Pätzold et al. 2008] and the methods of Zajic and Stuber [2008b, 2009], IMMEA provides better approximations to the desired properties of the reference model. By allowing at least one parameter (frequencies and/or gains) to be a random variable, the wideband deterministic model can be further modified to become a wideband stochastic model. Numerical results validate the utility of the resulting stochastic model. It is worth stressing that the proposed wideband reference model can be easily reduced to the narrowband model in Section 9.4.1 by removing the frequency-selectivity. Therefore, the corresponding narrowband simulation models, whose reference model is the narrowband MIMO V2V RS-GBSM in Section 9.4.1, can be obtained by removing the frequency-selectivity from the proposed wideband simulation models.

9.4.2.1 A Wideband MIMO V2V Reference Model

Let us now consider a wideband MIMO V2V communication system with M_T transmit and M_R receive omnidirectional antenna elements. Both the Tx and Rx are equipped with low-elevation antennas. Figure 9.18 illustrates the geometry of the proposed RS-GBSM, which is the combination of a two-ring model and a multiple confocal ellipse model incorporating LoS, single- and double-bounced rays. Note that in Figure 9.18, we used uniform linear antenna arrays with $M_T = M_R = 2$ as an example.

The two-ring model defines two rings of effective scatterers, one around the Tx and the other around the Rx. Suppose there are $N_{1,1}$ effective scatterers around the Tx lying on a ring of radius R_T and the $n_{1,1}$th ($n_{1,1} = 1, \ldots, N_{1,1}$) effective scatterer is denoted by $s^{(n_{1,1})}$. Similarly, assume there are $N_{1,2}$ effective scatterers around the Rx lying on a ring of radius R_R and the $n_{1,2}$th ($n_{1,2} = 1, \ldots, N_{1,2}$) effective scatterer is denoted by $s^{(n_{1,2})}$. The multiple confocal ellipses model with the Tx and Rx located at the foci represents the TDL structure and has $N_{l,3}$ effective scatterers on the lth ellipse (that is, the lth tap), where $l = 1, 2, \ldots, L$ with L being the total number of ellipses or taps. The semi-major axis of the lth ellipse and the $n_{l,3}$th ($n_{l,3} = 1, \ldots, N_{l,3}$) effective scatterer are denoted by a_l and $s^{(n_{l,3})}$, respectively. The distance between the Tx and Rx is $D = 2f$ with f denoting the half length of the distance between the two focal points of ellipses. The antenna element spacings at the Tx and Rx are designated by δ_T and δ_R, respectively. The multi-element antenna tilt angles are denoted by β_T and β_R. The Tx and Rx

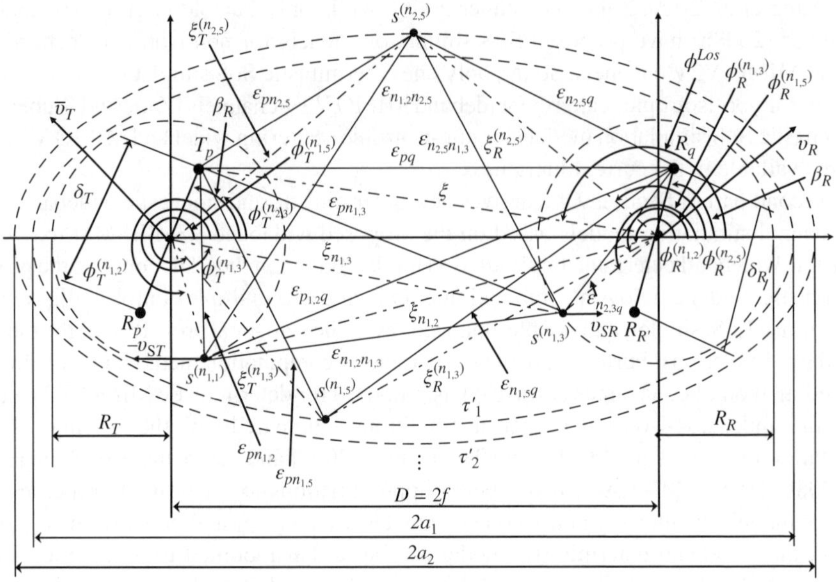

Figure 9.18 A geometry-based stochastic channel model combining a two-ring model and a multiple confocal ellipse model with LoS, single- and double-bounced rays for a wideband MIMO V2V channel

move with speeds v_T and v_R in directions determined by the angles of motion γ_T and γ_R, respectively. The AoA of the wave traveling from an effective scatterer $s^{(n_{1,1})}$, $s^{(n_{1,2})}$, and $s^{(n_{l,3})}$ toward the Rx are denoted by $\varphi_R^{(n_{1,1})}$, $\varphi_R^{(n_{1,2})}$, and $\varphi_R^{(n_{l,3})}$, respectively. The AoD of the wave that impinges on the effective scatterers $s^{(n_{1,1})}$, $s^{(n_{1,2})}$, and $s^{(n_{l,3})}$ are designated by $\varphi_T^{(n_{1,1})}$, $\varphi_T^{(n_{1,2})}$, and $\varphi_T^{(n_{l,3})}$, respectively. Note that φ^{LoS} denotes the AoA of a LoS path.

The MIMO fading channel can be described by a matrix $\mathbf{H}(t) = \left[h_{oq}(t, \tau')\right]_{M_R \times M_T}$ of size $M_R \times M_T$. According to the TDL concept, the complex impulse response between the oth ($o = 1, \ldots, M_T$) Tx, T_o, and the qth ($q = 1, \ldots, M_R$) Rx, R_q, can be expressed as $h_{oq}(t, \tau') = \sum_{l=1}^{L} c_l h_{l,oq}(t)\,\delta(\tau' - \tau_l')$ where c_l represents the gain of the lth tap, $h_{l,oq}(t)$ and τ_l' denote the complex time-variant tap coefficient and the discrete propagation delay of the lth tap, respectively. From the above RS-GBSM, the complex tap coefficient for the first tap of the $T_o - R_q$ link is a superposition of the LoS, single- and double-bounced components, and can be expressed as

$$h_{1,oq}(t) = h_{1,oq}^{LoS}(t) + \sum_{i=1}^{I} h_{1,oq}^{SB_i}(t) + h_{1,oq}^{DB}(t) \tag{9.77}$$

with

$$h_{1,oq}^{LoS}(t) = \sqrt{\frac{K_{oq}}{K_{oq}+1}}\, e^{-j2\pi f_c \tau_{oq}} e^{j\left[2\pi f_{T_{max}} t \cos\left(\pi - \varphi^{LoS} + \gamma_T\right) + 2\pi f_{R_{max}} t \cos\left(\varphi^{LoS} - \gamma_R\right)\right]} \tag{9.78a}$$

$$h_{1,oq}^{SB_i}(t) = \sqrt{\frac{\eta_{SB_{1,i}}}{K_{oq}+1}} \lim_{N_{1,i}\to\infty} \sum_{n_{1,i}=1}^{N_{1,i}} \frac{1}{\sqrt{N_{1,i}}} e^{j\left(\psi_{n_{1,i}} - 2\pi f_c \tau_{oq,n_{1,i}}\right)}$$

$$\times e^{j\left[2\pi f_{T_{max}} t\cos\left(\varphi_T^{(n_{1,i})}-\gamma_T\right)+2\pi f_{R_{max}} t\cos\left(\varphi_R^{(n_{1,i})}-\gamma_R\right)\right]} \tag{9.78b}$$

$$h_{1,oq}^{DB}(t)=\sqrt{\frac{\eta_{DB_1}}{K_{oq}+1}}\lim_{N_{1,1},N_{1,2}\to\infty}\sum_{n_{1,1},n_{1,2}=1}^{N_{1,1},N_{1,2}}\frac{1}{\sqrt{N_{1,1}N_{1,2}}}e^{j\left(\psi_{n_{1,1},n_{1,2}}-2\pi f_c\tau_{oq,n_{1,1},n_{1,2}}\right)}$$

$$\times e^{j\left[2\pi f_{T_{max}} t\cos\left(\varphi_T^{(n_{1,1})}-\gamma_T\right)+2\pi f_{R_{max}} t\cos\left(\varphi_R^{(n_{1,2})}-\gamma_R\right)\right]} \tag{9.78c}$$

where $\tau_{oq}=\varepsilon_{oq}/c$, $\tau_{oq,n_{1,i}}=(\varepsilon_{on_{1,i}}+\varepsilon_{n_{1,i}q})/c$, and $\tau_{oq,n_{1,1},n_{1,2}}=(\varepsilon_{on_{1,1}}+\varepsilon_{n_{1,1}n_{1,2}}+\varepsilon_{n_{1,2}q})/c$ are the travel times of the waves through the link T_o-R_q, $T_o-s^{(n_{1,i})}-R_q$, and $T_o-s^{(n_{1,1})}-s^{(n_{1,2})}-R_q$, respectively, as shown in Figure 9.18. The symbol $I=3$, and c and K_{oq} designate the speed of light and the Ricean factor, respectively. The complex tap coefficient for other taps ($l'>1$) of the T_o-R_q link is a superposition of the single- and double-bounced components, and can be expressed as

$$h_{l',oq}(t)=h_{l',oq}^{SB_3}(t)+h_{l',oq}^{DB_1}(t)+h_{l',oq}^{DB_2}(t) \tag{9.79}$$

with

$$h_{l',oq}^{SB_3}(t)=\sqrt{\eta_{SB_{l',3}}}\lim_{N_{l',3}\to\infty}\sum_{n_{l',3}=1}^{N_{l',3}}\frac{1}{\sqrt{N_{l',3}}}e^{j\left(\psi_{n_{l',3}}-2\pi f_c\tau_{oq,n_{l',3}}\right)}$$

$$\times e^{j\left[2\pi f_{T_{max}} t\cos\left(\varphi_T^{(n_{l',3})}-\gamma_T\right)+2\pi f_{R_{max}} t\cos\left(\varphi_R^{(n_{l',3})}-\gamma_R\right)\right]} \tag{9.80a}$$

$$h_{l',oq}^{DB_{1(2)}}(t)=\sqrt{\eta_{DB_{l',1(2)}}}\lim_{N_{1,1(2)},N_{l',3}\to\infty}\sum_{n_{1,1(2)},n_{l',3}=1}^{N_{1,1(2)},N_{l',3}}$$

$$\times\frac{1}{\sqrt{N_{1,1(2)}N_{l',3}}}e^{j\left(\psi_{n_{1,1(2)},n_{l',3}}-2\pi f_c\tau_{oq,n_{1,1(2)},n_{l',3}}\right)}$$

$$\times e^{j2\pi t\left[f_{T_{max}}\cos\left(\varphi_T^{(n_{1(l'),1(3)})}-\gamma_T\right)+f_{R_{max}}\cos\left(\varphi_R^{(n_{l'(1),3(2)})}-\gamma_R\right)\right]} \tag{9.80b}$$

$\tau_{oq,n_{l',3}}=(\varepsilon_{on_{l',3}}+\varepsilon_{n_{l',3}q})/c$ and $\tau_{oq,n_{1,1(2)},n_{l',3}}=(\varepsilon_{on_{1(l'),1(3)}}+\varepsilon_{n_{1(l'),1(3)}n_{l'(1),3(2)}}+\varepsilon_{n_{l'(1),3(2)}q})/c$ are the travel times of the waves through the link $T_o-s^{(n_{l',3})}-R_q$ and $T_o-s^{(n_{1,1})}(s^{(n_{l',3})})-s^{(n_{l',3})}(s^{(n_{1,2})})-R_q$, respectively, as illustrated in Figure 9.18. Energy-related parameters $\eta_{SB_{1,i}}$, η_{DB_1} and $\eta_{SB_{l',3}}$, $\eta_{DB_{l',1(2)}}$ specify how much the single- and double-bounced rays contribute to the total scattered power of the first tap and other taps, respectively. Note that these energy-related parameters satisfy $\sum_{i=1}^{I}\eta_{SB_{1,i}}+\eta_{DB_1}=1$ and $\eta_{SB_{l',3}}+\eta_{DB_{l',1}}+\eta_{DB_{l',2}}=1$. The phases $\psi_{n_{1,i}}$, $\psi_{n_{1,1},n_{1,2}}$, $\psi_{n_{l',3}}$, and $\psi_{n_{1,1(2)},n_{l',3}}$ are i.i.d. random variables with uniform distributions over $[-\pi,\pi)$, $f_{T_{max}}$ and $f_{R_{max}}$ are the maximum Doppler frequencies with respect to the Tx and Rx, respectively.

VTD significantly affects statistical properties at all taps of a wideband V2V channel [Sen and Matolak 2008]. To take the impact of VTD into account, we first distinguish between the moving cars around the Tx and Rx and the stationary roadside environment, using the two-ring

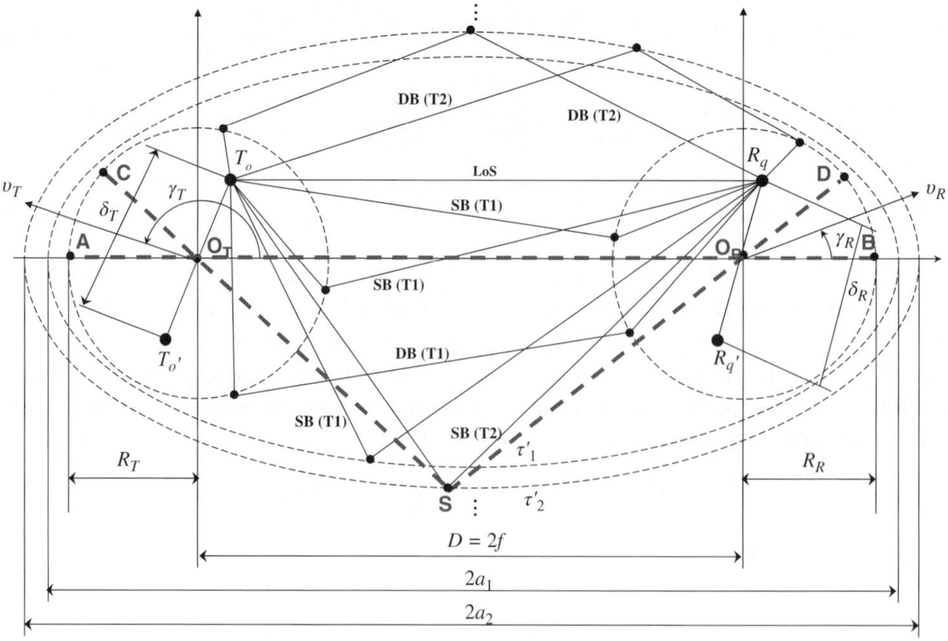

Figure 9.19 Geometrical description of the LoS, single-, and double-bounced rays for different taps in the proposed wideband MIMO V2V GBSM. SB, single-bounce; DB, double-bounce; T1, tap 1; T2, tap 2

model to mimic the moving cars and the multiple confocal ellipse model to depict the stationary roadside environment. For the first tap, the single-bounced rays are generated from the scatterers located on either of the two rings or the first ellipse, while the double-bounced rays are produced from the scatterers located on both rings. This means that the first tap contains a LoS component, a two-ring model with single- and double-bounced rays, and an ellipse model with single-bounced rays, as shown in Figure 9.19. For a low VTD, the value of K_{oq} is large since the LoS component can bear a significant amount of power. Also, the received scattered power is mainly from waves reflected by the stationary roadside environment, as described by the scatterers located on the first ellipse. The moving cars represented by the scatterers located on the two rings are sparse and thus more likely to be single-bounced, rather than double-bounced. This indicates that $\eta_{SB_{1,3}} > \max\{\eta_{SB_{1,1}}, \eta_{SB_{1,2}}\} > \eta_{DB_1}$. For a high VTD, the value of K_{oq} is smaller than the one in the low-VTD scenario. Also, due to the large number of moving cars, the double-bounced rays of the two-ring model bear more energy than the single-bounced rays of the two-ring and ellipse models: $\eta_{DB_1} > \max\{\eta_{SB_{1,1}}, \eta_{SB_{1,2}}, \eta_{SB_{1,3}}\}$.

For other taps, we assume that the single-bounced rays are generated only from the scatterers located on the corresponding ellipse, while the double-bounced rays are caused by the scatterers from the combined one ring (either of the two rings) and the corresponding ellipse, as illustrated in Figure 9.19. Note that according to the TDL structure, the double-bounced rays in one tap must be smaller in distance than the single-bounced rays of the ellipse on the next tap. As shown in Appendix A.8.8, this is valid only if the condition $\max\{R_T, R_R\} < \min\{a_l - a_{l-1}\}$ is fulfilled. For many current V2V channel measurement campaigns [Acosta et al. 2004, Acosta-Marum and Brunelli 2007, Sen and Matolak 2008], the resolution in delay is 100 ns. Then, the above condition can be modified as $\max\{R_T, R_R\} \leq 15$ m by calculating the equality $2(a_l - a_{l-1}) = c \cdot \tau'$ with $c = 3 \times 10^8$ m/s and $\tau' = 100$ ns. This indicates that the maximum acceptable width of the road is 30 m, which is sufficiently large to cover most roads in reality. In other words, the proposed wideband model with the specified TDL structure is valid for different scenarios. For low VTD, the received scattered power is mainly from waves reflected by the stationary roadside environment, as described by the scatterers located on the ellipse. This indicates that $\eta_{SB_{l',3}} > \max\{\eta_{DB_{l',1}}, \eta_{DB_{l',2}}\}$. For high VTD, due to the large number of moving cars, the double-bounced rays from the combined one-ring and ellipse models bear more energy than the single-bounced rays of the ellipse model: $\min\{\eta_{DB_{l',1}}, \eta_{DB_{l',2}}\} > \eta_{SB_{l',3}}$.

From Figure 9.18, based on the application of the law of cosines in triangles, the assumptions $\min\{R_T, R_R, a - f\} \gg \max\{\delta_T, \delta_R\}$ and $D \gg \max\{R_T, R_R\}$, and using the approximation $\sqrt{1+x} \approx 1 + x/2$ for small x, we have

$$\varepsilon_{oq} \approx D - k_o \delta_T \cos \beta_T - k_q \delta_R \cos(\varphi^{LoS} - \beta_R) \tag{9.81a}$$

$$\varepsilon_{on_{1,1}} \approx R_T - k_o \delta_T \cos(\varphi_T^{(n_{1,1})} - \beta_T) \tag{9.81b}$$

$$\varepsilon_{n_{1,1}q} \approx D - R_T \cos \varphi_T^{(n_{1,1})} - k_q \delta_R \cos(\varphi_R^{(n_{1,1})} - \beta_R) \tag{9.81c}$$

$$\varepsilon_{on_{1,2}} \approx D + R_R \cos \varphi_R^{(n_{1,2})} - k_o \delta_T \cos\left(\varphi_T^{(n_{1,2})} - \beta_T\right) \tag{9.81d}$$

$$\varepsilon_{n_{1,2}q} \approx R_R - k_q \delta_R \cos\left(\varphi_R^{(n_{1,2})} - \beta_R\right) \tag{9.81e}$$

$$\varepsilon_{n_{1,1}n_{1,2}} \approx D - R_T \cos \varphi_T^{(n_{1,1})} - R_R \cos\left(\varphi_R^{(n_{1,1})} - \varphi_R^{(n_{1,2})}\right) \tag{9.81f}$$

$$\varepsilon_{on_{l,3}} \approx \xi_T^{(l,3)} - k_o \delta_T \cos\left(\varphi_T^{(n_{l,3})} - \beta_T\right) \tag{9.81g}$$

$$\varepsilon_{n_{l,3}q} \approx \xi_R^{(l,3)} - k_q \delta_R \cos\left(\varphi_R^{(n_{l,3})} - \beta_R\right) \tag{9.81h}$$

$$\varepsilon_{n_{1,1}n_{l',3}} = \sqrt{\left(\xi_T^{(l',3)}\right)^2 + R_T^2 - 2\xi_T^{(l',3)} R_T \cos\left(\varphi_T^{(n_{l,1})} - \varphi_T^{(n_{l',3})}\right)} \tag{9.81i}$$

$$\varepsilon_{n_{l',3}n_{1,2}} = \sqrt{\left(\xi_R^{(l',3)}\right)^2 + R_R^2 - 2\xi_R^{(l',3)} R_R \cos\left(\varphi_R^{(n_{l,2})} - \varphi_R^{(n_{l',3})}\right)} \tag{9.81j}$$

where $\varphi^{LoS} \approx \pi$, $k_o = (M_T - 2o + 1)/2$, $k_q = (M_R - 2q + 1)/2$, $\xi_T^{(n_{l,3})} = \left(a_l^2 + f^2 + 2a_l f \times \cos \varphi_R^{(n_{l,3})}\right) / \left(a_l + f \cos \varphi_R^{(n_{l,3})}\right)$, and $\xi_R^{(n_{l,3})} = b_l^2 / \left(a_l + f \cos \varphi_R^{(n_{l,3})}\right)$ with b_l denoting the semi-minor axis of the lth ellipse. Note that the AoD $\varphi_T^{(n_{1,i})}$, $\varphi_T^{(n_{l',3})}$ and AoA $\varphi_R^{(n_{1,i})}$, $\varphi_R^{(n_{l',3})}$ are independent for double-bounced rays, while they are interdependent for single-bounced rays. Using the results in Section 9.4.1, we can express the relationships between the AoD and AoA for the single-bounced two-ring model as $\varphi_R^{(n_{1,1})} \approx \pi - \Delta_T \sin \varphi_T^{(n_{1,1})}$ and $\varphi_T^{(n_{1,2})} \approx \Delta_R \sin \varphi_R^{(n_{1,2})}$ with $\Delta_T \approx R_T/D$ and $\Delta_R \approx R_R/D$ and for the multiple confocal ellipse model as $\varphi_T^{(n_{l,3})} = \arcsin[b_l^2 \sin \varphi_R^{(n_{l,3})} / (a_l^2 + f^2 + 2a_l f \cos \varphi_R^{(n_{l,3})})]$.

Since the numbers of effective scatterers are assumed to be infinite – that is, $N_{1,i}, N_{l',3} \rightarrow \infty$ – the proposed model is actually a mathematical reference model and results in either Ricean PDF (the first tap) or Rayleighan PDF (other taps). For our reference model, the discrete AoDs $\varphi_T^{(n_{1,i})}$, $\varphi_T^{(n_{l',3})}$ and AoAs $\varphi_R^{(n_{1,i})}$, $\varphi_R^{(n_{l',3})}$, can be replaced by continuous expressions $\varphi_T^{(1,i)}$, $\varphi_T^{(l',3)}$ and $\varphi_R^{(1,i)}$, $\varphi_R^{(l',3)}$, respectively. To characterize these AoDs and AoAs, we use the von Mises PDF given in Eq. (9.7). Here, we give the expressions of the the von Mises PDF for the AoD $\varphi_T^{(1,1)}$ and AoAs $\varphi_R^{(1,2)}$, $\varphi_R^{(l,3)}$, which will be used later. Applying the von Mises PDF to the two-ring model, we obtain

- $f\left(\varphi_T^{(1,1)}\right) = \exp\left[k_T^{(1,1)} \cos\left(\varphi_T^{(1,1)} - \mu_T^{(1,1)}\right)\right] / \left[2\pi I_0\left(k_T^{(1,1)}\right)\right]$ for the AoD $\varphi_T^{(1,1)} \in [-\pi, \pi)$
- $f\left(\varphi_R^{(1,2)}\right) = \exp\left[k_R^{(1,2)} \cos\left(\varphi_R^{(1,2)} - \mu_R^{(1,2)}\right)\right] / \left[2\pi I_0\left(k_R^{(1,2)}\right)\right]$ for the AoA $\varphi_R^{(1,2)} \in [-\pi, \pi)$

Similarly, applying the von Mises PDF to the multiple confocal ellipses model, we get

- $f\left(\varphi_R^{(l,3)}\right) = \exp\left[k_R^{(l,3)} \cos\left(\varphi_R^{(l,3)} - \mu_R^{(l,3)}\right)\right] / \left[2\pi I_0\left(k_R^{(l,3)}\right)\right]$ with $\varphi_R^{(l,3)} \in [-\pi, \pi)$

Statistical Properties of the Proposed Model

In this section, based on the proposed wideband channel model, we will derive the ST CF, the corresponding SD PSD, the FCF, and the corresponding PDP for a non-isotropic scattering environment.

Space–Time CF Under the WSSUS condition, the correlation properties of two arbitrary channel impulse responses $h_{oq}(t, \tau')$ and $h_{o'q'}(t, \tau')$ of a MIMO V2V channel are completely determined by the correlation properties of $h_{l,oq}(t)$ and $h_{l,o'q'}(t)$ in each tap since no correlations exist between the underlying processes in different taps. Therefore, we can restrict our investigations to the following ST CF:

$$\rho_{h_{l,oq} h_{l,o'q'}}(\tau) = \mathbf{E}\left[h_{l,oq}(t)\, h^*_{l,o'q'}(t - \tau)\right] \tag{9.82}$$

where $(\cdot)^*$ denotes the complex conjugate operation and $\mathbf{E}[\cdot]$ designates the statistical expectation operator. Since the LoS, single-, and double-bounced components are independent of each other, based on Eq. (9.77) we have the following ST CF for the first tap

$$\rho_{h_{1,oq} h_{1,o'q'}}(\tau) = \rho_{h_{1,oq}^{LoS} h_{1,o'q'}^{LoS}}(\tau) + \sum_{i=1}^{I} \rho_{h_{1,oq}^{SB_i} h_{1,o'q'}^{SB_i}}(\tau) + \rho_{h_{1,oq}^{DB} h_{1,o'q'}^{DB}}(\tau) \tag{9.83}$$

For other taps, according to Eq. (9.79) we have the ST CF as

$$\rho_{h_{l',oq} h_{l',o'q'}}(\tau) = \rho_{h_{l',oq}^{SB_3} h_{l',o'q'}^{SB_3}}(\tau) + \rho_{h_{l',oq}^{DB_1} h_{l',o'q'}^{DB_1}}(\tau) + \rho_{h_{l',oq}^{DB_2} h_{l',o'q'}^{DB_2}}(\tau) \tag{9.84}$$

Applying the corresponding von Mises distribution, trigonometric transformations, and the equality $\int_{-\pi}^{\pi} \exp(a \sin c + b \cos c)\, dc = 2\pi I_0\left(\sqrt{a^2 + b^2}\right)$ [I.S.Gradshteyn and Ryzhik 2000], and following similar reasoning to Section 9.4.1, we can obtain the ST CF of the LoS, single-, and double-bounced components as follows.

1. In the case of the LoS component,

$$\rho_{h^{LoS}_{1,oq} h^{LoS}_{1,o'q'}}(\tau) = \sqrt{\frac{K_{oq} K_{o'q'}}{(K_{oq}+1)(K_{o'q'}+1)}} e^{j2\pi G + j2\pi\tau H} \tag{9.85}$$

where $G = O\cos\beta_T - Q\cos\beta_R$ and $H = f_{T_{max}}\cos\gamma_T - f_{R_{max}}\cos\gamma_R$ with $O = (o'-o)\,\delta_T/\lambda$, $Q = (q'-q)\,\delta_R/\lambda$.

2. In terms of the single-bounce two-ring model

$$\rho_{h^{SB_{1(2)}}_{1,oq} h^{SB_{1(2)}}_{1,o'q'}}(\tau) = \eta_{SB_{1,1(2)}} e^{jC^{SB_{1(2)}}_{T(R)}} \frac{I_0\left\{\sqrt{\left(A^{SB_{1(2)}}_{T(R)}\right)^2 + \left(B^{SB_{1(2)}}_{T(R)}\right)^2}\right\}}{\sqrt{(K_{oq}+1)(K_{o'q'}+1)}\, I_0\left(k^{(1,1(2))}_{T(R)}\right)} \tag{9.86}$$

where

$$A^{SB_{1(2)}}_{T(R)} = k^{(1,1(2))}_{T(R)} \cos\mu^{(1,1(2))}_{T(R)} + j2\pi\tau f_{T(R)_{max}} \cos\gamma_{T(R)} + j2\pi O(Q)\cos\beta_{T(R)} \tag{9.87a}$$

$$B^{SB_{1(2)}}_{T(R)} = k^{(1,1(2))}_{T(R)} \sin\mu^{(1,1(2))}_{T(R)} + j2\pi\tau(f_{T(R)_{max}} \sin\gamma_{T(R)} + f_{R(T)_{max}}\Delta_{T(R)}\sin\gamma_{R(T)}) + j2\pi(O(Q)\sin\beta_{T(R)} + Q(O)\Delta_{T(R)}\sin\beta_{R(T)}) \tag{9.87b}$$

$$C^{SB_{1(2)}}_{T(R)} = -2\pi\tau f_{R(T)_{max}} \cos\gamma_{R(T)} - 2\pi Q(O)\cos\beta_{R(T)} \tag{9.87c}$$

3. In the case of the single-bounce multiple confocal ellipses model

$$\rho_{h^{SB_3}_{l,oq} h^{SB_3}_{l,o'q'}}(\tau) = \frac{\eta_{SB_{l,3}}}{2\pi I_0\left(k^{(l,3)}_R\right) U} \int_{-\pi}^{\pi} e^{j2\pi\left[O\cos\left(\varphi^{(l,3)}_T - \beta_T\right) + Q\cos\left(\varphi^{(l,3)}_R - \beta_R\right)\right]} \times e^{k^{(l,3)}_R \cos\left(\varphi^{(l,3)}_R - \mu^{(l,3)}_R\right)} \times e^{j2\pi\tau\left[f_{T_{max}}\cos\left(\varphi^{(l,3)}_T - \gamma_T\right) + f_{R_{max}}\cos\left(\varphi^{(l,3)}_R - \gamma_R\right)\right]} d\varphi^{(l,3)}_R \tag{9.88}$$

where $\varphi^{(n_l,3)}_T = \arcsin[b_l^2 \sin\varphi^{(n_l,3)}_R / (a_l^2 + f^2 + 2a_l f\cos\varphi^{(n_l,3)}_R)]$ and $U = \sqrt{(K_{oq}+1)(K_{o'q'}+1)}$ only appears for the first tap.

4. In terms of the double-bounce component for the first tap

$$\rho_{h^{DB}_{1,oq} h^{DB}_{1,o'q'}}(\tau) = \eta_{DB_1} \frac{I_0\left\{\sqrt{\left(A^{DB_1}_T\right)^2 + \left(B^{DB_1}_T\right)^2}\right\} I_0\left\{\sqrt{\left(A^{DB_1}_R\right)^2 + \left(B^{DB_1}_R\right)^2}\right\}}{\sqrt{(K_{oq}+1)(K_{o'q'}+1)}\, I_0\left(k^{(1,1)}_T\right) I_0\left(k^{(1,2)}_R\right)} \tag{9.89}$$

where

$$A^{DB_1}_{T(R)} = k^{(1,1(2))}_{T(R)} \cos\mu^{(1,1(2))}_{T(R)} + j2\pi\tau f_{T(R)_{max}} \cos\gamma_{T(R)} + j2\pi O(Q)\cos\beta_{T(R)} \tag{9.90a}$$

$$B^{DB_1}_{T(R)} = k^{(1,1(2))}_{T(R)} \sin\mu^{(1,1(2))}_{T(R)} + j2\pi\tau f_{T(R)_{max}} \sin\gamma_{T(R)} + j2\pi O(Q)\sin\beta_{T(R)}. \tag{9.90b}$$

5. In terms of the double-bounce component for other taps

$$\rho_{h_{l',oq}^{DB_{1(2)}} h_{l',o'q'}^{DB_{1(2)}}}(\tau) = \eta_{DB_{l',1(2)}} \frac{I_0\left\{\sqrt{\left(A_T^{DB_{l',1(2)}}\right)^2 + \left(B_T^{DB_{l',1(2)}}\right)^2}\right\} \times I_0\left\{\sqrt{\left(A_R^{DB_{l',1(2)}}\right)^2 + \left(B_R^{DB_{l',1(2)}}\right)^2}\right\}}{I_0\left(k_T^{(1(l'),1(3))}\right) I_0\left(k_R^{(l'(1),3(2))}\right)} \tag{9.91}$$

where

$$A_T^{DB_{l',1(2)}} = k_T^{(1(l'),1(3))} \cos\mu_T^{(1(l'),1(3))} + j2\pi\tau f_{T_{max}} \cos\gamma_T + j2\pi O\cos\beta_T \tag{9.92a}$$

$$B_T^{DB_{l',1(2)}} = k_T^{(1(l'),1(3))} \sin\mu_T^{(1(l'),1(3))} + j2\pi\tau f_{T_{max}} \sin\gamma_T + j2\pi O\sin\beta_T \tag{9.92b}$$

$$A_R^{DB_{l',1(2)}} = k_R^{(l'(1),3(2))} \cos\mu_R^{(1(l'),1(3))} + j2\pi\tau f_{R_{max}} \cos\gamma_R + j2\pi Q\cos\beta_R \tag{9.92c}$$

$$B_R^{DB_{l',1(2)}} = k_R^{(l'(1),3(2))} \sin\mu_R^{(1(l'),1(3))} + j2\pi\tau f_{R_{max}} \sin\gamma_R + j2\pi Q\sin\beta_R. \tag{9.92d}$$

Note that since $k_T^{(l',3)}$, $\mu_T^{(l',3)}$ and $k_R^{(l',3)}$, $\mu_R^{(l',3)}$ refer to the same ellipse, $k_T^{(l',3)} = k_R^{(l',3)}$ and $\mu_T^{(l',3)} = \arcsin[b_{l'}^2 \sin\mu_R^{(l',3)}/(a_{l'}^2 + f^2 + 2a_{l'} f\cos\mu_R^{(l',3)})]$ hold. Finally, the ST CF of the channel impulse responses $h_{oq}(t,\tau')$ and $h_{o'q'}(t,\tau')$ can be expressed as $\rho_{h_{oq}h_{o'q'}}(\tau) = \sum_{l=1}^{L} c_l^2 \rho_{h_{l,oq}h_{l,o'q'}}(\tau)$.

Space–Doppler PSD Applying the Fourier transform to the ST CF $\rho_{h_{l,oq}h_{l,o'q'}}(\tau)$ in terms of τ, we can obtain the corresponding space-Doppler (SD) PSD as

$$S_{h_{l,oq}h_{l,o'q'}}(f_D) = \int_{-\infty}^{\infty} \rho_{h_{l,oq}h_{l,o'q'}}(\tau) e^{-j2\pi f_D\tau} d\tau \tag{9.93}$$

where f_D is the Doppler frequency. Therefore, SD PSD for the first tap is

$$S_{h_{1,oq}h_{1,o'q'}}(f_D) = S_{h_{1,oq}^{LoS} h_{1,o'q'}^{LoS}}(f_D) + \sum_{i=1}^{I} S_{h_{1,oq}^{SB_i} h_{1,o'q'}^{SB_i}}(f_D) + S_{h_{1,oq}^{DB} h_{1,o'q'}^{DB}}(f_D) \tag{9.94}$$

For other taps, we have the SD PSD as

$$S_{h_{l',oq}h_{l',o'q'}}(f_D) = S_{h_{l',oq}^{SB_3} h_{l',o'q'}^{SB_3}}(f_D) + S_{h_{l',oq}^{DB_1} h_{l',o'q'}^{DB_1}}(f_D) + S_{h_{l',oq}^{DB_2} h_{l',o'q'}^{DB_2}}(f_D) \tag{9.95}$$

Note that except for the SD PSD of the single-bounce multiple confocal ellipses model $S_{h_{l',oq}^{SB_3} h_{l',o'q'}^{SB_3}}(f_D)$, which must be evaluated numerically; for other SD PSDs, using the equality $\int_0^\infty I_0\left(j\alpha\sqrt{x^2+y^2}\right)\cos(\beta x)\ dx = \cos\left(y\sqrt{\alpha^2-\beta^2}\right)/\sqrt{\alpha^2-\beta^2}$ [I.S.Gradshteyn and Ryzhik 2000] and following similar reasoning to Section 9.4.1, we can derive the following closed-form solutions:

1. In the case of the LoS component,

$$S_{h_{1,oq}^{LoS} h_{1,o'q'}^{LoS}}(f_D) = \sqrt{\frac{K_{oq}K_{o'q'}}{(K_{oq}+1)(K_{o'q'}+1)}} e^{j2\pi G}\delta(f_D - H) \tag{9.96}$$

where $\delta(\cdot)$ denotes the Dirac delta function.

2. In terms of the single-bounce two-ring model

$$S_{h_{1,oq}^{SB_{1(2)}} h_{1,o'q'}^{SB_{1(2)}}}(f_D) = \frac{\eta_{SB_{1,1(2)}} 2e^{jU_{T(R)}^{SB_{1(2)}} + j2\pi O_{T(R)}^{SB_{1(2)}} \frac{D_{T(R)}^{SB_{1(2)}}}{W_{T(R)}^{SB_{1(2)}}}} \times \cos\left[\frac{E_{T(R)}^{SB_{1(2)}}}{W_{T(R)}^{SB_{1(2)}}}\sqrt{W_{T(R)}^{SB_{1(2)}} - 4\pi^2\left(O_{T(R)}^{SB_{1(2)}}\right)^2}\right]}{\sqrt{(K_{oq}+1)(K_{o'q'}+1)} I_0\left(k_{T(R)}^{(1,1(2))}\right)\sqrt{W_{T(R)}^{SB_{1(2)}} - 4\pi^2\left(O_{T(R)}^{SB_{1(2)}}\right)^2}} \tag{9.97}$$

where $O_{T(R)}^{SB_{1(2)}} = 2\pi\left(f_D \pm f_{R(T)_{max}}\cos\gamma_{R(T)}\right)$, $U_{T(R)}^{SB_{1(2)}} = \mp 2\pi Q(O)\cos\beta_{R(T)}$

$$W_{T(R)}^{SB_{1(2)}} = 4\pi^2 f_{T(R)_{max}}^2 + 4\pi^2 f_{R(T)_{max}}^2 \Delta_{T(R)}^2 \sin^2\gamma_{R(T)} + 8\pi^2 f_{T_{max}} f_{R_{max}} \Delta_{T(R)} \sin\gamma_T \sin\gamma_R \tag{9.98a}$$

$$\begin{aligned} D_{T(R)}^{SB_{1(2)}} &= -j2\pi k_{T(R)}^{(1,1(2))} J_{T(R)} + 4\pi^2 O(Q)(f_{T(R)_{max}} \cos\left(\beta_{T(R)} - \gamma_{T(R)}\right) \\ &+ \Delta_{T(R)} f_{R(T)_{max}} \sin\beta_{T(R)} \sin\gamma_{R(T)}) + 4\pi^2 Q(O)(\Delta_{T(R)} f_{T(R)_{max}} \\ &\times \sin\beta_{R(T)}\sin\gamma_{T(R)} + \Delta_{T(R)}^2 f_{R(T)_{max}} \sin\beta_{R(T)}\sin\gamma_{R(T)}) \end{aligned} \tag{9.98b}$$

$$\begin{aligned} E_{T(R)}^{SB_{1(2)}} &= j2\pi k_{T(R)}^{(1,1(2))}(f_{T(R)_{max}} \sin\left(\gamma_{T(R)} - \mu_{T(R)}^{(1,1(2))}\right) + f_{R(T)_{max}}\Delta_{T(R)}\sin\gamma_{R(T)} \\ &\times \cos\mu_{T(R)}^{(1,1(2))}) + 4\pi^2 O(Q)(f_{T(R)_{max}} \sin\left(\beta_{T(R)} - \gamma_{T(R)}\right) - \Delta_{T(R)} f_{R(T)_{max}} \\ &\times \cos\beta_{T(R)}\sin\gamma_{R(T)}) + 4\pi^2 Q(O)\Delta_{T(R)} f_{T(R)_{max}} \sin\beta_{R(T)}\cos\gamma_{T(R)} \end{aligned} \tag{9.98c}$$

with $J_{T(R)} = f_{T(R)_{max}}\cos\left(\gamma_{T(R)} - \mu_{T(R)}^{(1,1(2))}\right) - f_{R(T)_{max}}\Delta_{T(R)}\sin\gamma_{R(T)}\sin\mu_{T(R)}^{(1,1(2))}$. For the Doppler PSD in Eq. (9.97), the range of Doppler frequency is limited by $\left|f_D \pm f_{R(T)_{max}}\cos\gamma_{R(T)}\right| \le \sqrt{W_{T(R)}^{SB_{1(2)}}}/(2\pi)$.

3. In terms of the double-bounce component for the first tap

$$\begin{aligned} S_{h_{1,oq}^{DB} h_{1,o'q'}^{DB}}(f_D) &= \frac{\eta_{DB_1}}{\sqrt{(K_{oq}+1)(K_{o'q'}+1)} I_0\left(k_T^{(1,1)}\right) I_0\left(k_R^{(1,2)}\right)} 2e^{j2\pi f \frac{D_T^{DB}}{W_T^{DB}}} \\ &\times \frac{\cos\left(\frac{E_T^{DB}}{W_T^{DB}}\sqrt{W_T^{DB} - 4\pi^2 f^2}\right)}{\sqrt{W_T^{DB} - 4\pi^2 f^2}} \odot 2e^{j2\pi f \frac{D_R^{DB}}{W_R^{DB}}} \\ &\times \frac{\cos\left(\frac{E_R^{DB}}{W_R^{DB}}\sqrt{W_R^{DB} - 4\pi^2 f^2}\right)}{\sqrt{W_R^{DB} - 4\pi^2 f^2}} \end{aligned} \tag{9.99}$$

where $\odot$ denotes convolution, $W_{T(R)}^{DB} = 4\pi^2 f_{T(R)_{max}}^2$

$$D_{T(R)}^{DB} = 4\pi^2 O(Q) f_{T(R)_{max}} \cos\left(\beta_{T(R)} - \gamma_{T(R)}\right) - j2\pi k_{T(R)}^{(1,1(2))} f_{T(R)_{max}} \cos\left(\gamma_{T(R)} - \mu_{T(R)}^{(1,1(2))}\right) \quad (9.100a)$$

$$E_{T(R)}^{DB} = 4\pi^2 O(Q) f_{T(R)_{max}} \sin\left(\beta_{T(R)} + \gamma_{T(R)}\right) - j2\pi k_{T(R)}^{(1,1(2))} f_{T(R)_{max}} \sin\left(\gamma_{T(R)} - \mu_{T(R)}^{(1,1(2))}\right). \quad (9.100b)$$

For the Doppler PSD in Eq. (9.99), the range of Doppler frequency is limited by $|f_D| \leq f_{T_{max}} + f_{R_{max}}$.

4. In the case of the double-bounce component for other taps

$$S_{h_{l',oq}^{DB_{1(2)}} h_{l',o'q'}^{DB_{1(2)}}}(f_D) = \frac{\eta_{DB_{l',1(2)}}}{I_0\left(k_T^{(1(l'),1(3))}\right) I_0\left(k_R^{(l'(1),3(2))}\right)} 2e^{j2\pi f \frac{D_T^{DB_{l',1(2)}}}{W_T^{DB_{l',1(2)}}}} \times \frac{\cos\left(\frac{E_T^{DB_{l',1(2)}}}{W_T^{DB_{l',1(2)}}} \sqrt{W_T^{DB_{l',1(2)}} - 4\pi^2 f^2}\right)}{\sqrt{W_T^{DB_{l',1(2)}} - 4\pi^2 f^2}} \odot 2e^{j2\pi f \frac{D_R^{DB_{l',1(2)}}}{W_R^{DB_{l',1(2)}}}} \times \frac{\cos\left(\frac{E_R^{DB_{l',1(2)}}}{W_R^{DB_{l',1(2)}}} \sqrt{W_R^{DB_{l',1(2)}} - 4\pi^2 f^2}\right)}{\sqrt{W_R^{DB_{l',1(2)}} - 4\pi^2 f^2}} \quad (9.101)$$

where $W_T^{DB_{l',1(2)}} = 4\pi^2 f_{T_{max}}^2$, $W_R^{DB_{l',1(2)}} = 4\pi^2 f_{R_{max}}^2$

$$D_T^{DB_{l',1(2)}} = 4\pi^2 O f_{T_{max}} \cos(\beta_T - \gamma_T) - j2\pi k_T^{(1(l'),1(3))} f_{T_{max}} \cos\left(\gamma_T - \mu_T^{(1(l'),1(3))}\right) \quad (9.102a)$$

$$E_T^{DB_{l',1(2)}} = 4\pi^2 O f_{T_{max}} \sin(\beta_T - \gamma_T) + j2\pi k_T^{(1(l'),1(3))} f_{T_{max}} \sin\left(\gamma_T - \mu_T^{(1(l'),1(3))}\right) \quad (9.102b)$$

$$D_R^{DB_{l',1(2)}} = 4\pi^2 Q f_{R_{max}} \cos(\beta_R - \gamma_R) - j2\pi k_R^{(l'(1),3(2))} f_{R_{max}} \cos\left(\gamma_R - \mu_R^{(l'(1),3(2))}\right) \quad (9.102c)$$

$$E_R^{DB_{l',1(2)}} = 4\pi^2 Q f_{R_{max}} \sin(\beta_R - \gamma_R) + j2\pi k_R^{(l'(1),3(2))} f_{R_{max}} \sin\left(\gamma_R - \mu_R^{(l'(1),3(2))}\right). \quad (9.102d)$$

For the Doppler PSD in Eq. (9.101), the range of Doppler frequency is limited by $|f_D| \leq f_{T_{max}} + f_{R_{max}}$. Similar to the ST CF, the SD PSD of the channel impulse responses $h_{oq}(t, \tau')$ and $h_{o'q'}(t, \tau')$ can be expressed as $S_{h_{oq} h_{o'q'}}(f_D) = \sum_{l=1}^{L} c_l^2 S_{h_{l,oq} h_{l,o'q'}}(f_D)$.

Frequency Correlation Function and Power Delay Profile The FCF $\rho_{H_{oq}H_{o'q'}}(\Delta f')$ of the proposed wideband MIMO V2V GBSM is defined as $\rho_{H_{oq}H_{oq}}(\Delta f') = \mathbf{E}\left[H_{oq}(t, f') H^*_{oq}(t, f' - \Delta f')\right]$. Here, $H_{oq}(t, f')$ denotes the time-variant transfer function, which is the Fourier transform of the channel impulse response $h_{oq}(t, \tau')$ and can be expressed as $H_{oq}(t, f') = \sum_{l=1}^{L} c_l h_{l,oq}(t) e^{-j2\pi f'\tau'}$. Therefore, the FCF can be derived as

$$\rho_{H_{oq}H_{o'q'}}(\Delta f') = \sum_{l=1}^{L} c_l^2 e^{-j2\pi f'\tau_l'} \tag{9.103}$$

Applying the inverse Fourier transform to the FCF $\rho_{H_{oq}H_{o'q'}}(\Delta f')$ in Eq. (9.103), we can obtain the corresponding PDP as $S_{H_{oq}H_{o'q'}}(\tau') = \sum_{l=1}^{L} c_l^2 \delta(\tau' - \tau_l')$. It is obvious that the FCF and the PDP are completely determined by the number of propagation paths L, the path gains c_l, and the propagation delays τ_l'. Appropriate values for these parameters can be found in many measurement campaigns for wideband V2V channels [Acosta et al. 2004, Acosta-Marum and Brunelli 2007, Paier et al. 2009, Sen and Matolak 2008]. This allows us to fit the FCF $\rho_{H_{oq}H_{o'q'}}(\Delta f')$ and the corresponding PDP $S_{H_{oq}H_{o'q'}}(\tau')$ of the proposed model to any specified or measured FCF and PDP characterized by the sets $\{c_l\}_{l=1}^{L}$ and $\{\tau_l'\}_{l=1}^{L}$.

9.4.2.2 MIMO V2V Simulation Models

The reference models for non-isotropic scattering narrowband and wideband MIMO V2V channels proposed in Sections 9.4.1 and 9.4.2, respectively, assume an infinite number of effective scatterers, and thus cannot be implemented in practice. Comparing the narrowband reference model in Eq. (9.33) and the wideband reference model in Eqs (9.77) and (9.79), it is clear that the proposed wideband model can be easily reduced to our narrowband model by only considering the complex fading envelope for the first tap (that is, by removing the frequency selectivity) and assuming the power $\Omega_{oq} = 1$. Therefore, based on the proposed wideband reference model, this section will design realisable wideband simulation models that have reasonable complexity. Similarly, the narrowband simulation models can be obtained from the wideband simulation models by removing the frequency-selectivity.

Wideband Simulation Models

Based on the TDL structure, the complex impulse response of our simulation models is again composed of L discrete taps according to $\tilde{h}_{oq}(t, \tau') = \sum_{l=1}^{L} c_l \tilde{h}_{l,oq}(t)\, \delta(\tau' - \tau_l')$, where the complex time-variant tap coefficient $\tilde{h}_{l,oq}(t)$ is modeled using a finite number of effective scatterers. Assuming 2D non-isotropic scattering and using our reference model of Eqs (9.77) and (9.79), we propose the time-variant tap coefficient for the first tap as

$$\tilde{h}_{1,oq}(t) = \tilde{h}_{1,oq}^{LoS}(t) + \sum_{i=1}^{I} \tilde{h}_{1,oq}^{SB_i}(t) + \tilde{h}_{1,oq}^{DB}(t) \tag{9.104}$$

where

$$\tilde{h}_{1,oq}^{LoS}(t) = \sqrt{\frac{K_{oq}}{K_{oq}+1}}\, e^{-j2\pi f_c \tau_{oq}} e^{j\left[2\pi f_{T_{max}} t \cos\left(\pi - \phi^{LoS} + \gamma_T\right) + 2\pi f_{R_{max}} t \cos\left(\phi^{LoS} - \gamma_R\right)\right]} \tag{9.105a}$$

$$\tilde{h}_{1,oq}^{SB_i}(t) = \sqrt{\frac{\eta_{SB_{1,i}}}{K_{oq}+1}} \sum_{n_{1,i}=1}^{N_{1,i}} \frac{1}{\sqrt{N_{1,i}}} e^{j\left(\tilde{\psi}_{n_{1,i}} - 2\pi f_c \tau_{oq,n_{1,i}}\right)}$$

$$\times e^{j\left[2\pi f_{T_{max}} t \cos\left(\tilde{\phi}_T^{(n_{1,i})} - \gamma_T\right) + 2\pi f_{R_{max}} t \cos\left(\tilde{\phi}_R^{(n_{1,i})} - \gamma_R\right)\right]} \tag{9.105b}$$

$$\tilde{h}_{1,oq}^{DB}(t) = \sqrt{\frac{\eta_{DB_1}}{K_{oq}+1}} \sum_{n_{1,1},n_{1,2}=1}^{N_{1,1},N_{1,2}} \frac{1}{\sqrt{N_{1,1}N_{1,2}}} e^{j\left(\tilde{\psi}_{n_{1,1},n_{1,2}} - 2\pi f_c \tau_{oq,n_{1,1},n_{1,2}}\right)}$$

$$\times e^{j\left[2\pi f_{T_{max}} t \cos\left(\tilde{\phi}_T^{(n_{1,1})} - \gamma_T\right) + 2\pi f_{R_{max}} t \cos\left(\tilde{\phi}_R^{(n_{1,2})} - \gamma_R\right)\right]} \tag{9.105c}$$

and the time-variant tap coefficient for other taps as

$$\tilde{h}_{l',oq}(t) = \tilde{h}_{l',oq}^{SB_3}(t) + \tilde{h}_{l',oq}^{DB_1}(t) + \tilde{h}_{l',oq}^{DB_2}(t) \tag{9.106}$$

where

$$\tilde{h}_{l',oq}^{SB_3}(t) = \sqrt{\eta_{SB_{l',3}}} \sum_{n_{l',3}=1}^{N_{l',3}} \frac{1}{\sqrt{N_{l',3}}} e^{j\left(\tilde{\psi}_{n_{l',3}} - 2\pi f_c \tau_{oq,n_{l',3}}\right)}$$

$$\times e^{j\left[2\pi f_{T_{max}} t \cos\left(\tilde{\phi}_T^{(n_{l',3})} - \gamma_T\right) + 2\pi f_{R_{max}} t \cos\left(\tilde{\phi}_R^{(n_{l',3})} - \gamma_R\right)\right]} \tag{9.107a}$$

$$\tilde{h}_{l',oq}^{DB_{1(2)}}(t) = \sqrt{\eta_{DB_{l',1(2)}}} \sum_{n_{1,1(2)},n_{l',3}=1}^{N_{1,1(2)},N_{l',3}} \frac{1}{\sqrt{N_{1,1(2)}N_{l',3}}} e^{j\left(\tilde{\psi}_{n_{1,1(2)},n_{l',3}} - 2\pi f_c \tau_{oq,n_{1,1(2)},n_{l',3}}\right)}$$

$$\times e^{j2\pi t\left[f_{T_{max}} \cos\left(\tilde{\phi}_T^{(n_{1(l'),1(3)})} - \gamma_T\right) + f_{R_{max}} \cos\left(\tilde{\phi}_R^{(n_{l'(1),3(2)})} - \gamma_R\right)\right]} \tag{9.107b}$$

In Eqs (9.105) and (9.107), the AoDs $\tilde{\phi}_T^{(n_{1,i})}$ and $\tilde{\phi}_T^{(n_{l',3})}$, and AoAs $\tilde{\phi}_R^{(n_{1,i})}$ and $\tilde{\phi}_R^{(n_{l',3})}$ are discrete realizations of the random variables $\phi_T^{(1,i)}$ and $\phi_T^{(l',3)}$, and $\phi_R^{(1,i)}$ and $\phi_R^{(l',3)}$, respectively. Note that from our reference model in Section 6.2, we know that only the AoD $\tilde{\phi}_T^{(n_{1,1})}$ and AoAs $\tilde{\phi}_R^{(n_{1,2})}$ and $\tilde{\phi}_R^{(n_{l,3})}$ need to be generated based on the corresponding von Mises PDFs. The phases $\tilde{\psi}_{n_{1,i}}$, $\tilde{\psi}_{n_{1,1},n_{1,2}}$, $\tilde{\psi}_{n_{l',3}}$, and $\tilde{\psi}_{n_{1,1(2)},n_{l',3}}$ are i.i.d. random variables with uniform distributions over $[-\pi, \pi)$.

New Deterministic Simulation Model We first propose a new deterministic simulation model. This has parameters that are kept constant during simulation and thus only one simulation trial is required to obtain the desired statistical properties. The complex impulse response is $\tilde{h}_{oq}(t,\tau') = \sum_{l=1}^{L} c_l \tilde{h}_{l,oq}(t)\,\delta(\tau' - \tau_l')$, where the complex time-variant tap coefficient $\tilde{h}_{l,oq}(t)$ is a deterministic function and defined as in Eqs (9.104) and (9.106). Therefore, the essential issue in the design of our deterministic simulation model is to find the sets of AoD $\left\{\tilde{\phi}_T^{(n_{1,1})}\right\}_{n=1}^{N_{1,1}}$ and AoAs $\left\{\tilde{\phi}_R^{(n_{1,2})}\right\}_{n=1}^{N_{1,2}}$, $\left\{\tilde{\phi}_R^{(n_{l,3})}\right\}_{n=1}^{N_{l,3}}$ that make the simulation model reproduce the desired statistical properties of the reference model as faithfully as possible with reasonable complexity; that is, with a finite number of $N_{1,i}$ and $N_{l',3}$. To this end, the following conditions should be met, as previously mentioned in Chapter 5:

$\tilde{\phi}_{T(R)}^{(n_{1,1(2)})} \neq \pm\tilde{\phi}_{T(R)}^{(m_{1,1(2)})}$, $n_{1,1(2)} \neq m_{1,1(2)}$ and $\tilde{\phi}_{R}^{(n_{l,3})} \neq \pm\tilde{\phi}_{R}^{(m_{l,3})}$, $n_{l,3} \neq m_{l,3}$. However, in Chapter 5 we found that, unlike isotropic scattering environments, for non-isotropic scattering environments it is difficult to find the sets of AoDs and AoAs to meet these conditions. To the best of the authors' knowledge, so far only two parameter computation methods are available [Pätzold et al. 2008, Zajic and Stuber 2008b, 2009] for non-isotropic scattering MIMO V2V channels. However, neither of these solved the problem of finding sets of AoDs and AoAs to meet the conditions. This motivates us to propose a new parameter computation method, named IMMEA, to solve this difficulty. For easy comparison, we first give a brief description of these two existing methods.

The first method is MMEA [Pätzold et al. 2008], which was first proposed by de Leon and Patzold [2007] for F2M cellular channels. The MMEA originated from MEA [Pätzold 2002], used for isotropic scattering F2M channels. Based on MMEA, the AoD and AoA of our model can be designed as

$$\frac{n_{1,1(2)} - 1/4}{N_{1,1(2)}} = \int_{\mu_{T(R)}^{(1,1(2))}-\pi}^{\tilde{\phi}_{T(R)}^{(n_{1,1(2)})}} f\left(\phi_{T(R)}^{(1,1(2))}\right) d\phi_{T(R)}^{(1,1(2))}, \quad n_{1,1(2)} = 1, 2, \ldots, N_{1,1(2)} \tag{9.108a}$$

$$\frac{n_{l,3} - 1/4}{N_{l,3}} = \int_{\mu_{R}^{(l,3)}-\pi}^{\tilde{\phi}_{R}^{(n_{l,3})}} f\left(\phi_{R}^{(l,3)}\right) d\phi_{R}^{(l,3)}, \quad n_{l,3} = 1, 2, \ldots, N_{l,3} \tag{9.108b}$$

where $f\left(\phi_{T(R)}^{(1,1(2))}\right)$ ($\phi_{T(R)}^{(1,1(2))} \in [-\pi, \pi)$) and $f\left(\phi_{R}^{(l,3)}\right)$ ($\phi_{R}^{(l,3)} \in [-\pi, \pi)$) are von Mises PDFs with mean angles $\mu_{T(R)}^{(1,1(2))}$ and $\mu_{R}^{(l,3)}$, respectively. If the mean angles $\mu_{T(R)}^{(1,1(2))}$ and $\mu_{R}^{(l,3)}$ are equal to or less than zero, Eqs (9.108a) and (9.108b) become

$$\tilde{\phi}_{T(R)}^{(n_{1,1(2)})} = F_{T(R)}^{-1}\left(\frac{n_{1,1(2)} - 1/4}{N_{1,1(2)}}\right), \quad \tilde{\phi}_{T(R)}^{(n_{1,1(2)})} \in [-\pi, \pi) \tag{9.109a}$$

$$\tilde{\phi}_{R}^{(n_{l,3})} = F_{R_l}^{-1}\left(\frac{n_{l,3} - 1/4}{N_{l,3}}\right), \quad \tilde{\phi}_{R}^{(n_{l,3})} \in [-\pi, \pi) \tag{9.109b}$$

where $F_{T(R)}^{-1}(\cdot)$ and $F_{R_l}^{-1}(\cdot)$ denote the inverse function of the von Mises CDF for $\phi_{T(R)}^{(1,1(2))}$ and $\phi_{R}^{(l,3)}$, respectively. For $\mu_{T(R)}^{(1,1(2))} > 0$ and $\mu_{R}^{(l,3)} > 0$, it is clear that the $\tilde{\phi}_{T(R)}^{(n_{1,1}(m_{1,2}))}$ and $\tilde{\phi}_{R}^{(n_{l,3})}$ are designed over the ranges $\left[\mu_{T(R)}^{(1,1(2))} - \pi, \pi\right)$ and $\left[\mu_{R}^{(l,3)} - \pi, \pi\right)$, respectively, rather than over the whole range $[-\pi, \pi)$, which is necessary to the design of simulation model under the condition of non-isotropic scattering as addressed in Chapter 5 and de Leon and Patzold 2007. Note that as mentioned by de Leon and Patzold [2007], the MMEA can only meet the above conditions when $\mu_{T(R)}^{(1,1(2))} = 0$ and $\mu_{R}^{(l,3)} = 0$.

The other parameter computation method is the method proposed by Zajic and Stuber [2008b; 2009]. To distinguish this from MMEA, we call this method MMEA2 since it actually originated from the MEA of Pätzold [2002] as well. Following MMEA2, we have the AoD and AoA of our model as

$$\tilde{\phi}_{T(R)}^{(n_{1,1(2)})} = F_{T(R)}^{-1}\left(\frac{n_{1,1(2)} - 1/2}{N_{1,1(2)}}\right), \quad \tilde{\phi}_{T(R)}^{(n_{1,1(2)})} \in [-\pi, \pi) \tag{9.110a}$$

$$\tilde{\phi}_R^{(n_{l,3})} = F_{R_l}^{-1}\left(\frac{n_{l,3}-1/2}{N_{l,3}}\right), \qquad \tilde{\phi}_R^{(n_{l,3})} \in [-\pi, \pi) \tag{9.110b}$$

It is obvious that the MMEA2 chooses a value of $\frac{1}{2}$ rather than the $\frac{1}{4}$ chosen in MMEA. This is inspired by the modified MEDS of Patel et al. [2005] for the simulation of isotropic scattering V2V channels. However, MMEA2 cannot meet the above conditions.

In the following, we derive IMMEA, a method which *can* meet the above conditions for any non-isotropic scattering V2V channel. To this end, we first define the random variables $\phi'^{(1,1(2))}_{T(R)}$, which fulfill the von Mises distribution with the same mean angle $\mu^{(1,1(2))}_{T(R)}$ and parameter $k^{(1,1(2))}_{T(R)}$ as the random variables $\phi^{(1,1(2))}_{T(R)}$, over the range $\left[\mu^{(1,1(2))}_{T(R)} - \pi, \mu^{(1,1(2))}_{T(R)} + \pi\right)$. Similarly, we define the random variables $\phi'^{(l,3)}_R$ over the range $\left[\mu^{(l,3)}_R - \pi, \mu^{(l,3)}_R + \pi\right)$ that fulfill the von Mises distribution with the same mean angle $\mu^{(l,3)}_R$ and parameter $k^{(l,3)}_R$ as the random variables $\phi^{(l,3)}_R$. The discrete realizations of the new defined random variables $\phi'^{(1,1(2))}_{T(R)}$ and $\phi'^{(l,3)}_R$ can be designed as

$$\tilde{\phi}'^{(n_{1,1(2)})}_{T(R)} = F'^{-1}_{T(R)}\left(\frac{n_{1,1(2)}-1/4}{N_{1,1(2)}}\right), \; \tilde{\phi}'^{(n_{1,1(2)})}_{T(R)} \in \left[\mu^{(1,1(2))}_{T(R)} - \pi, \mu^{(1,1(2))}_{T(R)} + \pi\right) \tag{9.111a}$$

$$\tilde{\phi}'^{(n_{l,3})}_R = F'^{-1}_{R_l}\left(\frac{n_{l,3}-1/4}{N_{l,3}}\right), \qquad \tilde{\phi}'^{(n_{l,3})}_R \in \left[\mu^{(l,3)}_R - \pi, \mu^{(l,3)}_R + \pi\right) \tag{9.111b}$$

where $F'^{-1}_{T(R)}(\cdot)$ and $F'^{-1}_{R_l}(\cdot)$ denote the inverse functions of the von Mises CDF for $\phi'^{(1,1(2))}_{T(R)}$ and $\phi'^{(l,3)}_R$, respectively. The AoD and AoAs of our simulation model can be obtained by mapping $\tilde{\phi}'^{(n_{1,1(2)})}_{T(R)}$ and $\tilde{\phi}'^{(n_{l,3})}_R$ to the range $[-\pi, \pi)$ as

$$\tilde{\phi}^{(n_{1,1(2)})}_{T(R)} = \begin{cases} \tilde{\phi}'^{(n_{1,1(2)})}_{T(R)} + 2\pi, & \text{if} \quad \tilde{\phi}'^{(n_{1,1(2)})}_{T(R)} < -\pi \\ \tilde{\phi}'^{(n_{1,1(2)})}_{T(R)} - 2\pi, & \text{if} \quad \tilde{\phi}'^{(n_{1,1(2)})}_{T(R)} \geq \pi \\ \tilde{\phi}'^{(n_{1,1(2)})}_{T(R)}, & \text{else} \end{cases} \tag{9.112a}$$

$$\tilde{\phi}^{(n_{l,3})}_R = \begin{cases} \tilde{\phi}'^{(n_{l,3})}_R + 2\pi, & \text{if} \quad \tilde{\phi}'^{(n_{l,3})}_R < -\pi \\ \tilde{\phi}'^{(n_{l,3})}_R - 2\pi, & \text{if} \quad \tilde{\phi}'^{(n_{l,3})}_R \geq \pi \\ \tilde{\phi}'^{(n_{l,3})}_R, & \text{else} \end{cases} \tag{9.112b}$$

It is clear that IMMEA corresponds to MMEA when $\mu^{(1,1(2))}_{T(R)} = 0$ and $\mu^{(l,3)}_R = 0$. Also, IMMEA can meet the conditions above and thus should outperform the MMEA and MMEA2.

The correlation properties of our simulation model must be analyzed using time averages rather than statistical averages. The ST CF has to be computed according to $\tilde{\rho}_{\tilde{h}_{l,oq}\tilde{h}_{l,o'q'}}(\tau) = \left\langle \tilde{h}_{l,oq}(t)\, \tilde{h}^*_{l,o'q'}(t-\tau)\right\rangle$, where $\langle\cdot\rangle$ denotes the time-average operator. For brevity, the straightforward derivation of $\tilde{\rho}_{\tilde{h}_{l,oq}\tilde{h}_{l,o'q'}}(\tau)$ is omitted. Note that for $\{N_{1,i}, N_{l',3}\} \to \infty$, the ST CF of our simulation model matches that of the reference model; that is,

$\tilde{\rho}_{\tilde{h}_{l,oq}\tilde{h}_{l,o'q'}}(\tau) = \rho_{h_{l,oq}h_{l,o'q'}}(\tau)$ holds. The FCF of our simulation model can be computed in terms of $\tilde{\rho}_{\tilde{H}_{oq}\tilde{H}_{oq}}(\Delta f') = \left\langle \tilde{H}_{oq}(t, f')\, \tilde{H}^*_{oq}(t, f' - \Delta f') \right\rangle$, where $\tilde{H}_{oq}(t, f')$ is the Fourier transform of the complex impulse response $\tilde{h}_{oq}(t, \tau')$ with respect to the propagation delay τ'. Therefore, it is straightforward to obtain the FCF as $\tilde{\rho}_{\tilde{H}_{oq}\tilde{H}_{o'q'}}(\Delta f') = \sum_{l=1}^{L} c_l^2 e^{-j2\pi f'\tau_l'}$, which is equal to the FCF $\rho_{H_{oq}H_{o'q'}}(\Delta f')$ of our reference model in Eq. (9.103).

New Stochastic Simulation Model The deterministic model can be further modified to a stochastic simulation model by allowing both the phases and frequencies to be random variables. In contast to the deterministic model, the properties of the stochastic model vary for each simulation trial, but will converge to the desired ones when averaged over a sufficient number of simulation trials. To distinguish from our deterministic model, we define the complex impulse response of the proposed stochastic model as $\hat{h}_{oq}(t, \tau') = \sum_{l=1}^{L} c_l \hat{h}_{l,oq}(t)\, \delta(\tau' - \tau_l')$, where the complex time-variant tap coefficient $\hat{h}_{l,oq}(t)$ is a non-ergodic random process and defined as in Eqs (9.104) and (9.106). Unlike the deterministic simulation model, the AoD $\hat{\phi}_T^{(n_{1,1})}$ and AoAs $\hat{\phi}_R^{(n_{1,2})}$, $\hat{\phi}_R^{(n_{l,3})}$ of our stochastic model are random variables and thus vary for different simulation trials. Therefore, the fundamental issue for the design of the sets of AoD $\left\{\hat{\phi}_T^{(n_{1,1})}\right\}_{n=1}^{N_{1,1}}$ and AoAs $\left\{\hat{\phi}_R^{(n_{1,2})}\right\}_{n=1}^{N_{1,2}}$, $\left\{\hat{\phi}_R^{(n_{l,3})}\right\}_{n=1}^{N_{l,3}}$ is how to incorporate a random term into the AoA and AoD. To deal with this fundamental issue, we apply the method proposed by Zheng and Xiao [2002] for isotropic scattering F2M Rayleigh fading channels and design the AoD and AoAs as

$$\hat{\phi}_{T(R)}^{(n_{1,1(2)})} = F_{T(R)}^{-1}\left(\frac{n_{1,1(2)} - 1/2 + \theta_{T(R)}}{N_{1,1(2)}}\right), \quad \hat{\phi}_{T(R)}^{(n_{1,1(2)})} \in [-\pi, \pi) \tag{9.113a}$$

$$\hat{\phi}_R^{(n_{l,3})} = F_{R_l}^{-1}\left(\frac{n_{l,3} - 1/2 + \theta_{R_l}}{N_{l,3}}\right), \quad \hat{\phi}_R^{(n_{l,3})} \in [-\pi, \pi) \tag{9.113b}$$

where $\theta_{T(R)}$ and θ_{R_l} are random variables uniformly distributed in the interval $[-1/2, 1/2)$ and independent of each other. As mentioned in Chapter 4, the interval $[-1/2, 1/2)$ and the constant value $1/2$ are chosen to guarantee that the design of the AoA and AoDs is based on the desired range (here, $[-\pi, \pi)$). Due to the introduction of random variables $\theta_{T(R)}$ and θ_{R_l}, the sets of AoDs and AoAs vary for different simulations.

Unlike the deterministic simulation model, the ST CF of our stochastic model should be computed according to $\hat{\rho}_{\hat{h}_{l,oq}\hat{h}_{l,o'q'}}(\tau) = \mathbf{E}\left[\hat{h}_{l,oq}(t)\, \hat{h}^*_{l,o'q'}(t - \tau)\right]$. It can be shown that the ST CF $\hat{\rho}_{\hat{h}_{l,oq}\hat{h}_{l,o'q'}}(\tau)$ of our stochastic model matches the ST CF $\rho_{h_{l,oq}h_{l,o'q'}}(\tau)$ of the reference model irrespective of the values of $N_{1,i}$ and $N_{l',3}$; that is, for any $N_{1,i}$ and $N_{l',3}$. The derivation of $\hat{\rho}_{\hat{h}_{l,oq}\hat{h}_{l,o'q'}}(\tau)$ is complicated and lengthy; the details are omitted here for brevity. We invite interested readers to refer to Appendix A.5

The FCF of our simulation model can be computed in terms of $\hat{\rho}_{\hat{\mathrm{H}}_{oq}\hat{\mathrm{H}}_{oq}}(\Delta f') = \mathbf{E}\left[\hat{\mathrm{H}}_{oq}(t, f')\, \hat{\mathrm{H}}^*_{oq}(t, f' - \Delta f')\right]$, where $\hat{\mathrm{H}}_{oq}(t, f')$ is the Fourier transform of the complex impulse response $\hat{h}_{oq}(t, \tau')$ with respect to the propagation delay τ'. Therefore, the FCF is $\hat{\rho}_{\hat{\mathrm{H}}_{oq}\hat{\mathrm{H}}_{o'q'}}(\Delta f') = \sum_{l=1}^{L} c_l^2 e^{-j2\pi f'\tau_l'}$, which is equal to the FCF $\rho_{H_{oq}H_{o'q'}}(\Delta f')$ of our reference model in Eq. (9.103).

Narrowband Simulation Models

If we only consider the first tap (that is, we impose $L = 1$) on the wideband simulation models developed above, they reduce to narrowband MIMO V2V channel simulators. It follows that $\tilde{h}_{oq}(t, \tau') = \tilde{h}_{oq}(t)\,\delta(\tau')$ holds. Note that due to the similar relationship between the wideband reference model in Section 9.4.2 and the narrowband reference model in Section 9.4.1, the reduced narrowband MIMO V2V channel simulators are actually the corresponding simulation models for the proposed narrowband reference in Eqs (9.33) and (9.34). The complex fading envelope $\tilde{h}_{oq}(t)$ of the narrowband simulation model can be obtained by removing the subscript $(\cdot)_1$ from Eqs (9.104) and (9.105)

$$\tilde{h}_{oq}(t) = \tilde{h}_{oq}^{LoS}(t) + \sum_{i=1}^{I} \tilde{h}_{oq}^{SB_i}(t) + \tilde{h}_{oq}^{DB}(t) \tag{9.114}$$

where

$$\tilde{h}_{oq}^{LoS}(t) = \sqrt{\frac{K_{oq}}{K_{oq}+1}}\, e^{-j2\pi f_c \tau_{oq}} e^{j\left[2\pi f_{T_{max}} t \cos\left(\pi - \phi^{LoS} + \gamma_T\right) + 2\pi f_{R_{max}} t \cos\left(\phi^{LoS} - \gamma_R\right)\right]} \tag{9.115a}$$

$$\tilde{h}_{oq}^{SB_i}(t) = \sqrt{\frac{\eta_{SB_i}}{K_{oq}+1}} \sum_{n_i=1}^{N_i} \frac{1}{\sqrt{N_i}} e^{j\left(\tilde{\psi}_{n_i} - 2\pi f_c \tau_{oq,n_i}\right)} \times e^{j\left[2\pi f_{T_{max}} t \cos\left(\tilde{\phi}_T^{(n_i)} - \gamma_T\right) + 2\pi f_{R_{max}} t \cos\left(\tilde{\phi}_R^{(n_1,i)} - \gamma_R\right)\right]} \tag{9.115b}$$

$$\tilde{h}_{oq}^{DB}(t) = \sqrt{\frac{\eta_{DB_1}}{K_{oq}+1}} \sum_{n_1,n_2=1}^{N_1,N_2} \frac{1}{\sqrt{N_1 N_2}} e^{j\left(\tilde{\psi}_{n_1,n_2} - 2\pi f_c \tau_{oq,n_1,n_2}\right)} \times e^{j\left[2\pi f_{T_{max}} t \cos\left(\tilde{\phi}_T^{(n_1)} - \gamma_T\right) + 2\pi f_{R_{max}} t \cos\left(\tilde{\phi}_R^{(n_2)} - \gamma_R\right)\right]} \tag{9.115c}$$

In Eq. (9.105), the AoDs $\tilde{\phi}_T^{(n_i)}$ and AoAs $\tilde{\phi}_R^{(n_i)}$ are discrete realizations of the random variables $\phi_T^{SB_i}$ and $\phi_R^{SB_i}$ of the narrowband reference model in Section 9.4.1, respectively.

We first propose a deterministic simulation model. Analogous to the wideband deterministic simulation model proposed above, the key issue in the design of this narrowband deterministic simulation model is to properly design the AoD $\tilde{\phi}_T^{(n_1)}$ and the AoAs $\tilde{\phi}_R^{(n_2)}$ and $\tilde{\phi}_R^{(n_3)}$. Applying IMMEA, as shown in Eqs (9.111) and (9.112), we design the AoD and AoAs of our deterministic simulation model as

$$\tilde{\phi}_T^{(n_1)} = \begin{cases} \tilde{\phi}_T^{\prime(n_1)} + 2\pi, & \text{if} \quad \tilde{\phi}_T^{\prime(n_1)} < -\pi \\ \tilde{\phi}_T^{\prime(n_1)} - 2\pi, & \text{if} \quad \tilde{\phi}_T^{\prime(n_1)} \geq \pi \\ \tilde{\phi}_T^{\prime(n_1)}, & \text{else} \end{cases} \tag{9.116a}$$

$$\tilde{\phi}_R^{(n_{2(3)})} = \begin{cases} \tilde{\phi}_R^{\prime(n_{2(3)})} + 2\pi, & \text{if} \quad \tilde{\phi}_R^{\prime(n_{2(3)})} < -\pi \\ \tilde{\phi}_R^{\prime(n_{2(3)})} - 2\pi, & \text{if} \quad \tilde{\phi}_R^{\prime(n_{2(3)})} \geq \pi \\ \tilde{\phi}_R^{\prime(n_{2(3)})}, & \text{else} \end{cases} \tag{9.116b}$$

where $\tilde{\phi}_T^{(n_1)} \in [-\pi, \pi)$, $\tilde{\phi}_R^{(n_{2(3)})} \in [-\pi, \pi)$, and

$$\tilde{\phi}_T^{\prime(n_1)} = F_T^{\prime -1}\left(\frac{n_1 - 1/4}{N_1}\right), \quad \tilde{\phi}_T^{\prime(n_1)} \in \left[\mu_T^{TR} - \pi, \mu_T^{TR} + \pi\right) \tag{9.117a}$$

$$\tilde{\phi}_R^{\prime(n_{2(3)})} = F_{R-2(3)}^{\prime -1}\left(\frac{n_{2(3)} - 1/4}{N_{2(3)}}\right), \quad \tilde{\phi}_R^{\prime(n_{2(3)})} \in \left[\mu_R^{TR(EL)} - \pi, \mu_R^{TR(EL)} + \pi\right) \tag{9.117b}$$

where $F_T^{\prime -1}(\cdot)$ and $F_{R-2(3)}^{\prime -1}(\cdot)$ denote the inverse functions of the von Mises CDF for $\phi_T^{\prime(SB_1)}$ and $\phi_R^{\prime(SB_{2(3)})}$, respectively. The parameters $\phi_T^{\prime(SB_1)}$ and $\phi_R^{\prime(SB_{2(3)})}$ fulfill the same von Mises distributions as $\phi_T^{(SB_1)}$ and $\phi_R^{(SB_{2(3)})}$, respectively, but with different ranges: $\phi_T^{\prime(SB_1)} \in \left[\mu_T^{TR} - \pi, \mu_T^{TR} + \pi\right)$ and $\phi_R^{\prime(SB_{2(3)})} \in \left[\mu_R^{TR(EL)} - \pi, \mu_R^{TR(EL)} + \pi\right)$, while $\phi_T^{(SB_1)} \in [-\pi, \pi)$ and $\phi_R^{(SB_{2(3)})} \in [-\pi, \pi)$.

By allowing both the phases and frequencies to be random variables, Our deterministic model $\tilde{h}_{oq}(t)$ can be further modified to become a stochastic simulation model $\hat{h}_{oq}(t)$. The AoD $\hat{\phi}_T^{(n_1)}$, and AoAs $\hat{\phi}_R^{(n_2)}$ and $\hat{\phi}_R^{(n_3)}$ can be designed using the method described in Eq. (9.113)

$$\hat{\phi}_T^{(n_1)} = F_T^{-1}\left(\frac{n_1 - 1/2 + \theta_T}{N_1}\right), \quad \hat{\phi}_T^{(n_1)} \in [-\pi, \pi) \tag{9.118a}$$

$$\hat{\phi}_R^{(n_{2(3)})} = F_{R-2(3)}^{-1}\left(\frac{n_{2(3)} - 1/2 + \theta_{R-2(3)}}{N_{2(3)}}\right), \quad \hat{\phi}_R^{(n_{2(3)})} \in [-\pi, \pi) \tag{9.118b}$$

where $F_T^{-1}(\cdot)$ and $F_{R-2(3)}^{-1}(\cdot)$ denote the inverse function of the von Mises CDF for $\phi_T^{SB_1}$ and $\phi_R^{SB_{2(3)}}$, respectively, and θ_T and $\theta_{R-2(3)}$ are random variables uniformly distributed in the interval $[-1/2, 1/2)$ and independent to each other.

9.4.2.3 Numerical Results and Analysis

Unless otherwise specified, all the results presented in this section are obtained using $f_c = 5.9\,\text{GHz}$, $f_{T_{max}} = f_{R_{max}} = 570\,\text{Hz}$, $D = 300\,\text{m}$, $a_1 = 160\,\text{m}$, $a_2 = 180\,\text{m}$, $R_T = R_R = 10\,\text{m}$, $\beta_T = 60°$, $\beta_R = 45°$ $N_{1,1} = N_{1,2} = N_{1,3} = N_{2,3} = 30$ for the deterministic model, $N_{1,1} = N_{1,2} = N_{1,3} = N_{2,3} = 20$ for the stochastic model, and normalized sampling period $f_{T_{max}} T_s = 0.005$, where T_s is the sampling period.

Figures 9.20–9.22 compare the difference in the ST CF for the deterministic model obtained using different parameter computation methods: MMEA, MMEA2, and IMMEA. Without any loss of generality, here we choose the double-bounce two-ring model as the reference model for further investigation. Therefore, in Figures 9.20–9.22 we compare the difference in the ST CF $\tilde{\rho}_{h_{1,oq}^{DB} h_{1,o'q'}^{DB}}(\tau)$ from the desired $\rho_{h_{1,oq}^{DB} h_{1,o'q'}^{DB}}(\tau)$ by using the squared error $\left|\tilde{\rho}_{h_{1,oq}^{DB} h_{1,o'q'}^{DB}}(\tau) - \rho_{h_{1,oq}^{DB} h_{1,o'q'}^{DB}}(\tau)\right|^2$ for different non-isotropic scattering scenarios. From Figures 9.20–9.22, it is clear that IMMEA outperforms MMEA and MMEA2 for the different non-isotropic scattering scenarios. Comparing MMEA and MMEA2, we can conclude that neither method consistently outperforms the other for non-isotropic scattering scenarios. Figure 9.22 also shows that the MMEA results in a relatively large difference, even at shorter

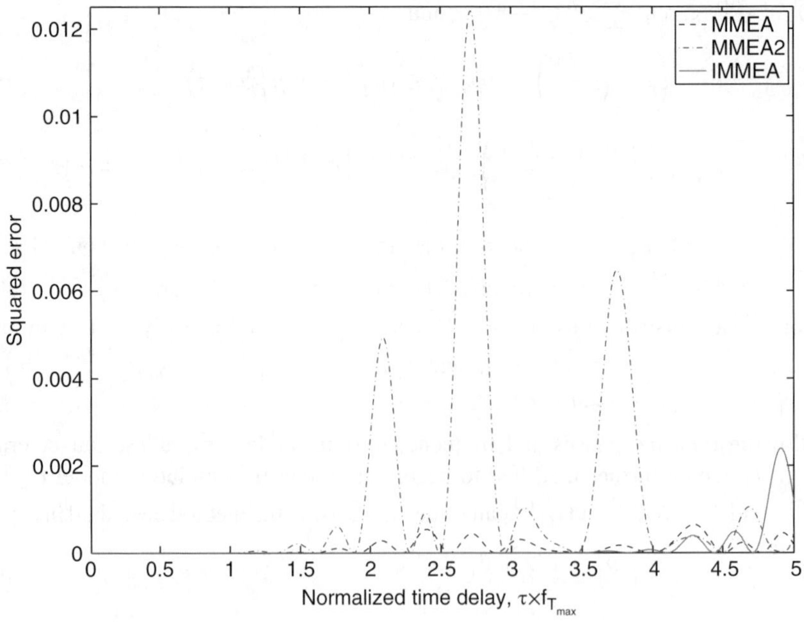

Figure 9.20 Squared error in the ST CF of the deterministic simulation model obtained using MMEA, MMEA2, and IMMEA for a non-isotropic scattering MIMO V2V channel: $\beta_T = \beta_R = \gamma_T = \gamma_R = \mu_T^{(1,1)} = \mu_R^{(1,2)} = -40°$, $k_T^{(1,1)} = k_R^{(1,2)} = 1$, and $\delta_T = \delta_R = 0.5\lambda$

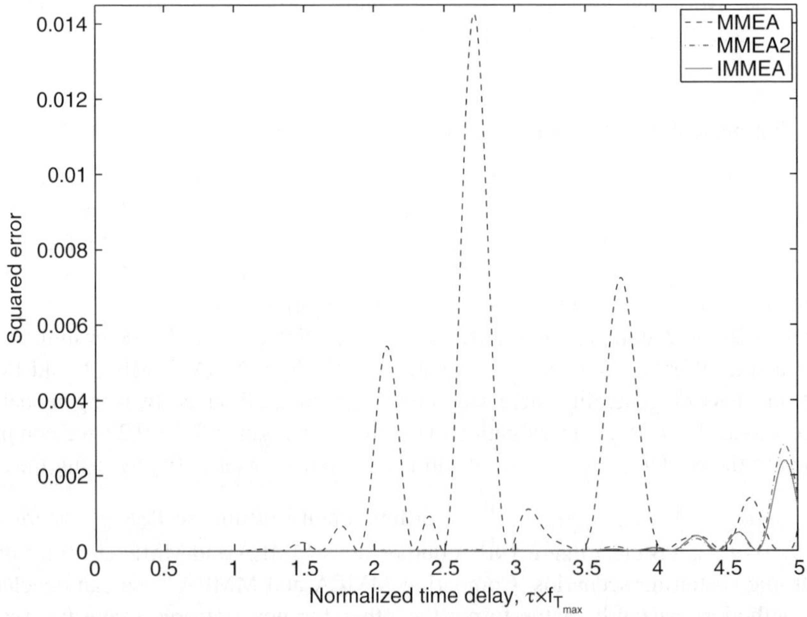

Figure 9.21 Squared error in the ST CF of the deterministic simulation model obtained using MMEA, MMEA2, and IMMEA for a non-isotropic scattering MIMO V2V channel: $\beta_T = \beta_R = \gamma_T = \gamma_R = \mu_T^{(1,1)} = \mu_R^{(1,2)} = -30°$, $k_T^{(1,1)} = k_R^{(1,2)} = 1$, and $\delta_T = \delta_R = 0.5\lambda$

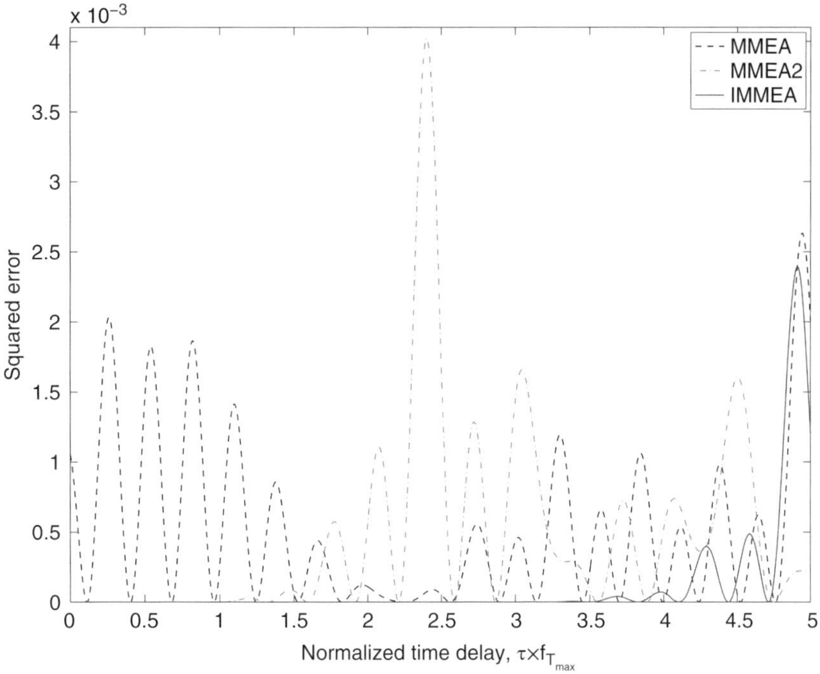

Figure 9.22 Squared error in the ST CF of the deterministic simulation model obtained using MMEA, MMEA2, and IMMEA for a non-isotropic scattering MIMO V2V channel: $\beta_T = \beta_R = \gamma_T = \gamma_R = \mu_T^{(1,1)} = \mu_R^{(1,2)} = 50°$, $k_T^{(1,1)} = k_R^{(1,2)} = 1$, and $\delta_T = \delta_R = 0.5\lambda$

time delays, which are of most interest for most communication systems [Patel et al. 2005]. This is because MMEA designs the AoDs/AoAs in a non-sufficient range when the mean AoDs/AoAs are larger than zero.

Figure 9.23 shows the ST CF of the reference model, the proposed IMMEA deterministic simulation model, and the proposed stochastic model. For simplicity, we only consider the double-bounced rays of the first tap in our reference model. In such a case, our model reduces to a double-bounce two-ring reference model. The results obtained for the stochastic model are averaged over $N_{sto} = 50$ trials. From Figure 9.23, it is obvious that the IMMEA model provides a good approximation to the ST CF of the reference model over a short normalized time delay of $0 \leq \tau f_{T_{max}} \leq 3$, which is the range typically of interest for communication systems [Patel et al. 2005]. The stochastic model presents a good approximation to the desired ST CF over a wider range of normalized time delays with an even smaller number of complex harmonic functions $N_{1,1}$ and $N_{1,2}$.

For further comparison, the ST CF of the MMEA and MMEA2 deterministic models are also plotted. Again, it is clear that the IMMEA model outperforms the MMEA and MMEA2 models. In addition, comparing Figures 9.23(a) and (b), we find that an increase in the antenna element spacing (that is, the value of $\delta_T$and δ_R) increases the difficulty in approximating the desired ST CF of the reference model.

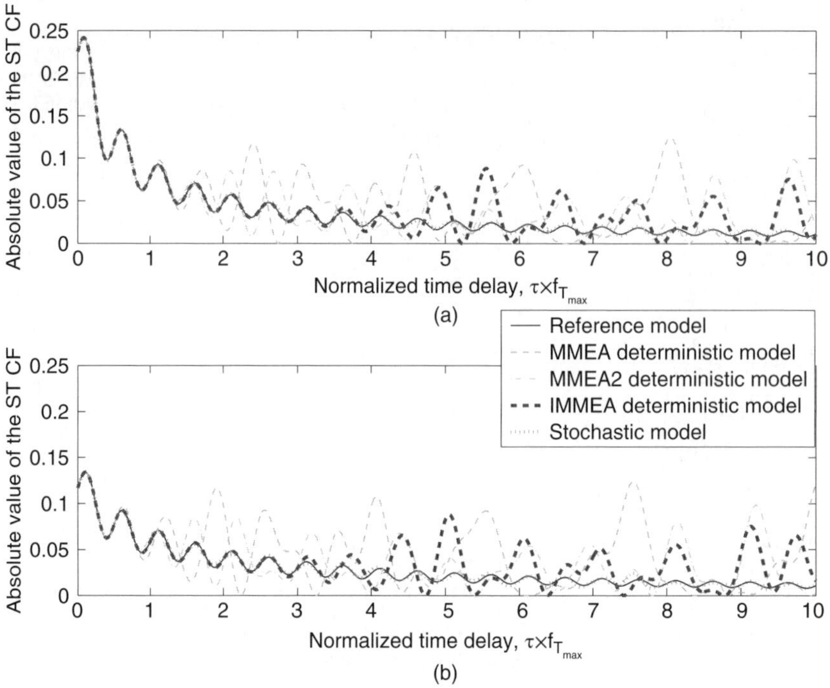

Figure 9.23 Absolute value of the ST CF of the reference model, deterministic simulation models (MMEA model, MMMEA2 model, and IMMEA model), and stochastic simulation model for a non-isotropic scattering MIMO V2V channel; ($\beta_T = \beta_R = \gamma_T = \gamma_R = \mu_T^{(1,1)} = \mu_R^{(1,2)} = -70°$ and $k_T^{(1,1)} = k_R^{(1,2)} = 1$), with antenna spacings (a) $\delta_T = \delta_R = 0.5\lambda$ and (b) $\delta_T = \delta_R = 1\lambda$

Figures 9.24 and 9.25 show the theoretical SD PSDs of the proposed V2V model for the first and second tap with different VTDs (low and high) and different antenna separations ($\delta_T = \delta_R = 0$ or 3λ) when the Tx and Rx are moving in opposite directions and the same direction, respectively. Note that when $\delta_T = \delta_R = 0$, the SD PSDs actually reduce to Doppler PSDs. For comparison purposes, the measured Doppler PSDs taken from the paper by Acosta-Marum and Brunelli [2007, Figs 4(a)–(d)] are also plotted in Figures 9.24(a) and (b) and Figures 9.25(a) and (b), respectively. Acosta-Marum and Brunelli [2007] peformed their measurement campaigns at a carrier frequency of 5.9 GHz on an expressway with low VTD. The distance between the Tx and Rx was approximately $D = 300$ m and the directions of movement were $\gamma_T = 0$, $\gamma_R = \pi$ (opposite direction) and $\gamma_T = \gamma_R = 0$ (same direction). Both the Tx and Rx were equipped with one omnidirectional antenna: the SISO case. Based on the scenarios measured by Acosta-Marum and Brunelli, we chose the following environment-related parameters:

- $k_T^{(1,1)} = 6.6$, $k_R^{(1,2)} = 8.3$, $k_R^{(1,3)} = 5.5$, $k_R^{(2,3)} = 7.7$, $\mu_T^{(1,1)} = 12.8°$, $\mu_R^{(1,2)} = 178.7°$, $\mu_R^{(1,3)} = 131.6°$, and $\mu_R^{(2,3)} = 31.3°$ for Figure 9.24
- $k_T^{(1,1)} = 9.6$, $k_R^{(1,2)} = 3.6$, $k_R^{(1,3)} = 11.5$, $k_R^{(2,3)} = 11.7$, $\mu_T^{(1,1)} = 21.7°$, $\mu_R^{(1,2)} = 147.8°$, $\mu_R^{(1,3)} = 171.6°$, and $\mu_R^{(2,3)} = 177.6°$ for Figure 9.25.

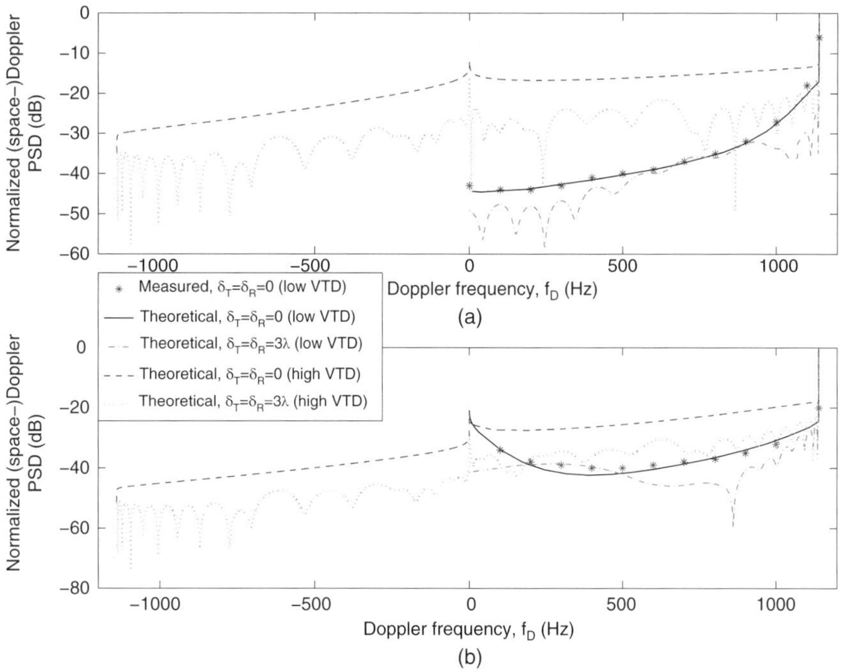

Figure 9.24 Normalized (space-)Doppler PSDs of (a) first tap and (b) second tap of the proposed wideband MIMO V2V channel model with low and high VTDs when the Tx and Rx are moving in opposite directions on an expressway

Considering the constraints of the Ricean factor and energy-related parameters for different taps, we choose the following parameters in order to fit the measured Doppler PSDs reported by Acosta-Marum and Brunelli under the condition of low VTD:

- $K_{oq} = 2.186$, $\eta_{DB_1} = 0.005$, $\eta_{SB_{1,1}} = 0.252$, $\eta_{SB_{1,2}} = 0.262$, and $\eta_{SB_{1,3}} = 0.481$ for Figure 9.24(a)
- $\eta_{DB_{2,1}} = \eta_{DB_{2,2}} = 0.119$ and $\eta_{SB_{2,3}} = 0.762$ for Figure 9.24(b)
- $K_{oq} = 3.786$, $\eta_{DB_1} = 0.051$, $\eta_{SB_{1,1}} = 0.335$, $\eta_{SB_{1,2}} = 0.203$, and $\eta_{SB_{1,3}} = 0.411$ for Figure 9.25(a)
- $\eta_{DB_{2,1}} = \eta_{DB_{2,2}} = 0.121$ and $\eta_{SB_{2,3}} = 0.758$ for Figure 9.25(b).

The excellent agreement between the theoretical and measured Doppler PSDs confirms the utility of the proposed wideband model. The environment-related parameters for high VTD in Figures 9.24 and 9.25 are the same as those for low VTD except $k_T^{(1,1)} = 0.6$ and $k_R^{(1,2)} = 1.3$, which are related to the distribution of moving scatterers. The SD PSDs for high VTD shown in Figures 9.24 and 9.25 were obtained with the following parameters

- $K_{oq} = 0.156$, $\eta_{DB_1} = 0.685$, $\eta_{SB_{1,1}} = \eta_{SB_{1,2}} = 0.126$, $\eta_{SB_{1,3}} = 0.063$ for the first tap
- $\eta_{DB_{2,1}} = \eta_{DB_{2,2}} = 0.456$, $\eta_{SB_{2,3}} = 0.088$ for the second tap.

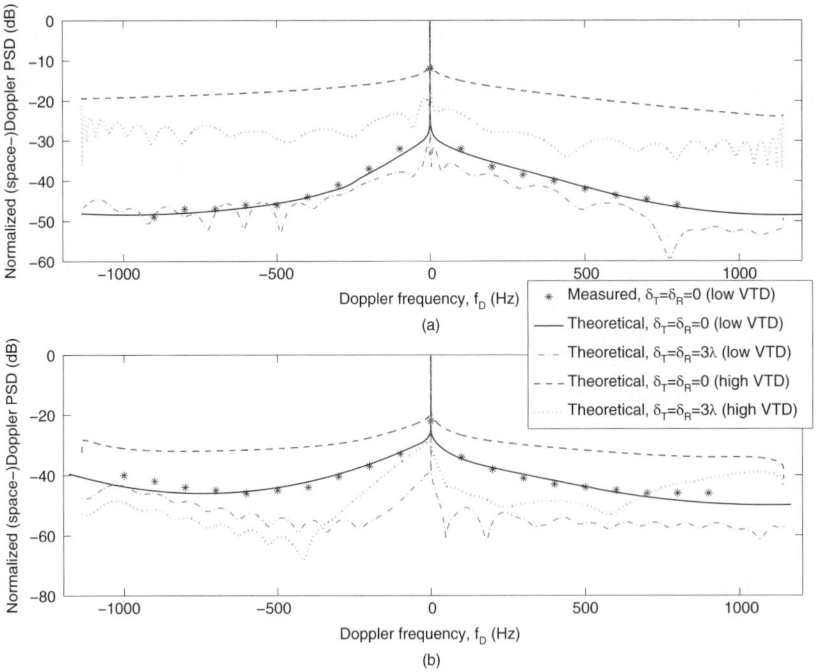

Figure 9.25 Normalized (space-)Doppler PSDs of (a) first tap and (b) second tap of the proposed wideband MIMO V2V channel model with low and high VTDs when the Tx and Rx are moving in the same direction on an expressway

Unfortunately, to the best of the authors' knowledge, no measurement results[5] are available regarding the impact of the high VTD (for example, a traffic jam) on the SD PSD. Comparing the theoretical SD PSDs at different VTDs in Figures 9.24 and 9.25, we see that VTD significantly affects the SD PSDs at different taps in V2V channels. Moreover, the space separation results in fluctuations in the SD PSDs.

Using the same parameters as Figures 9.24(a) and (b), Figures 9.26(a) and (b) depict the corresponding ST CFs for the first and second tap, respectively. Again, the VTD greatly affects the ST correlation properties at different taps. Higher VTDs lead to lower correlation properties. The ST CFs of the proposed IMMEA deterministic and stochastic simulation models are also plotted. The results obtained for the stochastic model are averaged over $N_{sto} = 30$ trials. It is obvious that the deterministic model provides a fairly good approximation to the ST CFs of the reference model, while the stochastic model gives a much better approximation with an even smaller number of complex harmonic functions.

Figure 9.27 shows the FCF of our reference model, and the IMMEA deterministic and stochastic simulation models. The results are obtained by using the parameters reported by Acosta-Marum and Brunelli [2007]:

[5] See, for example Acosta et al. [2004], Zajic et al. [2009], Sen and Matolak [2008], Paier et al. [2009], Cheng et al. [2007], Acosta-Marum and Brunelli [2007], Maurer et al. [2002].

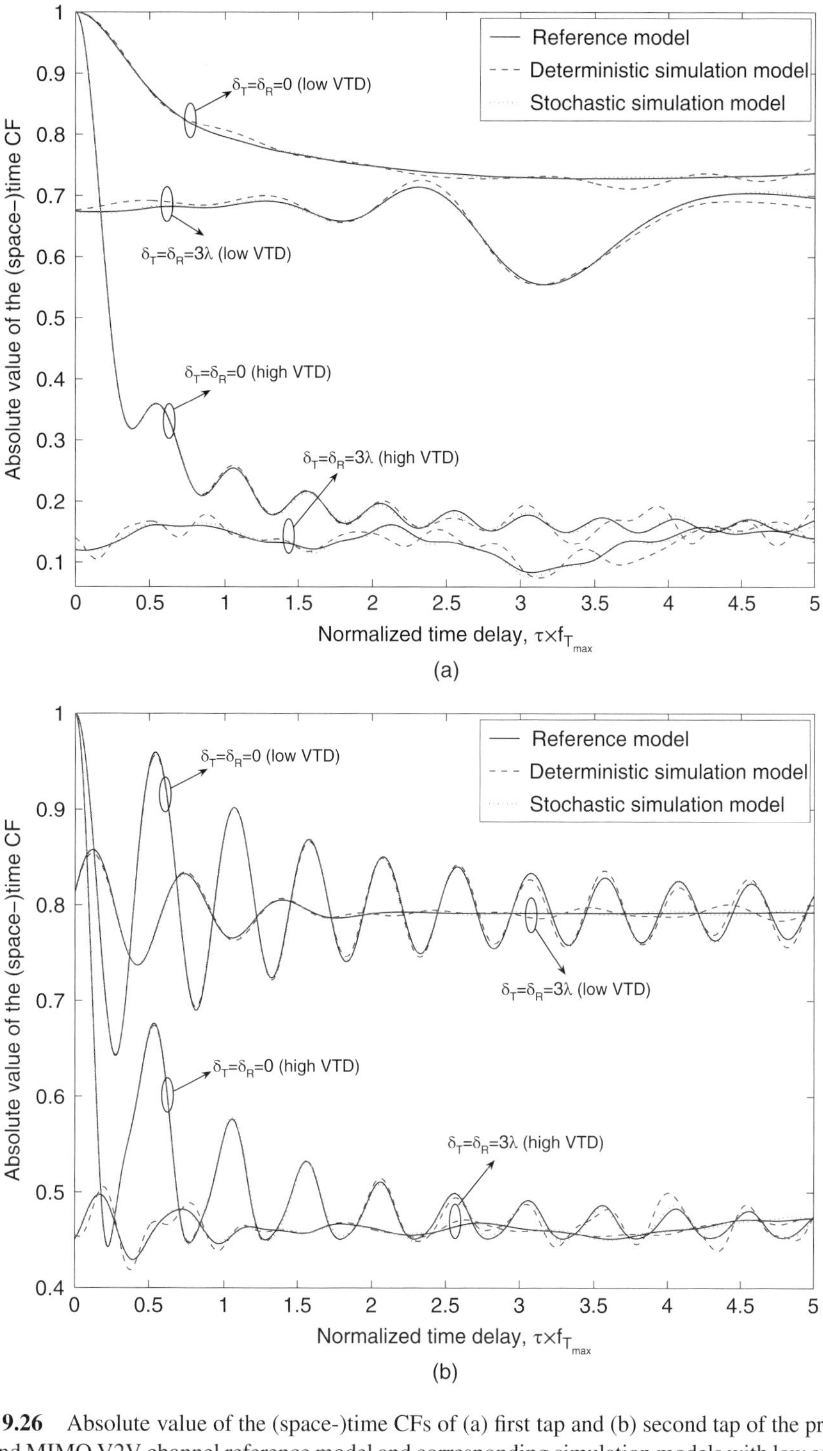

Figure 9.26 Absolute value of the (space-)time CFs of (a) first tap and (b) second tap of the proposed wideband MIMO V2V channel reference model and corresponding simulation models with low and high VTDs when the Tx and Rx are moving in opposite directions on an expressway

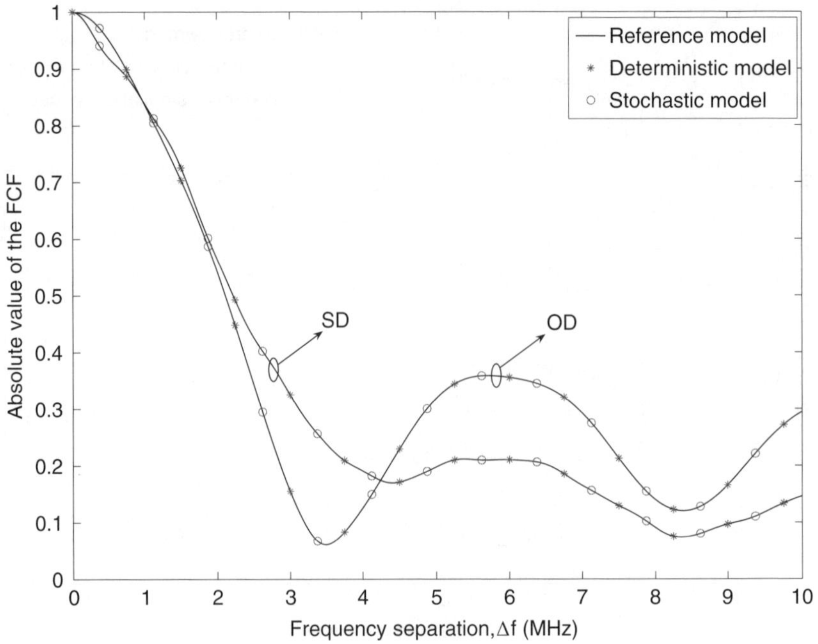

Figure 9.27 Absolute value of the FCF for the reference model and the deterministic and stochastic simulation models. SD: same direction; OD: opposite directions

- the propagation delays $\{\tau'\}_{l=1}^{4} = \{0, 0.1, 0.2, 0.3\}\,\mu s$ and the tap powers $\{c_l^2\}_{l=1}^{4} = \{0, -6.3, -25.1, -22.7\}$ dB when the Tx and Rx are moving in opposite directions
- the propagation delays $\{\tau'\}_{l=1}^{8} = \{0, 0.1, 0.2, 0.3, 0.4, 0.5, 0.6, 0.7\}\,\mu s$ and the tap powers $\{c_l^2\}_{l=1}^{8} = \{0, -11.2, -19, -21.9, -25.3, -24.4, -28.0, -26.1\}$ dB when the Tx and Rx are moving in the same direction.

It is clear that the FCFs of the simulation models are identical to those of the reference model.

Summary

Section 9.4.2 has described a wideband MIMO V2V RS-GBSM for the study of channel statistics at different time delays, taking into account the impact of VTD on channel statistics. From the proposed reference model, the ST CF, SD PSD, FCF, and PDP have been derived. Numerical results have demonstrated that VTD has a great impact on the resulting CFs and PSDs. It has been shown that the theoretical per-tap Doppler PSDs closely match the measured data, which validates the utility of the proposed reference model. Based on our reference model, we have proposed new deterministic and stochastic SoS simulation models for wideband MIMO V2V Ricean fading channels. We have demonstrated that by removing the frequency-selectivity, the wideband simulation models can be reduced to the narrowband ones, which are actually the corresponding simulation models for the proposed narrowband MIMO V2V reference model in Section 9.4.1. Furthermore, a new parameter computation method for a deterministic simulation model, named IMMEA, has been derived for non-isotropic scattering MIMO V2V channels. It has been demonstrated that IMMEA outperforms other existing methods, such as MMEA and MMEA2. Numerical results have shown that the proposed deterministic model with IMMEA gives a fairly good approximation to the desired properties of our reference model, while the proposed stochastic model provides better approximations over a wider range of normalized time delays with even small numbers of complex harmonic functions.

9.5 Scattering Theoretical Channel Models for Cooperative MIMO Systems

As described in Section 9.2.2, although the correlations between different links are important for the investigation of cooperative MIMO channels, the investigation of multi-link correlation is rare in the literature. Section 9.5 will present a detailed study of multi-link spatial correlations based on the general cooperative MIMO scattering theoretical channel models developed.

The multi-link correlation consists of large-scale fading correlation and small-scale fading correlation. Only a few papers have analyzed and modeled large-scale fading correlations, such as shadow-fading correlation, delay-spread correlation, and azimuth correlation. The 3rd Generation Partnership Project (3GPP) Spatial Channel Model (SCM) [3GPP 2003], the Wireless World Initiative New Radio Phase II (WINNER II) channel model [Kyösti et al. 2007], and the IEEE 802.16j channel model [IEEE 2007a] all investigated and modeled large-scale fading correlations of different links for multiple scenarios. However, as mentioned by Wang et al. [2010], these correlation models are not consistent and a unified correlation model for large-scale fading is necessary. Recently, Oestges et al. [2010] proposed a unified framework for investigation of both static and dynamic shadow-fading correlations for indoor and outdoor-to-indoor scenarios.

There are even fewer papers on investigation of small-scale fading correlations. In Xu et al. [2009], the authors proposed a multiuser MIMO channel model focusing on the investigation of the impact of surface roughness on spatial correlations. Wang et al. [2010] reported a preliminary investigation of spatial correlations for coordinated multi-point (CoMP) transmissions. The investigation of spatial correlations of multi-link propagation channels in amplify-and-forward (AF) relay systems was reported by Yin [2010]. However, all of these investigations of multi-link spatial correlations are scenario-specific. For example, Xu et al. [2009] only modeled the scenario where scatterers are located in streets, Wang et al. [2010] only focused on the CoMP scenario, and Yin [2010] only investigated the AF relay scenario. A unified channel model framework to investigate multi-link small-scale fading correlations for different scenarios is therefore highly desirable.

To fill this gap, Section 9.5 proposes a unified channel model framework for cooperative MIMO systems and investigates spatial correlations of different links in multiple scenarios. The main content of this section are listed as follows.

1. We will propose a wideband unified channel model framework that is suitable to mimic different links in cooperative MIMO systems, such as the BS–BS/RS/MS link, RS–RS/MS link, and MS–MS link. Due to different local scattering environments around BSs, RSs, and MSs, a high degree of link heterogeneity or variation is expected in cooperative

MIMO systems. Here, we are interested in 12 various cooperative MIMO environments as introduced in Section 9.1.2. Therefore, the proposed framework can be adapted to the 12 scenarios by simply adjusting key model parameters.
2. Taking a cooperative relay system, which includes three links (BS–RS, RS–MS, and BS–MS), as an example, we show how to apply the proposed channel model framework and derive a novel geometry-based stochastic model (GBSM) for multiple physical scenarios. The proposed GBSM is the first cooperative MIMO channel model that has the ability to mimic the impact of local scattering density (LSD) on channel characteristics.
3. From the proposed GBSMs, we further derive the multi-link spatial correlation functions that can significantly affect the performance of cooperative MIMO communication systems.
4. The impact of some important parameters, such as antenna element spacings and the LSDs, on multi-link spatial correlations in different scenarios is then investigated. Some interesting observations and conclusions are obtained, which can help give a better understanding of cooperative MIMO channels and thus better designs for cooperative MIMO systems.

9.5.1 A Unified Cooperative MIMO Channel-model Framework

Cooperative MIMO channel measurements [Czink et al. 2008, Oestges et al. 2010, Zhang et al. 2009] have clearly demonstrated that the degree of link heterogeneity in cooperative MIMO systems is highly related to local scattering environments around the devices. Therefore, the cooperative MIMO model framework needs to reflect the influence of different local scattering environments on link heterogeneity for different scenarios, while retaining an acceptable level of model complexity.

i	LoS component	$A_p \to B_q : h_{pq}^{LOS}(t,\tau)$
1	Single-bounced	$A_p \to S_B \to B_q : h_{pq}^{11}(t,\tau)$ $A_p \to S_c \to B_q : h_{pq}^{12}(t,\tau)$ $A_p \to S_D \to B_q : h_{pq}^{13}(t,\tau)$ $A_p \to S_A \to B_q : h_{pq}^{14}(t,\tau)$
2	Double-bounced	$A_p \to S_A \to S_B \to B_q : h_{pq}^{21}(t,\tau)$ $A_p \to S_A \to S_C \to B_q : h_{pq}^{22}(t,\tau)$ $A_p \to S_A \to S_D \to B_q : h_{pq}^{23}(t,\tau)$ $A_p \to S_D \to S_B \to B_q : h_{pq}^{24}(t,\tau)$ $A_p \to S_D \to S_C \to B_q : h_{pq}^{25}(t,\tau)$ $A_p \to S_C \to S_B \to B_q : h_{pq}^{26}(t,\tau)$
3	Triple-bounced	$A_p \to S_A \to S_D \to S_B \to B_q : h_{pq}^{31}(t,\tau)$ $A_p \to S_A \to S_D \to S_C \to$ $B_q : h_{pq}^{32}(t,\tau)$ $A_p \to S_A \to S_C \to S_B \to B_q : h_{pq}^{33}(t,\tau)$ $A_p \to S_D \to S_C \to$ $B_q : h_{pq}^{34}(t,\tau)$
4	Quadruple-bounced	$A_p \to S_A \to S_D \to S_C \to S_B \to B_q : h_{pq}^{41}(t,\tau)$

Let us now consider a general wideband cooperative MIMO system in which all nodes are surrounded by local scatterers and a link between Node A and Node B is presented as shown in Figure 9.28. It is assumed that each node can be in motion and is equipped with

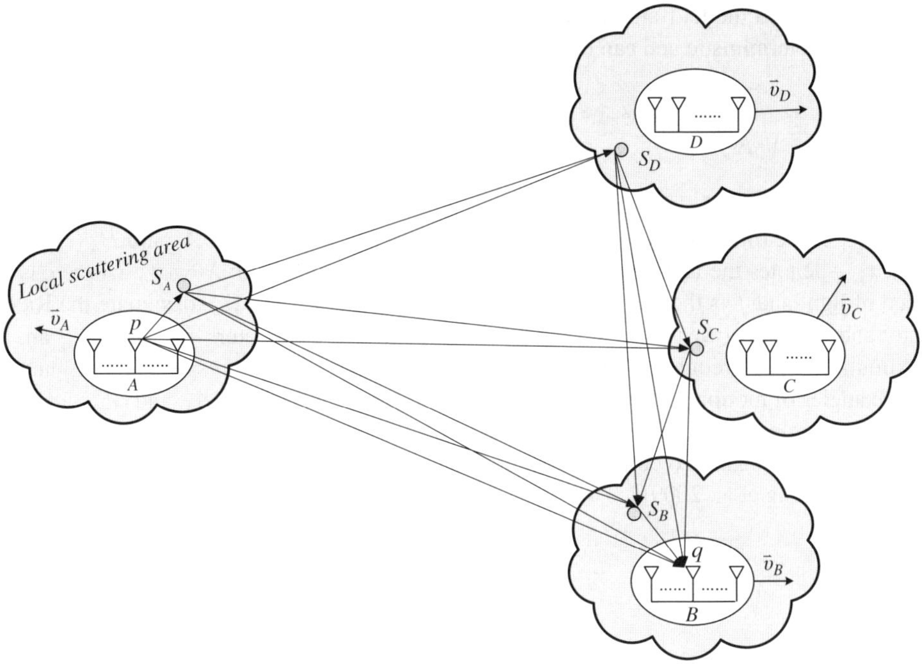

Figure 9.28 Geometry of a unified cooperative MIMO channel model framework

L antenna elements. The proposed unified channel model framework expresses the channel impulse response between the pth antenna in Node A and the qth antenna in Node B as the superposition of line-of-sight (LoS) and scattered rays

$$h_{pq}(t,\tau) = h_{pq}^{LoS}(t,\tau) + \sum_{i=1}^{I} \sum_{g=1}^{f_I(i)} h_{pq}^{ig}(t,\tau) \tag{9.119}$$

where $I \geq 1$ is the number of related local scattering areas, $f_I(i) = \frac{I!}{(I-i)! \cdot i!}$ denotes the total number of i-bounced components, and $h_{pq}^{ig}(t,\tau)$ represents the gth scattered component consisting of i-bounced rays. For example, $h_{pq}^{21}(t,\tau)$ denotes the first double-bounced component. It is worth noting that the parameter $f_I(i)$ is obtained not purely from the number of related local scattering areas, but also according to the following practical criterion: the i-bounced waves are always bounced by i scatterers located in different local scattering areas, from far to near relative to the receiver. Based on this practical criterion, some i-bounced components need not be considered, which makes the proposed model more practical. For the cooperative communication environment shown in Figure 9.28 with $I = 4$, the proposed model framework consists of the LoS component, $f_4(1) = 4$ single-bounced components, $f_4(2) = 6$ double-bounced components, $f_4(3) = 4$ triple-bounced components, and $f_4(4) = 1$ quadruple-bounced component. Multi-bounced components that violate the practical criterion, for example the $A_p - S_B - S_C - B_q$ double-bounced component, are not included.

In the proposed model framework Eq. (9.119), the LoS component of the channel impulse response is deterministic and can be expressed as [Cheng et al. 2009]:

$$h_{pq}^{LoS}(t,\tau) = \sqrt{\frac{K_{pq}\Omega_{pq}}{K_{pq}+1}} e^{-j2\pi\lambda^{-1}\chi_{pq}} e^{j\left[2\pi f_{\max}^{A} t\cos\left(\alpha_{pq}^{LoS}-\gamma_A\right)+2\pi f_{\max}^{B} t\cos\left(\phi_{pq}^{LoS}-\gamma_B\right)\right]} \delta(\tau-\tau_{LoS}) \tag{9.120}$$

where χ_{pq} is the travel path of the LoS waves through the link between A_p and B_q ($A_p - B_q$ link), τ_{LoS} denotes the LoS time delay, and λ is the wavelength with $\lambda = c/f$ where c is the speed of light and f is the carrier frequency. The symbols K_{pq} and Ω_{pq} designate the Ricean factor and the total power of the $A_p - B_q$ link, respectively. Parameters $f_{\max}^{A}$ and $f_{\max}^{B}$ are the maximum Doppler frequencies with respect to Node A and Node B, respectively, γ_A and γ_B are the angles of motion with respect to Node A and Node B, respectively, and α_{pq}^{LoS} and ϕ_{pq}^{LoS} denote the angles of arrival/departure of the LoS path with respect to Node A and Node B, respectively. The scattered component of the channel impulse response in Eq. (9.119) can be shown as [Cheng et al. 2009]:

$$h_{pq}^{ig}(t,\tau) = \sqrt{\frac{\eta_{pq}^{ig}\Omega_{pq}}{K_{pq}+1}} \lim_{\{N_k^g\}_{k=1}^{i}\to\infty} \sum_{\{n_k^g\}_{k=1}^{i}=1}^{\{N_k^g\}_{k=1}^{i}} \frac{1}{\sqrt{\prod_{k=1}^{i} N_k^g}} e^{j\left(\psi_{\{n_k^g\}_{k=1}^{i}} - 2\pi\lambda^{-1}\chi_{pq,\{n_k^g\}_{k=1}^{i}}\right)}$$
$$\times e^{j\left[2\pi f_{\max}^{A} t\cos\left(\alpha_{pq,\{n_k^g\}_{k=1}^{i}}-\gamma_A\right)+2\pi f_{\max}^{B} t\cos\left(\phi_{pq,\{n_k^g\}_{k=1}^{i}}-\gamma_B\right)\right]}$$
$$\delta(\tau-\tau_{\{n_k^g\}_{k=1}^{i}}) \tag{9.121}$$

where N_k^g is the number of effective scatterers in the kth local scattering area with respect to the gth i-bounced component, $\{\chi_{pq,n_k^g}\}_{k=1}^{i}$ is the travel path of the gth i-bounced waves through the $A_p - B_q$ link, $\{\tau_{n_k^g}\}_{k=1}^{i}$ denotes the time delay of the multipath components. The phases $\{\psi_{n_k}\}_{k=1}^{i}$ are independent and identically distributed (i.i.d.) random variables with uniform distributions over $[-\pi,\pi)$ and determined by scatterers $\{S_{n_K}\}_{k=1}^{i}$, $\{X_{n_k}\}_{k=1}^{i}$ represents $X_{n_1}, X_{n_2}, X_{n_3}, \ldots, X_{n_i}$, and $\{\alpha_{pq,n_k^g}\}_{k=1}^{i}$ and $\{\phi_{pq,n_k^g}\}_{k=1}^{i}$ denote angles of arrival/departure of a i-bounced path with respect to Node A and Node B, respectively. Here, η_{pq}^{ig} is a energy-related parameter specifying how much the gth i-bounced rays contribute to the total scattered power $\Omega_{pq}/(K_{pq}+1)$. Note that energy-related parameters satisfy $\sum_{i=1}^{I}\sum_{g=1}^{f_I(i)} \eta_{pq}^{ig} = 1$.

It is clear that the proposed unified channel model framework in Eq. (9.119) can naturally include the impact of the local scattering area on channel characteristics through choosing suitable $f_I(i)$ i-bounced components. Note that parameter $f_I(i)$ is related to the number of related local scattering areas I, which is determined by the physical environment: outdoor macrocell, microcell, picocell, and indoor. This means by properly adjusting the parameter I, the proposed model framework is suitable for different basic cooperative environments. Furthermore, the proposed model framework can model multiple links with different degrees of link heterogeneity due to different application scenes, such as BS cooperation, MS cooperation, and relay cooperation, for a typical cooperative environment simply by adjusting the Ricean factor K_{pq} and energy-related parameters η_{pq}^{ig}. How to properly set the key model parameters, I, K_{pq}, η_{pq}^{ig}, will be explained in the next subsection where the proposed unified cooperative channel model framework is implemented for a typical cooperative MIMO application scenario.

9.5.2 A New MIMO GBSM for Cooperative Relay Systems

Without loss of generality, Section 9.5.2 considers a wideband cooperative relay communication environment that includes three different links: BS–RS, RS–MS, and BS–MS, to implement the proposed cooperative MIMO channel model framework. Note that this cooperative MIMO GBSM can easily be adapted to other cooperative MIMO scenarios with multiple relays: the RS can be another BS for BS cooperation or another MS for MS cooperation.

In order to propose a generic cooperative MIMO GBSM that is suitable for the aforementioned 12 cooperative scenarios, we assume that the BS, RS, and MS are all surrounded by local scatterers. Figure 9.29 shows the geometry of the proposed cooperative MIMO GBSM, combining the LoS components and scattered components. To keep the readability of Figure 9.29, the LoS components are not shown. It is assumed that the BS, RS, and MS are all equipped with $A_B = A_R = A_M = 2$ uniform linear antenna arrays. The local scattering environment is characterized by the effective scatterers located on circular rings. Suppose there are N_1 effective scatterers around the MS and that these lie on a circular ring of radius $R_{1n_1} \leq \xi_{n_1}^M \leq R_{1n_2}$; the n_1th ($n_1 = 1, \ldots, N_1$) effective scatterer is denoted by S_{n_1}. Similarly, assume there are N_2 effective scatterers around the RS, and that these lie on a circular ring of radius $R_{2n_1} \leq \xi_{n_2}^R \leq$

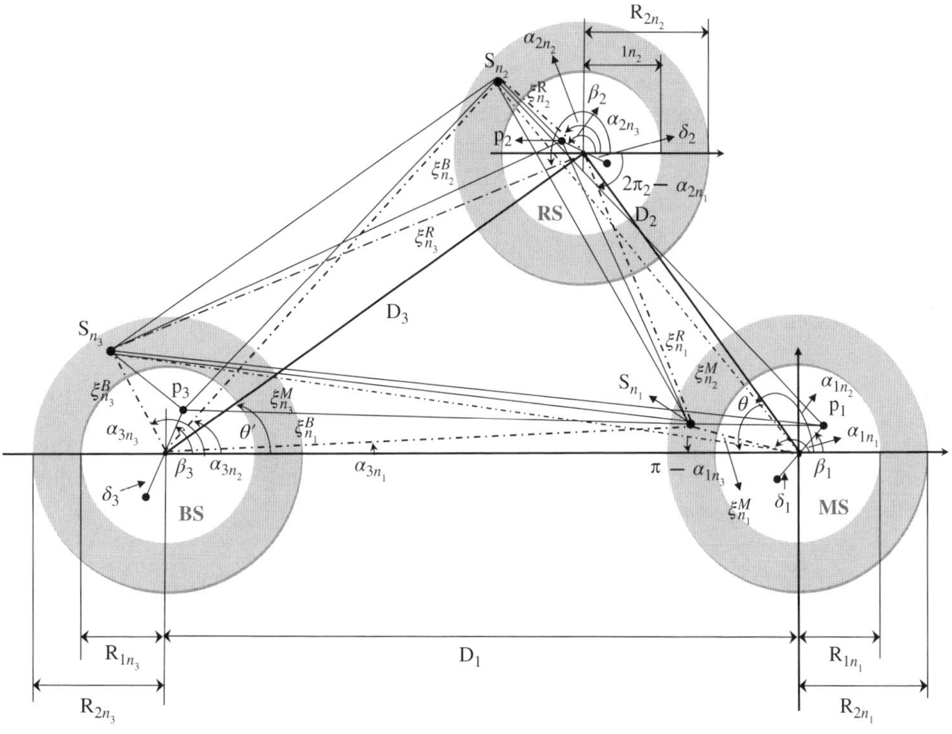

Figure 9.29 The proposed cooperative MIMO GBSM

Table 9.2 Definition of parameters in Figure 9.29

Parameter	Description
D_1, D_2, D_3	Distances of BS-MS, RS-MS, and BS-RS, respectively
$R_{1n_1}, R_{2n_1}; R_{1n_2}, R_{2n_2}; R_{1n_3}, R_{2n_3}$	Min and max radii of the circular rings around the MS, RS and BS, respectively
θ, θ'	Angles between the RS-MS link and BS-MS link, and between the BS-RS link and BS-MS link, respectively
$\delta_1, \delta_2, \delta_3$	Antenna element spacings of MS, RS and BS, respectively
$\beta_1, \beta_2, \beta_3$	Orientations of the MS, RS and RS antenna arrays in the x-y plane (relative to the x-axis), respectively
$\alpha_{1n_i}, \alpha_{2n_i}$, and α_{3n_i}	Azimuth angles of S_{n_i}-MS, S_{n_i}-RS, and S_{n_i}-BS links in the x-y plane (relative to the x-axis), respectively
$\xi^B_{n_1}, \xi^B_{n_2}, \xi^B_{n_3}$	Distances d(BS,S_{n_1}), d(BS,S_{n_2}), and d(BS,S_{n_3}), respectively
$\xi^R_{n_1}, \xi^R_{n_2}, \xi^R_{n_3}$	Distances d(RS,S_{n_1}), d(RS,S_{n_2}), and d(RS,S_{n_3}), respectively
$\xi^M_{n_1}, \xi^M_{n_2}, \xi^M_{n_3}$	Distances d(MS,S_{n_1}), d(MS,S_{n_2}), and d(MS,S_{n_3}), respectively
$\varepsilon_{p_i n_g}(\varepsilon_{n_g p_i}), \varepsilon_{p_i p_j}$, and $\varepsilon_{n_g n_k}$	Distances $d(p_i, S_{n_g})$, $d(p_i, p_j)$, and $d(S_{n_g}, S_{n_k})$, respectively

R_{2n_2}; the n_2th ($n_2 = 1, \ldots, N_2$) effective scatterer is denoted by S_{n_2}. For the local scattering area around the BS, N_3 effective scatterers lie on a circular ring of radius $R_{3n_1} \leq \xi^B_{n_3} \leq R_{3n_2}$ and the n_3th ($n_3 = 1, \ldots, N_3$) effective scatterer is denoted by S_{n_3}. The parameters in Figure 9.29 are defined in Table 9.2.

As this paper only focuses on the investigation of multi-link spatial correlations (not time or frequency correlations), we will neglect t and τ in Eq. (9.119) for the proposed channel model framework to simplify notations. In the following, we will show the channel gains of the three different links for the proposed cooperative MIMO GBSM.

9.5.2.1 BS–RS link

The channel gain of the BS-RS link between Antenna p_3 at the BS and antenna p_2 at the RS can be expressed as

$$h_{p_3p_2} = h^{LoS}_{p_3p_2} + \sum_{i=1}^{3}\sum_{g=1}^{f_3(i)} h^{ig}_{p_3p_2} \tag{9.122}$$

where $h^{LoS}_{p_3p_2}$ denotes the LoS component and $h^{ig}_{p_3p_2}$ represents the gth i-bounced component with the following expressions

$$h^{LoS}_{p_3p_2} = \sqrt{\frac{K_{p_3p_2}\Omega_{p_3p_2}}{K_{p_3p_2}+1}} e^{-j2\pi\lambda^{-1}\chi_{p_3p_2}} \tag{9.123}$$

$$h^{1g}_{p_3p_2} = \sqrt{\frac{\eta^{1g}_{p_3p_2}\Omega_{p_3p_2}}{K_{p_3p_2}+1}} \lim_{N_g\to\infty} \sum_{n_g=1}^{N_g} \frac{1}{\sqrt{N_g}} e^{j\left(\psi_{n_g} - 2\pi\lambda^{-1}\chi_{p_3p_2,n_g}\right)} \tag{9.124}$$

$$h_{p_3p_2}^{2g} = \sqrt{\frac{\eta_{p_3p_2}^{2g}\Omega_{p_3p_2}}{K_{p_3p_2}+1}} \lim_{N_{g_1},N_{g_2}\to\infty} \sum_{n_{g_1},n_{g_2}=1}^{N_{g_1},N_{g_2}} \frac{1}{\sqrt{N_{g_1}N_{g_2}}} e^{j\left(\psi_{n_{g_1},n_{g_2}} - 2\pi\lambda^{-1}\chi_{p_3p_2,n_{g_1},n_{g_2}}\right)} \tag{9.125}$$

$$h_{p_3p_2}^{31} = \sqrt{\frac{\eta_{p_3p_2}^{31}\Omega_{p_3p_2}}{K_{p_3p_2}+1}} \lim_{N_1,N_2,N_3\to\infty} \sum_{n_1,n_2,n_3=1}^{N_1,N_2,N_3} \frac{1}{\sqrt{N_1N_2N_3}} e^{j\left(\psi_{n_1,n_2,n_3} - 2\pi\lambda^{-1}\chi_{p_3p_2,n_1,n_2,n_3}\right)} \tag{9.126}$$

where $g = 1, 2, 3$, $\{g_1, g_2\} = \{3, 2\}$ for $g = 1$, $\{g_1, g_2\} = \{3, 1\}$ for $g = 2$, and $\{g_1, g_2\} = \{1, 2\}$ for $g = 3$. In Eqs (9.123)–(9.126), $\chi_{p_3p_2} = \varepsilon_{p_3p_2}$, $\chi_{p_3p_2,n_g} = \varepsilon_{p_3n_g} + \varepsilon_{n_gp_2}$, $\chi_{p_3p_2,n_{g_1},n_{g_2}} = \varepsilon_{p_3n_{g_1}} + \varepsilon_{n_{g_1}n_{g_2}} + \varepsilon_{n_{g_2}p_2}$, and $\chi_{p_3p_2,n_1,n_2,n_3} = \varepsilon_{p_3n_3} + \varepsilon_{n_3n_1} + \varepsilon_{n_1n_2} + \varepsilon_{n_2p_2}$ are the travel times of the waves through the link s$B_{p_3} - R_{p_2}$, $B_{p_3} - S_{n_g} - R_{p_2}$, $B_{p_3} - S_{n_{g_1}} - S_{n_{g_2}} - R_{p_2}$, and $B_{p_3} - S_{n_3} - S_{n_1} - S_{n_2} - R_{p_2}$, respectively. The symbols $K_{p_3p_2}$ and $\Omega_{p_3p_2}$ designate the Ricean factor and the total power of the BS–RS link, respectively. Parameters $\eta_{p_3p_2}^{1g}$, $\eta_{p_3p_2}^{2g}$, and $\eta_{p_3p_2}^{31}$ specify how much the single-, double-, and triple-bounced rays contribute to the total scattered power $\Omega_{p_3p_2}/(K_{p_3p_2}+1)$ with $\sum_{g=1}^{3}(\eta_{p_3p_2}^{1g} + \eta_{p_3p_2}^{2g}) + \eta_{p_3p_2}^{31} = 1$. The phases ψ_{n_g}, $\psi_{n_{g_1},n_{g_2}}$, and ψ_{n_1,n_2,n_3} are i.i.d. random variables with uniform distributions over $[-\pi, \pi)$.

From Figure 9.29 and based on the normally used assumption $\min\{D_1, D_2, D_3\} \gg \max\{\delta_1, \delta_2, \delta_3\}$ [Cheng et al. 2009] and the application of the law of cosines in appropriate triangles, the distances $\varepsilon_{p_3p_2}$, $\varepsilon_{p_3n_g}$, $\varepsilon_{n_gp_2}$, $\varepsilon_{n_1n_2}$, $\varepsilon_{n_3n_2}$, and $\varepsilon_{n_3n_1}$ in Eqs (9.123)–(9.126) can be expressed as

$$\varepsilon_{p_3p_2} \approx D_3 - \frac{\delta_3}{2}\cos(\beta_3 - \theta') + \frac{\delta_2}{2}\cos(\beta_2 - \theta') \tag{9.127}$$

$$\varepsilon_{p_3n_g} \approx \xi_{n_g}^{B} - \frac{\delta_3}{2}\cos(\beta_3 - \alpha_{1n_g}) \tag{9.128}$$

$$\varepsilon_{n_gp_2} \approx \xi_{n_g}^{R} - \frac{\delta_2}{2}\cos(\beta_2 - \alpha_{2n_g}) \tag{9.129}$$

$$\varepsilon_{n_1n_2} = [(\xi_{n_1}^{R})^2 + (\xi_{n_2}^{R})^2 - 2\xi_{n_1}^{R}\xi_{n_2}^{R}\cos(\alpha_{2n_1} - \alpha_{2n_2})]^{1/2} \tag{9.130}$$

$$\varepsilon_{n_3n_\varrho} = [(\xi_{n_3}^{B})^2 + (\xi_{n_\varrho}^{B})^2 - 2\xi_{n_3}^{B}\xi_{n_\varrho}^{B}\cos(\alpha_{3n_3} - \alpha_{1n_\varrho})]^{1/2} \tag{9.131}$$

where $\xi_{n_1}^{B} = \sqrt{D_1^2 + (\xi_{n_1}^{M})^2 + 2D_1\xi_{n_1}^{M}\cos\alpha_{1n_1}}$, $\xi_{n_1}^{R} = \sqrt{D_2^2 + (\xi_{n_1}^{M})^2 + 2D_2\xi_{n_1}^{M}\cos(\alpha_{1n_1} + \theta)}$, $\xi_{n_2}^{B} = \sqrt{D_3^2 + (\xi_{n_2}^{R})^2 + 2D_3\xi_{n_2}^{R}\cos(\alpha_{2n_2} - \theta')}$, $\xi_{n_3}^{R} = \sqrt{D_3^2 + (\xi_{n_3}^{B})^2 - 2D_3\xi_{n_3}^{B}\cos(\alpha_{3n_3} - \theta')}$, $\xi_{n_2}^{R} \in [R_{1n_2}, R_{2n_2}]$, $\xi_{n_3}^{B} \in [R_{1n_3}, R_{2n_3}]$, $\varrho = 1, 2$, and $g = 1, 2, 3$. Note that the AoD α_{3n_1}, α_{3n_2}, α_{3n_3} and AoA α_{2n_1}, α_{2n_2}, α_{2n_3} are independent for double- and triple-bounced rays, while they are interdependent for single-bounced rays. It is worth highlighting that scatterers S_{n_g} around MS, RS, and BS are relevant to the angles α_{1n_1}, α_{2n_2}, and α_{3n_3}, respectively. Therefore, all other AoDs and AoAs have to be related to the aforementioned three key angles. By following the general method given by Cheng et al. [2009], the relationship of the key angles with other AoAs and AoDs of BS-RS link can be obtained as: $\sin\alpha_{3n_1} = \frac{\xi_{n_1}^{M}}{\xi_{n_1}^{B}}\sin\alpha_{1n_1}$, $\sin(\alpha_{2n_1} + \theta) = \frac{\xi_{n_1}^{M}}{\xi_{n_1}^{R}}\sin(\theta + \alpha_{1n_1})$, $\sin(\alpha_{3n_2} - \theta') = \frac{\xi_{n_2}^{R}}{\xi_{n_2}^{B}}\sin(\alpha_{2n_2} - \theta')$, and $\sin(\alpha_{2n_3} - \theta') = \frac{\xi_{n_3}^{B}}{\xi_{n_3}^{R}}\sin(\alpha_{3n_3} - \theta')$.

Note that the distances and angles derived above are general expressions and thus are suitable for various basic scenarios. For outdoor macrocell and microcell scenarios, the assumption $\min\{D_1, D_2, D_3\} \gg \max\{\xi_{n_1}^M, \xi_{n_2}^R, \xi_{n_3}^B\}$, which is invalid for the outdoor picocell and indoor scenarios, is fulfilled. Therefore, for outdoor macrocell and microcell scenarios, we have the following reduced expressions: $\varepsilon_{n_1n_2} \approx D_2$, $\varepsilon_{n_3n_2} \approx D_3$, $\varepsilon_{n_3n_1} \approx D_1$, $\xi_{n_1}^B \approx D_1 + \xi_{n_1}^M \cos\alpha_{1n_1}$, $\xi_{n_1}^R \approx D_2 + \xi_{n_1}^M \cos(\alpha_{1n_1} + \theta)$, $\xi_{n_2}^B \approx D_3 + \xi_{n_2}^R \cos(\alpha_{2n_2} - \theta')$, $\xi_{n_3}^R \approx D_3 - \xi_{n_3}^B \cos(\alpha_{3n_3} - \theta')$, $\alpha_{3n_1} \approx \frac{\xi_{n_1}^M}{D_1} \sin\alpha_{1n_1}$, $\alpha_{2n_1} \approx 2\pi - \theta + \frac{\xi_{n_1}^M}{D_2} \sin(\theta + \alpha_{1n_1})$, $\alpha_{3n_2} \approx \theta' - \frac{\xi_{n_2}^R}{D_3} \sin(\alpha_{2n_2} - \theta')$, and $\alpha_{2n_3} \approx \pi + \theta' - \frac{\xi_{n_3}^B}{D_3} \sin(\alpha_{3n_3} - \theta')$.

9.5.2.2 BS–MS link

The channel gain of the BS–MS link between antenna p_3 at BS and antenna p_1 at the MS can be expressed as

$$h_{p_3p_1} = h_{p_3p_1}^{LoS} + \sum_{i=1}^{3} \sum_{g=1}^{f_3(i)} h_{p_3p_1}^{ig} \tag{9.132}$$

where $h_{p_3p_1}^{LoS}$ denotes the LoS component and $h_{p_3p_1}^{ig}$ represents the gth i-bounced component, with the following expressions

$$h_{p_3p_1}^{LoS} = \sqrt{\frac{K_{p_3p_1}\Omega_{p_3p_1}}{K_{p_3p_1}+1}} e^{-j2\pi\lambda^{-1}\chi_{p_3p_1}} \tag{9.133}$$

$$h_{p_3p_1}^{1g} = \sqrt{\frac{\eta_{p_3p_1}^{1g}\Omega_{p_3p_1}}{K_{p_3p_1}+1}} \lim_{N_g\to\infty} \sum_{n_g=1}^{N_g} \frac{1}{\sqrt{N_g}} e^{j\left(\psi_{n_g} - 2\pi\lambda^{-1}\chi_{p_3p_1,n_g}\right)} \tag{9.134}$$

$$h_{p_3p_1}^{2g} = \sqrt{\frac{\eta_{p_3p_1}^{2g}\Omega_{p_3p_1}}{K_{p_3p_1}+1}} \lim_{N_{g_1},N_{g_2}\to\infty} \sum_{n_{g_1},n_{g_2}=1}^{N_{g_1},N_{g_2}} \frac{1}{\sqrt{N_{g_1}N_{g_2}}} e^{j\left(\psi_{n_{g_1},n_{g_2}} - 2\pi\lambda^{-1}\chi_{p_3p_1,n_{g_1},n_{g_2}}\right)} \tag{9.135}$$

$$h_{p_3p_1}^{31} = \sqrt{\frac{\eta_{p_3p_1}^{31}\Omega_{p_3p_1}}{K_{p_3p_1}+1}} \lim_{N_1,N_2,N_3\to\infty} \sum_{n_1,n_2,n_3=1}^{N_1,N_2,N_3} \frac{1}{\sqrt{N_1N_2N_3}} e^{j\left(\psi_{n_1,n_2,n_3} - 2\pi\lambda^{-1}\chi_{p_3p_1,n_1,n_2,n_3}\right)} \tag{9.136}$$

where parameters g, g_1, and g_2 are the same as the ones in Eqs (9.123)–(9.126). In Eqs (9.133)–(9.136), $\chi_{p_3p_1} = \varepsilon_{p_3p_1}$, $\chi_{p_3p_1,n_g} = \varepsilon_{p_3n_g} + \varepsilon_{n_gp_1}$, $\chi_{p_3p_1,n_{g_1},n_{g_2}} = \varepsilon_{p_3n_{g_1}} + \varepsilon_{n_{g_1}n_{g_2}} + \varepsilon_{n_{g_2}p_1}$, and $\chi_{p_3p_1,n_1,n_2,n_3} = \varepsilon_{p_3n_3} + \varepsilon_{n_3n_2} + \varepsilon_{n_2n_1} + \varepsilon_{n_1p_1}$ are the travel times of the waves through the links $B_{p_3} - M_{p_1}$, $B_{p_3} - S_{n_g} - M_{p_1}$, $B_{p_3} - S_{n_{g_1}} - S_{n_{g_2}} - M_{p_1}$, and $B_{p_3} - S_{n_3} - S_{n_2} - S_{n_1} - M_{p_1}$, respectively. The symbols $K_{p_3p_1}$ and $\Omega_{p_3p_1}$ designate the Ricean factor and the total power of the BS–MS link, respectively. Parameters $\eta_{p_3p_1}^{1g}$, $\eta_{p_3p_1}^{2g}$, and $\eta_{p_3p_1}^{31}$ specify how much the single-, double-, and triple-bounced rays contribute to the total scattered power $\Omega_{p_3p_1}/(K_{p_3p_1}+1)$ with $\sum_{g=1}^{3}(\eta_{p_3p_1}^{1g} + \eta_{p_3p_1}^{2g}) + \eta_{p_3p_1}^{31} = 1$. The phases ψ_{n_1,n_3} are i.i.d. random variables with uniform distributions over $[-\pi, \pi)$.

Similar to the BS–RS link, by applying of the law of cosines in appropriate triangles, we have, with the help of the normally used assumption $\min\{D_1, D_2, D_3\} \gg \max\{\delta_1\delta_2, \delta_3\}$, the following expressions of the desired distances:

$$\varepsilon_{p_3p_1} \approx D_1 - \frac{\delta_3}{2}\cos\beta_3 + \frac{\delta_1}{2}\cos(\beta_1) \tag{9.137}$$

$$\varepsilon_{n_gp_1} \approx \xi^M_{n_g} - \frac{\delta_1}{2}\cos(\beta_1 - \phi_{n_g}) \tag{9.138}$$

where $g = 1, 2, 3$, $\xi^M_{n_1} \in [R_{1n_1}, R_{2n_1}]$, $\xi^M_{n_2} = \sqrt{D_2^2 + (\xi^R_{n_2})^2 - 2D_2\xi^R_{n_2}\cos\phi_{n_2}}$ with $\phi_{n_2} = 2\pi - \theta' - \alpha_{2n_2}$, and $\xi^M_{n_3} = \sqrt{D_1^2 + (\xi^B_{n_3})^2 - 2D_1\xi^B_{n_3}\cos\alpha_{3n_3}}$. The expressions of other distances of interest $\varepsilon_{p_3n_1}$, $\varepsilon_{p_3n_2}$, $\varepsilon_{p_3n_3}$, $\varepsilon_{n_3n_1}$, $\varepsilon_{n_3n_2}$, and $\varepsilon_{n_2n_1} = \varepsilon_{n_1n_2}$ have been given previously in the BS–RS link subsection. Similar to the BS–RS link, angles α_{1n_2} and α_{1n_3} need to be related to any one of three key angles as $\sin(\alpha_{1n_2} + \theta) = \frac{\xi^R_{n_2}}{\xi^M_{n_2}}\sin(\alpha_{2n_2} + \theta)$ and $\sin\alpha_{1n_3} = \frac{\xi^B_{n_3}}{\xi^M_{n_3}}\sin\alpha_{3n_3}$.

Also similar to the BS–RS link, the above derived expressions of the distances and angles are applicable to various basic scenarios. For outdoor macrocell and microcell scenarios, by using the assumption $\min\{D_1, D_2, D_3\} \gg \max\{\xi^M_{n_1}, \xi^R_{n_2}, \xi^B_{n_3}\}$, the following reduced expressions can be obtained: $\xi^M_{n_2} \approx D_2 - \xi^R_{n_2}\cos\phi_{n_2}$ with $\phi_{n_2} = 2\pi - (\alpha_{2n_2} + \theta)$, $\xi^M_{n_3} \approx D_1 - \xi^B_{n_3}\cos\alpha_{3n_3}$, $\alpha_{1n_2} \approx \pi - \theta - \frac{\xi^R_{n_2}}{D_2}\sin(\theta + \alpha_{2n_2})$, and $\alpha_{1n_3} \approx \pi - \frac{\xi^B_{n_3}}{D_1}\sin\alpha_{3n_3}$.

9.5.2.3 RS–MS link

The channel gain of RS-MS link between antenna p_2 at the BS and antenna p_1 at the MS can be expressed as

$$h_{p_2p_1} = h^{LoS}_{p_2p_1} + \sum_{i=1}^{3}\sum_{g=1}^{f_3(i)} h^{ig}_{p_2p_1} \tag{9.139}$$

where $h^{LoS}_{p_2p_1}$ denotes the LoS component and $h^{ig}_{p_2p_1}$ represents the gth i-bounced component, with the following expressions

$$h^{LoS}_{p_2p_1} = \sqrt{\frac{K_{p_2p_1}\Omega_{p_2p_1}}{K_{p_2p_1}+1}}e^{-j2\pi\lambda^{-1}\chi_{p_2p_1}} \tag{9.140}$$

$$h^{1g}_{p_2p_1} = \sqrt{\frac{\eta^{1g}_{p_2p_1}\Omega_{p_2p_1}}{K_{p_2p_1}+1}}\lim_{N_g\to\infty}\sum_{n_g=1}^{N_g}\frac{1}{\sqrt{N_g}}e^{j\left(\psi_{n_g}-2\pi\lambda^{-1}\chi_{p_2p_1,n_g}\right)} \tag{9.141}$$

$$h^{2g}_{p_2p_1} = \sqrt{\frac{\eta^{2g}_{p_2p_1}\Omega_{p_2p_1}}{K_{p_2p_1}+1}}\lim_{N_{g_1},N_{g_2}\to\infty}\sum_{n_{g_1},n_{g_2}=1}^{N_{g_1},N_{g_2}}\frac{1}{\sqrt{N_{g_1}N_{g_2}}}e^{j\left(\psi_{n_{g_1},n_{g_2}}-2\pi\lambda^{-1}\chi_{p_2p_1,n_{g_1},n_{g_2}}\right)} \tag{9.142}$$

$$h^{31}_{p_2p_1} = \sqrt{\frac{\eta^{31}_{p_2p_1}\Omega_{p_2p_1}}{K_{p_2p_1}+1}}\lim_{N_1,N_2,N_3\to\infty}\sum_{n_1,n_2,n_3=1}^{N_1,N_2,N_3}\frac{1}{\sqrt{N_1N_2N_3}}e^{j\left(\psi_{n_1,n_2,n_3}-2\pi\lambda^{-1}\chi_{p_2p_1,n_1,n_2,n_3}\right)} \tag{9.143}$$

where parameters g, g_1, and g_2 are the same as the ones in Eq. (9.123)–(9.126). In Eqs (9.140)–(9.143), $\chi_{p_2p_1} = \varepsilon_{p_2p_1}$, $\chi_{p_2p_1,n_g} = \varepsilon_{p_2n_g} + \varepsilon_{n_gp_1}$, $\chi_{p_2p_1,n_{g_1},n_{g_2}} = \varepsilon_{p_2n_{g_1}} + \varepsilon_{n_{g_1}n_{g_2}} + \varepsilon_{n_{g_2}p_1}$, and $\chi_{p_2p_1,n_1,n_2,n_3} = \varepsilon_{p_2n_2} + \varepsilon_{n_2n_3} + \varepsilon_{n_3n_1} + \varepsilon_{n_1p_1}$ are the travel times of the waves through the links $R_{p_2} - M_{p_1}$, $R_{p_2} - S_{n_g} - M_{p_1}$, $R_{p_2} - S_{n_{g_1}} - S_{n_{g_2}} - M_{p_1}$, and $R_{p_2} - S^{(n_2)} - S^{(n_3)} - S^{(n_1)} - M_{p_1}$, respectively. The symbols $K_{p_2p_1}$ and $\Omega_{p_2p_1}$ designate the Ricean factor and the total power of the RS–MS link, respectively. Parameters $\eta^{1g}_{p_2p_1}$, $\eta^{2g}_{p_2p_1}$, and $\eta^{31}_{p_2p_1}$ specify how much the single-, double-, and triple-bounced rays contribute to the total scattered power $\Omega_{p_2p_1}/(K_{p_2p_1}+1)$ with $\sum_{g=1}^{3}(\eta^{1g}_{p_2p_1} + \eta^{2g}_{p_2p_1}) + \eta^{31}_{p_2p_1} = 1$.

From Figure 9.29, it is clear that all the expressions of the desired distances have been given previously in the subsections on the BS–RS and BS–MS links, except the distance $\varepsilon_{p_2p_1}$, which has the following expression: $\varepsilon_{p_2p_1} \approx D_2 - \frac{\delta_2}{2}\cos(\beta_2+\theta) - \frac{\delta_1}{2}\cos(\beta_1-\theta)$, where the assumption $D_2 \gg \max\{\delta_2, \delta_1\}$ is utilized.

To characterize the AoDs and AoAs, we use the von Mises PDF given in Eq. (9.7). For the key angles, α_{1n_1}, α_{2n_2}, and α_{3n_3}, we use appropriate parameters (μ and k) of the von Mises PDF as μ_1 and k_1, μ_2 and k_2, and μ_3 and k_3, respectively.

9.5.2.4 Adjustment of Key Model Parameters

The proposed cooperative MIMO GBSM is adaptable to the above mentioned 12 cooperative scenarios for this interested typical cooperative MIMO environment by adjusting key model parameters. From previous section, we know that these important model parameters are the number of local scattering environment I, Ricean factors $K_{p_3p_2}$, $K_{p_3p_1}$, $K_{p_2p_1}$, and energy-related parameters $\eta^{ig}_{p_3p_2}$, $\eta^{ig}_{p_3p_1}$, and $\eta^{ig}_{p_2p_1}$. The paramter setting of I is basically based on basic scenario. For outdoor microcell, picocell, and indoor scenarios, we assume that the BS, RS, and MS are all surrounded by local scattering area as shown in Figure 9.29 and thus $I = 3$ in this case. For outdoor Marco-cell scenario, the BS is free of scatterers and thus $I = 2$. In this case, the channel model can also be obtained from the proposed model in Eq. (9.122), Eq. (9.132), and Eq. (9.139) by setting the energy-related parameters related to the local scatterers around BS equal to zero, e.g., for BS-RS link, the channel model can be obtained from Eq. (9.122) by setting $\eta^{13}_{p_3p_2} = \eta^{21}_{p_3p_2} = \eta^{22}_{p_3p_2} = \eta^{31}_{p_3p_2} = 0$. For outdoor macrocell BS cooperation scenario, RS actually represents the other BS, symbolled as BS2, and thus is free of scatterers as well. In this case, we have the currently most mature cooperative MIMO scheme: CoMP and the number of local scattering area $I = 1$. Similarly, the channel model with $I = 1$ can also be obtained from the proposed model in Eq. (9.122), Eq. (9.132), and Eq. (9.139) by setting the energy-related parameters related to the local scatterers around BS and RS(BS2) equal to zero. It is clear that the proposed GBSM can be adaptable to different basic scenarios by setting relevant energy-related parameters equal to zero. Therefore, the key model parameters of the proposed GBSM actually are reduced as the Ricean factors and energy-related parameters. The basic criterion of setting these key model parameters is summarized as following: the longer distance of the link and/or the higher the LSD, the smaller the Ricean factors and the larger the energy-related parameters of multi-bounced components, i.e., the multi-bounced components bear more energy than single-bounced components. Since the local scattering area is highly related to the degree of link heterogeneity in cooperative MIMO systems as presented in [Czink et al. 2008, Oestges et al. 2010, Zhang et al. 2009], the LSD significantly affects the channel characteristics and should be investigated. In general, the higher the LSD, the lower the possibility that the devices (BS/MS/RS) share the same scatterers. In this case, the cooperative MIMO environments present lower environment similarity. Therefore, the higher the LSD, the lower the environment similarity.

For macrocell scenarios, the Ricean factor $K_{p_3p_1}$ is very small or even close to zero due to the large distance D_1. Under the condition of BS cooperation scenes, Ricean factor $K_{p_2p_1}$ is similar to $K_{p_3p_1}$ due to the similar distances of D_1 and D_2. While the BS-RS link is disappeared and replaced by wired link. In this case, we only have single bounced components, i.e., $\eta^{11}_{p_3p_1}$ and $\eta^{11}_{p_2p_1}$. For MS/RS cooperation scenes, RS/MS actually represents the other MS/RS and is symbolled as MS2/RS2. In this case, Ricean factor $K_{p_3p_2}$ is similar to $K_{p_3p_1}$ due to the similar distances of D_1 and D_3. Since the large value of distances D_1 and D_3, for the BS-MS/BR2 link and BS-MS2/RS link, the impact of LSD on channel characteristics is small and in general the double-bounced rays bear more energy than single-bounced rays, i.e., $\{\eta^{21}_{p_3p_1}, \eta^{21}_{p_3p_2}\} > \{\eta^{11}_{p_3p_1}, \eta^{12}_{p_3p_1}, \eta^{11}_{p_3p_2}, \eta^{12}_{p_3p_2}\}$. While for the MS/RS2-MS2/RS link, due to the small distance of D_2 the impact of LSD is significant. For a low LSD, the scatterers are sparse and thus more likely single-bounced rays rather than double-bounced rays, i.e., $\{\eta^{11}_{p_2p_1}, \eta^{12}_{p_2p_1}\} > \eta^{21}_{p_2p_1}$, and Ricean factor $K_{p_2p_1}$ is large. For a high LSD, the double-bounced components bear more energy than single-bounced components, i.e., $\eta^{21}_{p_2p_1} > \{\eta^{11}_{p_2p_1}, \eta^{12}_{p_2p_1}\}$ and Ricean factor $K_{p_2p_1}$ is smaller than that in the low LSD. Similar to macrocell scenarios, the key model parameters setting for other 9 cooperative scenarios with the consideration of different LSDs can be easily obtained by following the aforementioned basic criterion and thus omits here for brevity. The main features of the proposed cooperative MIMO GBSM have been summarized in Table 9.3.

9.5.3 *Multi-link Spatial Correlation Functions*

In this section, based on the proposed cooperative MIMO GBSM in Section III, we will derive the multi-link spatial correlation functions for non-isotropic scattering cooperative MIMO environments. The spatial correlation properties between any two of the aforementioned three links – the BS–RS link, BS–MS link, and RS–MS link – will be investigated. The normalized spatial correlation function between any two links characterized by channel gains h_{pq} and $h_{p'q'}$, respectively, is defined as

$$\rho_{pq,p'q'} = \frac{\mathbf{E}\left[h_{pq}h^*_{p'q'}\right]}{\sqrt{\Omega_{pq}\Omega_{p'q'}}} \tag{9.144}$$

where $(\cdot)^*$ denotes the complex conjugate operation, $\mathbf{E}[\cdot]$ is the statistical expectation operator, $p, p' \in \{1, 2, \ldots, M_T\}$, and $q, q' \in \{1, 2, \ldots, M_R\}$. Substituting Eq. (9.122) and Eq. (9.132) into Eq. (9.144), we have the correlation function between the BS–RS link and BS–MS link as

$$\rho_{p_3p_2,p'_3p_1} = \rho^{LoS}_{p_3p_2,p'_3p_1} + \sum_{g=1}^{3}(\rho^{1g}_{p_3p_2,p'_3p_1} + \rho^{2g}_{p_3p_2,p'_3p_1}) + \rho^{31}_{p_3p_2,p'_3p_1} \tag{9.145}$$

with

$$\rho^{LoS}_{p_3p_2,p'_3p_1} = \sqrt{\frac{K_{p_3p_2}K_{p'_3p_1}}{\left(K_{p_3p_2}+1\right)\left(K_{p'_3p_1}+1\right)}}e^{j2\pi\lambda^{-1}\left(\chi_{p'_3p_1}-\chi_{p_3p_2}\right)} \tag{9.146}$$

Table 9.3 Main features of the proposed cooperative mimo GBSM

Links	BS–RS, RS–MS, and BS–MS links. Easily extended to include more links		
Scenarios	*Physical scenarios* Outdoor macrocell Outdoor microcell Outdoor picocell Indoor scenarios		*Application scenarios* BS cooperation MS cooperation Relay cooperation
	I	$k_{p_3p_2}$ $k_{p_3p_1}$ $k_{p_2p_1}$	$\eta^{1g}_{p_3p_2/p_3p_1/p_2p_1}$ $\eta^{2g}_{p_3p_2/p_3p_1/p_2p_1}$ $\eta^{31}_{p_3p_2/p_3p_1/p_2p_1}$ $(g = 1, 2, 3)$
Key parameters	The number of local scattering areas.	Ricean factor of the BS–RS link, BS–MS link, and RS–MS link, respectively.	Energy-related parameters that specify how much the single-, double-, and triple-bounced rays contribute to the total scattered power of the BS–RS, BS–MS, RS–MS link, respectively.
	By properly adjusting the key parameters, the proposed cooperative MIMO GBSM is suitable for 12 cooperation scenarios.		

$$\rho^{1g}_{p_3p_2,p'_3p_1} = \sqrt{\frac{\eta^{1g}_{p_3p_2}\eta^{1g}_{p'_3p_1}}{\left(K_{p_3p_2}+1\right)\left(K_{p'_3p_1}+1\right)}}\int_{-\pi}^{\pi}\int_{R_{1n_g}}^{R_{2n_g}} \times e^{j2\pi\lambda^{-1}\left(\chi_{p'_3p_1,g}-\chi_{p_3p_2,g}\right)} Q_g f(\emptyset_g)d\emptyset_g d\mathcal{I}_g \tag{9.147}$$

$$\rho^{2g}_{p_3p_2,p'_3p_1} = \sqrt{\frac{\eta^{2g}_{p_3p_2}\eta^{2g}_{p'_3p_1}}{\left(K_{p_3p_2}+1\right)\left(K_{p'_3p_1}+1\right)}}\int_{-\pi}^{\pi}\int_{-\pi}^{\pi}\int_{R_{1n_{g_1}}}^{R_{2n_{g_1}}}\int_{R_{1n_{g_2}}}^{R_{2n_{g_2}}} Q_{g_1g_2} f(\emptyset_{g_1})f(\emptyset_{g_2}) \times e^{j2\pi\lambda^{-1}\left(\chi_{p'_3p_1,g_1,g_2}-\chi_{p_3p_2,g_1,g_2}\right)} d\emptyset_{g_1}d\emptyset_{g_2}d\mathcal{I}_{g_1}d\mathcal{I}_{g_2} \tag{9.148}$$

$$\rho^{31}_{p_3p_2,p'_3p_1} = \sqrt{\frac{\eta^{31}_{p_3p_2}\eta^{31}_{p'_3p_1}}{\left(K_{p_3p_2}+1\right)\left(K_{p'_3p_1}+1\right)}}\int_{-\pi}^{\pi}\int_{-\pi}^{\pi}\int_{-\pi}^{\pi}\int_{R_{1n_1}}^{R_{2n_1}}\int_{R_{1n_2}}^{R_{2n_2}}\int_{R_{1n_3}}^{R_{2n_3}} Q_{123} \times e^{j2\pi\lambda^{-1}\left(\chi_{p'_3p_1,1,2,3}-\chi_{p_3p_2,1,2,3}\right)} f(\emptyset_1)f(\emptyset_2)f(\emptyset_3)d\emptyset_1 d\emptyset_2 d\emptyset_3 d\mathcal{I}_1 d\mathcal{I}_2 d\mathcal{I}_3 \tag{9.149}$$

where $\{\emptyset_g\}_{g=1}^{3} = \{\phi_1, \alpha_{2,2}, \alpha_{1,3}\}$ are the continuous expressions of the discrete expressions of angles α_{1n_1}, α_{2n_2}, α_{3n_3}, respectively, and $\varepsilon_{p_3'n_g} \approx \xi_{n_g}^{B} + \frac{\delta_3}{2}\cos(\beta_3 - \alpha_{1n_g})$, $\chi_{p_3'p_1,g} = \chi_{p_3'p_1,n_g}$, $\chi_{p_3p_2,g} = \chi_{p_3p_2,n_g}$, $\chi_{p_3'p_1,g_1,g_2} = \chi_{p_3'p_1,n_{g_1},n_{g_2}}$, $\chi_{p_3p_2,g_1,g_2} = \chi_{p_3p_2,n_{g_1},n_{g_2}}$, $\chi_{p_3'p_1,1,2,3} = \chi_{p_3'p_1,n_1,n_2,n_3}$, and $\chi_{p_3p_2,1,2,3} = \chi_{p_3p_2,n_1,n_2,n_3}$ with α_{1n_1}, α_{2n_2}, and α_{3n_3} being replaced by ϕ_1, $\alpha_{2,2}$, and $\alpha_{1,3}$, respectively. Parameters $f(\emptyset_g) = \exp\left[k_g \cos\left(\emptyset_g - \mu_g\right)\right] / \left[2\pi I_0\left(k_g\right)\right]$, $\{\mathcal{I}_g\}_{g=1}^{3} = \{\xi_{n_1}^{M}, \ \xi_{2n_2}^{R}, \xi_{1n_3}^{B}\}$, $Q_g = \frac{2\mathcal{I}_{n_g}}{R_{2n_g}^2 - R_{1n_g}^2}$, $Q_{g_1g_2} = \frac{4\mathcal{I}_{n_{g_1}}\mathcal{I}_{n_{g_2}}}{(R_{2n_{g_1}}^2 - R_{1n_{g_1}}^2)(R_{2n_{g_2}}^2 - R_{1n_{g_2}}^2)}$, $Q_{123} = \frac{8\mathcal{I}_{n_1}\mathcal{I}_{n_2}\mathcal{I}_{n_3}}{(R_{2n_1}^2 - R_{1n_1}^2)(R_{2n_2}^2 - R_{1n_2}^2)(R_{2n_3}^2 - R_{1n_3}^2)}$, and parameters g, g_1, and g_2 are the same as those in Eqs (9.123)–(9.126). Note that in Eq. (9.145) other correlation terms are equal to zero and thus omitted. These omitted correlation terms contain the integral of random phases ψ_{n_g}, $\psi_{n_{g_1},n_{g_2}}$, or ψ_{n_1,n_2,n_3}. Since the random phases fulfill a uniform distribution over the range of $[\pi, -\pi)$, the integral of the random phases in the range of $[\pi, -\pi)$ is equal to zero. Therefore, other correlation terms with the value of zero are omitted in Eq. (9.145).

Performing the substitution of Eqs (9.122) and (9.139) into Eq. (9.144), we can obtain the correlation function between the BS–RS link and RS–MS links as

$$\rho_{p_3p_2,p_2'p_1} = \rho_{p_3p_2,p_2'p_1}^{LoS} + \sum_{g=1}^{3}(\rho_{p_3p_2,p_2'p_1}^{1g} + \rho_{p_3p_2,p_2'p_1}^{2g}) + \rho_{p_3p_2,p_2'p_1}^{31} \tag{9.150}$$

with

$$\rho_{p_3p_2,p_2'p_1}^{LoS} = \sqrt{\frac{K_{p_3p_2}K_{p_2'p_1}}{\left(K_{p_3p_2}+1\right)\left(K_{p_2'p_1}+1\right)}}e^{j2\pi\lambda^{-1}\left(\chi_{p_2'p_1} - \chi_{p_3p_2}\right)} \tag{9.151}$$

$$\rho_{p_3p_2,p_2'p_1}^{1g} = \sqrt{\frac{\eta_{p_3p_2}^{1g}\eta_{p_2'p_1}^{1g}}{\left(K_{p_3p_2}+1\right)\left(K_{p_2'p_1}+1\right)}}\int_{-\pi}^{\pi}\int_{R_{1n_g}}^{R_{2n_g}} \times e^{j2\pi\lambda^{-1}\left(\chi_{p_2'p_1,g} - \chi_{p_3p_2,g}\right)} Q_g f(\emptyset_g) d\emptyset_g d\mathcal{I}_g \tag{9.152}$$

$$\rho_{p_3p_2,p_2'p_1}^{2g} = \sqrt{\frac{\eta_{p_3p_2}^{2g}\eta_{p_2'p_1}^{2g}}{\left(K_{p_3p_2}+1\right)\left(K_{p_2'p_1}+1\right)}}\int_{-\pi}^{\pi}\int_{-\pi}^{\pi}\int_{R_{1n_{g_1}}}^{R_{2n_{g_1}}}\int_{R_{1n_{g_2}}}^{R_{2n_{g_2}}} Q_{g_1g_2} f(\emptyset_{g_1}) f(\emptyset_{g_2}) \times e^{j2\pi\lambda^{-1}\left(\chi_{p_2'p_1,g_1,g_2} - \chi_{p_3p_2,g_1,g_2}\right)} d\emptyset_{g_1} d\emptyset_{g_2} d\mathcal{I}_{g_1} d\mathcal{I}_{g_2} \tag{9.153}$$

$$\rho_{p_3p_2,p_2'p_1}^{31} = \sqrt{\frac{\eta_{p_3p_2}^{31}\eta_{p_2'p_1}^{31}}{\left(K_{p_3p_2}+1\right)\left(K_{p_2'p_1}+1\right)}}\int_{-\pi}^{\pi}\int_{-\pi}^{\pi}\int_{-\pi}^{\pi}\int_{R_{1n_1}}^{R_{2n_1}}\int_{R_{1n_2}}^{R_{2n_2}}\int_{R_{1n_3}}^{R_{2n_3}} Q_{123} \times e^{j2\pi\lambda^{-1}\left(\chi_{p_2'p_1,1,2,3} - \chi_{p_3p_2,1,2,3}\right)} f(\emptyset_1) f(\emptyset_2) f(\emptyset_3) d\emptyset_1 d\emptyset_2 d\emptyset_3 d\mathcal{I}_1 d\mathcal{I}_2 d\mathcal{I}_3 \tag{9.154}$$

where $\varepsilon_{p_2'n_g} \approx \xi_{n_g}^{R} + \frac{\delta_2}{2}\cos(\beta_2 - \alpha_{2n_g})$, $\chi_{p_2'p_1,g} = \chi_{p_2'p_1,n_g}$, $\chi_{p_2'p_1,g_1,g_2} = \chi_{p_2'p_1,n_{g_1},n_{g_2}}$, and $\chi_{p_2'p_1,1,2,3} = \chi_{p_2'p_1,n_1,n_2,n_3}$ with α_{1n_1}, α_{2n_2}, and α_{3n_3} being replaced by ϕ_1, $\alpha_{2,2}$, and $\alpha_{1,3}$, respectively.

The substitution of Eqs (9.132) and (9.139) into Eq. (9.144) results in the correlation function between the BS–MS link and RS–MS link as

$$\rho_{p_3p_1,p_2p_1'} = \rho^{LoS}_{p_3p_1,p_2p_1'} + \sum_{g=1}^{3}(\rho^{1g}_{p_3p_1,p_2p_1'} + \rho^{2g}_{p_3p_1,p_2p_1'}) + \rho^{31}_{p_3p_1,p_2p_1'} \tag{9.155}$$

with

$$\rho^{LoS}_{p_3p_1,p_2p_1'} = \sqrt{\frac{K_{p_3p_1}K_{p_2p_1'}}{(K_{p_3p_1}+1)\left(K_{p_2p_1'}+1\right)}}e^{j2\pi\lambda^{-1}\left(\chi_{p_2p_1'}-\chi_{p_3p_1}\right)} \tag{9.156}$$

$$\rho^{1g}_{p_3p_1,p_2p_1'} = \sqrt{\frac{\eta^{1g}_{p_3p_1}\eta^{1g}_{p_2p_1'}}{(K_{p_3p_1}+1)\left(K_{p_2p_1'}+1\right)}}\int_{-\pi}^{\pi}\int_{R_{1n_g}}^{R_{2n_g}} \times e^{j2\pi\lambda^{-1}\left(\chi_{p_2p_1',g}-\chi_{p_3p_1,g}\right)}Q_g f(\emptyset_g)d\emptyset_g d\mathcal{I}_g, \tag{9.157}$$

$$\rho^{2g}_{p_3p_1,p_2p_1'} = \sqrt{\frac{\eta^{2g}_{p_3p_1}\eta^{2g}_{p_2p_1'}}{(K_{p_3p_1}+1)\left(K_{p_2p_1'}+1\right)}}\int_{-\pi}^{\pi}\int_{-\pi}^{\pi}\int_{R_{1n_{g1}}}^{R_{2n_{g1}}}\int_{R_{1n_{g2}}}^{R_{2n_{g2}}} Q_{g_1g_2}f(\emptyset_{g_1})f(\emptyset_{g_2}) \times e^{j2\pi\lambda^{-1}\left(\chi_{p_2p_1',g_1,g_2}-\chi_{p_3p_1,g_1,g_2}\right)}d\emptyset_{g_1}d\emptyset_{g_2}d\mathcal{I}_{g_1}d\mathcal{I}_{g_2} \tag{9.158}$$

$$\rho^{31}_{p_3p_1,p_2p_1'} = \sqrt{\frac{\eta^{31}_{p_3p_1}\eta^{31}_{p_2p_1'}}{(K_{p_3p_1}+1)\left(K_{p_2p_1'}+1\right)}}\int_{-\pi}^{\pi}\int_{-\pi}^{\pi}\int_{-\pi}^{\pi}\int_{R_{1n_1}}^{R_{2n_1}}\int_{R_{1n_2}}^{R_{2n_2}}\int_{R_{1n_3}}^{R_{2n_3}} Q_{123} \times e^{j2\pi\lambda^{-1}\left(\chi_{p_2p_1',1,2,3}-\chi_{p_3p_1,1,2,3}\right)}f(\emptyset_1)f(\emptyset_2)f(\emptyset_3)d\emptyset_1 d\emptyset_2 d\emptyset_3 d\mathcal{I}_1 d\mathcal{I}_2 d\mathcal{I}_3 \tag{9.159}$$

where $\varepsilon_{p_1'n_g} \approx \xi^M_{n_g} + \frac{\delta_1}{2}\cos(\beta_1 - \phi_{n_g})$ with α_{1n_1}, α_{2n_2}, and α_{3n_3} being replaced by ϕ_1, $\alpha_{2,2}$, and $\alpha_{1,3}$, respectively.

For outdoor macrocell and microcell BS cooperation and RS cooperation scenarios, based on the assumption $\min\{D_1, D_2, D_3\} \gg \max\{\xi^M_{n_1}, \xi^R_{n_2}, \xi^B_{n_3}\}$, the correlation functions in Eqs (9.147)–(9.149), Eqs (9.152)–(9.154), and Eqs (9.157)–(9.159) can be further simplified as expressed in Appendix A.8.9. Based on the derived general spatial correlation functions, the spatial correlation functions for other scenarios can be easily obtained by using the criterion for key model parameter setting explained in the last subsection.

9.5.4 Numerical Results and Analysis

In Section 9.5.4, the multi-link spatial correlation functions derived in Section 9.5.3 will be numerically analyzed in detail. Without loss of generality, the spatial correlation properties between the BS–RS and BS–MS links are chosen for further investigation. The parameters are as listed here or specified otherwise: $f = 2.4\,\text{GHz}$, $D_1 = D_3 = 100\,\text{m}$, $R_{1n_1} = R_{1n_2} = R_{1n_3} = 5\,\text{m}$, $R_{2n_1} = R_{2n_2} = R_{2n_3} = 50\,\text{mm}$, $\delta_3 = \delta_2 = \delta_1 = 0$, $\beta_3 = 30°$,

$\beta_2 = \beta_1 = 60°$, $K_{p_3p_2} = K_{p'_3p_1} = 0$, $k_1 = k_2 = k_3 = 10$, $\mu_1 = 120°$, $\mu_2 = 300°$, and $\mu_3 = 60°$.

9.5.4.1 Impact of Key Parameters on Multi-Link Spatial Correlation

Figure 9.30 illustrates the spatial correlation properties of different components in Eq. (9.145) as a function of θ' and D_3. It is obvious that high spatial correlations between the BS–RS link and BS–MS link can occur at certain distances D_3 and certain values of angle θ' for different components. This again demonstrates that small-scale spatial fading correlation should not simply be neglected, a possibility addressed by [Wang et al. 2010], and also highlights the importance of the work presented in this paper.

In Figure 9.31, we present the spatial correlation properties of all scattered components in Eq. (9.145) with parameters $\delta_3 = \delta_2 = \delta_1 = 3\lambda$ and $k_1 = k_2 = k_3 = 3$. Fig.9.31 clearly shows that the spatial correlation properties vary significantly for different scattered components. More importantly, we notice that the scattered component that includes more bounced rays has lower spatial correlation properties. This is because with more bounced rays, the component is related to more local scattering areas and thus experiences a higher degree of link heterogeneity resulting in lower link similarity.

Figures 9.32 and 9.33 compare the spatial correlation properties of different scattered components for different values of environment parameters D_g, R_{1n_g}, R_{2n_g}, k_g, and μ_g with $g = 1, 2, 3$. These environment parameters determine the distances between BS, RS, and MS, and determine the size and distribution of the local scattering areas. It is clear

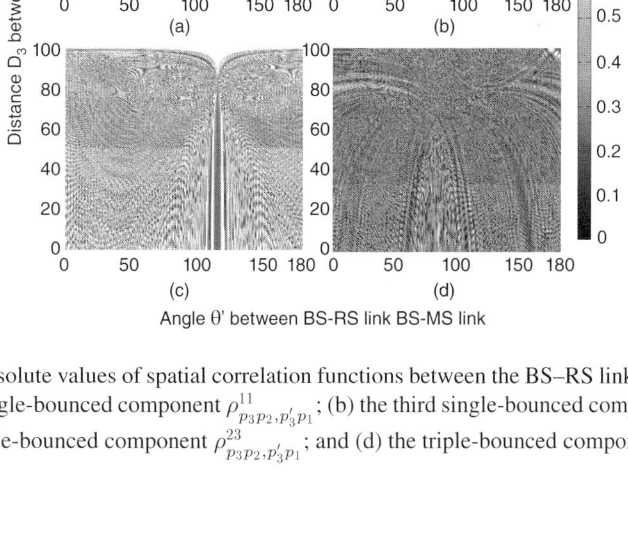

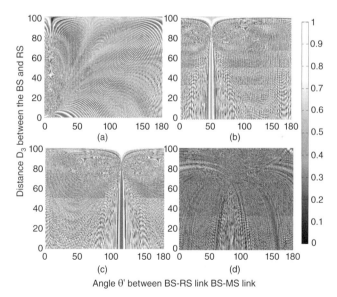

Figure 9.30 Absolute values of spatial correlation functions between the BS–RS link and BS–MS link for (a) the first single-bounced component $\rho^{11}_{p_3p_2,p'_3p_1}$; (b) the third single-bounced component $\rho^{13}_{p_3p_2,p'_3p_1}$; (c) the third double-bounced component $\rho^{23}_{p_3p_2,p'_3p_1}$; and (d) the triple-bounced component $\rho^{31}_{p_3p_2,p'_3p_1}$

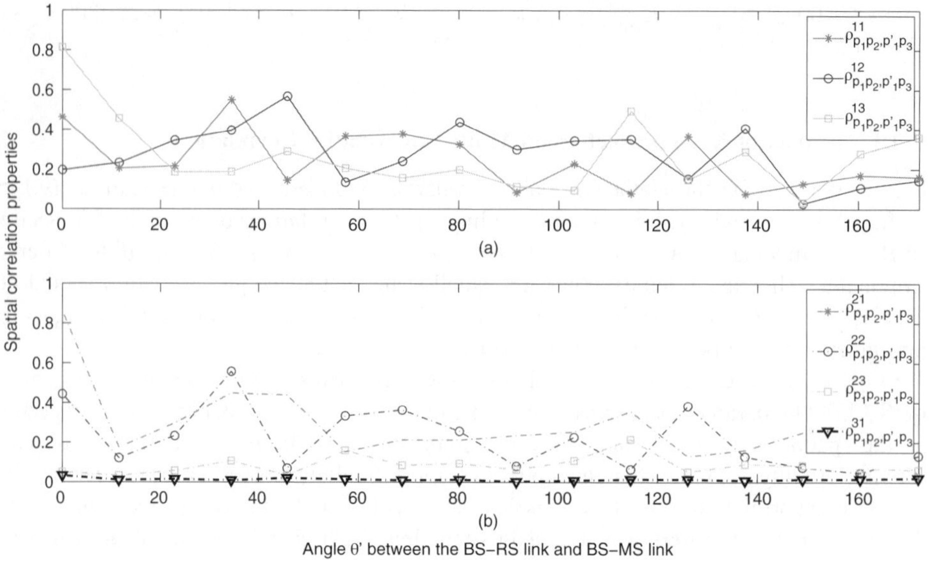

Figure 9.31 Absolute values of spatial correlation functions between the BS–RS link and BS–MS link for (a) the single-bounced components and (b) the double- and triple-bounce components

that these environment parameters significantly affect the spatial correlation properties of different scattered components. From Figure 9.32, we can also see that an increase in value k_g enhances spatial correlation. With larger values of k_g, the scatterers in the local scattering area are more concentrated and the received power mainly comes from a certain direction determined by μ_g. Therefore, in this case, the spatial correlation tends to be larger, a finding which also agrees with the conclusions of Xu et al. [2009]. Figure 9.33 also shows that a smaller local scattering area leads to higher spatial correlation properties, because the effective scatterers are more concentrated. It also allows us to conclude that compared to a narrowband cooperative MIMO system, a wideband system has a high possibility of expressing lower spatial correlation properties because the wider the system band, the more likely the system is to experience large local scattering areas.

Figure 9.34 shows the spatial correlation properties of different scattered components for different values of antenna element spacing δ_g and antenna array tilt angles β_g with $g = 1, 2, 3$ and parameters $k_1 = k_2 = k_3 = 3$. It is shown that both antenna element spacing and antenna array tilt angle affect the spatial correlation properties of different scattered components and that an increase of antenna spacing δ_g will decrease the spatial correlations. However, the impact of parameters δ_g and β_g on the spatial correlation properties tends to be marginal for scattered components with more bounced rays.

9.5.4.2 Validation of the Proposed Cooperative MIMO GBSM

So far, based on Figures 9.30–9.34 we have investigated in more detail the spatial correlation properties of different scattered components separately. In general, we find that the

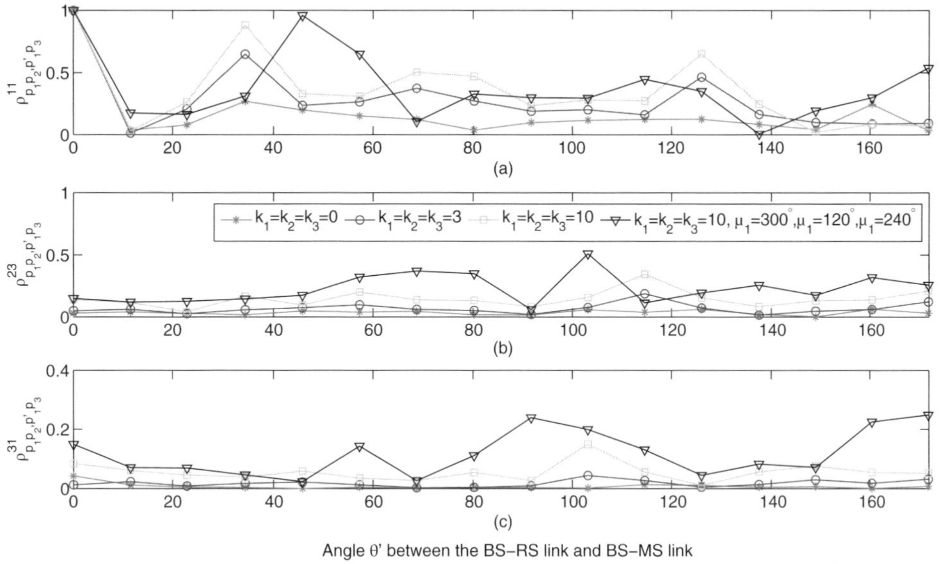

Figure 9.32 Absolute values of spatial correlation functions between the BS–RS link and BS–MS link for (a) the first single-bounced component $\rho^{11}_{p_3p_2,p_3'p_1}$; (b) the third double-bounced component $\rho^{23}_{p_3p_2,p_3'p_1}$; and (c) the triple-bounced component $\rho^{31}_{p_3p_2,p_3'p_1}$ with different values of parameters k_g and μ_g ($g = 1, 2, 3$)

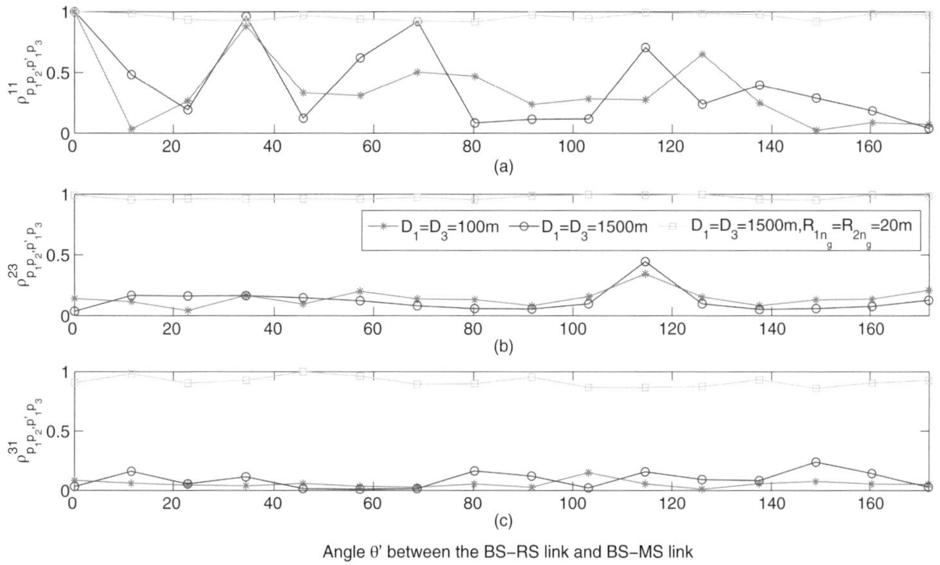

Figure 9.33 Absolute values of spatial correlation functions between the BS–RS link and BS–MS link for (a) the first single-bounced component $\rho^{11}_{p_3p_2,p_3'p_1}$; (b) the third double-bounced component $\rho^{23}_{p_3p_2,p_3'p_1}$; and (c) the triple-bounced component $\rho^{31}_{p_3p_2,p_3'p_1}$ with different values of parameters D_3, R_{1n_g}, and R_{2n_g} ($g = 1, 2, 3$)

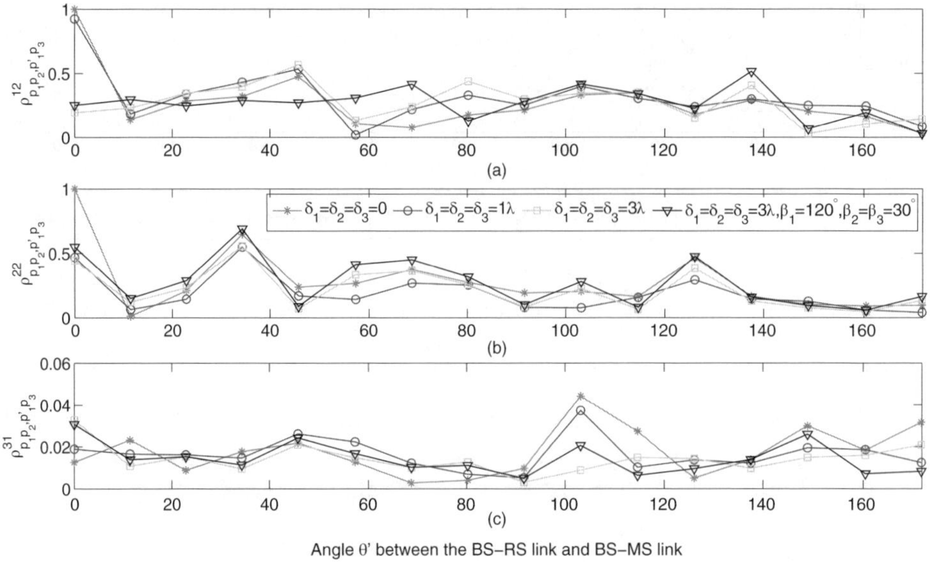

Figure 9.34 Absolute values of spatial correlation functions between the BS–RS link and BS–MS link for (a) the second single-bounced component $\rho^{12}_{p_3p_2,p'_3p_1}$; (b) the second double-bounced component $\rho^{22}_{p_3p_2,p'_3p_1}$; and (c) the triple-bounced component $\rho^{31}_{p_3p_2,p'_3p_1}$ with different values of parameters δ_g and β_g $(g = 1, 2, 3)$

multi-link spatial correlation increases with the increase of the environment parameter k_g, with the decrease of the size of the local scattering area, with the decrease of the value i for i-bounced rays, and/or with the decrease of the antenna spacing δ_g. Using these observations and conclusions, we will now investigate the spatial correlation properties of the proposed cooperative MIMO GBSM in Eq. (9.145) and thus validate the utility of the proposed cooperative MIMO channel model. Without loss of any generality, the outdoor macrocell MS cooperation scenario and indoor MS cooperation scenario are chosen for further investigation. As shown in Figure 9.35, three different LSD conditions are considered, with parameters $\delta_3 = \delta_2 = \delta_1 = 3\lambda$; that is, high LSD, low LSD, and mixed LSD.

For the outdoor macrocell MS cooperation scenario, the BS is free of scatterers and the RS actually represents the other MS, denoted as MS2. Therefore, the energy-related parameters related to the local scatterers around the BS are equal to zero: $\eta^{13}_{p_3p_2} = \eta^{13}_{p'_3p_1} = \eta^{21}_{p_3p_2} = \eta^{21}_{p'_3p_1} = \eta^{22}_{p_3p_2} = \eta^{22}_{p'_3p_1} = \eta^{31}_{p_3p_2} = \eta^{31}_{p'_3p_1} = 0$. We also assume $D_1 = D_3 = 1500$m and $K_{p_3p_2} = K_{p'_3p_1} = 0$ due to the large distance between BS, MS, and MS2. Considering the basic criterion of setting key model parameters expressed in Section 9.5.2, we choose the other energy-related parameters as:

- $\eta^{11}_{p_3p_2} = \eta^{11}_{p'_3p_1} = \eta^{12}_{p_3p_2} = \eta^{12}_{p'_3p_1} = 0.05$ and $\eta^{23}_{p_3p_2} = \eta^{23}_{p'_3p_1} = 0.9$ for high LSD, and $\eta^{11}_{p_3p_2} = \eta^{11}_{p'_3p_1} = \eta^{12}_{p_3p_2} = \eta^{12}_{p'_3p_1} = 0.2$
- $\eta^{23}_{p_3p_2} = \eta^{23}_{p'_3p_1} = 0.6$ for low LSD.

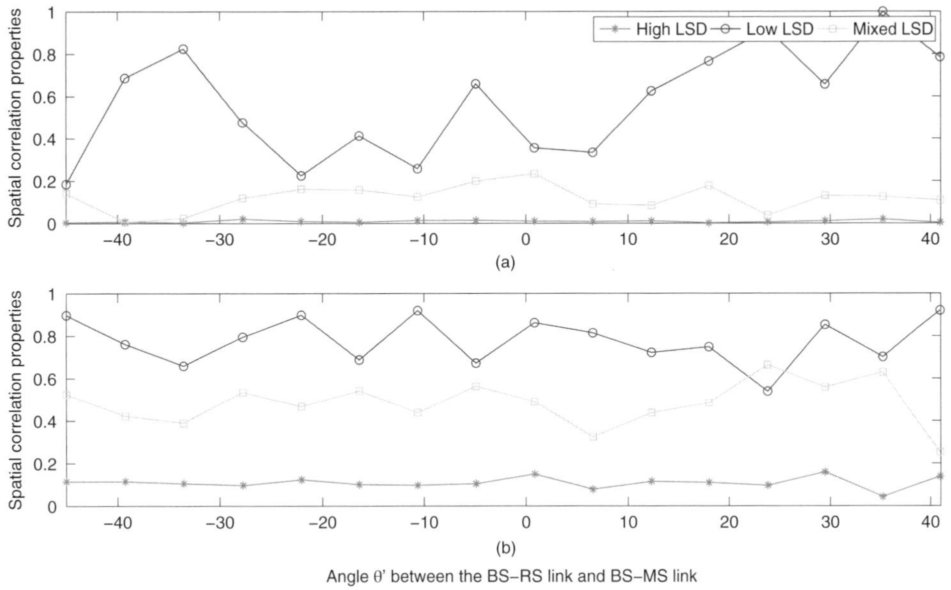

Figure 9.35 Absolute values of spatial correlation functions between the BS–RS link and BS–MS link for (a) the outdoor macrocell MS cooperation scenario and (b) the indoor MS cooperation scenario with different LSDs

For the mixed LSD case, we assume that the local scattering area around the MS presents low LSD and the one around MS2 shows high LSD, and thus assume energy-related parameters of $\eta^{11}_{p'_3p_1} = \eta^{12}_{p'_3p_1} = 0.2$, $\eta^{23}_{p'_3p_1} = 0.6$, $\eta^{11}_{p_3p_2} = \eta^{12}_{p_3p_2} = 0.1$, and $\eta^{23}_{p_3p_2} = 0.8$. In general, the higher the LSD, the larger and more distributed the local scattering area, and thereby the smaller the value of k_g and the larger the value of $R_{2n_g} - R_{1n_g}$ $(g = 1, 2, 3)$. Therefore, we have the following environment parameters:

- $k_1 = k_2 = 1$, $R_{1n_1} = R_{1n_2} = 5$ m, and $R_{2n_1} = R_{2n_2} = 200$ m for high LSD
- $k_1 = k_2 = 10$, $R_{1n_1} = R_{1n_2} = 5$m, and $R_{2n_1} = R_{2n_2} = 20$m for low LSD
- $k_1 = 10$, $k_2 = 2$, $\mu_1 = 60°$, $\mu_2 = 120°$, $R_{1n_1} = R_{1n_2} = 5$m, $R_{2n_1} = 20$m, and $R_{2n_2} = 100$m for mixed LSD.

Similarly, for indoor MS cooperation scenario, the RS actually represents the other MS, symbolled as MS2. Considering the small distance among BS, MS, and MS2, we assume $D_1 = D_3 = 50$ m. For mixed LSD case, we consider the scenario that the local scattering areas around BS and MS present low LSDs and the one around MS2 shows high LSD. The key model parameters are chosen as follows:

- $K_{p_3p_2} = K_{p'_3p_1} = 0.1$, $\eta^{11}_{p_3p_2} = \eta^{11}_{p'_3p_1} = \eta^{12}_{p_3p_2} = \eta^{12}_{p'_3p_1} = \eta^{13}_{p_3p_2} = \eta^{13}_{p'_3p_1} = 0.05$, $\eta^{21}_{p_3p_2} = \eta^{21}_{p'_3p_1} = \eta^{22}_{p_3p_2} = \eta^{22}_{p'_3p_1} = \eta^{23}_{p_3p_2} = \eta^{23}_{p'_3p_1} = 0.2$, and $\eta^{31}_{p_3p_2} = \eta^{31}_{p'_3p_1} = 0.25$ for high LSD
- $K_{p_3p_2} = K_{p'_3p_1} = 3$, $\eta^{11}_{p_3p_2} = \eta^{11}_{p'_3p_1} = \eta^{12}_{p_3p_2} = \eta^{12}_{p'_3p_1} = \eta^{13}_{p_3p_2} = \eta^{13}_{p'_3p_1} = 0.3$, and $\eta^{21}_{p_3p_2} = \eta^{21}_{p'_3p_1} = \eta^{22}_{p_3p_2} = \eta^{22}_{p'_3p_1} = \eta^{23}_{p_3p_2} = \eta^{23}_{p'_3p_1} = \eta^{31}_{p_3p_2} = \eta^{31}_{p'_3p_1} = 0.025$ for low LSD

- $K_{p_3p_2} = 0.5$, $K_{p_3'p_1} = 2.5$, $\eta^{11}_{p_3'p_1} = \eta^{12}_{p_3'p_1} = \eta^{13}_{p_3'p_1} = 0.25$, $\eta^{21}_{p_3'p_1} = 0.1$, $\eta^{22}_{p_3'p_1} = \eta^{23}_{p_3'p_1} = \eta^{31}_{p_3'p_1} = 0.05$, $\eta^{11}_{p_3p_2} = \eta^{12}_{p_3p_2} = \eta^{13}_{p_3p_2} = 0.05$, $\eta^{21}_{p_3p_2} = \eta^{23}_{p_3p_2} = 0.3$, $\eta^{22}_{p_3p_2} = 0.1$, and $\eta^{31}_{p_3p_2} = 0.15$ for mixed LSD.

The environment parameters are selected as:

- $k_1 = k_2 = k_3 = 1$, $R_{1n_1} = R_{1n_2} = R_{1n_3} = 2$m, and $R_{2n_1} = R_{2n_2} = R_{2n_3} = 25$m for high LSD
- $k_1 = k_2 = k_3 = 10$, $R_{1n_1} = R_{1n_2} = R_{1n_3} = 2$ m, and $R_{2n_1} = R_{2n_2} = R_{2n_3} = 8$ m for low LSD
- $k_1 = 6$, $k_2 = 2$, $k_3 = 15$, $\mu_1 = 60°$, $\mu_2 = 120°$, $\mu_3 = 240°$, $R_{1n_1} = R_{1n_2} = R_{1n_3} = 2$ m, $R_{2n_1} = 12$m, $R_{2n_2} = 20$ m, and $R_{2n_3} = 5$ m for mixed LSD.

Figure 9.35 shows that the LSD significantly affects the spatial correlation properties. It is observed that the higher the LSD, the lower the spatial correlation properties. This is because with a higher LSD, the local scattering area is larger and more distributed, resulting in the received power coming from many different directions. Figure 9.35 also illustrates that the indoor MS cooperation scenario has larger spatial correlation than the outdoor macrocell MS cooperation scenario. This results from the appearance of a LoS component in the indoor MS cooperation scenario due to the smaller distance between BS, MS, and MS2. Therefore, we can conclude that a high multi-link spatial correlation normally appears in a scenario with lower LSDs and LoS components. More importantly, from the observation in Figure 9.31 and based on the constraints of the energy-related parameters for cooperative scenarios with different LSDs, we know that with a higher LSD, the multi-bounced components bear more energy than the single-bounced ones and thus the corresponding cooperative environment has a higher possibility of a high degree of link heterogeneity (or a low degree of environment similarity). Therefore, the conclusion based on Figure 9.35 is consistent with our intuition that a low degree of environment similarity results in low multi-link spatial correlation.

9.5.5 Summary

Section 9.5 has proposed a novel unified cooperative MIMO channel model framework, from which a novel GBSM has been derived. The proposed multiple-ring GBSM is generic and adaptable to a wide variety of cooperative MIMO propagation scenarios. More importantly, the proposed GBSM is the first model that is capable of supporting investigations into the impact of LSD on channel statistics. From the proposed GBSM, the multi-link spatial correlations have been derived and numerically evaluated. Numerical results have shown that the antenna element spacings, environment parameters, and LSD have great impacts on the multi-link spatial correlation properties. It has also been demonstrated that high multi-link spatial correlation may exist if a cooperative communication system has a relatively narrow bandwidth and the underlying propagation environments have low LSDs and LoS components.

Bibliography

3GPP 2003 TR 25.996: Spatial channel model for multiple input multiple output (mimo) simulations (rel. 6). Technical report.

Abdi A and Kaveh M 2002 A space-time correlation model for multielement antenna systems in mobile fading channels. *IEEE Journal on Selected Areas in Communications* **20**(3), 550–560.

Abdi A, Barger J and Kaveh M 2002 A parametric model for the distribution of the angle of arrival and the associated correlation function and power spectrum at the mobile station. *IEEE Transactions on Vehicular Technology* **51**(3), 425–434.

Acosta G, Tokuda K and Ingram M 2004 Measured joint Doppler-delay power profiles for vehicle-to-vehicle communications at 2.4 GHz *Global Telecommunications Conference, 2004. GLOBECOM '04. IEEE*, vol. 6, pp. 3813–3817.

Acosta-Marum G and Brunelli D 2007 Six time- and frequency- selective empirical channel models for vehicular wireless LANs. *IEEE Vehicular Technology Magazine* **2**(4), 4–11.

Adachi F, Feeney M, Williamsona A and Parsons J 1986 Cross-correlation between the envelopes of 900 MHz signals received at a mobile radio base station site. *IEE Proceedings – Communications, Radar & Signal Processing* **133**, 506–512.

Ahumada L, Feick R, Valenzuela R and Morales C 2005 Measurement and characterization of the temporal behavior of fixed wireless links. *IEEE Transactions on Vehicular Technology* **54**(6), 1913–1922.

Akki A 1994 Statistical properties of mobile-to-mobile land communication channels. *IEEE Transactions on Vehicular Technology* **43**(4), 826–831.

Akki A and Haber F 1986 A statistical model of mobile-to-mobile land communication channel. *IEEE Transactions on Vehicular Technology* **35**(1), 2–7.

Almers P, Bonek E, Burr A, Czink N, Debbah M, Degli-Esposti, Hofstetter H, Kyösti P, Laurenson D, Matz G, Molisch AF, Oestges C and Özcelik H 2007 Survey of channel and radio propagation models for wireless MIMO systems. *EURASIP Journal on Wireless Communications and Networking* **Article ID 19070**, 1–19.

Bolcskei H, Borgmann M and Paulraj A 2003 Impact of the propagation environment on the performance of space-frequency coded MIMO-OFDM. *IEEE Journal on Selected Areas in Communications* **21**(3), 427–439.

Byers G and Takawira F 2004 Spatially and temporally correlated MIMO channels: modeling and capacity analysis. *IEEE Transactions on Vehicular Technology* **53**(3), 634–643.

Chen P and Li H 2007 Modeling and applications of space-time correlation for MIMO fading signals. *IEEE Transactions on Vehicular Technology* **56**(4), 1580–1590.

Chen TA, Fitz M, Kuo WY, Zoltowski M and Grimm J 2000 A space-time model for frequency nonselective Rayleigh fading channels with applications to space-time modems. *IEEE Journal on Selected Areas in Communications* **18**(7), 1175–1190.

Cheng L, Henty B, Stancil D, Bai F and Mudalige P 2007 Mobile vehicle-to-vehicle narrow-band channel measurement and characterization of the 5.9 GHz dedicated short range communication (DSRC) frequency band. *IEEE Journal on Selected Areas in Communications* **25**(8), 1501–1516.

Cheng X, Wang CX, Laurenson D, Salous S and Vasilakos A 2009 An adaptive geometry-based stochastic model for non-isotropic MIMO mobile-to-mobile channels. *IEEE Transactions on Wireless Communications* **8**(9), 4824–4835.

Chong CC, Tan CM, Laurenson D, McLaughlin S, Beach M and Nix A 2003 A new statistical wideband spatio-temporal channel model for 5-GHz band WLAN systems. *IEEE Journal on Selected Areas in Communications* **21**(2), 139–150.

Czink N, Bandemer B and Vilar G 2008 Spatial separation of multi-user MIMO channels *COST 2100 TD (08) 622, Lille.*

de Leon CGD and Patzold M 2007 Sum-of-sinusoids-based simulation of flat fading wireless propagation channels under non-isotropic scattering conditions *Global Telecommunications Conference, 2007. GLOBECOM '07. IEEE*, pp. 3842–3846.

Foersler J 2003 P802.15-02/490r1-SG3a: Channel modeling sub-committee report (final). Technical report, IEEE.

Fulghum T, Molnar K, Furht B and Ahson SA 1998 The Jakes fading model incorporating angular spread for a disk of scatterers *Vehicular Technology Conference, 1998. VTC 98. 48th IEEE*, vol. 1, pp. 489–493.

Gradshteyn IS and Ryzhik IM *Table of Integrals, Series, and Products*. London, UK: Academic press, 6th edn., 2000.

Hogstad BO, Pätzold M, Chopra A, Kim D and Yeom KB (2005, December). A wideband MIMO channel simulation model based on the geometrical elliptical scattering model. In Proc. 15th Meeting of the Wireless World Research Forum (WWRF).

IEEE 2007a Multi-hop relay system evaluation methodology (channel model and performance metric). Technical report, IEEE 802.16 Broadband Wireless Access Working Group.

IEEE 2007b P802.11p/D2.01: Standard for wireless local area networks providing wireless communications while in vehicular environment. Technical report, IEEE.

Jakes WC 1994 *Microwave Mobile Communications* 2nd edn. Wiley-IEEE Press.

Kalkan M and Clarke R 1997 Prediction of the space-frequency correlation function for base station diversity reception. *IEEE Transactions on Vehicular Technology* **46**(1), 176–184.

Kaltenberger F, Gesbert D, Knopp R and Kountouris M 2008 Correlation and capacity of measured multi-user MIMO channels *Personal, Indoor and Mobile Radio Communications, 2008. PIMRC 2008. IEEE 19th International Symposium on*, pp. 1–5 IEEE.

Karedal J, Tufvesson F, Czink N, Paier A, Dumard C, Zemen T, Mecklenbrauker C and Molisch A 2009 A geometry-based stochastic MIMO model for vehicle-to-vehicle communications. *IEEE Transactions on Wireless Communications* **8**(7), 3646–3657.

Kosch T and Franz W 2005 Technical concept and prerequisites of car-to-car communication *The 5th European Congress and Exhibition on Intelligent Transport Systems and Services (ITS2005).*

Kyösti P, Meinilä J, Hentilä L, Zhao X, Jämsä T, Schneider C, Narandzić M, Milojević M, Hong A, Ylitalo J, Holappa VM, Alatossava M, Bultitude R, de Jong Y and Rautiainen T 2007 WINNER II Channel Models D1.1.2 V1.1.

Latinovic Z, Abdi A and Bar-Ness Y 2003 A wideband space-time model for MIMO mobile fading channels *Wireless Communications and Networking, 2003*, vol. 1, pp. 338–342.

Lee W 1970 Level crossing rates of an equal-gain predetection diversity combiner. *IEEE Transactions on Communication Technology* **18**(4), 417–426.

Matolak D 2008 Channel modeling for vehicle-to-vehicle communications. *IEEE Communications Magazine* **46**(5), 76–83.

Maurer J, Fugen T and Wiesbeck W 2002 Narrow-band measurement and analysis of the inter-vehicle transmission channel at 5.2 GHz *Vehicular Technology Conference, 2002. VTC Spring 2002. IEEE 55th*, vol. 3, pp. 1274–1278.

Medbo J, Berg JE and Harrysson F 2004 Temporal radio channel variations with stationary terminal *Vehicular Technology Conference, 2004. VTC2004-Fall. 2004 IEEE 60th*, vol. 1, pp. 91–95.

Oestges C, Czink N, Bandemer B, Castiglione P, Kaltenberger F and Paulraj A 2010 Experimental characterization and modeling of outdoor-to-indoor and indoor-to-indoor distributed channels. *IEEE Transactions on Vehicular Technology* **PP**(99), 1–1.

Paier A, Karedal J, Czink N, Dumard CC, Zemen T, Tufvesson F, Molisch AF and Mecklenbräuker CF 2009 Characterization of vehicle-to-vehicle radio channels from measurements at 5.2 GHz. *Wireless Personal Communications* **50**(1), 19–32.

Papoulis A and Pillai SU 2002 *Probability, Random Variables and Stochastic Processes* 4th edn. McGraw-Hill, NJ.

Patel C, Stuber S and Pratt T 2005 Simulation of Rayleigh-faded mobile-to-mobile communication channels. *IEEE Transactions on Communications* **53**(10), 1773.

Pätzold M 2002 *Mobile Fading Channels*. John Wiley and Sons.

Pätzold M and Hogstad B 2006 A wideband space-time MIMO channel simulator based on the geometrical one-ring model *IEEE 64th Vehicular Technology Conference, 2006. VTC-2006 Fall*, pp. 1–6.

Pätzold M and Laue F 1998 Statistical properties of Jakes' fading channel simulator. *Proc. IEEE VTC'98-Spring* **2**, 712–718.

Patzold M and Youssef N 2001 Modeling and simulation of direction-selective and frequency-selective mobile radio channels. *International Journal of Electronics and Communications* **55**(6), 433–442.

Pätzold M, Hogstad B and Youssef N 2008 Modeling, analysis, and simulation of MIMO mobile-to-mobile fading channels. *IEEE Transactions on Wireless Communications* **7**(2), 510–520.

Pätzold M, Killat U and Laue F 1998 An extended Suzuki model for land mobile satellite channels and its statistical properties. *IEEE Transactions on Vehicular Technology* **47**(2), 617–630.

Rice SO 1958 Distribution of the duration of fades in radio transmission: Gaussian noise model. *Bell System Technical Journal* **37**, 581–635.

Salz J and Winters J 1994 Effect of fading correlation on adaptive arrays in digital mobile radio. *IEEE Transactions on Vehicular Technology* **43**(4), 1049–1057.

Schumacher L, Pedersen K and Mogensen P 2002 From antenna spacings to theoretical capacities – guidelines for simulating MIMO systems *Personal, Indoor and Mobile Radio Communications, 2002. The 13th IEEE International Symposium on*, vol. 2, pp. 587–592.

Sen I and Matolak D 2008 Vehicle-vehicle channel models for the 5-GHz band. *IEEE Transactions on Intelligent Transportation Systems* **9**(2), 235–245.

Soma P, Baum D, Erceg V, Krishnamoorthy R and Paulraj A 2002 Analysis and modeling of multiple-input multiple-output (MIMO) radio channel based on outdoor measurements conducted at 2.5 GHz for fixed BWA applications *Communications, 2002. ICC 2002. IEEE International Conference on*, vol. 1, pp. 272–276.

Stüber GL 2001 *Principles of Mobile Communications* 2nd edn. Kluwer Academic.

Wang CX, Hong X, Ge X, Cheng X, Zhang G and Thompson J 2010 Cooperative MIMO channel models: A survey. *IEEE Communications Magazine* **48**(2), 80–87.

Wang LC, Liu WC and Cheng YH 2009 Statistical analysis of a mobile-to-mobile Rician fading channel model. *IEEE Transactions on Vehicular Technology* **58**(1), 32–38.

Wang S, Abdi A, Salo J, El-Sallabi H, Wallace J, Vainikainen P and Jensen M 2007 Time-varying MIMO channels: Parametric statistical modeling and experimental results. *IEEE Transactions on Vehicular Technology* **56**(4), 1949–1963.

Wang Y and Zoubir A 2007 Some new techniques of localization of spatially distributed source *Signals, Systems and Computers, 2007. ACSSC 2007*, pp. 1807–1811.

Xu W, Zekavat S and Tong H 2009 A novel spatially correlated multiuser MIMO channel modeling: Impact of surface roughness. *IEEE Transactions on Antennas and Propagation* **57**(8), 2429–2438.

Yin X 2010 Spatial cross-correlation of multilink propagation channels in amplify-and-forward relay systems *Proceedings of the 4th International ICST Workshop on Channel Measurement and Modeling (IwonCMM2010)*.

Youssef N, Wang CX and Pätzold M 2005 A study on the second order statistics of Nakagami-Hoyt mobile fading channels. *IEEE Transactions on Vehicular Technology* **54**(4), 1259–1265.

Zajic A and Stuber G 2006 A new simulation model for mobile-to-mobile Rayleigh fading channels *Wireless Communications and Networking Conference, 2006. WCNC 2006. IEEE*, vol. 3, pp. 1266–1270.

Zajic A and Stuber G 2008a Space-time correlated mobile-to-mobile channels: Modelling and simulation. *IEEE Transactions on Vehicular Technology* **57**(2), 715–726.

Zajic A and Stuber G 2008b Three-dimensional modeling, simulation, and capacity analysis of space–time correlated mobile-to-mobile channels. *IEEE Transactions on Vehicular Technology* **57**(4), 2042–2054.

Zajic A and Stuber G 2009 Three-dimensional modeling and simulation of wideband MIMO mobile-to-mobile channels. *IEEE Transactions on Wireless Communications* **8**(3), 1260–1275.

Zajic A, Stiiber G, Pratt T and Nguyen S 2008 Envelope level crossing rate and average fade duration in mobile-to-mobile fading channels *Communications, 2008. ICC '08. IEEE International Conference on*, pp. 4446–4450.

Zajic A, Stuber G, Pratt T and Nguyen S 2009 Wideband MIMO mobile-to-mobile channels: Geometry-based statistical modeling with experimental verification. *IEEE Transactions on Vehicular Technology* **58**(2), 517–534.

Zhang J, Dong D, Nie X, Liu G and Dong W 2009 Propagation characteristics of wideband relay channels in urban environment *Proceedings of The 3rd International Workshop on Broadband MIMO Channel Measurement and Modeling, August 25*, pp. 1–6.

Zheng Y and Xiao C 2002 Improved models for the generation of multiple uncorrelated Rayleigh fading waveforms. *IEEE Communications Letters* **6**(6), 256–258.

Appendix A

A.1 Influence of Neglecting Doppler Shift within the Sensing Periods

The objective function derived under the assumption that all parameters are known except Doppler frequency and the complex amplitude, is given in Eq. (5.45). This expression is derived without considering the phase shift caused by the Doppler frequency in each sensing period $T_{\rm sc}$. In this appendix, we show that neglecting this phase shift results in a decrease of the SNR by a level that is a function of the Doppler frequency. We show that the SNR decrease resulting is negligible provided the ratio between the sampling rate and the switching rate is sufficiently high.

The sounding signal $u(t)$ can be written as

$$u(t) = \sum_{k=0}^{K-1} a_k p(t - kT_{\rm p}) \tag{A.1}$$

where $a_k, k = 0, \ldots, K-1$ and $p(t)$ denote, respectively, the (possibly complex) sounding sequence of length K and the shaping pulse whose duration T_p is related to the duration of the burst signal T_a according to $T_a = KT_p$. Each shaping pulse is sampled with a sampling rate of N_s. So the pulse shape can be written as

$$p(t) = \sum_{s=0}^{N_s-1} p_s h(t - sT_s) \tag{A.2}$$

with T_s denoting the sample duration. Finally, we can rewrite $u(t)$ to be

$$u(t) = \sum_{k=0}^{K-1} \sum_{s=0}^{N_s-1} a_k p_s h(t - kT_{\rm p} - sT_s). \tag{A.3}$$

Propagation Channel Characterization, Parameter Estimation and Modelling for Wireless Communications, First Edition.
Xuefeng Yin and Xiang Cheng.

This expression indicates that $u(t)$ consists of a total of KN_s samples. For simplicity, we can write $u(t)$ as

$$u(t) = \sum_{n=1}^{KN_s} u(t_n) \tag{A.4}$$

with $u(t_n)$ denoting the envelope of the signal transmitted at t_n and $t_n = (n-1) \cdot T_s, n = 1, \ldots, KN_s$.

Inserting Eq. (A.4) into Eq. (5.45), we can rewrite the integral operation as a summation over discrete samples:

$$\begin{aligned} z(\nu; y) = &\sum_{m_2=1}^{M_2} \sum_{m_1=1}^{M_1} \tilde{c}_{1,m_1}(\mathbf{\Omega}'_1)^* \tilde{c}_{2,m_2}(\mathbf{\Omega}'_2)^* \sum_{i=1}^{I} \exp\{-j2\pi\nu t_{i,m_2,m_1}\} \\ &\cdot \sum_{n=1}^{KN_s} u(t_n - \tau)^* y(t_{i,m_2,m_1} + t_n) \end{aligned} \tag{A.5}$$

The signal $u(t)$ has power P_u. The observed signal $y(t)$ at time $t = t_{i,m_2,m_1} + t_n$ can be described according to

$$\begin{aligned} y(t_{i,m_2,m_1} + t_n) = &\alpha_\ell c_{1,m_1}(\mathbf{\Omega}'_1) c_{2,m_2}(\mathbf{\Omega}'_2) \exp\{j2\pi\nu(t_{i,m_2,m_1} + t_n)\} u(t_n - \tau) \\ &+ \sqrt{\frac{N_0}{2}} q_2(t_n + t_{i,m_2,m_1}) W(t_n + t_{i,m_2,m_1}) \end{aligned}$$

where ν' denotes the true Doppler frequency of the wave. Substituting the above expression into Eq. (A.5) yields

$$\begin{aligned} z(\nu; y) = &\sum_{m_2=1}^{M_2} \sum_{m_1=1}^{M_1} \alpha_\ell \tilde{c}_{1,m_1}(\mathbf{\Omega}'_1)^* c_{1,m_1}(\mathbf{\Omega}'_1) \tilde{c}_{2,m_2}(\mathbf{\Omega}'_2)^* c_{2,m_2}(\mathbf{\Omega}'_2) \sum_{i=1}^{I} \exp\{-j2\pi\nu t_{i,m_2,m_1}\} \\ &\cdot \sum_{n=1}^{KN_s} \exp\{j2\pi\nu' \cdot (t_{i,m_2,m_1} + t_n)\} \underbrace{u(t_n - \tau)^* u(t_n - \tau)}_{=P_u} \\ &+ \sum_{m_2=1}^{M_2} \sum_{m_1=1}^{M_1} \tilde{c}_{1,m_1}(\mathbf{\Omega}'_1)^* \tilde{c}_{2,m_2}(\mathbf{\Omega}'_2)^* \sum_{i=1}^{I} \exp\{-j2\pi\nu t_{i,m_2,m_1}\} \\ &\cdot \sum_{n=1}^{KN_s} u(t_n - \tau)^* \sqrt{\frac{N_0}{2}} q_2(t_n + t_{i,m_2,m_1}) W(t_n + t_i, m_2, m_1) \end{aligned}$$

Replacing $c_{1,m_1}(\mathbf{\Omega}'_1)$ by $\|\mathbf{c}_1(\mathbf{\Omega}'_1)\| \tilde{c}_{1,m_1}(\mathbf{\Omega}'_1)$, and $c_{2,m_2}(\mathbf{\Omega}'_2)$ by $\|\mathbf{c}_2(\mathbf{\Omega}'_2)\| \tilde{c}_{2,m_2}(\mathbf{\Omega}'_2)$, we obtain

$$\begin{aligned} z(\nu; y) = &\|\mathbf{c}_1(\mathbf{\Omega}'_1)\| \|\mathbf{c}_2(\mathbf{\Omega}'_2)\| \alpha_\ell \\ &\times \sum_{i=1}^{I} \sum_{m_2=1}^{M_2} \sum_{m_1=1}^{M_1} |\tilde{c}_{1,m_1}(\mathbf{\Omega}'_1)^*|^2 |\tilde{c}_{2,m_2}(\mathbf{\Omega}'_2)^*|^2 \exp\{-j2\pi\nu t_{i,m_2,m_1}\} \end{aligned}$$

$$\sum_{n=1}^{KN_s} \exp\{j2\pi\nu' \cdot (t_{i,m_2,m_1} + t_n)\} P_u + \sqrt{\frac{N_0}{2}} \sum_{m_2=1}^{M_2} \sum_{m_1=1}^{M_1} \tilde{c}_{1,m_1}(\mathbf{\Omega}_1')^* \tilde{c}_{2,m_2}(\mathbf{\Omega}_2')^*$$

$$\cdot \sum_{i=1}^{I} \exp\{-j2\pi\nu t_{i,m_2,m_1}\} \sum_{n=1}^{KN_s} u(t_n - \tau)^* W(t_n + t_i, m_2, m_1)$$

Under Assumption (A) in Section 5.5.2, namely that the field patterns of the Tx and Rx antenna elements are respectively identical at $\mathbf{\Omega}_1'$ and $\mathbf{\Omega}_2'$, $|\tilde{c}_{1,m_1}(\mathbf{\Omega}_1')^*|^2$ and $|\tilde{c}_{2,m_2}(\mathbf{\Omega}_2')^*|^2$ can be calculated to be $1/\|\mathbf{c}_1(\mathbf{\Omega}_1')\|^2$ and $1/\|\mathbf{c}_2(\mathbf{\Omega}_2')\|^2$, respectively. Applying this result in the above equation, we obtain

$$z(\nu; y) = \frac{\alpha_\ell}{\|\mathbf{c}_1(\mathbf{\Omega}_1')\|\|\mathbf{c}_2(\mathbf{\Omega}_2')\|} \sum_{i=1}^{I} \sum_{m_2=1}^{M_2} \sum_{m_1=1}^{M_1} \exp\{j2\pi(\nu_\ell' - \nu) t_{i,m_2,m_1}\} \sum_{n=1}^{KN_s} \exp\{j2\pi\nu_\ell' t_n\} P_u$$

$$+ \sqrt{\frac{N_0}{2}} \sum_{m_2=1}^{M_2} \sum_{m_1=1}^{M_1} \tilde{c}_{1,m_1}(\mathbf{\Omega}_1')^* \tilde{c}_{2,m_2}(\mathbf{\Omega}_2')^* \sum_{i=1}^{I} \exp\{-j2\pi\nu t_{i,m_2,m_1}\}$$

$$\cdot \sum_{n=1}^{KN_s} u(t_n - \tau)^* W(t_n + t_i, m_2, m_1)$$

Replacing t_n by $(n-1)T_s$, we obtain

$$z(\nu; y) = \frac{\alpha_\ell P_u}{\|\mathbf{c}_1(\mathbf{\Omega}_1')\|\|\mathbf{c}_2(\mathbf{\Omega}_2')\|} \left[\sum_{n=1}^{KN_s} \exp\{j2\pi\nu_\ell'(n-1)T_s\} \right.$$

$$\left. \cdot \sum_{i=1}^{I} \sum_{m_2=1}^{M_2} \sum_{m_1=1}^{M_1} \exp\{j2\pi(\nu_\ell' - \nu) t_{i,m_2,m_1}\} + V(\nu) \right]$$

Similar to the derivation in Appendix A.2, the noise term $V(\nu)$ can be calculated to be

$$V(\nu) = \frac{1}{\sqrt{\gamma_s}} \sum_{i,m_2,m_1,n} W_{i,m_2,m_1,n}' \exp\{-j2\pi\nu t_{i,m_2,m_1}\} \tag{A.6}$$

where $\gamma_s \doteq \frac{|\alpha_\ell|^2 P_u}{N_0}$ denotes the signal-to-noise ratio at each sample and $W_{i,m_2,m_1,n}'$ represents complex circularly symmetric white Gaussian noise with unit variance. Hence,

$$|z(\nu; y)| = \frac{|\alpha_\ell| P_u}{\|\mathbf{c}_1(\mathbf{\Omega}_1')\|\|\mathbf{c}_2(\mathbf{\Omega}_2')\|} \cdot \left| \left[\sum_{n=1}^{KN_s} \exp\{j2\pi\nu_\ell'(n-1)T_s\} \right. \right.$$

$$\sum_{i=1}^{I} \sum_{m_2=1}^{M_2} \sum_{m_1=1}^{M_1} \exp\{j2\pi(\nu_\ell' - \nu) t_{i,m_2,m_1}\}$$

$$\left. \left. + \frac{1}{\sqrt{\gamma_s}} \sum_{i,m_2,m_1,n} W_{i,m_2,m_1,n}' \exp\{-j2\pi\nu t_{i,m_2,m_1}\} \right] \right|$$

Noticing that

$$\sum_{n=1}^{KN_s} \exp\{j2\pi\nu'_\ell(n-1)T_s\} = \exp\{j\pi\nu'_\ell(KN_s-1)T_s\}\frac{\sin(\pi\nu'_\ell KN_sT_s)}{\sin(\pi\nu'_\ell T_s)}$$

$|z(\nu;y)|$ becomes

$$|z(\nu;y)| = \frac{|\alpha_\ell|P_u}{\|\boldsymbol{c}_1(\boldsymbol{\Omega}'_1)\|\|\boldsymbol{c}_2(\boldsymbol{\Omega}'_2)\|}\cdot\left|\left[\exp\{j\pi\nu'_\ell(KN_s-1)T_s\}\frac{\sin(\pi\nu'_\ell KN_sT_s)}{\sin(\pi\nu'_\ell T_s)}\right.\right.$$
$$\cdot\sum_{i=1}^{I}\sum_{m_2=1}^{M_2}\sum_{m_1=1}^{M_1}\exp\{j2\pi(\nu'_\ell-\nu)t_{i,m_2,m_1}\}$$
$$\left.\left.+\frac{1}{\sqrt{\gamma_s}}\sum_{i,m_2,m_1,n}W'_{i,m_2,m_1,n}\exp\{-j2\pi\nu t_{i,m_2,m_1}\}\right]\right|$$

We may rewrite $z(\nu;y)$ after some algebraic manipulations according to

$$|z(\nu;y)| = \frac{|\alpha_\ell|P_u}{\|\boldsymbol{c}_1(\boldsymbol{\Omega}'_1)\|\|\boldsymbol{c}_2(\boldsymbol{\Omega}'_2)\|}\cdot\left|\left[\left|\frac{\sin(\pi\nu'_\ell KN_sT_s)}{\sin(\pi\nu'_\ell T_s)}\right|\cdot\left|\frac{\sin(\pi(\nu'_\ell-\nu)IT_{\mathrm{cy}})}{\sin(\pi(\nu'_\ell-\nu)T_{\mathrm{cy}})}\right|\right.\right.$$
$$\cdot\left|\frac{\sin(\pi(\nu'_\ell-\nu)M_1T_{\mathrm{t}})}{\sin(\pi(\nu'_\ell-\nu)T_{\mathrm{t}})}\right|\cdot\left|\frac{\sin(\pi(\nu'_\ell-\nu)M_2T_{\mathrm{r}})}{\sin(\pi(\nu'_\ell-\nu)T_{\mathrm{r}})}\right|$$
$$\left.\left.+\frac{1}{\sqrt{\gamma_s}}\sum_{i,m_2,m_1,n}W''_{i,m_2,m_1,n}\exp\{-j2\pi\nu t_{i,m_2,m_1}\}\right]\right|$$
$$=\frac{|\alpha_\ell|P_uF_{\mathrm{n}}(\nu'_\ell)IM_2M_1}{\|\boldsymbol{c}_1(\boldsymbol{\Omega}'_1)\|\|\boldsymbol{c}_2(\boldsymbol{\Omega}'_2)\|}\cdot[F_{\mathrm{cy}}(\nu)\cdot F_{\mathrm{t}}(\nu)\cdot F_{\mathrm{r}}(\nu)$$
$$+\frac{1}{F_{\mathrm{n}}(\nu'_\ell)IM_2M_1\sqrt{\gamma_s}}\sum_{i,m_2,m_1,n}W''_{i,m_2,m_1,n}\exp\{-j2\pi\nu t_{i,m_2,m_1}\}] \tag{A.7}$$

where $F_{\mathrm{cy}}(\nu)$, $F_{\mathrm{t}}(\nu)$ and $F_{\mathrm{r}}(\nu)$ are given by Eq. (5.49),

$$F_{\mathrm{n}}(\nu'_\ell) \doteq \left|\frac{\sin(\pi\nu'_\ell KN_sT_{\mathrm{n}})}{\sin(\pi\nu'_\ell T_{\mathrm{n}})}\right|$$

and $W''_{i,m_2,m_1,n}$ is a complex random variable with same property as $W'_{i,m_2,m_1,n}$. We can recast Eq. (A.7) to be

$$|z(\nu;y)| = \frac{|\alpha_\ell|P_uF_{\mathrm{n}}(\nu'_\ell)IM_2M_1}{\|\boldsymbol{c}_1(\boldsymbol{\Omega}'_1)\|\|\boldsymbol{c}_2(\boldsymbol{\Omega}'_2)\|}\cdot\left[F_{\mathrm{cy}}(\nu)F_{\mathrm{t}}(\nu)F_{\mathrm{r}}(\nu)\right.$$
$$\left.+\frac{1}{F_{\mathrm{n}}(\nu'_\ell)IM_2M_1\sqrt{\gamma_s}}\sum_{i,m_2,m_1,n}W''_{i,m_2,m_1,n}\exp\{-j2\pi\nu t_{i,m_2,m_1}\}\right] \tag{A.8}$$

Table A.1 The values of β versus ν'_ℓ

ν'_ℓ in Hz	β in dB
0	0
24,509	−0.056
49,019	−0.225
73,529	−0.508
98,039	−0.912

The variance of the noise term in the above expression is calculated to be

$$\frac{1}{F_{\mathrm{n}}^2(\nu'_\ell)I^2M_2^2M_1^2\gamma_{\mathrm{s}}}IM_2M_1KN_s = \frac{(KN_s)^2}{F_{\mathrm{n}}^2(\nu'_\ell)IM_2M_1KN_s\gamma_{\mathrm{s}}}$$
$$= \frac{(KN_s)^2}{F_{\mathrm{n}}^2(\nu'_\ell)\gamma_{\mathrm{O}}}$$
$$= \frac{1}{\left(\frac{F_{\mathrm{n}}(\nu'_\ell)}{KN_s}\right)^2\gamma_{\mathrm{O}}}$$
$$= \frac{1}{\beta(\nu'_\ell)\gamma_{\mathrm{O}}}$$

where $\beta(\nu'_\ell) \doteq (\frac{F_{\mathrm{n}}(\nu'_\ell)}{KN_s})^2 = \left|\frac{\sin(\pi\nu'_\ell KN_sT_n)}{\sin(\pi\nu'_\ell T_n)KN_s}\right|^2$ represents the factor with respect to ν'_ℓ, and γ_{O} denotes the signal-to-noise ratio at the output of the estimation scheme.

Finally, Eq. (A.8) is of the form

$$|z(\nu;y)| = \frac{|\alpha_\ell|P_uF_{\mathrm{n}}(\nu'_\ell)IM_2M_1}{\|\mathbf{c}_1(\mathbf{\Omega}'_1)\|\|\mathbf{c}_2(\mathbf{\Omega}'_2)\|}\cdot F_{\mathrm{n}}(\nu'_\ell)\cdot\left[F_{\mathrm{cy}}(\nu)F_{\mathrm{t}}(\nu)F_{\mathrm{r}}(\nu) + \frac{1}{\sqrt{\beta(\nu'_\ell)\gamma_{\mathrm{O}}}}W\right] \quad \text{(A.9)}$$

The function $\beta(\nu'_\ell)$ decreases when ν'_ℓ increases; in other words, the output SNR γ_{O} is lowered by $|10\log 10(\beta(\nu'_\ell))|$ dB. Let us take an example with the parameters given in Table 5.3. The values of $\beta(\nu'_\ell)$ under different values of ν'_ℓ are shown in Table A.1. The valid Doppler frequency range according to the intracycle observations T_r is calculated to be $[-98{,}039$ Hz, $+98{,}039$ Hz$]$. From the table we observe that $\beta(\nu'_\ell)$ has its lowest value when the true Doppler frequency equals the highest valid value for estimation.

A.2 Simplification of the Noise Component in an Objective Function

The noise term arising in the z function $V(\nu)$ can be written as

$$V(\nu) = \frac{|\mathbf{c}_1(\mathbf{\Omega}'_1)\|\mathbf{c}_2(\mathbf{\Omega}'_2)|}{P_uT_{\mathrm{sc}}}\sum_{i=1}^{I}\sum_{m_2=1}^{M_2}\sum_{m_1=1}^{M_1}\tilde{c}_{1,m_1}(\mathbf{\Omega}'_1)^*\tilde{c}_{2,m_2}(\mathbf{\Omega}'_2)^*$$
$$\exp(-j2\pi\nu t_{i,m_2,m_1})\sqrt{\frac{N_0}{2}\frac{1}{\alpha_1}}\int_0^{T_{\mathrm{sc}}}u(t-\tau)^*q_2(t)W(t+t_{i,m_2,m_1})\mathrm{d}t$$

Denoting the integral by W'_{i,m_2,m_1}, where W'_{i,m_2,m_1} is a complex circularly Gaussian random variable with variance $2P_uT_{sc}$, $V(\nu)$ can be recast as

$$V(\nu) = \frac{|\boldsymbol{c}_1(\boldsymbol{\Omega}'_1)\|\boldsymbol{c}_2(\boldsymbol{\Omega}'_2)|}{P_uT_{sc}} \sum_{i=1}^{I} \sum_{m_2=1}^{M_2} \sum_{m_1=1}^{M_1} \tilde{c}_{1,m_1}(\boldsymbol{\Omega}'_1)^* \tilde{c}_{2,m_2}(\boldsymbol{\Omega}'_2)^*$$

$$\exp(-j2\pi\nu t_{i,m_2,m_1})\sqrt{\frac{N_0}{2}\frac{1}{\alpha_1}}W'_{i,m_2,m_1} \tag{A.10}$$

The right-hand side in Eq. (A.10) can be written as

$$V(\nu) = \sum_{i=1}^{I} \sum_{m_2=1}^{M_2} \sum_{m_1=1}^{M_1} \sqrt{\frac{N_0}{2}} \frac{\|\boldsymbol{c}_1(\boldsymbol{\Omega}'_1)\|\|\boldsymbol{c}_2(\boldsymbol{\Omega}'_2)\|}{P_uT_{sc}\alpha_\ell} \tilde{c}_{1,m_1}(\boldsymbol{\Omega}'_1)^* \tilde{c}_{2,m_2}(\boldsymbol{\Omega}'_2)^*$$

$$W'_{i,m_2,m_1}\exp(-j2\pi\nu t_{i,m_2,m_1}) \tag{A.11}$$

The individual element in the summation $V_{i,m_2,m_1}(\nu)$ reads

$$V_{i,m_2,m_1}(\nu) \doteq \sqrt{\frac{N_0}{2}} \frac{\|\boldsymbol{c}_1(\boldsymbol{\Omega}'_1)\|\|\boldsymbol{c}_2(\boldsymbol{\Omega}'_2)\|}{P_uT_{sc}\alpha_\ell} \tilde{c}_{1,m_1}(\boldsymbol{\Omega}'_1)^* \tilde{c}_{2,m_2}(\boldsymbol{\Omega}'_2)^*$$
$$W'_{i,m_2,m_1}\exp(-j2\pi\nu t_{i,m_2,m_1})$$

$V(\nu)$ is obtained as

$$V(\nu) = \sum_{i=1}^{I} \sum_{m_2=1}^{M_2} \sum_{m_1=1}^{M_1} W''_{i,m_2,m_1} \cdot \exp(-j2\pi\nu t_{i,m_2,m_1})$$

where

$$W''_{i,m_2,m_1} \doteq \sqrt{\frac{N_0}{2}} \frac{\|\boldsymbol{c}_1(\boldsymbol{\Omega}'_1)\|\|\boldsymbol{c}_2(\boldsymbol{\Omega}'_2)\|}{P_uT_{sc}\alpha_\ell} \tilde{c}_{1,m_1}(\boldsymbol{\Omega}'_1)^* \tilde{c}_{2,m_2}(\boldsymbol{\Omega}'_2)^* W'_{i,m_2,m_1}$$

The variance of W''_{i,m_2,m_1} can be calculated according to

$$\begin{aligned}
\mathrm{Var}(W''_{i,m_2,m_1}) &= \left|\sqrt{\frac{N_0}{2}} \frac{\|\boldsymbol{c}_1(\boldsymbol{\Omega}'_1)\|\|\boldsymbol{c}_2(\boldsymbol{\Omega}'_2)\|}{P_uT_{sc}\alpha_\ell} \tilde{c}_{1,m_1}(\boldsymbol{\Omega}'_1)^* \tilde{c}_{2,m_2}(\boldsymbol{\Omega}'_2)^*\right|^2 \cdot \mathrm{Var}(W'_{i,m_2,m_1}) \\
&= \frac{N_0}{2} \frac{\|\boldsymbol{c}_1(\boldsymbol{\Omega}'_1)||^2\|\boldsymbol{c}_2(\boldsymbol{\Omega}'_2)||^2}{P_u^2T_{sc}^2|\alpha_\ell|^2} |\tilde{c}_{1,m_1}(\boldsymbol{\Omega}'_1)^*|^2 |\tilde{c}_{2,m_2}(\boldsymbol{\Omega}'_2)^*|^2 \cdot 2P_uT_{sc} \\
&= \frac{N_0}{2} \frac{\|\boldsymbol{c}_1(\boldsymbol{\Omega}'_1)||^2\|\boldsymbol{c}_2(\boldsymbol{\Omega}'_2)||^2}{P_u^2T_{sc}^2 \parallel \alpha_\ell|^2} \frac{1}{\|\boldsymbol{c}_1(\boldsymbol{\Omega}'_1)||^2} \frac{1}{\|\boldsymbol{c}_2(\boldsymbol{\Omega}'_2)||^2} \cdot 2P_uT_{sc} \\
&= \frac{N_0}{P_uT_{sc}|\alpha_\ell|^2}
\end{aligned} \tag{A.12}$$

Defining $\gamma_I \doteq P_u|\alpha_\ell|^2/(N_0/T_{sc})$, $V(\nu)$ can be written as

$$V(\nu) = \sum_{i,m_2,m_1} \frac{1}{\sqrt{\gamma_I}} N_{i,m_2,m_1} \cdot \exp(-j2\pi\nu t_{i,m_2,m_1})$$

where N_{i,m_2,m_1} denotes a complex circularly symmetrical Gaussian variable with unit variance.

A.3 Derivations of Equations (7.6)–(7.8)

In a single-SDS scenario, the ML estimate $\hat{\phi}$ of the NA is a possibly non-unique value of the argument of

$$\Lambda(\phi) \doteq \sum_{t=t_1}^{t_N} \| \tilde{\boldsymbol{c}}(\phi)^{\mathrm{H}} \boldsymbol{y}(t) \|_{\mathrm{F}}^2 \tag{A.13}$$

for which this function is maximum. In Eq. (A.13), $\tilde{\boldsymbol{c}}(\phi) \doteq \| \boldsymbol{c}(\phi) \|^{-1} \boldsymbol{c}(\phi)$, with $\| \cdot \|$ denoting the Euclidean norm of the vector given as an argument, is the normalized array response.

Hence, assuming that $\Lambda(\phi)$ fulfils the necessary regularity conditions, $\hat{\phi}$ satisfies the identity $\Lambda'(\hat{\phi}) = 0$ with $\Lambda'(\hat{\phi}) \doteq \frac{\mathrm{d}\Lambda(\phi)}{\mathrm{d}\phi}|_{\phi=\hat{\phi}}$. Let us write $\hat{\phi} = \bar{\phi} + \check{\phi}$, with $\bar{\phi}$ denoting the true NA and $\check{\phi}$ representing the estimation error. We assume that the estimation error $\check{\phi}$ is small, so that $\Lambda'(\hat{\phi})$ can be approximated by its first-order Taylor series expansion around $\bar{\phi}$; that is, $\Lambda'(\hat{\phi}) \approx \Lambda'(\bar{\phi}) + \check{\phi} \cdot \Lambda''(\bar{\phi})$, where $\Lambda''(\bar{\phi}) \doteq \frac{\mathrm{d}^2\Lambda(\phi)}{\mathrm{d}\phi^2}|_{\phi=\bar{\phi}}$. Inserting this approximation in the identity $\Lambda'(\hat{\phi}) = 0$ yields:

$$\Lambda'(\bar{\phi}) + \check{\phi} \cdot \Lambda''(\bar{\phi}) \approx 0 \tag{A.14}$$

In the sequel we consider the case where the observation is noiseless ($\sigma_w^2 = 0$ in Eq. (7.9)). Furthermore, the array elements are assumed to be isotropic. In this case, the following equalities hold

$$\begin{aligned}
\mathrm{Re}\{\boldsymbol{c}(\bar{\phi})^{\mathrm{H}} \boldsymbol{c}'(\bar{\phi})\} &= 0 \\
\mathrm{Re}\{(\boldsymbol{c}^{\mathrm{H}}(\bar{\phi}) \boldsymbol{c}'(\bar{\phi}))^2\} &= -\, |\tilde{\boldsymbol{c}}^{\mathrm{H}}(\bar{\phi}) \boldsymbol{c}'(\bar{\phi})|^2 \\
\mathrm{Re}\{\boldsymbol{c}^{\mathrm{H}}(\bar{\phi}) \boldsymbol{c}''(\bar{\phi})\} &= -\, \| \boldsymbol{c}'(\bar{\phi}) \|^2
\end{aligned}$$

where $\boldsymbol{c}''(\bar{\phi}) \doteq \frac{\mathrm{d}^2\boldsymbol{c}(\phi)}{\mathrm{d}\phi^2}|_{\phi=\bar{\phi}}$. Making use of the approximation $\boldsymbol{y}(t) \approx \boldsymbol{y}_{\mathrm{GAM}}(t)$ in Eq. (A.13), we calculate $\Lambda'(\bar{\phi})$ and $\Lambda''(\bar{\phi})$ in Eq. (A.14) to be:

$$\Lambda'(\bar{\phi}) = 2 \cdot \sum_{t=t_1}^{t_N} |\alpha(t)|^2 \cdot \mathrm{Re}\left\{ \frac{\beta(t)}{\alpha(t)} \right\} \cdot \left(\| \boldsymbol{c}'(\bar{\phi}) \|^2 + \mathrm{Re}\left\{ (\tilde{\boldsymbol{c}}(\bar{\phi})^{\mathrm{H}} \boldsymbol{c}'(\bar{\phi}))^2 \right\} \right) \tag{A.15}$$

$$\Lambda''(\bar{\phi}) = -2 \cdot \sum_{t=t_1}^{t_N} |\alpha(t)|^2 \cdot (\| \boldsymbol{c}'(\bar{\phi}) \|^2 + \mathrm{Re}\{(\tilde{\boldsymbol{c}}(\bar{\phi})^{\mathrm{H}} \boldsymbol{c}'(\bar{\phi}))^2\}) + g(t, \bar{\phi}) \tag{A.16}$$

where

$$\begin{aligned}
g(t, \bar{\phi}) \doteq& -\mathrm{Re}\{\alpha^*(t)\beta(t) \cdot [\boldsymbol{c}''(\bar{\phi})^{\mathrm{H}} \tilde{\boldsymbol{c}}'(\bar{\phi}) + 2\tilde{\boldsymbol{c}}(\bar{\phi})^{\mathrm{H}} \tilde{\boldsymbol{c}}'(\bar{\phi}) \| \boldsymbol{c}'(\bar{\phi}) \|^2 + \boldsymbol{c}(\bar{\phi})^{\mathrm{H}} \tilde{\boldsymbol{c}}(\bar{\phi}) \boldsymbol{c}''(\bar{\phi})^{\mathrm{H}} \tilde{\boldsymbol{c}}(\bar{\phi})]\} \\
& - |\beta(t)|^2 (\mathrm{Re}\{\boldsymbol{c}'(\bar{\phi})^{\mathrm{H}} \tilde{\boldsymbol{c}}(\bar{\phi}) \boldsymbol{c}''(\bar{\phi})^{\mathrm{H}} \tilde{\boldsymbol{c}}'(\bar{\phi})\} + \| \boldsymbol{c}'(\bar{\phi}) \|^4)
\end{aligned} \tag{A.17}$$

Under Assumptions (1)–(3) in Section 7.2.1, the random processes $\alpha^*(t)\beta(t)$ and $|\beta(t)|^2$ in Eq. (A.17) have expectation 0 and $\sigma_\alpha^2 \sigma_{\tilde{\phi}}^2$, respectively, and variance $4\sigma_\alpha^4 \sigma_{\tilde{\phi}}^2$ and $10\sigma_\alpha^4 \sigma_{\tilde{\phi}}^4$, respectively. Thus, in the case where $\sigma_{\breve{\phi}}$ is small, $g(t,\bar{\phi})$ is much smaller than the other terms in $\Lambda''(\bar{\phi})$. By inserting Eq. (A.16) into Eq. (A.14) and neglecting the term $g(t,\bar{\phi})$, $\Lambda''(\bar{\phi})$ can be approximated by

$$\Lambda''(\bar{\phi}) \approx \tilde{\Lambda}''(\bar{\phi}) \approx -2 \cdot \sum_{t=t_1}^{t_N} |\alpha(t)|^2 \cdot (\| \boldsymbol{c}'(\bar{\phi})\|^2 + \mathrm{Re}\{(\tilde{\boldsymbol{c}}(\bar{\phi})^{\mathrm{H}} \boldsymbol{c}'(\bar{\phi}))^2\}) \tag{A.18}$$

Making use of Eq. (A.15) and Eq. (A.18) in Eq. (A.14) we conclude that

$$\breve{\phi} \approx \breve{\phi} \doteq -\frac{\Lambda'(\bar{\phi})}{\tilde{\Lambda}''(\bar{\phi})} \tag{A.19}$$

Inserting Eq. (A.15) and Eq. (A.18) into Eq. (A.19) yields

$$\breve{\phi} = \frac{\sum_{t=t_1}^{t_N} |\alpha(t)|^2 \cdot \mathrm{Re}\left\{\frac{\beta(t)}{\alpha(t)}\right\}}{\sum_{t=t_1}^{t_N} |\alpha(t)|^2} \tag{A.20}$$

In the special case where $N = 1$, Eq. (A.20) reduces to

$$\breve{\phi} = \mathrm{Re}\left\{\frac{\beta(t)}{\alpha(t)}\right\} \tag{A.21}$$

It can be shown from Eq. (A.20) that the conditional expectation and conditional variance of $\breve{\phi}$ given $\boldsymbol{\alpha} = [\alpha(t_1), \ldots, \alpha(t_N)]$ read

$$\mathrm{E}[\breve{\phi}|\boldsymbol{\alpha}] = 0 \tag{A.22}$$

$$\mathrm{Var}[\breve{\phi}|\boldsymbol{\alpha}] = \frac{1}{2} \cdot \frac{\sigma_\beta^2}{\sum_{t=t_1}^{t_N} |\alpha(t)|^2} \tag{A.23}$$

Let us define $z \doteq \sum_{t=t_1}^{t_N} |\alpha(t)|^2$. The conditional PDF of $\breve{\phi}$ given z is of the form

$$f_{\breve{\phi}}(\phi|z) = \frac{\sqrt{z}}{\sqrt{\pi} \cdot \sigma_\beta} \cdot \exp\left\{-\frac{z}{\sigma_\beta^2} \cdot \phi^2\right\} \tag{A.24}$$

The random variable z is chi-squared distributed with $2N$ degrees of freedom. Its PDF reads

$$f(z) = \frac{1}{\Gamma(N)\sigma_\alpha^{2N}} \cdot z^{(N-1)} \cdot \exp\left\{-\frac{z}{\sigma_\alpha^2}\right\}, z > 0 \tag{A.25}$$

Combining Eq. (A.24) and Eq. (A.25) we obtain for the PDf of $\breve{\phi}$

$$\begin{aligned} f_{\breve{\phi}}(\phi) &= \int_0^\infty f_{\breve{\phi}}(\phi|z) f(z) \mathrm{d}z \\ &= \frac{\Gamma(N+\frac{1}{2})}{\sqrt{\pi}\Gamma(N)} \cdot \frac{1}{\sigma_{\breve{\phi}}} \cdot \frac{1}{(1+\frac{\breve{\phi}^2}{\sigma_{\breve{\phi}}^2})^{(N+\frac{1}{2})}} \end{aligned} \tag{A.26}$$

Making use of Eq. (A.26) we derive the variance of $\breve{\phi}$:

$$\begin{aligned}\mathrm{Var}[\breve{\phi}] &= \frac{\Gamma(N-1)}{2\Gamma(N)}\cdot\sigma_{\tilde{\phi}}^2 \\ &= \begin{cases}\infty, & N=1\\ \frac{1}{2(N-1)}\cdot\sigma_{\tilde{\phi}}^2, & N>1\end{cases}\end{aligned} \tag{A.27}$$

The conditional expectation of $|\breve{\phi}|$ given z reads

$$\mathrm{E}[|\breve{\phi}||z] = \frac{\sigma_\beta}{\sqrt{\pi\cdot z}}$$

From this result, the expectation of $|\breve{\phi}|$ is calculated to be

$$\begin{aligned}\mathrm{E}[|\breve{\phi}|] &= \int_0^\infty \frac{\sigma_\beta}{\sqrt{\pi\cdot z}} f(z)\mathrm{d}z \\ &= \frac{\Gamma(N-\frac{1}{2})}{\sqrt{\pi}\Gamma(N)}\cdot\sigma_{\tilde{\phi}}\end{aligned} \tag{A.28}$$

In the special case where $N=1$, Eq. (A.28) becomes

$$\mathrm{E}[|\breve{\phi}|] = \mathrm{E}\left[\left|\mathrm{Re}\left\{\frac{\beta(t)}{\alpha(t)}\right\}\right|\right] = \sigma_{\tilde{\phi}} \tag{A.29}$$

Making use of Eq. (A.19), the PDF, the variance and the absolute expectation of $\breve{\phi}$ can be approximated with the right-hand expressions in Eqs (A.26)–(A.28), respectively. The same approach can be used to approximate higher moments of $\breve{\phi}$.

A.4 Derivation of Eqs (4.20a) and (4.20b)

In this appendix, we first derive the time-average autocorrelation function $\tilde{\rho}_{\tilde{h}_i\tilde{h}_i}(\tau)$ of the in-phase component $\tilde{h}_i(t)$ Eq. (4.12) of the proposed deterministic simulation model:

$$\begin{aligned}\tilde{\rho}_{\tilde{h}_i\tilde{h}_i}(\tau) &= \langle\tilde{h}_i(t)\tilde{h}_i(t-\tau)\rangle = \lim_{T\to\infty}\frac{1}{2T}\int_{-T}^{T}\tilde{h}_i(t)\tilde{h}_i(t-\tau)\,dt \\ &= \lim_{T\to\infty}\frac{1}{2T}\int_{-T}^{T}\frac{1}{N_iM_i}\sum_{n_i,m_i=1}^{N_i,M_i}\sum_{n_i',m_i'=1}^{N_iM_i}\cos\Big[\tilde{\psi}_{n_im_i}+2\pi f_{T_{max}}t\cos(\tilde{\phi}_T^{m_i}-\gamma_T) \\ &\quad +2\pi f_{R_{max}}t\cos(\tilde{\phi}_R^{n_i}-\gamma_R)\Big]\cos\Big[\tilde{\psi}_{n_i'm_i'}+2\pi f_{T_{max}}(t-\tau)\cos(\tilde{\phi}_T^{m_i'}-\gamma_T) \\ &\quad +2\pi f_{R_{max}}(t-\tau)\cos(\tilde{\phi}_R^{n_i'}-\gamma_R)\Big]\,dt \\ &= \lim_{T\to\infty}\frac{1}{2T}\int_{-T}^{T}\frac{1}{N_iM_i}\sum_{n_i,m_i=1}^{N_i,M_i}\cos\Big[\tilde{\psi}_{n_im_i}+2\pi f_{T_{max}}t\cos(\tilde{\phi}_T^{m_i}-\gamma_T)+2\pi f_{R_{max}}t\end{aligned}$$

$$\times \cos(\tilde{\phi}_R^{n_i} - \gamma_R)\Big] \cos\Big[\tilde{\psi}_{n_i m_i} + 2\pi f_{T_{max}}(t-\tau)\cos(\tilde{\phi}_T^{m_i} - \gamma_T) + 2\pi f_{R_{max}}$$
$$\times (t-\tau)\cos(\tilde{\phi}_R^{n_i} - \gamma_R)\Big]\, dt$$
$$(\text{since } n_i \neq n_i' \text{ and/or } m_i \neq m_i', \tilde{\rho}_{\tilde{h}_i \tilde{h}_i}(\tau) = 0)$$
$$= \lim_{T\to\infty} \frac{1}{2T} \int_{-T}^{T} \frac{1}{2N_i M_i} \sum_{n_i, m_i = 1}^{N_i, M_i} \cos\Big[2\pi f_{T_{max}} \tau \cos(\tilde{\phi}_T^{m_i} - \gamma_T)$$
$$+ 2\pi f_{R_{max}} \tau \cos(\tilde{\phi}_R^{n_i} - \gamma_R) \Big]\, dt \tag{A.30}$$

The integral of Eq. (A.30) is trivial as the integrand does not contain the variable of integration. Therefore, the expression of $\tilde{\rho}_{\tilde{h}_i \tilde{h}_i}(\tau)$ in Eq. (4.20a) can be easily obtained from Eq. (A.30).

The derivation of the time-average cross-correlation function $\tilde{\rho}_{\tilde{h}_i \tilde{h}_q}(\tau)$ between the in-phase component $\tilde{h}_i(t)$ Eq. (4.12) and the quadrature component $\tilde{h}_q(t)$ Eq. (4.13) of the proposed deterministic simulation model is outlined as follows:

$$\tilde{\rho}_{\tilde{h}_i \tilde{h}_q}(\tau) = \left\langle \tilde{h}_i(t)\, \tilde{h}_q(t-\tau) \right\rangle = \lim_{T\to\infty} \frac{1}{2T} \int_{-T}^{T} \tilde{h}_i(t)\, \tilde{h}_q(t-\tau)\, dt$$
$$= \lim_{T\to\infty} \frac{1}{2T} \int_{-T}^{T} \frac{1}{\sqrt{N_i M_i}\sqrt{N_q M_q}} \sum_{n_i, m_i = 1}^{N_i, M_i} \sum_{n_q, m_q = 1}^{N_q, M_q}$$
$$\times \cos\Big[\tilde{\psi}_{n_i m_i} + 2\pi f_{T_{max}} t \cos(\tilde{\phi}_T^{m_i} - \gamma_T) + 2\pi f_{R_{max}} t \cos(\tilde{\phi}_R^{n_i} - \gamma_R)\Big]$$
$$\times \sin\Big[\tilde{\psi}_{n_q m_q} + 2\pi f_{T_{max}} (t-\tau) \cos(\tilde{\phi}_T^{m_q} - \gamma_T)$$
$$+ 2\pi f_{R_{max}} (t-\tau) \cos(\tilde{\phi}_R^{n_q} - \gamma_R)\Big]\, dt$$
$$= \begin{cases} -\lim_{T\to\infty} \dfrac{1}{2T} \displaystyle\int_{-T}^{T} \dfrac{1}{2NM} \sum_{n,m=1}^{N,M} \\ \quad \times \sin[2\pi f_{T_{max}} \tau \cos(\tilde{\phi}_T^{m} - \gamma_T) + 2\pi f_{R_{max}} \tau \cos(\tilde{\phi}_R^{n} - \gamma_R)]\, dt, \\ \qquad N_i = N_q = N \text{ and } M_i = M_q = M (Cases\ I \text{ and } III) \\ 0, \qquad N_i \neq N_q \text{ and } M_i \neq M_q (Case\ II) \end{cases} \tag{A.31}$$

where at the third equality of Eq. (A.31), setting $N_i \neq N_q$ and $M_i \neq M_q$ results in the integrand for *Case II* containing the variable of integration t, unlike the integrand for other cases.

A.5 Derivation of the CF $\hat{\rho}_{\hat{h}\hat{h}}(\tau)$

In this appendix, we derive the CF $\hat{\rho}_{\hat{h}\hat{h}}(\tau)$ for the stochastic simulation model in Eq. (4.21).

$$\hat{\rho}_{\hat{h}\hat{h}}(\tau) = E\left[\hat{h}(t)\,\hat{h}^*(t-\tau)\right]$$

$$= \begin{cases} \frac{1}{NM}\sum_{n,m=1}^{N,M}\int_{\frac{-1}{2}}^{\frac{1}{2}}\int_{\frac{-1}{2}}^{\frac{1}{2}} e^{j2\pi\tau\left\{f_{T_{max}}\cos\left[F^{-1}\left(\frac{m-1/2+\theta_T}{vM}\right)-\gamma_T\right]+f_{R_{max}}\cos\left[F^{-1}\left(\frac{n-1/2+\theta_R}{uN}\right)-\gamma_R\right]\right\}} \\ \times d\theta_T d\theta_R,\quad N_i = N_q = N \text{ and } M_i = M_q = M \qquad (Cases\ I \text{ and } III) \\ \frac{1}{2N_iM_i}\sum_{n_i,m_i=1}^{N_i,M_i}\int_{\frac{-1}{2}}^{\frac{1}{2}}\int_{\frac{-1}{2}}^{\frac{1}{2}}\cos\left\{2\pi\tau\left(f_{T_{max}}\cos\left[F^{-1}\left(\frac{m_i-1/2+\theta_T}{M_i}\right)-\gamma_T\right]\right.\right. \\ \left.\left.+f_{R_{max}}\cos\left[F^{-1}\left(\frac{n_i-1/2+\theta_R}{N_i}\right)-\gamma_R\right]\right)\right\}d\theta_T d\theta_R + \frac{1}{2N_qM_q}\sum_{n_q,m_q=1}^{N_q,M_q} \\ \times\int_{\frac{-1}{2}}^{\frac{1}{2}}\int_{\frac{-1}{2}}^{\frac{1}{2}}\cos\left\{2\pi\tau\left(f_{T_{max}}\cos\left[F^{-1}\left(\frac{m_q-1/2+\theta_T}{M_q}\right)-\gamma_T\right]\right.\right. \\ \left.\left.+f_{R_{max}}\cos\left[F^{-1}\left(\frac{n_q-1/2+\theta_R}{N_q}\right)-\gamma_R\right]\right)\right\}d\theta_T d\theta_R \\ \qquad (Case\ II) \end{cases}$$

$$= \begin{cases} \frac{NM}{NM}\sum_{n,m=1}^{N,M}\int_{F^{-1}\left(\frac{n-1}{uN}\right)}^{F^{-1}\left(\frac{n}{uN}\right)}\int_{F^{-1}\left(\frac{m-1}{vM}\right)}^{F^{-1}\left(\frac{m}{vM}\right)} e^{j2\pi\tau\left[f_{T_{max}}\cos(\beta_T^m-\gamma_T)+f_{R_{max}}\cos(\beta_R^n-\gamma_R)\right]} \\ \times\frac{e^{k_T\cos(\beta_T^m-\mu_T)+k_R\cos(\beta_R^n-\mu_R)}}{4\pi^2 I_0(k_T)I_0(k_R)} \\ \times d\beta_T^m d\beta_R^n\ N_i = N_q = N \text{ and } M_i = M_q = M \qquad (Cases\ I \text{ and } III) \\ \frac{N_iM_i}{2N_iM_i}\sum_{n_i,m_i=1}^{N_i,M_i}\int_{F^{-1}\left(\frac{n_i-1}{N_i}\right)}^{F^{-1}\left(\frac{n_i}{N_i}\right)}\int_{F^{-1}\left(\frac{m_i-1}{M_i}\right)}^{F^{-1}\left(\frac{m_i}{M_i}\right)} \\ \times\cos\left\{2\pi\tau\left[f_{T_{max}}\cos(\beta_T^{m_i}-\gamma_T)+f_{R_{max}}\cos\left(\beta_R^{n_i}-\gamma_R\right)\right]\right\} \\ \times\frac{e^{k_T\cos(\beta_T^{m_i}-\mu_T)+k_R\cos(\beta_R^{n_i}-\mu_R)}}{4\pi^2 I_0(k_T)I_0(k_R)}d\beta_T^{m_i}d\beta_R^{n_i} + \frac{N_qM_q}{2N_qM_q}\sum_{n_q,m_q=1}^{N_q,M_q}\int_{F^{-1}\left(\frac{n_q-1}{N_q}\right)}^{F^{-1}\left(\frac{n_q}{N_q}\right)}\int_{F^{-1}\left(\frac{m_q-1}{M_q}\right)}^{F^{-1}\left(\frac{m_q}{M_q}\right)} \\ \times\cos\left\{2\pi\tau\left[f_{T_{max}}\times\cos(\beta_T^{m_q}-\gamma_T)+f_{R_{max}}\cos(\beta_R^{n_q}-\gamma_R)\right]\right\} \\ \times\frac{e^{k_T\cos(\beta_T^{m_q}-\mu_T)+k_R\cos(\beta_R^{n_q}-\mu_R)}}{4\pi^2 I_0(k_T)I_0(k_R)}d\beta_T^{m_q}d\beta_R^{n_q} \\ \qquad (Case\ II) \end{cases}$$

$$
= \begin{cases} \dfrac{1}{2\pi^2 I_0(k_T) I_0(k_R)} \displaystyle\int_0^{\pi}\int_0^{\pi} e^{j2\pi\tau[f_{T_{max}}\cos(\beta_T-\gamma_T)+f_{R_{max}}\cos(\beta_R-\gamma_R)]} \\ \qquad \times e^{k_T\cos(\beta_T^m-\mu_T)+k_R\cos(\beta_R^n-\mu_R)} d\beta_T d\beta_R, \qquad (Case\ I) \\ \dfrac{1}{4\pi^2 I_0(k_T) I_0(k_R)} \displaystyle\int_{-\pi}^{\pi}\int_{-\pi}^{\pi} \cos\{2\pi\tau[f_{T_{max}}\cos(\beta_T-\gamma_T)+f_{R_{max}}\cos(\beta_R-\gamma_R)]\} \\ \qquad \times e^{k_T\cos(\beta_T-\mu_T)+k_R\cos(\beta_R-\mu_R)} d\beta_T d\beta_R, \qquad (Case\ II) \\ \dfrac{1}{4\pi^2 I_0(k_T) I_0(k_R)} \displaystyle\int_{-\pi}^{\pi}\int_{-\pi}^{\pi} e^{j2\pi\tau\ [f_{T_{max}}\cos(\beta_T-\gamma_T)+f_{R_{max}}\cos(\beta_R-\gamma_R)]} \\ \qquad \times e^{k_T\cos(\beta_T^m-\mu_T)+k_R\cos(\beta_R^n-\mu_R)} d\beta_T d\beta_R \qquad (Case\ III) \end{cases} \tag{A.32}
$$

where at the third equality of Eq. (A.32), the integration variables θ_T and θ_R were replaced by $\beta_T^{m_{i/q}} = F^{-1}\left(\frac{m_{i/q}-1/2+\theta_T}{M_{i/q}}\right)$, $d\theta_T = \frac{M_{i/q}e^{k_T\cos\left(\beta_T^{m_{i/q}}-\mu_T\right)}}{2\pi I_0(k_T)} d\beta_T^{m_{i/q}}$, $\beta_R^{n_{i/q}} = F^{-1}\left(\frac{n_{i/q}-1/2+\theta_R}{N_{i/q}}\right)$, $d\theta_R = \frac{N_{i/q}e^{k_R\cos\left(\beta_R^{n_{i/q}}-\mu_R\right)}}{2\pi I_0(k_R)} d\beta_R^{n_{i/q}}$, $\beta_T^m = F^{-1}\left(\frac{m-1/2+\theta_T}{M}\right)$, $d\theta_T = \frac{Me^{k_T\cos(\beta_T^m-\mu_T)}}{2\pi I_0(k_T)} d\beta_T^m$, $\beta_R^n = F^{-1}\left(\frac{n-1/2+\theta_R}{N}\right)$, and $d\theta_R = \frac{Ne^{k_R\cos(\beta_R^n-\mu_R)}}{2\pi I_0(k_R)} d\beta_R^n$. The two single definite integrals in the last equality of Eq. (A.32) can be solved by using the equality of Eq. $\int_{-\pi}^{\pi} e^{a\sin c+b\cos c}dc = 2\pi I_0(\sqrt{a^2+b^2})$ [I.S.Gradshteyn and Ryzhik 2000]. After some manipulation, the closed-form expression of the CF $\hat{\rho}_{\hat{h}\hat{h}}(\tau)$ can be obtained and is the same as $\rho_{hh}(\tau)$ in Eq. (4.2).

A.6 Probability Density Functions

Rician PDF

The Rician PDF is:

$$
f_g(r) = \frac{r}{\sigma^2}\exp\left\{-\frac{r^2+A^2}{2\sigma^2}\right\} I_o\left(\frac{rA}{\sigma^2}\right), r \geq 0 \tag{A.33}
$$

where for the fading analysis scenarios,

- A: the amplitude of the LoS component
- σ^2: the variance of faded component, and
- $I_0(\cdot)$: the zeroth-order modified Bessel function

Rayleigh PDF

The Rayleigh density function is written as

$$
f_g(r) = fracr\sigma^2\exp\{-fracr^2 2\sigma^2\}, r \geq 0 \tag{A.34}
$$

The Rayleigh PDF is usually used to describe the PDF of the gain coefficients when the LoS component does not exist.

Log-normal PDF

The log-normal density function is written as

$$f_g(r) = \frac{1}{(2\pi)^{\frac{1}{2}}\sigma r} \exp\left\{-\frac{(Lnr - \mu)^2}{2\sigma^2}\right\}, r > 0 \tag{A.35}$$

A.7 Computation of the Gerschgorin Radii

In this section, we first sketch the procedure for the calculation of the Gerschgorin Radii (GRs). The reader is referred to the paper by Wu et al. [1995] for the detailed derivation. Then, we show that the GRs computed using the projection onto the noise eigenvectors are zero, while the GRs calculated using the projection onto the signal eigenvectors are positive. These results are obtained in a scenario with D specular-scatterers. They can also be used in the single-SDS case where the signal subspace induced by the SDS has effective rank equal to D. Furthermore, we show that in an SDS scenario, the ratio between the two largest positive GRs is independent of the noise and is approximately proportional to the square root of the ratio between the corresponding signal eigenvalues.

We consider a scenario with D specular scatterers. The covariance matrix of the signals at the output of the Rx correlators reads

$$\boldsymbol{\Sigma_y} = \boldsymbol{A}(\boldsymbol{\phi})\boldsymbol{V}\boldsymbol{A}(\boldsymbol{\phi})^{\mathrm{H}} + \sigma_w^2 \boldsymbol{I}_M \tag{A.36}$$

where $\boldsymbol{\phi} = [\phi_1, \ldots, \phi_D]$, $\boldsymbol{A}(\boldsymbol{\phi}) = [\boldsymbol{c}(\phi_1), \ldots, \boldsymbol{c}(\phi_D)]$, and $\boldsymbol{V} = \mathrm{diag}(\sigma_{\alpha_1}^2, \sigma_{\alpha_2}^2, \ldots, \sigma_{\alpha_D}^2)$. We now rewrite $\boldsymbol{A}(\boldsymbol{\phi})$ as $\boldsymbol{A}_M(\boldsymbol{\phi}) = [\boldsymbol{a}_1(\boldsymbol{\phi}), \ldots, \boldsymbol{a}_m(\boldsymbol{\phi}), \ldots, \boldsymbol{a}_M(\boldsymbol{\phi})]^{\mathrm{T}}$, where $\boldsymbol{a}_m(\phi)^{\mathrm{T}}$ denotes the mth row of $\boldsymbol{A}(\boldsymbol{\phi})$.

Following the method for the computation of GRs described by Wu et al. [1995], the covariance matrix in Eq. (A.36) is partitioned according to

$$\boldsymbol{\Sigma_y} = \begin{bmatrix} \Sigma_1 & \vartheta \\ \vartheta^{\mathrm{H}} & \varepsilon \end{bmatrix} \tag{A.37}$$

where $\boldsymbol{\Sigma}_1$ is the $(M-1) \times (M-1)$ matrix

$$\begin{aligned} \Sigma_1 &= \boldsymbol{A}_{M-1}(\phi)\boldsymbol{V}\boldsymbol{A}_{M-1}(\phi)^{\mathrm{H}} + \sigma_w^2 \boldsymbol{I}_{M-1} \\ &= \Sigma_{1,\mathrm{sig}} + \sigma_w^2 \boldsymbol{I}_{M-1} \end{aligned} \tag{A.38}$$

the $(M \quad 1)$-element vector $\boldsymbol{\vartheta}$ is given by

$$\begin{aligned} \vartheta &= \boldsymbol{A}_{M-1}(\phi)\boldsymbol{V}\boldsymbol{a}_M(\phi)^* \\ &= \boldsymbol{A}_{M-1}(\phi) \cdot \mathrm{diag}(c_M^*(\phi_1), \ldots, c_M^*(\phi_L)) \cdot \boldsymbol{v} \\ &= \boldsymbol{A}'_{M-1}(\phi) \cdot \boldsymbol{v} \end{aligned} \tag{A.39}$$

and ε is the (M, M) element of $\boldsymbol{\Sigma_y}$. In the above expressions $\boldsymbol{\Sigma}_{1,\mathrm{sig}} = \boldsymbol{A}_{M-1}(\boldsymbol{\phi})\boldsymbol{V}\boldsymbol{A}_{M-1}(\boldsymbol{\phi})^{\mathrm{H}}$, $\boldsymbol{v} \doteq [\sigma_{a_1}^2, \ldots, \sigma_{a_D}^2]^{\mathrm{T}}$, and $\boldsymbol{A}'_{M-1}(\boldsymbol{\phi}) \doteq \boldsymbol{A}_{M-1}(\boldsymbol{\phi}) \cdot \mathrm{diag}(c_M^*(\phi_1), \ldots, c_M^*(\phi_D))$, with $(\cdot)^*$ denoting complex conjugation.

For simplicity, we assume that the array elements are isotropic and without loss of generality that $|c_M(\phi_d)| = 1$, $\ell = 1, \ldots, D$. As will be discussed later, these assumptions can be dropped in practise. Under these two assumptions, the diagonal matrix in Eq. (A.39) is a unitary matrix. As a consequence, $\boldsymbol{A}_{M-1}(\boldsymbol{\phi})\boldsymbol{V}\boldsymbol{A}_{M-1}(\boldsymbol{\phi})^{\mathrm{H}} = \boldsymbol{A}'_{M-1}(\boldsymbol{\phi})\boldsymbol{V}\boldsymbol{A}'_{M-1}(\boldsymbol{\phi})^{\mathrm{H}}$. Hence the eigenvectors of $\boldsymbol{\Sigma}_{1,\text{sig}}$ coincide with the eigenvectors of $\boldsymbol{A}'_{M-1}(\boldsymbol{\phi})\boldsymbol{V}\boldsymbol{A}'_{M-1}(\boldsymbol{\phi})^{\mathrm{H}}$. Thus the noise eigenvectors of $\boldsymbol{\Sigma}_1$ are orthogonal to the space spanned by the column vectors in $\boldsymbol{A}'_{M-1}(\boldsymbol{\phi})$.

To calculate the GRs, the eigenvalue decomposition of $\boldsymbol{\Sigma}_1$ is first performed:

$$\boldsymbol{\Sigma}_1 = \boldsymbol{U}\boldsymbol{\Lambda}\boldsymbol{U}^{\mathrm{H}} \tag{A.40}$$

The unitary matrix $\boldsymbol{U} = [\boldsymbol{e}_1, \ldots, \boldsymbol{e}_{M-1}]$ contains the eigenvectors of $\boldsymbol{\Sigma}_1$ and $\boldsymbol{\Lambda} = \mathrm{diag}(\lambda_1, \ldots, \lambda_{M-1})$ has diagonal elements equal to the eigenvalues of $\boldsymbol{\Sigma}_1$ ranked in decreasing order. The GRs are calculated to be

$$r_m = |\boldsymbol{e}_m^H \boldsymbol{\vartheta}|, \quad m = 1, \ldots, M-1 \tag{A.41}$$

Obviously when $\boldsymbol{e}_m$ is a noise eigenvector, $r_m = 0$ as $\boldsymbol{e}_m$ is orthogonal to the column space of $\boldsymbol{A}'_{M-1}(\boldsymbol{\phi})$. Furthermore, when $\boldsymbol{e}_m$ is a signal eigenvector, r_m will be non-zero and independent of the noise. This property is used by Wu et al. [1995] to estimate the number of sources. When $D > M-1$, all GRs are non-zero since the eigenvectors $\boldsymbol{e}_1, \ldots, \boldsymbol{e}_{M-1}$ corresponding to the $M-1$ dominant signal components belong to the column space of $\boldsymbol{A}'_{M-1}(\boldsymbol{\phi})$ [Besson and Stoica 1999].

We now derive analytical expressions of the GRs associated with the signal eigenvectors in a single-SDS scenario. To this end, we use the two-ray model [Bengtsson and Ottersten 2000] to approximate the received signal:

$$\boldsymbol{y}(t) \approx \boldsymbol{y}_{2SS}(t) = \alpha_1(t)\boldsymbol{c}(\bar{\phi}_1) + \alpha_2(t)\boldsymbol{c}(\bar{\phi}_2) + \boldsymbol{w}(t) \tag{A.42}$$

where $\bar{\phi}_1 = \bar{\phi} - \sigma_{\tilde{\phi}}$ and $\bar{\phi}_2 = \bar{\phi} + \sigma_{\tilde{\phi}}$. The random processes $\alpha_1(t)$ and $\alpha_2(t)$ are uncorrelated and white, with identical variances $\mathrm{E}[|\alpha_1(t)|^2] = \mathrm{E}[|\alpha_2(t)|^2] = P$. The eigenvalues λ_+ and λ_- associated with the signal eigenvectors $\boldsymbol{e}_+$ and $\boldsymbol{e}_-$ respectively of the covariance matrix of $\boldsymbol{y}_{2SS}(t)$ are calculated to be [Kaveh and Barabell 1986]

$$\lambda_{\pm} = PM \cdot (1 \pm |\zeta_M|) + \sigma_w^2$$

where $\zeta_M = \frac{1}{M}\boldsymbol{c}^{\mathrm{H}}(\bar{\phi}_1)\boldsymbol{c}(\bar{\phi}_2)$. The ratio between the largest signal eigenvalue $\lambda_+ - \sigma_w^2$ and the smallest signal eigenvalue $\lambda_- - \sigma_w^2$ reads

$$\frac{\lambda_+ - \sigma_w^2}{\lambda_- - \sigma_w^2} = \frac{1 + |\zeta_M|}{1 - |\zeta_M|}$$

To calculate the GRs we first partition the covariance matrix of $\boldsymbol{y}_{2SS}(t)$ according to Eq. (A.37). The signal eigenvectors $\boldsymbol{e}_+$ and $\boldsymbol{e}_-$ of $\boldsymbol{\Sigma}_1$ are obtained to be

$$\boldsymbol{e}_{\pm} = \frac{1}{\sqrt{2(M-1)(1 \pm |\zeta_{M-1}|)}}\bigg(\boldsymbol{c}_{M-1}(\bar{\phi}_1) \pm \frac{|\zeta_{M-1}|}{\zeta_{M-1}} \cdot \boldsymbol{c}_{M-1}(\bar{\phi}_2)\bigg)$$

with $\boldsymbol{c}_{M-1}(\phi) = [c_1(\phi), c_2(\phi), \ldots, c_{M-1}(\phi)]^{\mathrm{T}}$ and $\zeta_{M-1} = \frac{1}{M-1}\boldsymbol{c}^{\mathrm{H}}_{M-1}(\bar{\phi}_1)\boldsymbol{c}_{M-1}(\bar{\phi}_2)$. Inserting $\boldsymbol{e}_+$ and $\boldsymbol{e}_-$ in Eq. (A.41) yields

$$r_+ = P\sqrt{2(M-1)(1 \pm |\zeta_{M-1}|)}\left|\cos\left(\frac{M}{4}\pi(\cos(\bar{\phi}_2) - \cos(\bar{\phi}_1))\right)\right|$$

$$r_- = P\sqrt{2(M-1)(1 \pm |\zeta_{M-1}|)}\left|\sin\left(\frac{M}{4}\pi(\cos(\bar{\phi}_2) - \cos(\bar{\phi}_1))\right)\right|$$

It can be shown that when the AS $\sigma_{\tilde{\phi}}$ is small, r_+ is larger than r_-. The ratio of the largest GR r_+ to the second largest GR r_- can be calculated to be

$$\frac{r_+}{r_-} = \frac{\sqrt{1+|\zeta_{M-1}|}}{\sqrt{1-|\zeta_{M-1}|}} \cdot \left|\mathrm{ctg}\left(\frac{M}{4}\pi(\cos(\bar{\phi}_2) - \cos(\bar{\phi}_1))\right)\right| \tag{A.43}$$

In the case where $\sigma_{\tilde{\phi}}$ is small and the array elements are isotropic, $\frac{\sqrt{1+|\zeta_{M-1}|}}{\sqrt{1-|\zeta_{M-1}|}} \approx \frac{M}{M-1} \cdot \frac{\sqrt{1+|\zeta_M|}}{\sqrt{1-|\zeta_M|}}$. Hence, the ratio $\frac{r_+}{r_-}$ can be approximated in this case according to

$$\begin{aligned}\frac{r_+}{r_-} &\approx \frac{M}{M-1} \cdot \frac{\sqrt{1+|\zeta_M|}}{\sqrt{1-|\zeta_M|}} \cdot \left|\mathrm{ctg}\left(\frac{M}{4}\pi(\cos(\bar{\phi}_2) - \cos(\bar{\phi}_1))\right)\right| \\ &= \sqrt{\frac{\lambda_1 - \sigma_w^2}{\lambda_2 - \sigma_w^2}} \cdot \frac{M}{M-1}\left|\mathrm{ctg}\left(\frac{M}{4}\pi(\cos(\bar{\phi}_2) - \cos(\bar{\phi}_1))\right)\right| \end{aligned} \tag{A.44}$$

The right-hand term in Eq. (A.44) is independent of the noise and proportional to the square root of the ratio of the corresponding signal eigenvalues. Further calculation not presented here shows that Eq. (A.44) is valid provided the antenna elements have identical radiation pattern and, moreover, this pattern exhibits the same value at azimuths $\bar{\phi} - \sigma_{\tilde{\phi}}$ and $\bar{\phi} + \sigma_{\tilde{\phi}}$. Notice that the latter condition is usually fulfilled to a good approximation as the antenna radiation pattern changes slowly with respect to azimuth within the array beam-width. Hence, from a practical point of view the fluctuation of the pattern in small azimuth ranges – of the size of $2\sigma_{\tilde{\phi}}$ – is negligible.

A.8 Derivations for Chapter 9

A.8.1 Derivation of Eq. (9.11)

Substituting Eqs (9.5), (9.6), and (9.9) into Eq. (9.10) and after some mathematical manipulation, we have

$$\begin{aligned}\rho_{l,oq;l,o'q'}(\tau,\chi) &= \frac{1}{2\pi I_0(k)}\sum_{i=0}^{\Lambda_l-1}\sum_{r=1}^{R_c} Q_{l,i,r}e^{jC_{l,i}} \\ &\times \int_{\mu_{l,i,r}-\Delta\mu_{l,i}}^{\mu_{l,i,r}+\Delta\mu_{l,i}} e^{A_{l,i,r}\cos\phi^R_{l,i,r}}e^{B_{l,i,r}\sin\phi^R_{l,i,r}}\,d\phi^R_{l,i,r}.\end{aligned} \tag{A.45}$$

Considering the periodic property of trigonometric function and the relationship between the sine and cosine functions, the well-known series $e^{z\sin\theta} = I_0(z) + 2\sum_{k=0}^{\infty}(-)^k I_{2k+1}(z)$ $\sin[(2k+1)\theta] + 2\sum_{k=1}^{\infty}(-)^k I_{2k}(z)\cos(2k\theta)$ [Abramowitz and Stegun 1965] can be further simplified as:

$$e^{z\sin\theta} = I_0(z) + 2\sum_{\ell=1}^{\infty}(-1)^{\ell} I_{\ell}(z)\cos[\ell\theta + \ell\pi/2] \tag{A.46}$$

By making use of another well-known series $e^{z\cos\theta} = I_0(z) + 2\sum_{k=1}^{\infty} I_k(z)\cos(k\theta)$ in the book by Abramowitz and Stegun [1965] and the further simplified series in Eq. (A.46), $\rho_{l,oq;l,o'q'}(\tau,\chi)$ can be further expressed as:

$$\begin{aligned}\rho_{l,oq;l,o'q'}(\tau,\chi) = {} & \frac{1}{2\pi I_0(k)}\sum_{i=0}^{\Lambda_l-1}\sum_{r=1}^{R_c} Q_{l,i,r}e^{jC_{l,i}} \\ & \times \int_{\mu_{l,i,r}-\Delta e_{l,i}}^{\mu_{l,i,r}+\Delta e_{l,i}} \left[I_0(A_{l,i,r}) + 2\sum_{\ell=1}^{\infty} I_{\ell}(A_{l,i,r})\cos(\ell\phi_{l,i,r}^R)\right] \\ & \times \left[I_0(B_{l,i,r}) + 2\sum_{\ell'=1}^{\infty}(-1)^{\ell'} I_{\ell'}(B_{l,i,r})\cos(\ell'\phi_{l,i,r}^R + \ell'\pi/2)\right] d\phi_{l,i,r}^R\end{aligned} \tag{A.47}$$

To obtain the closed-form expression of Eq. (9.11), the following definite integral functions from I.S.Gradshteyn and Ryzhik 2000 are used:

$$\int_{\mu-\Delta\mu}^{\mu+\Delta\mu}\cos(\ell\phi)d\phi = \frac{2\sin(\ell\Delta\mu)\cos(\ell\mu)}{\ell} \tag{A.48a}$$

$$\int_{\mu-\Delta\mu}^{\mu+\Delta\mu}\cos(\ell'\phi + \ell'\pi/2)d\phi = \frac{2\sin(\ell'\Delta\mu)\cos(\ell'\mu + \ell'\pi/2)}{\ell'} \tag{A.48b}$$

$$\int_{\mu-\Delta\mu}^{\mu+\Delta\mu}\cos(\ell\phi + \ell\pi/2)\cos(\ell\phi)d\phi = \Delta\mu\cos(\ell\pi/2) + \frac{\sin(2\ell\Delta\mu)\cos(2\ell\mu + \ell\pi/2)}{2\ell} \tag{A.48c}$$

$$\begin{aligned}\int_{\mu-\Delta\mu}^{\mu+\Delta\mu}\sin[\ell'\phi + (\ell'-1)\pi/2]\cos(\ell\phi)d\phi = {} & \frac{\sin[(\ell+\ell')\Delta\mu]\cos[(\ell+\ell')\mu + \ell'\pi/2]}{\ell+\ell'} \\ & + \frac{\sin[(\ell-\ell')\Delta\mu]\cos[(\ell-\ell')\mu - \ell'\pi/2]}{\ell-\ell'}\end{aligned} \tag{A.48d}$$

A.8.2 Derivation of Eq. (9.16)

Substituting $\Delta\mu_{l,i} = \pi$ and $\Lambda_l = R_c = 1$ into Eq. (9.11), we have

$$\rho_{l,oq;l,o'q'}(\tau,\chi) = \frac{e^{jC}\left[I_0(A)I_0(B) + 2\sum_{\ell=1}^{\infty}(-1)^{\ell} I_{2\ell}(A)I_{2\ell}(B)\right]}{I_0(k)} \tag{A.49}$$

By making use of the Neumann's addition theorem [Watson 1952], the following equation can be obtained by further considering $J_\ell(jz) = j^\ell I_\ell(z)$, where ℓ is integral, and setting the angle $\phi = \pi/2$ in the theorem:

$$I_0[(Z^2 + z^2)] = I_0(Z)I_0(z) + 2\sum_{\ell=1}^{\infty}(-1)^\ell I_{2\ell}(Z)I_{2\ell}(z) \tag{A.50}$$

With the help of Eqs (A.49) and (A.50), Eq. (9.16) can be easily obtained.

A.8.3 Derivation of Eqs (9.43)–(9.48)

In this appendix, following the same derivation procedure (that is, the same newly proposed method), we will derive these general relationships for the two-ring model in Eqs (9.43)–(9.46) and the ellipse model in Eqs (9.47) and (9.48). In Figure A.1, applying the laws of cosines and sines to the triangle $O_T s^{(n_1)} O_R$, we obtain:

$$\xi_{n_1}^2 = R_T^2 + D^2 - 2DR_T \cos\phi_T^{(n_1)} \tag{A.51}$$

$$R_T^2 = \xi_{n_1}^2 + D^2 + 2D\xi_{n_1} \cos\phi_R^{(n_1)} \tag{A.52}$$

$$\frac{R_T}{\sin\phi_R^{(n_1)}} = \frac{\xi_{n_1}}{\sin\phi_T^{(n_1)}} \tag{A.53}$$

From the above expressions, we can easily obtain Eq. (9.43) and Eq. (9.44). Similarly, applying the laws of cosines and sines to the triangle $O_T s^{(n_2)} O_R$, we have

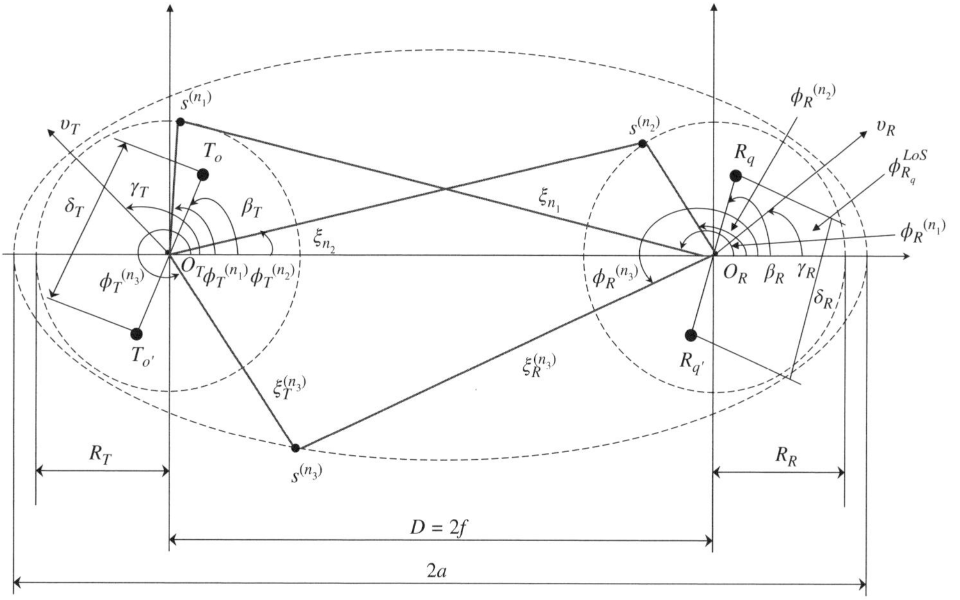

Figure A.1 Relationship between the AoA and AoD for single-bounced rays.

$$\xi_{n_2}^2 = R_R^2 + D^2 + 2DR_R \cos\phi_R^{(n_2)} \tag{A.54}$$

$$R_R^2 = \xi_{n_2}^2 + D^2 - 2D\xi_{n_2} \cos\phi_T^{(n_2)} \tag{A.55}$$

$$\frac{R_R}{\sin\phi_T^{(n_2)}} = \frac{\xi_{n_2}}{\sin\phi_R^{(n_2)}} \tag{A.56}$$

We can easily obtain Eqs (9.45) and (9.46) from these expressions. Analogously, applying the laws of cosines and sines to the triangle $O_T s^{(n_3)} O_R$, we get

$$\left(\xi_T^{(n_3)}\right)^2 = (\xi_R^{(n_3)})^2 + D^2 + 2D\xi_R^{(n_3)} \cos\phi_R^{(n_3)} \tag{A.57}$$

$$\left(\xi_R^{(n_3)}\right)^2 = (\xi_T^{(n_3)})^2 + D^2 - 2D\xi_T^{(n_3)} \cos\phi_T^{(n_3)} \tag{A.58}$$

$$\frac{\xi_R^{(n_3)}}{\sin\phi_T^{(n_3)}} = \frac{\xi_T^{(n_3)}}{\sin\phi_R^{(n_3)}} \tag{A.59}$$

Based on the above expressions, and the following equalities $D = 2f$ and $\xi_T^{(n_3)} + \xi_R^{(n_3)} = 2a$, we can get Eqs (9.47) and (9.48).

A.8.4 Derivation of Eq. (9.53)

Considering the von Mises PDF for the two-ring model, applying the following approximate relationships $\phi_R^{(n_1)} \approx \pi - \Delta_T \sin\phi_T^{(n_1)}$ and $\phi_T^{(n_2)} \approx \Delta_R \sin\phi_R^{(n_2)}$, and substituting Eq. (9.34b) and Eq. (9.36)–(9.39) into Eq. (9.49), we have

$$\rho_{h_{oq}^{SB_{1(2)}} h_{o'q'}^{\prime SB_{1(2)}}}(\tau,\chi) = \frac{\left[2\pi I_0\left(k_{T(R)}^{SB_{1(2)}}\right)\right]^{-1} e^{jC_{T(R)}^{SB_{1(2)}}}}{\sqrt{\left(K_{oq}+1\right)\left(K_{o'q'}+1\right)}} \times \int_{-\pi}^{\pi} e^{\left(A_{T(R)}^{SB_{1(2)}} \cos\phi_{T(R)}^{SB_{1(2)}} + B_{T(R)}^{SB_{1(2)}} \sin\phi_{T(R)}^{SB_{1(2)}}\right)} d\phi_{T(R)}^{SB_{1(2)}} \tag{A.60}$$

where $A_{T(R)}^{SB_{1(2)}}$, $B_{T(R)}^{SB_{1(2)}}$, and $C_{T(R)}^{SB_{1(2)}}$ have been given in Eqs (9.54a)–(9.54c). The definite integrals in the right-hand side of Eq. (A.60) can be solved using the equality $\int_{-\pi}^{\pi} e^{a\sin c + b\cos c} dc = 2\pi I_0(\sqrt{a^2+b^2})$ [I.S.Gradshteyn and Ryzhik 2000]. After some manipulation, we can get the closed-form expression Eq. (9.53).

A.8.5 Derivation of Eq. (9.60)

Given $a^2 + b^2 = c(d^2 + e^2)$, after some complex manipulation, we can rewrite $I_0\left[\sqrt{\left(A_{T(R)}^{SB_{1(2)}}\right)^2 + \left(B_{T(R)}^{SB_{1(2)}}\right)^2}\right]$ as

$$I_0\left[j\sqrt{W_{T(R)}^{SB_{1(2)}}}\sqrt{\left(\tau + \frac{D_{T(R)}^{SB_{1(2)}}}{W_{T(R)}^{SB_{1(2)}}}\right)^2 + \left(\frac{E_{T(R)}^{SB_{1(2)}}}{W_{T(R)}^{SB_{1(2)}}}\right)^2}\right] \tag{A.61}$$

where $W_{T(R)}^{SB_{1(2)}}$, $D_{T(R)}^{SB_{1(2)}}$, and $E_{T(R)}^{SB_{1(2)}}$ have been given in Eqs (9.61b)–(9.61d). Note that the expression Eq. (A.61) corrects the expressions in Zajic and Stuber [2008, Eqs(38),(39)]. By applying the Fourier transform to Eq. (9.53) in terms of the time separation τ, and using Eq. (A.61) and the equality $\int_0^{\infty} I_0(j\alpha\sqrt{x^2+y^2})\cos(\beta x)dx = \cos(y\sqrt{\alpha^2-\beta^2})/\sqrt{\alpha^2-\beta^2}$ [I.S. Gradshteyn and Ryzhik 2000], we can obtain Eq. (9.60).

A.8.6 Comparison of the Doppler PSDs with different CFs, Eqs (9.49) and (9.50)

To further clarify which CF definition, Eq. (9.49) or Eq. (9.50), results in the correct Doppler PSD to accurately reflect the underlying physical phenomena of real channels, we first derive the relationship between the Doppler PSD based on the CF of Eq. (9.49), $S_{h_{oq}h_{oq}}(f_D)$, and the Doppler PSD based on the CF of Eq. (9.50), $\tilde{S}_{h_{oq}h_{oq}}(f_D)$. Considering the equality $\tilde{\rho}_{h_{oq}h_{oq}}(\tau) = \rho^*_{h_{oq}h_{oq}}(\tau)$ and the Fourier transform relation between the CF and Doppler PSD, we have

$$\tilde{S}_{h_{oq}h_{oq}}(f_D) = S^*_{h_{oq}h_{oq}}(-f_D) \tag{A.62}$$

From Eq. (A.62), it is clear that only if $S_{h_{oq}h_{oq}}(f_D)$ is a real function and symmetrical to the origin does the equality $\tilde{S}_{h_{oq}h_{oq}}(f_D) = S_{h_{oq}h_{oq}}(f_D)$ hold. Note that due to the Fourier transform relationship, the equality $\tilde{S}_{h_{oq}h_{oq}}(f_D) = S_{h_{oq}h_{oq}}(f_D)$ leads to the equality $\tilde{\rho}_{h_{oq}h_{oq}}(\tau) = \rho_{h_{oq}h_{oq}}(\tau)$ and vice versa. We now proceed the comparison of $S_{h_{oq}h_{oq}}(f_D)$ and $\tilde{S}_{h_{oq}h_{oq}}(f_D)$ in the following two common scenarios.

- *Scenario 1* is a non-isotropic F2M macrocell propagation environment ($f_{T_{max}} = 0$), as shown in Figure A.2(a). We use a one-ring model to represent this scenario, where the ring of scatterers is around the Rx – that is, MS – and the MS moves toward the direction of the Tx: $\gamma_R = \pi$. Note that most of the scatterers are located in a small part of the ring facing the motion of the MS: $\mu_R = \pi$.
- *Scenario 2* is an isotropic M2M propagation environment ($k_T^{TR} = k_R^{TR} = 0$), where the Tx and Rx move in opposite directions ($\gamma_T = 0$ and $\gamma_R = \pi$), as shown in Figure A.2(b). Here, a single-bounce two-ring model is used.

For *Scenario 1*, based on Eq. (A.62), the opposite results for the Doppler PSD are expected as shown in Figure A.3, where $k_R^{TR} = 3$. Since the MS moves toward the majority of received signals, the maximum Doppler PSD should appear at $f_D = f_{R_{max}} = 570$ Hz. From Figure A.3, it is clear that the $S_{h_{oq}h_{oq}}(f_D)$ presents the underlying physical phenomena for *Scenario 1*. For *Scenario 2*, as expected from Eq. (A.62), the opposite results of the Doppler PSD with respect to the range of Doppler frequencies are illustrated in Figure A.3, where $f_{T_{max}} = f_{R_{max}} = 570$ Hz were used. Since the Tx and Rx are moving in opposite directions, the Doppler PSD should be limited to the range of Doppler frequencies $0 \leq f_D \leq 1140$ Hz, whereas the maximum Doppler PSD exists at $f_D = 0$ and $f_D = f_{T_{max}} + f_{R_{max}} = 1140$ Hz. Again, from Figure A.3, it is obvious that $S_{h_{oq}h_{oq}}(f_D)$ reflects the underlying physical phenomenon for *Scenario 2*. Therefore, we can conclude that $S_{h_{oq}h_{oq}}(f_D)$ is able to accurately capture the underlying physical phenomena of real channels for any scenario,

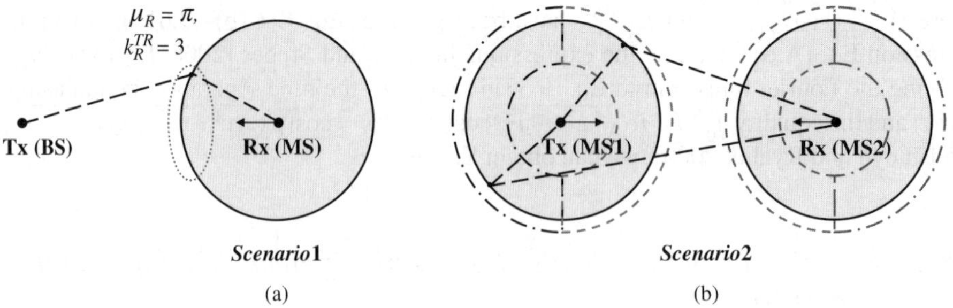

Figure A.2 Graphical description of (a) *Scenario1* and (b) *Scenario2*.

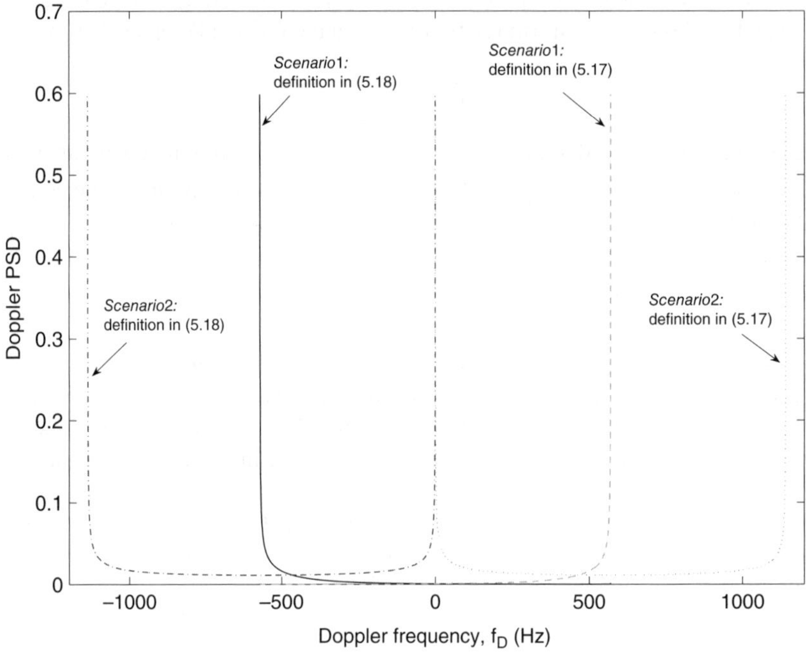

Figure A.3 Comparison of the Doppler PSDs of *Scenario1* and *Scenario2* based on the CF definitions in Eqs (9.49) and (9.50).

while $\tilde{S}_{h_{oq}h_{oq}}(f_D)$ cannot. It is worth stressing that for an isotropic F2M macro-cell scenario (Clarke's scenario), where no scatterers are around the Tx, we find that the difference of the Doppler PSD caused by two CF definitions vanishes; that is, $\tilde{S}_{h_{oq}h_{oq}}(f_D) = S_{h_{oq}h_{oq}}(f_D)$. This is because Clarke's scenario has the U-shaped Doppler PSD, which is a real function and symmetrical to the origin. This seems to be the reason why the CF of Eq. (9.50) was widely misapplied.

A.8.7 Derivation of Eq. (9.75b)

Note that Eq. (9.75b) includes two formulas regarding parameters $b_2^{SB_1}$ and $b_2^{SB_2}$. In this appendix, we only derive the parameter $b_2^{SB_1}$ since the derivation of the parameter $b_2^{SB_2}$ is exactly the same. From the general expression of $b_2^{SB_1}$ in Eq. (9.74a) with $m = 2$, we have

$$\begin{aligned}
b_2^{SB_1} =& \frac{b_0^{SB_1} 2\pi}{I_0(k_T^{TR})} \int_{-\pi}^{\pi} e^{k_T^{TR}\cos(\phi_T^{SB_1}-\mu_T^{TR})} \left(\frac{f_{T_{max}}^2}{2} + \frac{f_{R_{max}}^2 \Delta_T^2 \sin^2\gamma_R}{2} + f_{R_{max}}^2 \cos^2\gamma_R \right. \\
& \left. + f_{T_{max}} f_{R_{max}} \Delta_T \sin\gamma_T \sin\gamma_R \right) d\phi_T^{SB_1} - \left[\frac{b_0^{SB_1} 2\pi}{I_0(k_T^{TR})} \int_{-\pi}^{\pi} e^{k_T^{TR}\cos(\phi_T^{SB_1}-\mu_T^{TR})} 2 f_{T_{max}} f_{R_{max}} \right. \\
& \times \cos\gamma_R \cos(\phi_T^{SB_1} - \gamma_T) d\phi_T^{SB_1} + \frac{b_0^{SB_1} 2\pi}{I_0(k_T^{TR})} \int_{-\pi}^{\pi} e^{k_T^{TR}\cos(\phi_T^{SB_1}-\mu_T^{TR})} f_{R_{max}}^2 \Delta_T \sin(2\gamma_R) \\
& \left. \times \sin\phi_T^{SB_1} d\phi_T^{SB_1} \right] + \left[\frac{b_0^{SB_1} 2\pi}{I_0(k_T^{TR})} \int_{-\pi}^{\pi} e^{k_T^{TR}\cos(\phi_T^{SB_1}-\mu_T^{TR})} f_{T_{max}}^2 \frac{\cos(2(\phi_T^{SB_1} - \gamma_T))}{2} d\phi_T^{SB_1} \right. \\
& - \frac{b_0^{SB_1} 2\pi}{I_0(k_T^{TR})} \int_{-\pi}^{\pi} e^{k_T^{TR}} e^{\cos(\phi_T^{SB_1}-\mu_T^{TR})} \frac{f_{R_{max}}^2 \Delta_T^2 \sin^2\gamma_R \cos(2\phi_T^{SB_1})}{2} d\phi_T^{SB_1} + \frac{b_0^{SB_1} 2\pi}{I_0(k_T^{TR})} \\
& \left. \int_{-\pi}^{\pi} e^{k_T^{TR}\cos(\phi_T^{SB_1}-\mu_T^{TR})} f_{T_{max}} f_{R_{max}} \Delta_T \sin\gamma_R \sin(\gamma_T - 2\phi_T^{SB_1}) d\phi_T^{SB_1} \right] \\
=& X_1 - [X_2 + X_3] + [X_4 - X_5 + X_6]. \qquad \text{(A.63)}
\end{aligned}$$

For the above equation, the first term X_1 can be readily obtained as the following closed-form expression based on the equality $\int_{-\pi}^{\pi} e^{a\sin c + b\cos c} dc = 2\pi I_0(\sqrt{a^2+b^2})$ [I.S.Gradshteyn and Ryzhik 2000]:

$$X_1 = b_0^{SB_1} 4\pi^2 \left(\frac{f_{T_{max}}^2}{2} + \frac{f_{R_{max}}^2 \Delta_T^2 \sin^2\gamma_R}{2} + f_{R_{max}}^2 \cos^2\gamma_R + f_{T_{max}} f_{R_{max}} \Delta_T \sin\gamma_T \sin\gamma_R \right) \qquad \text{(A.64)}$$

The second term X_2 is given as follows

$$\begin{aligned}
X_2 =& \frac{b_0^{SB_1} 4\pi f_{T_{max}} f_{R_{max}} \cos\gamma_R}{I_0(k_T^{TR})} \int_{-\pi}^{\pi} e^{k_T^{TR}\cos(\phi_T^{SB_1}-\mu_T^{TR})} \frac{e^{j(\phi_T^{SB_1}-\gamma_T)} + e^{-j(\phi_T^{SB_1}-\gamma_T)}}{2} d\phi_T^{SB_1} \\
=& \frac{b_0^{SB_1} 2\pi f_{T_{max}} f_{R_{max}} \cos\gamma_R}{I_0(k_T^{TR})} \left[e^{j(\pi/2+\mu_T^{TR}-\gamma_T)} \int_{-\pi/2+\mu_T^{TR}}^{3\pi/2+\mu_T^{TR}} e^{-j\beta_T + k_T^{TR}\sin\beta_T} d\beta_T \right. \\
& \left. + e^{j(\pi/2-\mu_T^{TR}+\gamma_T)} \int_{-\pi/2-\mu_T^{TR}}^{3\pi/2-\mu_T^{TR}} e^{-j\theta_T + k_T^{TR}\sin\theta_T} d\theta_T \right] \qquad \text{(A.65)}
\end{aligned}$$

where in the second equality we have used the equalities $\phi_T^{SB_1} = \pi/2 - \beta_T + \mu_T^{TR}$ and $\phi_T^{SB_1} = -\pi/2 + \theta_T + \mu_T^{TR}$, so that $d\phi_T^{SB_1} = -d\beta_T$ and $d\phi_T^{SB_1} = d\theta_T$, respectively. Based on the modified equality $\int_{\alpha_l}^{\alpha_u} e^{-jV\theta+z\sin\theta}d\theta = 2\pi(-j)^V I_V(z)$ with $\alpha_u - \alpha_l = 2\pi$ obtained from I.S.Gradshteyn and Ryzhik [2000], Eq. (A.65) becomes

$$X_2 = b_0^{SB_1} 8\pi^2 f_{T_{max}} f_{R_{max}} \cos\gamma_R \cos(\gamma_T - \mu_T^{TR}) \frac{I_1(k_T^{TR})}{I_0(k_T^{TR})} \tag{A.66}$$

Using the same procedure shown above, the closed-form expressions of terms X_3, X_4, X_5, and X_6 can be obtained in a similar way. The substitution of $X_1, X_2, \ldots, X_6$ into Eq. (A.63) gives the final result of $b_2^{SB_1}$, as shown in Eq. (9.75b).

A.8.8 Derivation of the Condition max$\{R_T, R_R\} <$ min$\{a_l - a_{l-1}\}$ that Guarantees the TDL Structure of our Model

For the first tap, Figure 9.19 clearly shows that the longest distance caused by the double-bounced rays is the link $O_T - A - B - O_R$, which is equal to $2R_T + 2R_R + 2f$. According to the TDL structure, the inequality $2R_T + 2R_R + 2f < 2a_2$ should be fulfilled. Considering $R_T + R_R + 2f < 2a_1$ and based on the transitivity of inequalities, we know that if $R_T + R_R + 2a_1 < 2a_2$ then $2R_T + 2R_R + 2f < 2a_2$. Therefore, we can conclude that the condition $R_T + R_R < 2a_2 - 2a_1$ guarantees the fulfillment of the TDL structure for the first tap.

For other taps ($l' > 1$), since the derivations of the condition that guarantees the fulfillment of the TDL structure are the same, here we only detail the derivation of the condition for the second tap. From Figure 9.19, it is clear that the longest distance in the second tap caused by the double-bounced rays is either the link $O_T - C - S - O_R$, which is equal to $2R_T + 2a_2$, or the link $O_T - S - D - O_R$, which is equal to $2R_R + 2a_2$. In terms of the TDL structure, the inequality $\max\{R_T, R_R\} < \{a_2 - a_1\}$ should be fulfilled. Therefore, we can conclude that the condition $\max\{R_T, R_R\} < \min\{a_{l'} - a_{l'-1}\}$ guarantees the fulfillment of the TDL structure for other taps.

Since the condition $R_T + R_R < 2a_2 - 2a_1$ for the first tap can be rewritten as

- $R_T < a_2 - a_1$ if $R_T \geq R_R$
- $R_R < a_2 - a_1$ (if $R_R \geq R_T$)

we can obtain the general condition $\max\{R_T, R_R\} < \min\{a_l - a_{l-1}\}$ that guarantees the fulfillment of the TDL structure of our model.

A.8.9 The Reduced Expressions of Spatial Correlation

For outdoor macrocell and microcell BS cooperation and relay cooperation scenarios, the assumption $\min\{D_1, D_2, D_3\} \gg \max\{\xi_{n_1}^M, \xi_{n_2}^R, \xi_{n_3}^B\}$ is applicable. In this case, by using trigonometric transformations, the equality $\int_{-\pi}^{\pi} e^{a\sin c + b\cos c}dc = 2\pi I_0(\sqrt{a^2+b^2})$ [I.S.Gradshteyn and Ryzhik 2000], and the results of Cheng et al. [2009], the spatial correlation between BS-RS link and BS–MS link in Eqs (9.147)–(9.149) can be reduced to:

$$
\begin{aligned}
\rho^{1g}_{p_3p_2,p_3'p_1} &= \sqrt{\frac{\eta^{1g}_{p_3p_2}\eta^{1g}_{p_3'p_1}}{I_0\{k_g\}\left(K_{p_3p_2}+1\right)\left(K_{p_3'p_1}+1\right)}e^{jC^{11}}} \\
&\times \int_{R_{1ng}}^{R_{2ng}} e^{jE^{11}} I_0\left\{\sqrt{(A^{1g})^2+(B^{1g})^2}\right\} Q_g d\mathcal{I}_g
\end{aligned} \tag{A.67}
$$

$$
\begin{aligned}
\rho^{2g}_{p_3p_2,p_3'p_1} &= \sqrt{\frac{\eta^{2g}_{p_3p_2}\eta^{2g}_{p_3'p_1}}{I_0\{k_{g_1}\}I_0\{k_{g_2}\}(K_{p_3p_2}+1)(K_{p_3'p_1}+1)}e^{jC^{2g}}} \\
&\times \int_{R_{1ng_1}}^{R_{2ng_1}}\int_{R_{1ng_2}}^{R_{2ng_2}} e^{jE^{2g}} I_0\left\{\sqrt{(A_1^{2g})^2+(B_1^{2g})^2}\right\} \\
&\times I_0\left\{\sqrt{(A_2^{2g})^2+(B_2^{2g})^2}\right\} Q_{g_1g_2} d\mathcal{I}_{g_1} d\mathcal{I}_{g_2}
\end{aligned} \tag{A.68}
$$

$$
\begin{aligned}
\rho^{31}_{p_3p_2,p_3'p_1} &= \sqrt{\frac{\eta^{31}_{p_3p_2}\eta^{31}_{p_3'p_1}}{I_0\{k_1\}I_0\{k_2\}I_0\{k_3\}(K_{p_3p_2}+1)(K_{p_3'p_1}+1)}e^{jC^{31}}} \\
&\times \int_{R_{1n_1}}^{R_{2n_1}}\int_{R_{1n_2}}^{R_{2n_2}}\int_{R_{1n_3}}^{R_{2n_3}} e^{jE^{31}} I_0\left\{\sqrt{(A_M^{31})^2+(B_M^{31})^2}\right\} \\
&\times I_0\left\{\sqrt{(A_R^{31})^2+(B_R^{31})^2}\right\} I_0\left\{\sqrt{(A_B^{31})^2+(B_B^{31})^2}\right\} Q_{123} d\mathcal{I}_1 d\mathcal{I}_2 d\mathcal{I}_3
\end{aligned} \tag{A.69}
$$

where $C^{11} = C_p^{11} - 2\pi\lambda^{-1}\delta_3\cos\beta_3$ with $C_P^{11} = -2\pi\lambda^{-1}[D_2 + \frac{\delta_2}{2}\cos(\beta_2+\theta)]$, $E^{11} = E^{22} = 2\pi\lambda^{-1}\xi_{n_1}^M$, $A^{11} = A_1^{22} - j2\pi\lambda^{-1}\delta_3\sin\beta_3\frac{\xi_{n_1}^M}{D_1}$, $B^{11} = B_1^{22} = -j2\pi\lambda^{-1}[\frac{\delta_1}{2}\cos\beta_1 - \xi_{n_1}^M + \frac{\delta_2\xi_{n_1}^M}{2D_2}\sin(\beta_2+\theta)\sin\theta] + k_1\cos\mu_1$ $C^{12} = C_P^{12} + 2\pi\lambda^{-1}\delta_3\cos(\beta_3-\theta')]$ with $C_P^{22} = 2\pi\lambda^{-1}[D_2 + \frac{\delta_1}{2}\cos(\beta_1+\theta)]$, $E^{12} = E^{21} = -2\pi\lambda^{-1}\xi_{n_2}^R$, $A^{12} = A_1^{21} - j2\pi\lambda^{-1}\delta_3\sin(\beta_3 - \theta')\frac{\xi_{n_2}^R}{D_3}\cos\theta'$, $B^{12} = B_1^{21} + j2\pi\lambda^{-1}\delta_3\sin(\beta_3-\theta')\frac{\xi_{n_2}^R}{D_3}\sin\theta'$, $C^{13} = 2\pi\lambda^{-1}[\frac{\delta_1}{2}\cos\beta_1 - \frac{\delta_2}{2}\cos(\theta'-\beta_2)] - C^{31}$, $E^{13} = E^{21} = E^{22} = 0$, $A^{13} = A_B^{31} + j2\pi\lambda^{-1}[\xi_{n_3}^B\sin\theta' - \frac{\delta_1\xi_{n_3}^B}{2D_1}\sin\beta_1 + \frac{\delta_2\xi_{n_3}^B}{2D_3}\sin(\theta'-\beta_2)\cos\theta']$, $B^{13} = B_B^{31} + j2\pi\lambda^{-1}[\xi_{n_3}^B\cos\theta' - \xi_{n_3}^B - \frac{\delta_2\xi_{n_3}^B}{2D_3}\sin(\theta' - \beta_2)\sin\theta']$, $C^{21} = C_P^{12} - 2\pi\lambda^{-1}\delta_1\cos\beta_1$, $A_1^{21} = j2\pi\lambda^{-1}[\xi_{n_2}^R\sin\theta + \frac{\delta_2}{2}\sin\beta_2 + \frac{\delta_1\xi_{n_2}^R}{2D_2}\sin(\beta_1+\theta)\cos\theta] + k_2\sin\mu_2$, $B_1^{21} = j2\pi\lambda^{-1}[\frac{\delta_2}{2}\cos\beta_2 - \xi_{n_2}^R\cos\theta + \frac{\delta_1\xi_{n_2}^R}{2D_2}\sin(\beta_1+\theta)\sin\theta] + k_2\cos\mu_2$, $A_2^{21} = A_2^{22} = j2\pi\lambda^{-1}\delta_1\sin\beta_1\frac{\xi_{n_3}^B}{D_1} + k_3\sin\mu_3$, $B_2^{21} = B_2^{22} = k_3\cos\mu_3$, $C^{22} = C_P^{11} - 2\pi\lambda^{-1}\delta_1\cos\beta_1$, $A_1^{22} = -j2\pi\lambda^{-1}[\frac{\delta_1}{2}\sin\beta_1 + \frac{\delta_2\xi_{n_1}^M}{2D_2}\sin(\beta_2+\theta)\cos\theta] + k_1\sin\mu_1$, $C^{23} = c^{31} + 2\pi\lambda^{-1}[\frac{\delta_3}{2}\cos\beta_3 + \frac{\delta_3}{2}\cos(\beta_3-\theta')]$, $E^{23} = E^{11} + E^{12}$, $A_1^{23} = A_M^{31} + j2\pi\lambda^{-1}\frac{\delta_3}{2}\sin\beta_3\frac{\xi_{n_1}^M}{D_1}$, $B_1^{23} = B_M^{31} - j2\pi\lambda^{-1}\xi_{n_1}^M$, $A_2^{23} = A^{12} - j2\pi\lambda^{-1}\frac{\delta_1\xi_{n_2}^R}{2D_2}\sin(\beta_1+\theta)\cos\theta$, $B_2^{23} = A^{12} - j2\pi\lambda^{-1}\frac{\delta_1\xi_{n_2}^R}{2D_2}\sin(\beta_1+\theta)\sin\theta$, $C^{31} = 2\pi\lambda^{-1}[D_3 - D_1]$, $E^{31} = E^{11} - E^{12}$,

$A^{31}_{M(R)} = \mp j2\pi\lambda^{-1}\frac{\delta_{3(2)}}{2}\sin\beta_{3(2)} + k_{n_{1(2)}}\sin\mu_{n_{1(2)}}$, $B^{31}_{M(R)} = \mp j2\pi\lambda^{-1}\frac{\delta_{3(2)}}{2}\cos\beta_{3(2)} + k_{n_{1(2)}}$ $\cos\mu_{n_{1(2)}}$, $A^{31}_{B} = j2\pi\lambda^{-1}\delta_3\sin\beta_3 + k_3\sin\mu_3$, $B^{31}_{B} = j2\pi\lambda^{-1}\delta_3\cos\beta_3 + k_3\cos\mu_3$.

Similarly, we can reduced the spatial correlation between BS–RS link and RS–MS link in Eqs (9.152)–(9.154) to:

$$\rho^{1g}_{p_3p_2,p'_2p_1} = \sqrt{\frac{\eta^{1g}_{p_3p_2}\eta^{1g}_{p'_2p_1}}{I_0\{k_g\}(K_{p_3p_2}+1)(K_{p'_2p_1}+1)}}e^{j\tilde{C}^{11}} \times \int_{R_{1n_g}}^{R_{2n_g}} e^{j\tilde{E}^{11}} I_0\left\{\sqrt{(\tilde{A}^{1g})^2 + (\tilde{B}^{1g})^2}\right\} Q_g d\mathcal{I}_g \tag{A.70}$$

$$\rho^{2g}_{p_3p_2,p'_2p_1} = \sqrt{\frac{\eta^{2g}_{p_3p_2}\eta^{2g}_{p'_2p_1}}{I_0\{k_{g_1}\}I_0\{k_{g_2}\}(K_{p_3p_2}+1)(K_{p'_2p_1}+1)}}e^{j\tilde{C}^{21}} \times \int_{R_{1n_{g_1}}}^{R_{2n_{g_1}}}\int_{R_{1n_{g_2}}}^{R_{2n_{g_2}}} e^{j\tilde{E}^{21}} I_0\left\{\sqrt{(\tilde{A}^{2g}_1)^2 + (\tilde{B}^{2g}_1)^2}\right\} \times I_0\left\{\sqrt{(\tilde{A}^{2g}_2)^2 + (\tilde{B}^{2g}_2)^2}\right\} Q_{g_1g_2} d\mathcal{I}_{g_1} d\mathcal{I}_{g_2} \tag{A.71}$$

$$\rho^{31}_{p_3p_2,p'_2p_1} = \sqrt{\frac{\eta^{31}_{p_3p_2}\eta^{31}_{p'_2p_1}}{I_0\{k_1\}I_0\{k_2\}I_0\{k_3\}(K_{p_3p_2}+1)(K_{p'_2p_1}+1)}}e^{j\tilde{C}^{31}} \times \int_{R_{1n_1}}^{R_{2n_1}}\int_{R_{1n_2}}^{R_{2n_2}}\int_{R_{1n_3}}^{R_{2n_3}} e^{jE^{31}} I_0\left\{\sqrt{(\tilde{A}^{31}_M)^2 + (\tilde{B}^{31}_M)^2}\right\} \times I_0\left\{\sqrt{(\tilde{A}^{31}_R)^2 + (\tilde{B}^{31}_R)^2}\right\} I_0\left\{\sqrt{(\tilde{A}^{31}_B)^2 + (\tilde{B}^{31}_B)^2}\right\} Q_{123} d\mathcal{I}_1 d\mathcal{I}_2 d\mathcal{I}_3 \tag{A.72}$$

where $\tilde{C}^{11} = \tilde{C}^{11}_P - 2\pi\lambda^{-1}\delta_2\cos(\beta_2+\theta)$ with $\tilde{C}^{11}_P = -2\pi\lambda^{-1}[D_1 - \frac{\delta_3}{2}\cos\beta_3]$, $\tilde{E}^{11} = \tilde{E}^{22} = 2\pi\lambda^{-1}\xi^M_{n_1}$, $\tilde{A}^{11} = \tilde{A}^{22}_1 - j2\pi\lambda^{-1}\frac{\delta_2\xi^M_{n_1}}{D_2}\sin(\beta_2+\theta)\cos\theta$, $\tilde{B}^{11} = \tilde{B}^{22}_1 - j2\pi\lambda^{-1}$ $\frac{\delta_2\xi^M_{n_1}}{D_2}\sin(\beta_2+\theta)\sin\theta$, $\tilde{C}^{12} = 2\pi\lambda^{-1}[D_2 + \frac{\delta_1}{2}\cos(\beta_1+\theta) + \frac{\delta_3}{2}\cos(\beta_3-\theta')] - \tilde{C}^{31}$, $\tilde{E}^{12} = 0$, $\tilde{A}^{12} = \tilde{A}^{21}_2 + j2\pi\lambda^{-1}[\xi^R_{n_2}\sin\theta + \frac{\delta_1\xi^R_{n_2}}{2D_2}\sin(\beta_1+\theta)\cos\theta - \xi^R_{n_2}\sin\theta' - \frac{\delta_3}{2}\sin(\beta_3-\theta')\frac{\xi^R_{n_2}}{D_3}$ $\cos\theta']$, $\tilde{B}^{12} = \tilde{B}^{31}_R + j2\pi\lambda^{-1}[\xi^R_{n_2}\cos\theta + \frac{\delta_1\xi^R_{n_2}}{2D_2}\sin(\beta_1+\theta)\sin\theta - \xi^R_{n_2}\cos\theta' + \frac{\delta_3}{2}\sin(\beta_3 - \theta')\frac{\xi^R_{n_2}}{D_3}\sin\theta']$, $\tilde{C}^{13} = \tilde{C}^{21}_1 - 2\pi\lambda^{-1}\delta_2\cos(\theta'-\beta_2)$, $\tilde{E}^{13} = \tilde{E}^{21}_1 = -2\pi\lambda^{-1}\xi^B_{n_3}$, $\tilde{A}^{13} = \tilde{A}^{21}_1 + j2\pi\lambda^{-1}\frac{\delta_2\xi^B_{n_3}}{D_3}\sin(\theta'-\beta_2)\cos\theta'$, $\tilde{B}^{13} = \tilde{B}^{21}_1 - j2\pi\lambda^{-1}\frac{\delta_2\xi^B_{n_3}}{D_3}\sin(\theta'-\beta_2)\sin\theta'$, $\tilde{C}^{21} = 2\pi\lambda^{-1}[D_1 + \frac{\delta_1}{2}\cos\beta_1]$, $\tilde{A}^{21}_1 = j2\pi\lambda^{-1}[-\frac{\delta_1\xi^B_{n_3}}{2D_1}\sin\beta_1 - \frac{\delta_3}{2}\sin\beta_3] + k_3\sin\mu_3$, $\tilde{B}^{21}_1 = j2\pi\lambda^{-1}[-\xi^B_{n_3} - \frac{\delta_3}{2}\cos\beta_3] + k_3\cos\mu_3$, $\tilde{A}^{21}_2 = \tilde{A}^{31}_R = j2\pi\lambda^{-1}\delta_2\sin\beta_2 + k_2\sin\mu_2$, $\tilde{B}^{21}_2 = \tilde{B}^{31}_R = j2\pi\lambda^{-1}\delta_2\cos\beta_2 + k_2\cos\mu_2$, $\tilde{C}^{22} = \tilde{C}^{11}_P - 2\pi\lambda^{-1}\cos(\theta'-\beta_2)$, $\tilde{A}^{22}_1 = -j2\pi\lambda^{-1}$ $[\frac{\delta_1}{2}\sin\beta_1 + -\frac{\delta_3}{2}\sin\beta_3\frac{\xi^M_{n_1}}{D_1}] + k_1\sin\mu_1$, $\tilde{B}^{22}_1 = -j2\pi\lambda^{-1}[\frac{\delta_1}{2}\cos\beta_1 + \xi^M_{n_1}] + k_1\cos\mu_1$, $\tilde{A}^{22}_2 = j2\pi\lambda^{-1}\sin(\theta'-\beta_2)\frac{\xi^B_{n_3}}{D_3}\cos\theta + k_3\sin\mu_3$, $\tilde{B}^{22}_2 = -j2\pi\lambda^{-1}\sin(\theta'-\beta_2)\frac{\xi^B_{n_3}}{D_3}\sin\theta +$

$k_3 \cos \mu_3$, $\tilde{C}^{31} = 2\pi\lambda^{-1}[D_3 - D_2]$, $\tilde{E}^{31} = \tilde{E}^{11} - \tilde{E}^{13}$, $\tilde{A}^{31}_{M(B)} = \mp j2\pi\lambda^{-1}\frac{\delta_{3(1)}}{2}\sin\beta_{3(1)} + k_{n_{1(3)}} \sin\mu_{n_{1(3)}}$, $\tilde{B}^{31}_{M(B)} = \mp j2\pi\lambda^{-1}\frac{\delta_{3(1)}}{2}\cos\beta_{3(1)} + k_{n_{1(3)}}\cos\mu_{n_{1(3)}}$.

For BS–MS link and RS–MS link, the spatial correlation between them shown in Eqs (9.157)–(9.147) can be reduced to

$$\rho^{1g}_{p_3p_1,p_2p_1'} = \sqrt{\frac{\eta^{1g}_{p_3p_1}\eta^{1g}_{p_2p_1'}}{I_0\left\{k_g\right\}(K_{p_3p_1}+1)(K_{p_2p_1'}+1)}}e^{j\widehat{C}^{11}} \times \int_{R_{1n_g}}^{R_{2n_g}} e^{j\widehat{E}^{11}} I_0\left\{\sqrt{(\widehat{A}^{1g})^2+(\widehat{B}^{1g})^2}\right\} Q_g d\mathcal{I}_g \tag{A.73}$$

$$\rho^{2g}_{p_3p_1,p_2p_1'} = \sqrt{\frac{\eta^{2g}_{p_3p_1}\eta^{2g}_{p_2p_1'}}{I_0\{k_{g_1}\}I_0\{k_{g_2}\}(K_{p_3p_1}+1)(K_{p_2p_1'}+1)}}e^{j\widehat{C}^{2g}} \times \int_{R_{1n_{g_1}}}^{R_{2n_{g_1}}}\int_{R_{1n_{g_2}}}^{R_{2n_{g_2}}} e^{j\widehat{E}^{2g}} I_0\left\{\sqrt{(\widehat{A}^{2g}_1)^2+(\widehat{B}^{2g}_1)^2}\right\} \times I_0\left\{\sqrt{(\widehat{A}^{2g}_2)^2+(\widehat{B}^{2g}_2)^2}\right\} Q_{g_1g_2} d\mathcal{I}_{g_1} d\mathcal{I}_{g_2} \tag{A.74}$$

$$\rho^{31}_{p_3p_1,p_2p_1'} = \sqrt{\frac{\eta^{31}_{p_3p_1}\eta^{31}_{p_2p_1'}}{I_0\{k_1\}I_0\{k_2\}I_0\{k_3\}(K_{p_3p_1}+1)(K_{p_2p_1'}+1)}}e^{j\widehat{C}^{31}} \times \int_{R_{1n_1}}^{R_{2n_1}}\int_{R_{1n_2}}^{R_{2n_2}}\int_{R_{1n_3}}^{R_{2n_3}} e^{j\widehat{E}^{31}} I_0\left\{\sqrt{(\widehat{A}^{31}_M)^2+(\widehat{B}^{31}_M)^2}\right\} \times I_0\left\{\sqrt{(\widehat{A}^{31}_R)^2+(\widehat{B}^{31}_R)^2}\right\} I_0\left\{\sqrt{(\widehat{A}^{31}_B)^2+(\widehat{B}^{31}_B)^2}\right\} Q_{123} d\mathcal{I}_1 d\mathcal{I}_2 d\mathcal{I}_3 \tag{A.75}$$

where $\widehat{C}^{11} = 2\pi\lambda^{-1}[D_2 + \frac{\delta_2}{2}\cos(\beta_2+\theta) - D_1 + \frac{\delta_3}{2}\cos\beta_3]$, $\widehat{E}^{11} = 0$, $\widehat{A}^{11} = \widehat{A}^{22}_1 + j2\pi\lambda^{-1}[\frac{\delta_2\xi^M_{n_1}}{2D_2}\sin(\beta_2+\theta)\cos\theta + \frac{\delta_3}{2}\sin\beta_3\frac{\xi^M_{n_1}}{D_1}]$, $\widehat{B}^{11} = \widehat{B}^{22}_1 + j2\pi\lambda^{-1}[-2\xi^M_{n_1} + \frac{\delta_2\xi^M_{n_1}}{2D_2}\sin(\beta_2+\theta)\sin\theta]$, $\widehat{C}^{12} = \widehat{C}^{23} - 2\pi\lambda^{-1}\delta_1\cos(\beta_1+\theta)$, $\widehat{E}^{12} = \widehat{E}^{23} = 2\pi\lambda^{-1}\xi^R_{n_2}$, $\widehat{A}^{12} = \widehat{A}^{23}_2 - j2\pi\lambda^{-1}\frac{\delta_1\xi^R_{n_2}}{D_2}\sin(\beta_1+\theta)\cos\theta$, $\widehat{B}^{12} = \widehat{B}^{23}_2 - j2\pi\lambda^{-1}\frac{\delta_1\xi^R_{n_2}}{D_2}\sin(\beta_1+\theta)\sin\theta$, $\widehat{C}^{13} = \widehat{C}^{22} - 2\pi\lambda^{-1}\delta_1\cos\beta_1$, $\widehat{E}^{13} = \widehat{E}^{22} = -2\pi\lambda^{-1}\xi^B_{n_3}$, $\widehat{A}^{13} = \widehat{A}^{22}_2 + j2\pi\lambda^{-1}\frac{\delta_1\xi^B_{n_3}}{D_1}\sin\beta_1$, $\widehat{B}^{13} = \widehat{B}^{22}_2 = -j2\pi\lambda^{-1}[\xi^B_{n_3}\cos\theta' - \frac{\delta_2\xi^B_{n_3}}{2D_3}\sin(\theta'-\beta_2)\sin\theta' + \frac{\delta_3}{2}\cos\beta_3] + k_3\cos\mu_3$, $\widehat{C}^{21} = \widehat{C}^{31} - 2\pi\lambda^{-1}\frac{\delta_1}{2}(\cos\beta_1 + \cos(\beta_1+\theta))$, $\widehat{E}^{21} = \widehat{E}^{31} = \widehat{E}^{12} + \widehat{E}^{13}$, $\widehat{A}^{21}_1 = -j2\pi\lambda^{-1}[-\xi^R_{n_2}\sin\theta + \frac{\delta_1\xi^R_{n_2}}{2D_2}\sin(\beta_1+\theta)\cos\theta + \frac{\delta_2}{2}\sin\beta_2] + k_2\sin\mu_2$, $\widehat{B}^{21}_1 = -j2\pi\lambda^{-1}[-\xi^R_{n_2}\cos\theta + \frac{\delta_1\xi^R_{n_2}}{2D_2}\sin(\beta_1+\theta)\sin\theta + \frac{\delta_2}{2}\cos\beta_2] + k_2\cos\mu_2$, $\widehat{A}^{21}_2 = j2\pi\lambda^{-1}[\frac{\delta_1\xi^B_{n_3}}{2D_1}\sin\beta_1 + \frac{\delta_3}{2}\sin\beta_3] + k_3\sin\mu_3$, $\widehat{B}^{21}_2 = j2\pi\lambda^{-1}[-\xi^B_{n_3} + \frac{\delta_3}{2}\cos\beta_3] + k_3\cos\mu_3$, $\widehat{C}^{22} = 2\pi\lambda^{-1}[D_3 + \frac{\delta_2}{2}\cos(\theta'-\beta_2)]$, $\widehat{A}^{22}_1 = \widehat{A}^{23}_1 = \widehat{A}^{31}_M = j2\pi\lambda^{-1}\delta_1\sin\beta_1 + k_1\sin\mu_1$, $\widehat{B}^{22}_1 = \widehat{B}^{23}_1 = \widehat{B}^{31}_M = j2\pi\lambda^{-1}\delta_1\cos\beta_1 + k_1\cos\mu_1$, $\widehat{A}^{22}_2 = -j2\pi\lambda^{-1}[\frac{\delta_3}{2}\sin\beta_3 + \xi^B_{n_3}\sin\theta' + \frac{\delta_2\xi^B_{n_3}}{2D_3}\sin(\theta'-\beta_2)\cos\theta'] + k_3\sin\mu_3$, $\widehat{C}^{23} = 2\pi\lambda^{-1}[-D_3 + \frac{\delta_3}{2}\cos(\beta_3-\theta')]$, $\widehat{A}^{23}_2 = -j2\pi\lambda^{-1}[\frac{\delta_2}{2}\sin\beta_2 + \xi^R_{n_2}\sin\theta' + \frac{\delta_3}{2}\sin(\beta_3-\theta')\frac{\xi^R_{n_2}}{D_3}$

$\cos\theta'] + k_2 \sin\mu_2$, $\widehat{B}_2^{23} = -j2\pi\lambda^{-1}[\frac{\delta_2}{2}\cos\beta_2 + \xi_{n_2}^R \cos\theta' - \frac{\delta_3}{2}\sin(\beta_3 - \theta')\frac{\xi_{n_2}^R}{D_3}\sin\theta'] +$ $k_2 \cos\mu_2$, $\widehat{C}^{31} = 2\pi\lambda^{-1}[D_1 - D_2]$, $\widehat{A}_{R(B)}^{31} = -j2\pi\lambda^{-1}\frac{\delta_{2(1)}}{2}\sin\beta_{2(1)} + k_{n_{2(3)}}\sin\mu_{n_{2(3)}}$, $\widehat{B}_{R(B)}^{31} = -j2\pi\lambda^{-1}\frac{\delta_{2(1)}}{2}\cos\beta_{2(1)} + k_{n_{2(3)}}\cos\mu_{n_{2(3)}}$.

Bibliography

Bengtsson M and Ottersten B 2000 Low-complexity estimators for distributed sources. *IEEE Transactions on Signal Processing* **48**(8), 2185–2194.

Besson O and Stoica P 1999 Decoupled estimation of DoA and angular spread for spatially distributed sources. *IEEE Transactions on Signal Processing* **49**, 1872–1882.

Kaveh M and Barabell AJ 1986 The statistical performance of the MUSIC and the Minimum-Norm algorithms in resolving plane waves in noise. *IEEE Transactions on Acoustics, Speech, and Signal Processing* **34**(2), 331–341.

Wu HT, Yang JF and Chen FK 1995 Source number estimators using transformed Gerschgorin Radii. *IEEE Trans. Signal Processing* **43**, 1325–1333.

Abramowitz M and Stegun IA 1965 *Handbook of Mathematical Functions with Formulas, Graphs and Mathematical Tables*.

Gradshteyn I and Ryzhik I 2000 *Table of Integrals, Series, and Products* 6th edn. Academic.

Watson G 1952 *A Treatise of the Theory of Bessel Functions* 2nd edn. Cambridge.

Zajic A and Stuber G 2008 Three-dimensional modeling, simulation, and capacity analysis of space time correlated mobile-to-mobile channels. *IEEE Transactions on Vehicular Technology* **57**(4), 2042–2054.

Index

Propagation Channel Characterization, Parameter Estimation and Modelling for Wireless Communications, First Edition.
Xuefeng Yin and Xiang Cheng.